A Modern Introduction to Mathematical Analysis

Alessandro Fonda

A Modern Introduction to Mathematical Analysis

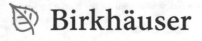 Birkhäuser

Alessandro Fonda
Dipartimento di Matematica e Geoscienze
Università degli Studi di Trieste
Trieste, Italy

ISBN 978-3-031-23715-7   ISBN 978-3-031-23713-3   (eBook)
https://doi.org/10.1007/978-3-031-23713-3

This book is published under the imprint Birkhäuser, www.birkhauser-science.com by the registered company Springer Nature Switzerland AG
The registered company address is: Gewerbestrasse 11, 6330 Cham, Switzerland

To my parents, Thea and Luciano

Preface

This book brings together the classical topics of mathematical analysis normally taught in the first two years of a university course. It is the outcome of the lessons I have been teaching for many years in the undergraduate courses in mathematics, physics, and engineering at my university.

Many excellent books on mathematical analysis have already been written, so a natural question to ask is: Why write another book on this subject? I will try to provide a brief answer to that question.

The main novelty of this book lies in the treatment of the theory of the integral. Kurzweil and Henstock's theory is presented here in Chaps. 7, 9, and 11. Compared to Riemann's theory, it requires modest additional effort from the student, but that effort will be repaid with significant benefits. Consider that it includes Lebesgue's theory itself, since a function is integrable according to Kurzweil and Henstock if and only if it is integrable according to Lebesgue, together with its absolute value. In this theory, the Fundamental Theorem turns out to be very general and natural, and it finds its generalization in Taylor's formula with an integral remainder, requiring only essential hypotheses. Moreover, the improper integral happens to be a normal integral.

Despite the modest additional effort required, no demands for preliminary knowledge about the integral will be made on the student for the purpose of reading this book. Students will be guided as they construct all the necessary mathematical tools, starting from the very beginning.

Indeed, the book starts with some preliminaries on logic and set theory. It is a short vademecum, without formal rigor, and will help readers orient themselves with respect to the notations to be used later on.

In Chap. 1, we introduce the main sets on which we base the rest of the theory. These are the numeric sets, mainly $\mathbb{R}$ and $\mathbb{C}$, the space $\mathbb{R}^N$, and the metric spaces. It is in this general context that the concepts of continuity and limit will be developed in Chaps. 2 and 3, respectively. The discussion of numerical series, together with the series of functions, will be postponed to Chap. 8.

Chapter 4 is dedicated to the notions of compactness and completeness. Although they seem to be rather abstract concepts, they happen to be necessary for a rigorous treatment of differential and integral calculus.

I would like to make special mention here of the original construction of the exponential function and the trigonometric functions, which I propose in Chap. 5.

They are introduced as particular cases of a function with complex values that is constructed with elementary geometric tools.

Differential calculus is first developed in Chap. 6 for functions of one real variable and, later in Chap. 10, for functions of several variables. Here the reader will find the implicit function theorem proved by induction, as in the original proofs by Dini and Genocchi–Peano.

As was already noted, integral calculus is presented in Chaps. 7, 9, and 11. This approach to the integral was introduced independently by J. Kurzweil [5] in 1957 and R. Henstock [3] in 1961.

In Chap. 12, the theory of differential forms and their integral on M-surfaces and on differentiable M-manifolds is developed in detail, following the approach of Spivak [6]. The Stokes–Cartan theorem, with its classic corollaries the curl and divergence theorems, is the final result. Besides being a fundamental tool in applications, it stands out for its elegance and formal perfection, like the most sublime works of art.

The book can be used at different teaching levels, in line with the preferences of the teacher. As mentioned earlier, I have proposed it as an early postsecondary text. However, it could also be used in an advanced course in analysis or by scholars wishing to understand the Kurzweil–Henstock integral starting with a simple approach.

Unlike the majority of books on these subjects, this one contains almost no exercises. The reason for this is that many textbooks containing only exercises have already been published, which fits perfectly with the arguments of this book. An example is *Solving Problems in Mathematical Analysis* by T. Radożycki, published by Springer in 2020, which is divided into three volumes:

- Part I. Sets, Functions, Limits, Derivatives, Integrals, Sequences, and Series;
- Part II. Definite, Improper and Multidimensional Integrals, Functions of Several Variables, and Differential Equations;
- Part III. Curves and Surfaces, Conditional Extremes, Curvilinear Integrals, Complex Functions, Singularities, and Fourier Series.

A list of other recent textbooks with solved exercises is provided in the bibliography.

Finally, I would like to mention that this book would never have written without the strong motivation provided by my students. I thank them all, and I hope that the book will encourage others in the future to become involved in such a beautiful and fruitful theory as mathematical analysis.

Trieste, Italy   Alessandro Fonda

Contents

Part I   The Basics of Mathematical Analysis

1   Sets of Numbers and Metric Spaces 3
    1.1   The Natural Numbers and the Induction Principle 3
           1.1.1   Recursive Definitions 4
           1.1.2   Proofs by Induction 6
           1.1.3   The Binomial Formula 9
    1.2   The Real Numbers 12
           1.2.1   Supremum and Infimum 14
           1.2.2   The Square Root 17
           1.2.3   Intervals 19
           1.2.4   Properties of $\mathbb{Q}$ and $\mathbb{R} \setminus \mathbb{Q}$ 20
    1.3   The Complex Numbers 23
           1.3.1   Algebraic Equations in $\mathbb{C}$ 25
           1.3.2   The Modulus of a Complex Number 26
    1.4   The Space $\mathbb{R}^N$ 28
           1.4.1   Euclidean Norm and Distance 30
    1.5   Metric Spaces 33

2   Continuity 41
    2.1   Continuous Functions 41
    2.2   Intervals and Continuity 50
    2.3   Monotone Functions 51
    2.4   The Exponential Function 53
    2.5   The Trigonometric Functions 56
    2.6   Other Examples of Continuous Functions 59

3   Limits 63
    3.1   The Notion of Limit 63
    3.2   Some Properties of Limits 65
    3.3   Change of Variables in the Limit 68
    3.4   On the Limit of Restrictions 70
    3.5   The Extended Real Line 72
    3.6   Some Operations with $-\infty$ and $+\infty$ 76
    3.7   Limits of Monotone Functions 83

3.8   Limits for Exponentials and Logarithms......................... 86
3.9   Liminf and Limsup 90

4   Compactness and Completeness 93
4.1   Some Preliminaries on Sequences 93
4.2   Compact Sets... 95
4.3   Compactness and Continuity 97
4.4   Complete Metric Spaces... 99
4.5   Completeness and Continuity 102
4.6   Spaces of Continuous Functions 103

5   Exponential and Circular Functions 107
5.1   The Construction.. 107
        5.1.1   Preliminaries for the Proof 108
        5.1.2   Definition on a Dense Set 113
        5.1.3   Extension to the Whole Real Line 114
5.2   Exponential and Circular Functions............................. 117
5.3   Limits for Trigonometric Functions............................. 119

Part II   Differential and Integral Calculus in $\mathbb{R}$

6   The Derivative 127
6.1   Some Differentiation Rules....................................... 130
6.2   The Derivative Function 135
6.3   Remarkable Properties of the Derivative 136
6.4   Inverses of Trigonometric and Hyperbolic Functions 139
6.5   Convexity and Concavity 141
6.6   L'Hôpital's Rules 146
6.7   Taylor Formula.. 154
6.8   Local Maxima and Minima.. 158
6.9   Analyticity of Some Elementary Functions 159

7   The Integral 161
7.1   Riemann Sums 161
7.2 δ-Fine Tagged Partitions................................. 163
7.3   Integrable Functions on a Compact Interval 165
7.4   Elementary Properties of the Integral 168
7.5   The Fundamental Theorem 170
7.6   Primitivable Functions.. 172
7.7   Primitivation by Parts and by Substitution...................... 177
7.8   The Taylor Formula with Integral Form Remainder 181
7.9   The Cauchy Criterion 182
7.10   Integrability on Subintervals 184
7.11 R-Integrable and Continuous Functions 187
7.12   Two Theorems Involving Limits 191
7.13   Integration on Noncompact Intervals 194
7.14   Functions with Vector Values 198

Part III Further Developments

8 Numerical Series and Series of Functions 203
 8.1 Introduction and First Properties 203
 8.2 Series of Real Numbers... 208
 8.3 Series of Complex Numbers.. 213
 8.4 Series of Functions.. 215
  8.4.1 Power Series 217
  8.4.2 The Complex Exponential Function 220
  8.4.3 Taylor Series 222
  8.4.4 Fourier Series 224
 8.5 Series and Integrals 231

9 More on the Integral 233
 9.1 Saks–Henstock Theorem 233
 9.2 *L*-Integrable Functions 236
 9.3 Monotone Convergence Theorem 241
 9.4 Dominated Convergence Theorem 245
 9.5 Hake's Theorem.. 247

Part IV Differential and Integral Calculus in $\mathbb{R}^N$

10 The Differential 255
 10.1 The Differential of a Scalar-Valued Function 256
 10.2 Some Computational Rules 259
 10.3 Twice Differentiable Functions 261
 10.4 Taylor Formula.. 263
 10.5 The Search for Maxima and Minima 267
 10.6 Implicit Function Theorem: First Statement...................... 269
 10.7 The Differential of a Vector-Valued Function 272
 10.8 The Chain Rule 274
 10.9 Mean Value Theorem 277
 10.10 Implicit Function Theorem: General Statement 280
 10.11 Local Diffeomorphisms 286
 10.12 *M*-Surfaces... 287
 10.13 Local Analysis of *M*-Surfaces 291
 10.14 Lagrange Multipliers.. 294
 10.15 Differentiable Manifolds 299

11 The Integral 301
 11.1 Integrability on Rectangles 301
 11.2 Integrability on a Bounded Set 305
 11.3 The Measure 307
 11.4 Negligible Sets 310
 11.5 A Characterization of Measurable Bounded Sets 313
 11.6 Continuous Functions and *L*-Integrable Functions............... 317

11.7   Limits and Derivatives under the Integration Sign 319
11.8   Reduction Formula.. 324
11.9   Change of Variables in the Integral 333
11.10   Change of Measure by Diffeomorphisms.......................... 340
11.11   The General Theorem on Change of Variables 342
11.12   Some Useful Transformations in $\mathbb{R}^2$ 344
11.13   Cylindrical and Spherical Coordinates in $\mathbb{R}^3$ 348
11.14   The Integral on Unbounded Sets 350
11.15   The Integral on M-Surfaces 360
11.16 M-Dimensional Measure 363
11.17   Length and Area 365
11.18   Approximation with Smooth M-Surfaces 369
11.19   The Integral on a Compact Manifold 370

12   Differential Forms 375
12.1   An Informal Definition 376
12.2   Algebraic Operations 378
12.3   The Exterior Differential 380
12.4   Differential Forms in $\mathbb{R}^3$ 382
12.5   The Integral on an M-Surface 384
12.6   Pull-Back Transformation 389
12.7   Oriented Boundary of a Rectangle 392
12.8   Gauss Formula 395
12.9   Oriented Boundary of an M-Surface.............................. 397
12.10   Stokes–Cartan Formula 401
12.11   Physical Interpretation of Curl and Divergence 406
12.12   The Integral on an Oriented Compact Manifold................... 408
12.13   Closed and Exact Differential Forms 411
12.14   On the Precise Definition of a Differential Form................. 418

Bibliography 425

Index 427

Preliminaries

This preliminary section has as its goal to introduce the main language and notations used in the book. Logic and set theory are treated in an informal way, without aiming for the highest mathematical rigor. Indeed, a rigorous treatment would require a solid background in mathematics, which students just starting out in their college career will not usually possess.

The Symbols of Logic

In mathematical language, we usually deal with propositions, indicated by $\mathcal{P}$, $\mathcal{Q}$, and so forth. Moreover, we are accustomed to combining them in different ways, for example,

$$\mathcal{P} \text{ and } \mathcal{Q}, \qquad \mathcal{P} \text{ or } \mathcal{Q}, \qquad \mathcal{P} \Rightarrow \mathcal{Q}, \qquad \mathcal{P} \Leftrightarrow \mathcal{Q}.$$

Let us explain the meaning of these. We start with

$$\mathcal{P} \text{ and } \mathcal{Q}.$$

It is true if both $\mathcal{P}$ and $\mathcal{Q}$ are true; otherwise, it is false. We can draw a table where all four cases are exemplified:[1]

$\mathcal{P}$	$\mathcal{Q}$	$\mathcal{P}$ and $\mathcal{Q}$
T	T	T
T	F	F
F	T	F
F	F	F

---

[1] In all tables, T means that a proposition is true, F that it is false.

Let us now consider

$$\mathcal{P} \text{ or } \mathcal{Q}.$$

It is true if at least one of the two is true and false when both $\mathcal{P}$ and $\mathcal{Q}$ are false. Here is the corresponding table:

$\mathcal{P}$	$\mathcal{Q}$	$\mathcal{P} \text{ or } \mathcal{Q}$
T	T	T
T	F	T
F	T	T
F	F	F

Let us now analyze

$$\mathcal{P} \Rightarrow \mathcal{Q}.$$

It is false only when $\mathcal{P}$ is true and $\mathcal{Q}$ is false; in all other cases it is true. Here is its table:

$\mathcal{P}$	$\mathcal{Q}$	$\mathcal{P} \Rightarrow \mathcal{Q}$
T	T	T
T	F	F
F	T	T
F	F	T

Let us conclude with

$$\mathcal{P} \Leftrightarrow \mathcal{Q}.$$

It is true if both are true or if both are false. Otherwise, it is false. And here is its table:

$\mathcal{P}$	$\mathcal{Q}$	$\mathcal{P} \Leftrightarrow \mathcal{Q}$
T	T	T
T	F	F
F	T	F
F	F	T

It is very important to be able to logically deny a proposition. The negation of $\mathcal{P}$ will be denoted by $\neg \mathcal{P}$ (read as "non $\mathcal{P}$"): it is true when $\mathcal{P}$ is false, and vice versa.

For example, we have the following *De Morgan rules*:

$$\neg\,(P \text{ and } Q) \quad \text{is equivalent to} \quad \neg P \text{ or } \neg Q,$$
$$\neg\,(P \text{ or } Q) \quad \text{is equivalent to} \quad \neg P \text{ and } \neg Q.$$

It is possible, moreover, to verify that

$$P \Rightarrow Q \quad \text{is equivalent to} \quad \neg P \text{ or } Q.$$

Consequently,

$$\neg\,(P \Rightarrow Q) \quad \text{is equivalent to} \quad P \text{ and } \neg Q.$$

Logical Propositions

Our propositions will often involve one or more "variables." For example, we could write them as follows: $P(x)$, which contains the variable x, in which case we will typically find the following two types of propositions. The first one,

$$\forall x : \; P(x),$$

means

"for every x one has that $P(x)$ is true."

The second,

$$\exists x : \; P(x),$$

means

"there exists at least one x for which $P(x)$ is true."

Let us see how their negation can be formulated. One has that

$$\neg\,(\forall x : \; P(x)) \quad \text{is equivalent to} \quad \exists x : \; \neg P(x)$$

and

$$\neg\,(\exists x : \; P(x)) \quad \text{is equivalent to} \quad \forall x : \; \neg P(x).$$

To be more precise, these x will be assumed to be the elements of some set. Thus, this leads us to a brief review of the theory of sets.

The Language of Set Theory

First Symbols

We are more or less familiar with some numerical sets like, for example,

$\mathbb{N}$, the set of natural numbers;
$\mathbb{Z}$, the set of integer numbers;
$\mathbb{Q}$, the set of rational numbers;
$\mathbb{R}$, the set of real numbers;
$\mathbb{C}$, the set of complex numbers.

Their nature will be further studied as we progress through the book, and several other sets will be introduced later. To treat sets correctly, we need to develop a proper language. This is why we will now introduce some symbols explaining their meaning.

Let us first introduce the symbol $\in$. Writing

$$a \in A$$

means "a belongs to the set A" or "a is an element of A." Its negation is written $a \notin A$ and reads "a does not belong to A" or "a is not an element of A."

For example, let $A = \{1, 2, 3\}$ be the set[2] whose elements are the three natural numbers 1, 2, and 3. We clearly have

$$1 \in A, \qquad 2 \in A, \qquad 3 \in A,$$

whereas

$$4 \notin A, \qquad \tfrac{1}{2} \notin A, \qquad \pi \notin A.$$

Let us now present the symbol $\subseteq$. We will write

$$A \subseteq B$$

and read "A is contained in B" whenever every element of A is also an element of B. In symbols,

$$x \in A \quad \Rightarrow \quad x \in B.$$

For example, if, as previously, $A = \{1, 2, 3\}$, we have that $A \subseteq \mathbb{N}$, but also $A \subseteq \mathbb{R}$.

---

[2] In this example, the set A is defined by listing its elements, which are finite in number.

If $A \subseteq B$, we also say that "A is a subset of B," and we can also write $B \supseteq A$. The negation of $A \subseteq B$ is written $A \nsubseteq B$ or $B \nsupseteq A$, and we read this as "A is not contained in B" or "B does not contain A."

We say that two sets A and B are "equal" if they coincide, i.e., if they have the same elements; in such a case we will write

$$A = B .$$

Therefore,

$$A = B \quad \Leftrightarrow \quad A \subseteq B \text{ and } B \subseteq A .$$

The negation of $A = B$ is written as $A \neq B$; in this case, we say that A and B are different, i.e., they do not coincide.

Let us emphasize the following "order relation" properties:

- $A \subseteq A$;
- $A \subseteq B$ and $B \subseteq A \Rightarrow A = B$;
- $A \subseteq B$ and $B \subseteq C \Rightarrow A \subseteq C$.

We end this section by introducing a very peculiar set, the "empty set," which is a set having no elements. It is denoted by the symbol

$$\emptyset .$$

It is convenient to consider $\emptyset$ as a subset of any other set, i.e.,

$$\emptyset \subseteq A , \quad \text{for any } A .$$

Some Examples of Sets

Let us begin with the simplest sets, those having a single element, for example,

$$A = \{3\} , \qquad A = \{\mathbb{N}\} , \qquad A = \{\emptyset\} .$$

The first one is a set having the number 3 as its only element. The second one has a single element $\mathbb{N}$, and the third one has only the element $\emptyset$. We thus observe that the elements of a set may be other sets as well. We could have sets of the type

$$A = \{\mathbb{N}, \mathbb{Q}\} , \qquad A = \{\emptyset, \mathbb{Z}, \mathbb{R}\} , \qquad A = \{\{\pi\}, \{1, 2, 3\}, \mathbb{N}\}$$

and of the type

$$A = \{3, \{3\}, \mathbb{N}, \{\mathbb{N}, \mathbb{Q}\}\} .$$

In this last case, one must be careful with symbols: we see that $3 \in A$, hence $\{3\} \subseteq A$, but also $\{3\} \in A$, with $\{3\}$ being an element of A.

Let us also consider, as a last example, the set

$$A = \{\emptyset, \{\emptyset\}\}.$$

We have that $\emptyset \in A$, since $\emptyset$ is one of the elements of A, and hence $\{\emptyset\} \subseteq A$. But we also have that $\{\emptyset\} \in A$, being $\{\emptyset\}$ an element of A, and hence $\{\{\emptyset\}\} \subseteq A$. We also recall that $\emptyset \subseteq A$, since this is true for every set.

Operations with Sets

It is normal practice to choose a "universal set" where we operate. We will denote it by E. All the objects we will speak of necessarily lie in this set.

We define the "intersection" of two sets A and B: it is the set[3]

$$A \cap B = \{x : x \in A \text{ and } x \in B\},$$

whose elements belong to both sets. Notice that the intersection could also be the empty set: in that case, we say that A and B are "disjoint."

On the other hand, the "union" of two sets A and B is the set

$$A \cup B = \{x : x \in A \text{ or } x \in B\},$$

whose elements belong to at least one of the two sets, and possibly also to both.

The "difference" of the two sets A and B is the set

$$A \setminus B = \{x : x \in A \text{ and } x \notin B\},$$

whose elements belong to the first set but not the second. In particular, the set $E \setminus A$ is said to be "complementary" to A and is denoted by $\mathcal{C}A$. Hence,

$$\mathcal{C}A = \{x : x \notin A\}.$$

The following *De Morgan rules* hold true:

$$\mathcal{C}(A \cap B) = \mathcal{C}A \cup \mathcal{C}B, \qquad \mathcal{C}(A \cup B) = \mathcal{C}A \cap \mathcal{C}B.$$

The "product" of the two sets A and B is the set

$$A \times B = \{(a, b) : a \in A, \ b \in B\},$$

---

[3] Here the sets are defined by specifying the properties that their elements must possess.

whose elements are the "ordered couples " (a, b), where at the first position we have an element of A and at the second position an element of B.

The Concept of Function

A "function" (sometimes also called "application") is defined by assigning three sets:

- set A, the "domain" of the function;
- set B, the "codomain" of the function;
- set $G \subseteq A \times B$, the "graph" of the function, having the following property: *for every $a \in A$ there is a unique $b \in B$ such that $(a, b) \in G$.*

A function defined in such a way is usually written $f : A \to B$ (read "f from A to B"). To each element a of the domain we have a well determined associated element b of the codomain: such a b will be denoted by $f(a)$, and we will write $a \mapsto f(a)$. We thus have that

$$(a, b) \in G \quad \Leftrightarrow \quad b = f(a),$$

i.e.,

$$G = \{(a, b) \in A \times B : b = f(a)\} = \{(a, f(a)) : a \in A\}.$$

For example, the function $f : \mathbb{N} \to \mathbb{R}$, defined as $f(n) = n/(n+1)$, associates to every $n \in \{0, 1, 2, 3, \dots\}$ the corresponding value $n/(n+1)$, i.e.,

$$n \mapsto \frac{n}{n+1}.$$

We will thus have that

$$0 \mapsto 0, \qquad 1 \mapsto \frac{1}{2}, \qquad 2 \mapsto \frac{2}{3}, \qquad 3 \mapsto \frac{3}{4}, \qquad \dots$$

Note that the values of this function are all rational numbers, so we could have defined using the same formula a function $f : \mathbb{N} \to \mathbb{Q}$. Such a function, however, is not the same as the previous one since they do not have the same codomain.

A function whose domain is the set $\mathbb{N}$ of natural numbers is also called a "sequence," and a different notation is usually preferred: if $s : \mathbb{N} \to B$ is such a sequence, instead of $s(n)$, it is customary to write s_n, and the sequence itself is denoted by $(s_n)_n$.

The function $f : \mathbb{R} \to \mathbb{R}$, defined by $f(x) = x^2$, associates to every $x \in \mathbb{R}$ its square. Notice that

$$f(-x) = f(x), \quad \text{for every } x \in \mathbb{R}.$$

We will say that such a function is "even." If, instead, as for the function $f : \mathbb{R} \to \mathbb{R}$ defined by $f(x) = x^3$, one has that

$$f(-x) = -f(x), \quad \text{for every } x \in \mathbb{R},$$

we will say that such a function is "odd." Clearly, a function could very well be neither even nor odd.

Sometimes it could be useful to use the notation $f(\cdot)$ instead of just f. For example, if $g : \mathbb{R} \times \mathbb{R} \to \mathbb{R}$ is a given function associating to each $(x, y) \in \mathbb{R}^2$ a real number $g(x, y)$, then by $g(\cdot, y) : \mathbb{R} \to \mathbb{R}$ we will denote the function $x \mapsto g(x, y)$ for any fixed $y \in \mathbb{R}$.

The "image" of the function $f : A \to B$ is the set

$$f(A) = \{f(a) : a \in A\},$$

and, in general, for every set $U \subseteq A$ we can write

$$f(U) = \{f(a) : a \in U\};$$

it is the image of the function $f|_U$, the restriction of f to the domain U.

The "composition" of two functions $f : A \to B$ and $g : B \to C$ is the function $g \circ f : A \to C$ defined as

$$(g \circ f)(a) = g(f(a)).$$

It could also be defined only assuming that the image of f is a subset of the domain of g.

A function $f : A \to B$ is said to be

- "injective" if $a_1 \neq a_2 \Rightarrow f(a_1) \neq f(a_2)$;
- "surjective" if $f(A) = B$;
- "bijective" if it is both injective and surjective.

If $f : A \to B$ is bijective, then for every $b \in B$ there is an $a \in A$ such that $f(a) = b$ (f is surjective), and such an element a is unique (f is injective). One can thus define a function from B to A that associates to every $b \in B$ the unique element $a \in A$ such that $f(a) = b$. This is the so-called "inverse function" of $f : A \to B$, and it is usually denoted by $f^{-1} : B \to A$. Thus,

$$f(a) = b \quad \Leftrightarrow \quad a = f^{-1}(b).$$

The word "bijective" will thus be synonymous with "invertible." Notice that, for every $a \in A$ and $b \in B$,

$$f^{-1}(f(a)) = a, \qquad f(f^{-1}(b)) = b.$$

For any function f, whether invertible or not, given a set $V \subseteq B$, it is common practice to write

$$f^{-1}(V) = \{a \in A : f(a) \in V\}.$$

This is the so-called "counterimage set" of V; it is composed of those elements a of A whose associated element $f(a)$ belongs to V.

To conclude this brief presentation, let us recall that, given two functions $f : A \to B$ and $g : A \to B$, if the codomain B has an addition operation, we can define the function $f + g : A \to B$ as follows:

$$(f + g)(a) = f(a) + g(a).$$

Similar definitions can be given for the difference, product, and quotient of two functions:

$$(f - g)(a) = f(a) - g(a),$$

$$(fg)(a) = f(a)\, g(a),$$

$$\left(\frac{f}{g}\right)(a) = \frac{f(a)}{g(a)}.$$

Part I

The Basics of Mathematical Analysis

Sets of Numbers and Metric Spaces

1

In this chapter, we introduce the main settings where all the theory will be developed. First, we discuss the sets of numbers $\mathbb{N}$, $\mathbb{Z}$, $\mathbb{Q}$, $\mathbb{R}$, and $\mathbb{C}$, then the space $\mathbb{R}^N$, and, finally, abstract metric spaces.

1.1   The Natural Numbers and the Induction Principle

In 1898 Giuseppe Peano, in his fundamental paper "Arithmetices principia: nova methodo exposita", provided an axiomatic description of the set of natural numbers $\mathbb{N}$. We briefly state those axioms as follows:

(a) There exists an element, called "zero," denoted by 0.
(b) Every element n has a "successor" n'.
(c) 0 is not the successor of any element.
(d) Different elements have different successors.
(e) **Induction principle:** If S is a subset of $\mathbb{N}$ such that
  (i) $0 \in S$,
  (ii) $n \in S \Rightarrow n' \in S$,
  then $S = \mathbb{N}$.

It is tacitly understood that condition *(ii)* must be verified for any $n \in \mathbb{N}$. We may therefore read it in the following way:

(ii) If, for some n, we have that $n \in S$, then also $n' \in S$.

We then introduce the familiar symbols $0' = 1$, $1' = 2$, $2' = 3$, and so on.
From these few axioms, making use of set theory, Peano showed how to recover all the properties of the natural numbers. In particular, we can define addition and

© The Author(s), under exclusive license to Springer Nature Switzerland AG 2023
A. Fonda, *A Modern Introduction to Mathematical Analysis*,
https://doi.org/10.1007/978-3-031-23713-3_1

multiplication such that

$$n' = n + 1.$$

Moreover, writing $m \leq n$ whenever there exists a $p \in \mathbb{N}$ such that $m + p = n$, we obtain an order relation. We will assume here that all the properties of addition, multiplication, and the order relation defined on $\mathbb{N}$ are well known.

1.1.1 Recursive Definitions

The induction principle can be used to define a sequence of objects

$$A_0, A_1, A_2, A_3, \ldots$$

We proceed in the following way:

(j) We define A_0.
(jj) Assuming that A_n has already been defined, for some n, we define A_{n+1}.

In this way, if we denote by S the set of those n for which A_n is well defined, it is easy to see that such a set S confirms (i) and (ii) in the induction principle. Hence, S coincides with $\mathbb{N}$, meaning that every A_n is well defined.

For example, we can define the "powers" a^n by setting

(j) $a^0 = 1$,
(jj) $a^{n+1} = a \cdot a^n$.

We then verify that[1]

$$a^1 = a \cdot a^0 = a \cdot 1 = a,$$
$$a^2 = a \cdot a^1 = a \cdot a,$$
$$a^3 = a \cdot a^2 = a \cdot a \cdot a$$
$$a^4 = a \cdot a^3 = a \cdot a \cdot a \cdot a$$

$$\ldots$$

Henceforth, we will assume that all elementary properties of powers are well known.

---

[1] If $a = 0$, it is sometimes a subtle matter to define 0^0. However, in this book we will always assume that $0^0 = 1$.

Let us now define the "factorial" $n!$ by setting

(j) $0! = 1$,
(jj) $(n + 1)! = (n + 1) \cdot n!$.

We then see that

$$1! = 1 \cdot 0! = 1 \cdot 1 = 1$$
$$2! = 2 \cdot 1! = 2 \cdot 1$$
$$3! = 3 \cdot 2! = 3 \cdot 2 \cdot 1$$
$$4! = 4 \cdot 3! = 4 \cdot 3 \cdot 2 \cdot 1$$

$$\ldots$$

Finally, let us define the "summation" of $\alpha_0, \alpha_1, \ldots, \alpha_n$ using the notation

$$\sum_{k=0}^{n} \alpha_k \,,$$

which reads "the sum of α_k when k goes from 0 to n." We set

$$\sum_{k=0}^{0} \alpha_k = \alpha_0 \,,$$

and, assuming that $\sum_{k=0}^{n} \alpha_k$ has been defined for some n, we set

$$\sum_{k=0}^{n+1} \alpha_k = \sum_{k=0}^{n} \alpha_k + \alpha_{n+1} \,.$$

In the preceding notation, an index appears, denoted by k; it takes all the integer values between 0 and n. Informally,

$$\sum_{k=0}^{n} \alpha_k = \alpha_0 + \alpha_1 + \alpha_2 + \cdots + \alpha_n \,.$$

The fact that the index is denoted by the letter k is unimportant; we can use any other letter or symbol to denote it, for instance

$$\sum_{j=0}^{n} \alpha_j \,, \qquad \sum_{\ell=0}^{n} \alpha_\ell \,, \qquad \sum_{m=0}^{n} \alpha_m \,, \qquad \sum_{\star=0}^{n} \alpha_\star \,, \quad \ldots$$

Notice that the same sum could be written

$$\sum_{k=1}^{n+1} \alpha_{k-1}, \qquad \sum_{k=2}^{n+2} \alpha_{k-2}, \qquad \cdots \qquad \sum_{k=m}^{n+m} \alpha_{k-m},$$

or even

$$\sum_{k=0}^{n} \alpha_{n-k}.$$

As you can see, the use of the summation symbol has many variants, and we will sometimes need them in what follows.

1.1.2 Proofs by Induction

The induction principle can also be used to prove a sequence of propositions

$$P_0, \ P_1, \ P_2, \ P_3, \ \ldots$$

We must proceed as follows:

(j) We verify P_0.
(jj) Assuming the truth of P_n for some n, we verify P_{n+1}.

In this way, denoting by S the set of those n for which P_n is true, then S verifies both (i) and (ii) in the induction principle. Hence, S coincides with $\mathbb{N}$, so all propositions P_n are true.

We now provide some examples.

Example 1 We want to prove the **Bernoulli inequality**

$$P_n: \qquad\qquad\qquad (1+a)^n \geq 1 + na.$$

We first see that P_0 is true since surely $(1+a)^0 \geq 1 + 0 \cdot a$. Let us now assume that P_n is true for some n; under this assumption, we need to verify P_{n+1}. Indeed, we have

$$(1+a)^{n+1} = (1+a)^n(1+a) \geq (1+na)(1+a) = 1+(n+1)a+na^2 \geq 1+(n+1)a,$$

hence, P_{n+1} is also true. In conclusion, P_n is true for every $n \in \mathbb{N}$.

Remark 1.1 In this section we are dealing with natural numbers; however, the Bernoulli inequality is true as well for any real number $a \geq -1$, and the proof is exactly the same. Similar remarks could be made for the other formulas in the following discussion.

Example 2 The following properties of summation can be proven by induction:

$$\sum_{k=0}^{n} (\alpha_k + \beta_k) = \sum_{k=0}^{n} \alpha_k + \sum_{k=0}^{n} \beta_k \,,$$

which informally reads

$$(\alpha_0 + \beta_0) + (\alpha_1 + \beta_1) + \cdots + (\alpha_n + \beta_n) = (\alpha_0 + \alpha_1 + \cdots + \alpha_n) + (\beta_0 + \beta_1 + \cdots + \beta_n) \,,$$

and

$$\sum_{k=0}^{n} (C\alpha_k) = C \left(\sum_{k=0}^{n} \alpha_k \right) ,$$

which informally reads

$$C\alpha_0 + C\alpha_1 + \cdots + C\alpha_n = C(\alpha_0 + \alpha_1 + \cdots + \alpha_n) \,.$$

Let us prove, for instance, the first one. We first verify that it holds for $n = 0$, with

$$\sum_{k=0}^{0} (\alpha_k + \beta_k) = \alpha_0 + \beta_0 = \sum_{k=0}^{0} \alpha_k + \sum_{k=0}^{0} \beta_k \,.$$

Assuming now that the formula is true for some n, we have

$$\sum_{k=0}^{n+1} (\alpha_k + \beta_k) = \sum_{k=0}^{n} (\alpha_k + \beta_k) + (\alpha_{n+1} + \beta_{n+1})$$

$$= \sum_{k=0}^{n} \alpha_k + \sum_{k=0}^{n} \beta_k + (\alpha_{n+1} + \beta_{n+1})$$

$$= \left(\sum_{k=0}^{n} \alpha_k + \alpha_{n+1} \right) + \left(\sum_{k=0}^{n} \beta_k + \beta_{n+1} \right) = \sum_{k=0}^{n+1} \alpha_k + \sum_{k=0}^{n+1} \beta_k \,,$$

so that the formula also holds for $n + 1$. The proof is thus complete.

Example 3 The following formula involves a "telescopic sum":

$$\sum_{k=0}^{n}(a_{k+1} - a_k) = a_{n+1} - a_0 \,,$$

which can be visualized as

$$(a_1 - a_0) + (a_2 - a_1) + (a_3 - a_2) + \cdots + (a_n - a_{n-1}) + (a_{n+1} - a_n) = a_{n+1} - a_0 \,.$$

It can also be proved by induction.

Example 4 Let us prove the identity

$$P_n : \qquad\qquad a^{n+1} - b^{n+1} = (a - b)\Big(\sum_{k=0}^{n} a^k b^{n-k}\Big) \,.$$

We first verify P_0, i.e.,

$$a^{0+1} - b^{0+1} = (a - b)a^0 b^{0-0} \,,$$

which is clearly true. Assume now P_n to be true for some $n \in \mathbb{N}$; then

$$(a - b)\Big(\sum_{k=0}^{n+1} a^k b^{n+1-k}\Big) = (a - b)\Big(\sum_{k=0}^{n} a^k b^{n+1-k}\Big) + (a - b)a^{n+1}b^0$$

$$= (a - b)\Big(\sum_{k=0}^{n} a^k b^{n-k}\Big)b + (a - b)a^{n+1}$$

$$= (a^{n+1} - b^{n+1})b + (a - b)a^{n+1} = a^{n+2} - b^{n+2} \,,$$

so that P_{n+1} is also true. Thus, we have proved that P_n is true for every $n \in \mathbb{N}$.

As particular cases of the preceding formula we have

$$a^2 - b^2 = (a - b)(a + b) \,,$$
$$a^3 - b^3 = (a - b)(a^2 + ab + b^2) \,,$$
$$a^4 - b^4 = (a - b)(a^3 + a^2 b + ab^2 + b^3) \,,$$
$$a^5 - b^5 = (a - b)(a^4 + a^3 b + a^2 b^2 + ab^3 + b^4) \,,$$

$$\cdots$$

Notice also the formula

$$\sum_{k=0}^{n} a^k = \frac{a^{n+1} - 1}{a - 1},$$

which holds for any $a \neq 1$, obtained from the preceding formula taking $b = 1$.

In some cases it could be useful to start the sequence of propositions, e.g., from P_1 instead of P_0 or from any other of them, say, $P_{\bar{n}}$. However, one can always reduce to the previous case by a shift of the indices, so that the principle of induction indeed remains of the same nature. Briefly, to prove the propositions

$$P_{\bar{n}}, \; P_{\bar{n}+1}, \; P_{\bar{n}+2}, \; P_{\bar{n}+3}, \ldots,$$

we verify the first proposition $P_{\bar{n}}$ and then, assuming that P_n is true for some $n \geq \bar{n}$, we verify P_{n+1}.

As an exercise, the reader could try to prove by induction the following identities:

$$1 + 2 + 3 + \cdots + n = \frac{n(n + 1)}{2},$$

$$1^2 + 2^2 + 3^2 + \cdots + n^2 = \frac{n(n + 1)(2n + 1)}{6},$$

$$1^3 + 2^3 + 3^3 + \cdots + n^3 = \frac{n^2(n + 1)^2}{4}.$$

Notice the beautiful equality

$$1^3 + 2^3 + 3^3 + \cdots + n^3 = (1 + 2 + 3 + \cdots + n)^2.$$

1.1.3 The Binomial Formula

Let us define, for any couple of natural numbers n, k such that $k \leq n$, the binomial coefficients

$$\binom{n}{k} = \frac{n!}{k!(n - k)!}.$$

The following identity holds:

$$\binom{n}{k - 1} + \binom{n}{k} = \binom{n + 1}{k}.$$

Indeed,

$$\binom{n}{k-1} + \binom{n}{k} = \frac{n!}{(k-1)!(n-k+1)!} + \frac{n!}{k!(n-k)!}$$

$$= \frac{n!k + n!(n-k+1)}{k!(n-k+1)!}$$

$$= \frac{n!(n+1)}{k!(n-k+1)!} = \frac{(n+1)!}{k!((n+1)-k)!}.$$

It is sometimes useful to represent the binomial coefficients in the so-called Pascal triangle

$$\binom{0}{0}$$

$$\binom{1}{0} \quad \binom{1}{1}$$

$$\binom{2}{0} \quad \binom{2}{1} \quad \binom{2}{2}$$

$$\binom{3}{0} \quad \binom{3}{1} \quad \binom{3}{2} \quad \binom{3}{3}$$

$$\binom{4}{0} \quad \binom{4}{1} \quad \binom{4}{2} \quad \binom{4}{3} \quad \binom{4}{4}$$

$$\binom{5}{0} \quad \binom{5}{1} \quad \binom{5}{2} \quad \binom{5}{3} \quad \binom{5}{4} \quad \binom{5}{5}$$

$$\cdots \quad \cdots \quad \cdots \quad \cdots \quad \cdots \quad \cdots \quad \cdots$$

which we can explicitly write as

$$1$$
$$1 \quad 1$$
$$1 \quad 2 \quad 1$$
$$1 \quad 3 \quad 3 \quad 1$$
$$1 \quad 4 \quad 6 \quad 4 \quad 1$$
$$1 \quad 5 \quad 10 \quad 10 \quad 5 \quad 1$$

$$\cdots \quad \cdots \quad \cdots \quad \cdots \quad \cdots$$

We will now prove, for every $n \in \mathbb{N}$, the **binomial formula**

$$P_n : \qquad (a+b)^n = \sum_{k=0}^{n} \binom{n}{k} a^{n-k} b^k .$$

It will be necessary to prove separately the case $n = 0$ and then start the induction from $n = 1$.

If $n = 0$, then

$$(a+b)^0 = \binom{0}{0} a^{0-0} b^0 ,$$

and the formula holds. Assuming now $n \geq 1$, we proceed by induction. We first see that it holds when $n = 1$:

$$(a+b)^1 = \binom{1}{0} a^{1-0} b^0 + \binom{1}{1} a^{1-1} b^1 .$$

Now, assuming that P_n is true for some $n \geq 1$, we prove that P_{n+1} is also true:

$$
\begin{aligned}
(a+b)^{n+1} &= (a+b)(a+b)^n \\
&= (a+b) \left(\sum_{k=0}^{n} \binom{n}{k} a^{n-k} b^k \right) \\
&= a \left(\sum_{k=0}^{n} \binom{n}{k} a^{n-k} b^k \right) + b \left(\sum_{k=0}^{n} \binom{n}{k} a^{n-k} b^k \right) \\
&= \sum_{k=0}^{n} \binom{n}{k} a^{n-k+1} b^k + \sum_{k=0}^{n} \binom{n}{k} a^{n-k} b^{k+1} \\
&= a^{n+1} + \sum_{k=1}^{n} \binom{n}{k} a^{n-k+1} b^k + \sum_{k=0}^{n-1} \binom{n}{k} a^{n-k} b^{k+1} + b^{n+1} \\
&= a^{n+1} + \sum_{k=1}^{n} \binom{n}{k} a^{n-k+1} b^k + \sum_{k=1}^{n} \binom{n}{k-1} a^{n-(k-1)} b^{(k-1)+1} + b^{n+1} \\
&= a^{n+1} + \sum_{k=1}^{n} \left[\binom{n}{k} + \binom{n}{k-1} \right] a^{n-k+1} b^k + b^{n+1}
\end{aligned}
$$

$$= a^{n+1} + \sum_{k=1}^{n} \binom{n+1}{k} a^{n-k+1} b^k + b^{n+1}$$

$$= \sum_{k=0}^{n+1} \binom{n+1}{k} a^{n+1-k} b^k .$$

We have thus proved by induction that P_n is true for every $n \in \mathbb{N}$.

As particular cases of the binomial formula we have

$$(a+b)^2 = a^2 + 2ab + b^2 ,$$

$$(a+b)^3 = a^3 + 3a^2b + 3ab^2 + b^3 ,$$

$$(a+b)^4 = a^4 + 4a^3b + 6a^2b^2 + 4ab^3 + b^4 ,$$

$$(a+b)^5 = a^5 + 5a^4b + 10a^3b^2 + 10a^2b^3 + 5ab^4 + b^5 ,$$

$$\dots$$

1.2   The Real Numbers

Starting from the set of natural numbers

$$\mathbb{N} = \{0, 1, 2, 3, \dots\} ,$$

by the use of set theory arguments it is possible first to construct the set of integer numbers

$$\mathbb{Z} = \{\dots, -3, -2, -1, 0, 1, 2, 3, \dots\}$$

and then the set of rational numbers

$$\mathbb{Q} = \left\{ \frac{m}{n} : m \in \mathbb{Z}, n \in \mathbb{N}, n \neq 0 \right\} .$$

This set has a lot of nice features from an algebraic point of view. Let us briefly review them.

1. An "order relation" $\leq$ is defined, with the following properties.
   For every choice of x, y, z
   (a) $x \leq x$.
   (b) $[x \leq y \text{ and } y \leq x] \Rightarrow x = y$.
   (c) $[x \leq y \text{ and } y \leq z] \Rightarrow x \leq z$.
   Moreover, such an order relation is "total" since any two elements x and y are comparable:

(d) $x \leq y$ or $y \leq x$.
If $x \leq y$, we will also write $y \geq x$. If $x \leq y$ and $y \neq x$, we will write $x < y$, or $y > x$.

2. An addition operation $+$ is defined, with the following properties.
   For any choice of x, y, z
   (a) (Associative) $x + (y + z) = (x + y) + z$.
   (b) There exists an "identity element" 0: we have $x + 0 = x = 0 + x$.
   (c) x has an "inverse element" $-x$: we have $x + (-x) = 0 = (-x) + x$.
   (d) (Commutative) $x + y = y + x$.
   (e) If $x \leq y$, then $x + z \leq y + z$.

3. A multiplication operation $\cdot$ is defined, with the following properties.
   For any choice of x, y, z
   (a) (Associative) $x \cdot (y \cdot z) = (x \cdot y) \cdot z$.
   (b) There exists an "identity element" 1: we have $x \cdot 1 = x = 1 \cdot x$.
   (c) If $x \neq 0$, then x has an "inverse element" x^{-1}: we have $x \cdot x^{-1} = 1 = x^{-1} \cdot x$.
   (d) (Commutative) $x \cdot y = y \cdot x$.
   (e) If $x \leq y$ and $z \geq 0$, then $x \cdot z \leq y \cdot z$,
   and a property involving both operations:
   (f) (Distributive) $x \cdot (y + z) = (x \cdot y) + (x \cdot z)$.

A set satisfying the foregoing properties is called an "ordered field." The set $\mathbb{Q}$ is, in some sense, the smallest ordered field.

We will often omit the symbol $\cdot$ in multiplication. Moreover, we adopt the usual notations, writing $z = y - x$ if $z + x = y$ and $z = \frac{y}{x}$ if $zx = y$, with $x \neq 0$. In particular, $x^{-1} = \frac{1}{x}$.

We rediscover the set $\mathbb{N}$ as a subset of $\mathbb{Q}$. Indeed, 0 and 1 are the identity elements of addition and multiplication, respectively, and then we have $2 = 1 + 1, 3 = 2 + 1$, and so on.

Besides its nice algebraic properties, the set of rational numbers $\mathbb{Q}$ is not rich enough to deal with such an elementary geometric problem as the measuring of the diagonal of a square whose side's length is 1, as the following theorem states.

Theorem 1.2 *There is no rational number x such that $x^2 = 2$.*

Proof By contradiction, assume that there exist $m, n \in \mathbb{N}$ different from 0 such that

$$\left(\frac{m}{n}\right)^2 = 2,$$

i.e., $m^2 = 2n^2$. Then m needs to be even, and so there exists a nonzero $m_1 \in \mathbb{N}$ such that $m = 2m_1$. We thus have $4m_1^2 = 2n^2$, i.e., $2m_1^2 = n^2$. But then n also needs to be even, and so there exists a nonzero $n_1 \in \mathbb{N}$ such that $2n_1 = n$. Hence,

$$\frac{m}{n} = \frac{m_1}{n_1} \quad \text{and} \quad \left(\frac{m_1}{n_1}\right)^2 = 2.$$

We can now repeat the argument as many times as we want, continuing the division by 2 of numerator and denominator:

$$\frac{m}{n} = \frac{m_1}{n_1} = \frac{m_2}{n_2} = \frac{m_3}{n_3} = \cdots = \frac{m_k}{n_k} = \ldots,$$

where m_k and n_k are nonzero natural numbers such that $m = 2^k m_k$, $n = 2^k n_k$. Then, since $n_k \geq 1$, we have that $n \geq 2^k$ for any natural number $k \geq 1$. In particular, $n \geq 2^n$. But the Bernoulli inequality tells us that $2^n = (1+1)^n \geq 1+n$, and all this implies that $n \geq 1 + n$, which is clearly false.   ∎

Therefore, one feels the need to further extend the set $\mathbb{Q}$ so as to be able to deal with this kind of problem. It is indeed possible to construct a set $\mathbb{R}$, containing $\mathbb{Q}$, which is an ordered field and, hence, satisfies properties (1), (2), and (3) and moreover satisfies the following property.

4. **Separation Property.** *Given two nonempty subsets A, B of $\mathbb{R}$ such that*

$$\forall a \in A \quad \forall b \in B \quad a \leq b,$$

there exists an element $c \in \mathbb{R}$ such that

$$\forall a \in A \quad \forall b \in B \quad a \leq c \leq b.$$

Mathematical Analysis is based on the set $\mathbb{R}$. We will assume that the reader is familiar with its elementary algebraic properties.

1.2.1 Supremum and Infimum

In this section we analyze some fundamental tools in $\mathbb{R}$. Let us start with some definitions.

A subset E of $\mathbb{R}$ is said to be "bounded from above" if there exists $\alpha \in \mathbb{R}$ such that, for every $x \in E$, we have $x \leq \alpha$; such a number α is then an "upper bound" of E. If, moreover, $\alpha \in E$, then we will say that α is the "maximum" of E, and we will write $\alpha = \max E$.

Analogously, the set E is said to be "bounded from below" if there exists $\beta \in \mathbb{R}$ such that, for every $x \in E$, we have $x \geq \beta$; such a number β is then a "lower bound" of E. If, moreover, we have that $\beta \in E$, then we will say that β is the "minimum" of E, and we will write $\beta = \min E$.

The set E is said to be "bounded" if it is both bounded from above and below.

Some remarks are in order. The maximum, when it exists, is unique. However, a set could be bounded from above without having a maximum, as the example $E = \{x \in \mathbb{R} : x < 0\}$ shows. Similar considerations can be made for the minimum.

Theorem 1.3 *If E is nonempty and bounded from above, then the set of all upper bounds of E has a minimum.*

Proof Let B be the set of all upper bounds of E. Then

$$\forall a \in E \quad \forall b \in B \quad a \leq b,$$

and by the separation property there exists a $c \in \mathbb{R}$ such that

$$\forall a \in E \quad \forall b \in B \quad a \leq c \leq b.$$

This means that c is an upper bound of E, and hence $c \in B$, and it is also a lower bound of B. Hence, $c = \min B$. ∎

If E is nonempty and bounded from above, the smallest upper bound of E is called the "supremum" of E: it is a real number $s \in \mathbb{R}$ that will be denoted by sup E. It is characterized by the following two properties:

(i) $\forall x \in E \quad x \leq s$.
(ii) $\forall s' < s \quad \exists x \in E: \quad x > s'$.

The two preceding properties can also be equivalently written as follows:

(i) $\forall x \in E \quad x \leq s$.
(ii) $\forall \varepsilon > 0 \quad \exists x \in E: \quad x > s - \varepsilon$.

In the second expression, we understand that the number $\varepsilon > 0$ can be arbitrarily small.

If the supremum sup E belongs to E, then sup $E = \max E$; as we saw earlier, however, this is not always the case.

We can state the following analogue of the preceding theorem.

Theorem 1.4 *If E is nonempty and bounded from below, then the set of all lower bounds of E has a maximum.*

If E is nonempty and bounded from below, the greatest lower bound of E is called the "infimum" of E: It is a real number $\overline{\imath} \in \mathbb{R}$ that will be denoted by inf E. It is characterized by the following two properties:

(i) $\forall x \in E \quad x \geq \overline{\imath}$.
(ii) $\forall \overline{\imath}' > \overline{\imath} \quad \exists x \in E: \quad x < \overline{\imath}'$.

The two foregoing properties can also be equivalently written as follows:

(i) $\forall x \in E \quad x \geq \bar{\iota}.$
(ii) $\forall \varepsilon > 0 \quad \exists x \in E : \quad x < \bar{\iota} + \varepsilon.$

If the infimum inf E belongs to E, then inf $E = \min E$; however, the minimum could not exist.

Notice that, defining the set

$$E^- = \{x \in \mathbb{R} : -x \in E\},$$

we have

$$E \text{ is bounded from above } \Leftrightarrow \ E^- \text{ is bounded from below,}$$

and in that case,

$$\sup E = -\inf E^-,$$

while

$$E \text{ is bounded from below } \Leftrightarrow \ E^- \text{ is bounded from above,}$$

and in that case,

$$\inf E = -\sup E^-.$$

In the case where E is *not* bounded from above, we will write

$$\sup E = +\infty.$$

Theorem 1.5 *The set* $\mathbb{N}$ *is not bounded from above, i.e.,* $\sup \mathbb{N} = +\infty$.

Proof Assume by contradiction that $\mathbb{N}$ is bounded from above. Then $s = \sup \mathbb{N}$ is a real number. By the properties of the supremum, there exists an $n \in \mathbb{N}$ such that $n > s - \frac{1}{2}$. But then $n + 1 \in \mathbb{N}$ and

$$n + 1 > s - \frac{1}{2} + 1 > s,$$

thereby contradicting the fact that s is an upper bound for $\mathbb{N}$.   ∎

In the case where E is *not* bounded from below, we will write

$$\inf E = -\infty.$$

For instance, we have $\inf \mathbb{Z} = -\infty$.

1.2.2   The Square Root

The following property will be used several times.

Lemma 1.6 *If* $0 \le \alpha < \beta$, *then* $\alpha^2 < \beta^2$.

Proof If $0 \le \alpha < \beta$, then $\alpha^2 = \alpha\alpha \le \alpha\beta < \beta\beta = \beta^2$.   ■

We will now prove that there exists a real number $c > 0$ such that $c^2 = 2$. Let us define the sets

$$A = \{x \in \mathbb{R} : x \ge 0 \text{ and } x^2 < 2\}, \quad B = \{x \in \mathbb{R} : x \ge 0 \text{ and } x^2 > 2\}.$$

Let us check that

$$\forall a \in A \quad \forall b \in B \quad a \le b.$$

Indeed, if not, it would be $0 \le b < a$, and, hence, by Lemma 1.6, $b^2 < a^2$. But we know that $a^2 < 2$ and $b^2 > 2$, hence $a^2 < b^2$, so we find a contradiction. By the separation property, there is an element $c \in \mathbb{R}$ such that

$$\forall a \in A \quad \forall b \in B \quad a \le c \le b.$$

Notice that, since $1 \in A$, it is surely the case that $c \ge 1$. We will now prove, by contradiction, that $c^2 = 2$.

If $c^2 > 2$, then, for $n \ge 1$,

$$\left(c - \frac{1}{n}\right)^2 = c^2 - \frac{2c}{n} + \frac{1}{n^2} \ge c^2 - \frac{2c}{n};$$

hence, if $n > 2c/(c^2 - 2)$, since $c \ge 1$ and $n \ge 1$, then

$$c - \frac{1}{n} \ge 0 \quad \text{and} \quad \left(c - \frac{1}{n}\right)^2 > 2,$$

so that $c - \frac{1}{n} \in B$. But then $c \le c - \frac{1}{n}$, which is clearly impossible.

If $c^2 < 2$, then, for $n \geq 1$,

$$\left(c + \frac{1}{n}\right)^2 = c^2 + \frac{2c}{n} + \frac{1}{n^2} \leq c^2 + \frac{2c}{n} + \frac{1}{n} = c^2 + \frac{2c+1}{n} \, ;$$

hence, if $n > (2c+1)/(2-c^2)$, then $(c + \frac{1}{n})^2 < 2$, and therefore $c + \frac{1}{n} \in A$. But then $c + \frac{1}{n} \leq c$, which is impossible.

Since both assumptions $c^2 > 2$ and $c^2 < 2$ lead to a contradiction, it must be that $c^2 = 2$.

Lemma 1.6 also tells us that there cannot exist any other positive solutions of the equation

$$x^2 = 2 \, ,$$

which therefore has exactly two solutions, $x = c$ and $x = -c$.

The same type of reasoning can be used to prove that, for any positive real number r, there exists a unique positive real number c such that $c^2 = r$. This number c is called the square root of r, and we write $c = \sqrt{r}$. Notice that the equation $x^2 = r$ has indeed two solutions, $x = \sqrt{r}$ and $x = -\sqrt{r}$. One also writes $\sqrt{0} = 0$, whereas the square root of a negative number remains undefined. This subject will be reconsidered in the framework of the complex numbers.

At this point we are ready to deal with the quadratic equation

$$ax^2 + bx + c = 0 \, ,$$

where a, b, and c are real numbers, with $a \neq 0$. It can be written equivalently as follows:

$$\left(x + \frac{b}{2a}\right)^2 = \frac{b^2 - 4ac}{(2a)^2} \, .$$

Thus, we see that the equation is solvable if and only if $b^2 - 4ac \geq 0$, in which case the solutions are

$$x = \frac{-b \pm \sqrt{b^2 - 4ac}}{2a} \, .$$

Let us now define the "absolute value" (or "modulus") of a real number x as

$$|x| = \sqrt{x^2} = \begin{cases} x & \text{if } x \geq 0 \, , \\ -x & \text{if } x < 0 \, . \end{cases}$$

The following properties may be easily verified. For every x_1, x_2 in $\mathbb{R}$,

$$|x_1 x_2| = |x_1|\,|x_2|\,,$$

whereas

$$|x_1 + x_2| \le |x_1| + |x_2|\,.$$

We will sometimes also need the inequality

$$\big||x_1| - |x_2|\big| \le |x_1 - x_2|$$

and the equivalence

$$|x| \le \alpha \quad \Leftrightarrow \quad -\alpha \le x \le \alpha\,.$$

1.2.3 Intervals

Let us explain what we mean by "interval."

Definition 1.7 An interval is a nonempty subset I of $\mathbb{R}$ having the following property: if α, β are two of its elements, then I contains all the numbers between them.

We will not exclude the case where I only has a single element.

Proposition 1.8 *Let I be an interval, define $a = \inf I$, $b = \sup I$ (possibly $a = -\infty$ or $b = +\infty$), and assume $a \ne b$. If $a < x < b$, then $x \in I$.*

Proof If $a < x < b$, then, by the properties of the infimum and supremum, we can always find α and β in I such that $a < \alpha < x < \beta < b$. Thus, by the preceding definition, I contains x. ∎

By the foregoing proposition, distinguishing the cases where a and b can be real numbers or not and whether or not they belong to I, we can conclude that any interval I must be among those in the following list:

$$[a, b] = \{x : a \le x \le b\}\,,$$

$$]a, b[= \{x : a < x < b\}\,,$$

$$[a, b[= \{x : a \le x < b\}\,,$$

$$]a, b] = \{x : a < x \le b\}\,,$$

$$[a, +\infty[= \{x : x \ge a\}\,,$$

$$]a, +\infty[= \{x : x > a\},$$

$$]-\infty, b] = \{x : x \leq b\},$$

$$]-\infty, b[= \{x : x < b\},$$

$$\mathbb{R}, \quad \text{sometimes denoted by }]-\infty, +\infty[.$$

Note that, when $a = b$, the interval $[a, a]$ reduces to a single point. In that case we say that the interval is "degenerate."

Theorem 1.9 (Cantor Theorem) *Let $(I_n)_n$ be a sequence of intervals of the type $I_n = [a_n, b_n]$, with $a_n \leq b_n$, such that*

$$I_0 \supseteq I_1 \supseteq I_2 \supseteq I_3 \supseteq \ldots$$

Then there is a $c \in \mathbb{R}$ that belongs to all the intervals I_n.

Proof Let us define the two sets

$$A = \{a_n : n \in \mathbb{N}\},$$

$$B = \{b_n : n \in \mathbb{N}\}.$$

For any a_n in A and any b_m in B (not necessarily having the same index), we have that $a_n \leq b_m$. Indeed, if $n \leq m$, then $I_n \supseteq I_m$, hence $a_n \leq a_m \leq b_m \leq b_n$. On the other hand, if $n > m$, then $I_m \supseteq I_n$, so that $a_m \leq a_n \leq b_n \leq b_m$. We have thus proved that

$$\forall a \in A \quad \forall b \in B \quad a \leq b.$$

By the separation property, there is a $c \in \mathbb{R}$ such that

$$\forall a \in A \quad \forall b \in B \quad a \leq c \leq b.$$

In particular, $a_n \leq c \leq b_n$, which means that $c \in I_n$ for every $n \in \mathbb{N}$. ∎

1.2.4 Properties of $\mathbb{Q}$ and $\mathbb{R} \setminus \mathbb{Q}$

We will now study the "density" of $\mathbb{Q}$ and $\mathbb{R} \setminus \mathbb{Q}$ in the set of real numbers $\mathbb{R}$.

Theorem 1.10 *Given two real numbers α, β, with $\alpha < \beta$, there always exists a rational number in between them.*

Proof Let us consider three different cases.

First Case: $0 \leq \alpha < \beta$. Choose $n \in \mathbb{N}$ such that

$$n > \frac{1}{\beta - \alpha} \,,$$

and let $m \in \mathbb{N}$ be the greatest natural number such that

$$m < n\beta \,.$$

Then clearly $\frac{m}{n} < \beta$, and we will now show that it must be that $\frac{m}{n} > \alpha$. By contradiction, assume that $\frac{m}{n} \leq \alpha$; then

$$\frac{m+1}{n} \leq \alpha + \frac{1}{n} < \alpha + (\beta - \alpha) = \beta \,,$$

meaning $m + 1 < n\beta$, in contradiction to the fact that m is the greatest natural number less than $n\beta$.

Second Case: $\alpha < 0 < \beta$. It is sufficient to choose 0, which is a rational number.

Third Case: $\alpha < \beta \leq 0$. We can reduce this case to the first case, changing signs: since $0 \leq -\beta < -\alpha$, there exists a rational number $\frac{m}{n}$ such that $-\beta < \frac{m}{n} < -\alpha$. Hence, $\alpha < -\frac{m}{n} < \beta$. ∎

Theorem 1.11 *Given two real numbers* α, β, *with* $\alpha < \beta$, *there always exists an irrational number in between them.*

Proof By the previous theorem, there exists a rational number $\frac{m}{n}$ such that

$$\alpha + \sqrt{2} < \frac{m}{n} < \beta + \sqrt{2}.$$

Consequently,

$$\alpha < \frac{m}{n} - \sqrt{2} < \beta \,,$$

with $\frac{m}{n} - \sqrt{2} \notin \mathbb{Q}$. ∎

We will now discover a crucial difference between the sets $\mathbb{Q}$ and $\mathbb{R} \setminus \mathbb{Q}$. Let us consider the following sequence of nonnegative rational numbers:

0	1	2	3	4	5	6	7	8	9	10	11	12	13	14	15	...
↓	↓	↓	↓	↓	↓	↓	↓	↓	↓	↓	↓	↓	↓	↓	↓	
$\frac{0}{1}$	$\frac{1}{1}$	$\frac{1}{2}$	$\frac{2}{1}$	$\frac{1}{3}$	$\frac{2}{2}$	$\frac{3}{1}$	$\frac{1}{4}$	$\frac{2}{3}$	$\frac{3}{2}$	$\frac{4}{1}$	$\frac{1}{5}$	$\frac{2}{4}$	$\frac{3}{3}$	$\frac{4}{2}$	$\frac{5}{1}$	...

As you can see, the sequence is built choosing rational numbers having the sum of their numerator and denominator equal to 1, then 2, then 3, and so on. In this way, all nonnegative rational numbers will sooner or later appear in the list. We now modify it in order to make it injective, checking from the beginning all the numbers in the list, one by one, and eliminating those that had already appeared previously:

$$
\begin{array}{cccccccccccccccccc}
0 & 1 & 2 & 3 & 4 & 5 & 6 & 7 & 8 & 9 & 10 & 11 & 12 & 13 & 14 & 15 & \dots \\
\downarrow & \downarrow & \downarrow & \downarrow & \downarrow & \downarrow & \downarrow & \downarrow & \downarrow & \downarrow & \downarrow & \downarrow & \downarrow & \downarrow & \downarrow & \downarrow & \\
\frac{0}{1} & \frac{1}{1} & \frac{1}{2} & \frac{2}{1} & \frac{1}{3} & \frac{3}{1} & \frac{1}{4} & \frac{2}{3} & \frac{3}{2} & \frac{4}{1} & \frac{1}{5} & \frac{5}{1} & \frac{1}{6} & \frac{2}{5} & \frac{3}{4} & \frac{4}{3} & \dots
\end{array}
$$

We are now ready to introduce the negative numbers as well, so as to obtain *all* the rationals:

$$
\begin{array}{cccccccccccccccccc}
0 & 1 & 2 & 3 & 4 & 5 & 6 & 7 & 8 & 9 & 10 & 11 & 12 & 13 & 14 & 15 & \dots \\
\downarrow & \downarrow & \downarrow & \downarrow & \downarrow & \downarrow & \downarrow & \downarrow & \downarrow & \downarrow & \downarrow & \downarrow & \downarrow & \downarrow & \downarrow & \downarrow & \\
\frac{0}{1} & \frac{1}{1} & -\frac{1}{1} & \frac{1}{2} & -\frac{1}{2} & \frac{2}{1} & -\frac{2}{1} & \frac{1}{3} & -\frac{1}{3} & \frac{3}{1} & -\frac{3}{1} & \frac{1}{4} & -\frac{1}{4} & \frac{2}{3} & -\frac{2}{3} & \frac{3}{2} & \dots
\end{array}
$$

In this way, we have constructed a bijective function $\varphi : \mathbb{N} \to \mathbb{Q}$. We will thus say that $\mathbb{Q}$ is a "countable" set.

Let us now prove that $\mathbb{R}$ is not countable, i.e., there cannot exist any bijective function $\varphi : \mathbb{N} \to \mathbb{R}$. We will show this assuming by contradiction that there exists a surjective function $\psi : \mathbb{N} \to [0, 1]$. Divide the interval $[0, 1]$ into three equal parts:

$$
\left[0, \tfrac{1}{3}\right], \qquad \left[\tfrac{1}{3}, \tfrac{2}{3}\right], \qquad \left[\tfrac{2}{3}, 1\right].
$$

Choose one of them, and denote it by I_0, with the property that $\psi(0) \notin I_0$. Now iterate this procedure: divide I_0 into three equal parts and denote by $I_1 = [a_1, b_1]$ one of these with the property that $\psi(1) \notin I_1$. Then divide I_1 into three equal parts and denote by $I_2 = [a_2, b_2]$ one of these with the property that $\psi(2) \notin I_2$, and so on. In this way, we have constructed a sequence of intervals $I_n = [a_n, b_n]$ with the property that $\psi(n) \notin I_n$ for every $n \in \mathbb{N}$ and

$$
I_0 \supseteq I_1 \supseteq I_2 \supseteq I_3 \supseteq \dots
$$

By Cantor's Theorem 1.9, there is a $c \in \mathbb{R}$ that belongs to all the intervals I_n. Hence, $c \neq \psi(n)$ for every $n \in \mathbb{N}$, contradicting the surjectivity of the function ψ.

We now claim that $\mathbb{R} \setminus \mathbb{Q}$ cannot be countable. Indeed, if it were countable, we would have an injective sequence $(\alpha_n)_n$ such that $\{\alpha_n : n \in \mathbb{N}\} = \mathbb{R} \setminus \mathbb{Q}$. On the other hand, since we know that $\mathbb{Q}$ is countable, there is an injective sequence $(\beta_n)_n$

such that $\{\beta_n : n \in \mathbb{N}\} = \mathbb{Q}$. Then the sequence defined as

$$\alpha_0 , \ \beta_0 , \ \alpha_1 , \ \beta_1 , \ \alpha_2 , \ \beta_2 , \ \alpha_3 , \ \beta_3 , \ \ldots$$

would contain all real numbers, and we know that this is impossible. Our claim is thus proved.

1.3   The Complex Numbers

Let us consider the set

$$\mathbb{R} \times \mathbb{R} = \{(a, b) : a \in \mathbb{R}, \ b \in \mathbb{R}\} ,$$

which is often denoted by $\mathbb{R}^2$. We define an addition operation as

$$(a, b) + (a', b') = (a + a', b + b') .$$

The following properties are readily verified. For any choice of (a, b), (a', b'), (a'', b''),

(a) (Associative) $(a, b) + [(a', b') + (a'', b'')] = [(a, b) + (a', b')] + (a'', b'')$.
(b) There exists an "identity element" $(0, 0)$: We have

$$(a, b) + (0, 0) = (a, b) = (0, 0) + (a, b) .$$

(c) (a, b) has an "inverse element" $-(a, b) = (-a, -b)$: We have

$$(a, b) + (-a, -b) = (0, 0) = (-a, -b) + (a, b) .$$

(d) (Commutative) $(a, b) + (a', b') = (a', b') + (a, b)$.

We also define a multiplication operation $\cdot$ as

$$(a, b) \cdot (a', b') = (aa' - bb', ab' + ba') .$$

We can then verify the following properties. For any choice of (a, b), (a', b'), (a'', b''),

(a) (Associative) $(a, b) \cdot [(a', b') \cdot (a'', b'')] = [(a, b) \cdot (a', b')] \cdot (a'', b'')$.
(b) There exists an "identity element" $(1, 0)$: We have

$$(a, b) \cdot (1, 0) = (a, b) = (1, 0) \cdot (a, b) .$$

(c) If $(a, b) \neq (0, 0)$, then (a, b) has an "inverse element"

$$(a, b)^{-1} = \left(\frac{a}{a^2 + b^2}, \frac{-b}{a^2 + b^2} \right).$$

We have

$$(a, b) \cdot \left(\frac{a}{a^2 + b^2}, \frac{-b}{a^2 + b^2} \right) = (1, 0) = \left(\frac{a}{a^2 + b^2}, \frac{-b}{a^2 + b^2} \right) \cdot (a, b).$$

(d) (Commutative) $(a, b) \cdot (a', b') = (a', b') \cdot (a, b)$.
(e) (Distributive) $(a, b) \cdot [(a', b') + (a'', b'')] = [(a, b) \cdot (a', b')] + [(a, b) \cdot (a'', b'')]$.

(Henceforth, we will often omit the sign "$\cdot$".) In this way, $(\mathbb{R}^2, +, \cdot)$ has the algebraic structure of a *field*; it is indicated by $\mathbb{C}$ and referred to as a *complex field*. Its elements will be called "complex numbers."

We can view $\mathbb{C}$ as an extension of $\mathbb{R}$ identifying each element of the type $(a, 0)$ with the corresponding real number a. The operations of addition and multiplication are indeed preserved:

$$(a, 0) + (a', 0) = (a + a', 0),$$
$$(a, 0) \cdot (a', 0) = (aa', 0).$$

We now focus on the identity

$$(a, b) = (a, 0) + (0, 1)(b, 0).$$

It is worth introducing a new symbol for the element $(0, 1)$. We will write

$$(0, 1) = i.$$

In this way, having identified $(a, 0)$ with a and $(b, 0)$ with b, we can write

$$(a, b) = a + ib.$$

For any complex number $z = a + ib$, the real numbers a and b are called the "real part" and "imaginary part" of z, respectively, and they are denoted by

$$a = \Re(z), \qquad b = \Im(z).$$

Now we present a crucial identity:

$$i^2 = (0, 1)(0, 1) = (-1, 0) = -1.$$

Using this simple information, we can verify that all the usual symbolic rules are satisfied; for instance,

$$(a + ib) + (a' + ib') = (a + a') + i(b + b').$$

$$(a + ib)(a' + ib') = (aa' - bb') + i(ab' + ba').$$

We are therefore allowed to manipulate complex numbers using all the algebraic rules we know well. In the next section we provide an example.

1.3.1   Algebraic Equations in $\mathbb{C}$

Let $z = a + ib$ be a fixed complex number, with $a, b \in \mathbb{R}$. We want to solve the equation

$$u^2 = z.$$

We will refer to the solutions $u \in \mathbb{C}$ as "complex square roots" of the number z (or "square roots" for short, being careful not to confuse them with the notion of square root already introduced in $\mathbb{R}$). If $b = 0$, then we find

$$u = \begin{cases} \pm\sqrt{a} & \text{if } a \geq 0, \\ \pm i\sqrt{-a} & \text{if } a < 0. \end{cases}$$

Otherwise, if $b \neq 0$, then let us write $u = x + iy$. Then

$$x^2 - y^2 = a, \qquad 2xy = b.$$

Since $b \neq 0$, we have $x \neq 0$ and $y \neq 0$. We can then write $y = \frac{b}{2x}$ and obtain

$$x^4 - ax^2 - \frac{b^2}{4} = 0,$$

whence

$$x^2 = \frac{a + \sqrt{a^2 + b^2}}{2}.$$

Thus, we have found the two solutions

$$u = \pm \left[\sqrt{\frac{a + \sqrt{a^2 + b^2}}{2}} + i\, \frac{b}{\sqrt{2(a + \sqrt{a^2 + b^2})}} \right].$$

We now turn to the quadratic equation

$$au^2 + bu + c = 0,$$

where a, b, and c are any fixed complex numbers, with $a \neq 0$. As we saw in the real case, this equation is equivalent to

$$\left(u + \frac{b}{2a}\right)^2 = \frac{b^2 - 4ac}{(2a)^2};$$

hence, setting

$$v = u + \frac{b}{2a}, \quad z = \frac{b^2 - 4ac}{(2a)^2},$$

we are led to $v^2 = z$, i.e., to the problem of finding the square roots of z, a problem we already know how to solve.

To conclude, for a more general polynomial equation

$$a_n u^n + a_{n-1} u^{n-1} + \cdots + a_1 u + a_0 = 0,$$

where $a_0, a_1, \ldots, a_n$ are any fixed complex numbers, with $a_n \neq 0$, we have the following theorem.

Theorem 1.12 (Fundamental Theorem of Algebra) *Any polynomial equation has, in the complex field, at least one solution.*

The problem of finding a general procedure to determine the solutions of the foregoing equation has troubled mathematicians for a very long time. We encountered it in the case $n = 2$, and it was also settled if $n = 3$ or 4. However, if $n \geq 5$, then it has finally been proved that such a general procedure does not exist.

1.3.2 The Modulus of a Complex Number

We now examine some additional properties of complex numbers. If $z = a + ib$, with $a, b \in \mathbb{R}$, then we define the "modulus" of z,

$$|z| = \sqrt{a^2 + b^2}.$$

Notice that, if $z = a \in \mathbb{R}$, then we recover the absolute value

$$|a| = \sqrt{a^2} = \begin{cases} a & \text{if } a \geq 0, \\ -a & \text{if } a < 0. \end{cases}$$

Given two complex numbers z_1 and z_2, let us verify the identity

$$|z_1 z_2| = |z_1|\,|z_2|\,.$$

Indeed, if $z_1 = a_1 + ib_1$ and $z_2 = a_2 + ib_2$, then

$$\begin{aligned}
|z_1 z_2|^2 &= (a_1 a_2 - b_1 b_2)^2 + (a_1 b_2 + b_1 a_2)^2 \\
&= a_1^2 a_2^2 - 2 a_1 a_2 b_1 b_2 + b_1^2 b_2^2 + a_1^2 b_2^2 + 2 a_1 b_2 b_1 a_2 + b_1^2 a_2^2 \\
&= a_1^2 a_2^2 + b_1^2 b_2^2 + a_1^2 b_2^2 + b_1^2 a_2^2 \\
&= (a_1^2 + b_1^2)(a_2^2 + b_2^2) = |z_1|^2 |z_2|^2\,.
\end{aligned}$$

In particular, if the two numbers coincide, then

$$|z^2| = |z|^2\,,$$

and it can be proved by induction that, for every $n \in \mathbb{N}$,

$$|z^n| = |z|^n\,.$$

Moreover, if $z \neq 0$, since $|z^{-1} z| = 1$, then we have

$$|z^{-1}| = |z|^{-1}\,.$$

Hence, for any positive integer n,

$$|z^{-n}| = |(z^{-1})^n| = |z^{-1}|^n = (|z|^{-1})^n = |z|^{-n}.$$

Thus, we have seen that the equality $|z^n| = |z|^n$ holds for every $n \in \mathbb{Z}$.

For any complex number $z = a + ib$ let us define its "complex conjugate" $z^* = a - ib$ (sometimes denoted by $\bar{z}$). The following properties hold:

$$(z_1 + z_2)^* = z_1^* + z_2^*\,,$$
$$(z_1 z_2)^* = z_1^* z_2^*\,,$$
$$z^{**} = z\,,$$
$$|z^*| = |z|\,,$$
$$z z^* = |z|^2\,,$$
$$\Re(z_1 + z_2) = \Re(z_1) + \Re(z_2)\,, \qquad \Im(z_1 + z_2) = \Im(z_1) + \Im(z_2)\,,$$
$$\Re(z) = \frac{1}{2}(z + z^*)\,, \qquad \Im(z) = \frac{1}{2i}(z - z^*)\,,$$
$$|\Re(z)| \leq |z|\,, \qquad |\Im(z)| \leq |z|\,,$$

and, if $z \neq 0$,

$$z^{-1} = \frac{z^*}{|z|^2}.$$

Let us now prove the following subadditivity property of the modulus:

$$|z_1 + z_2| \leq |z_1| + |z_2|.$$

Indeed,

$$
\begin{aligned}
|z_1 + z_2|^2 &= (z_1 + z_2)(z_1 + z_2)^* \\
&= (z_1 + z_2)(z_1^* + z_2^*) \\
&= z_1 z_1^* + z_1 z_2^* + z_2 z_1^* + z_2 z_2^* \\
&= |z_1|^2 + z_1 z_2^* + (z_1 z_2^*)^* + |z_2|^2 \\
&= |z_1|^2 + 2\Re(z_1 z_2^*) + |z_2|^2 \\
&\leq |z_1|^2 + 2|z_1 z_2^*| + |z_2|^2 \\
&= |z_1|^2 + 2|z_1| \, |z_2^*| + |z_2|^2 \\
&= |z_1|^2 + 2|z_1| \, |z_2| + |z_2|^2 = (|z_1| + |z_2|)^2,
\end{aligned}
$$

and Lemma 1.6 completes the proof.

1.4   The Space $\mathbb{R}^N$

Let us introduce the set $\mathbb{R}^N$, composed of the N-tuples $(x_1, x_2, \ldots, x_N)$, where x_1, $x_2, \ldots, x_N$ are real numbers. We will denote its elements by the symbols

$$\boldsymbol{x}, \boldsymbol{x}', \boldsymbol{x}'', \ldots$$

Let us start by defining an addition operation in $\mathbb{R}^N$. Given two elements

$$\boldsymbol{x} = (x_1, x_2, \ldots, x_N) \quad \text{and} \quad \boldsymbol{x}' = (x_1', x_2', \ldots, x_N'),$$

we set

$$\boldsymbol{x} + \boldsymbol{x}' = (x_1 + x_1', x_2 + x_2', \ldots, x_N + x_N').$$

The following properties hold. For any choice of $\boldsymbol{x}, \boldsymbol{x}', \boldsymbol{x}''$,

(a) (Associative) $(\boldsymbol{x} + \boldsymbol{x}') + \boldsymbol{x}'' = \boldsymbol{x} + (\boldsymbol{x}' + \boldsymbol{x}'')$.
(b) There exists an "identity element" $\boldsymbol{0} = (0, 0, \ldots, 0)$.
    We have $\boldsymbol{x} + \boldsymbol{0} = \boldsymbol{x} = \boldsymbol{0} + \boldsymbol{x}$.
(c) $\boldsymbol{x} = (x_1, x_2, \ldots, x_N)$ has an "inverse element" $(-\boldsymbol{x}) = (-x_1, -x_2, \ldots, -x_N)$. We have $\boldsymbol{x} + (-\boldsymbol{x}) = \boldsymbol{0} = (-\boldsymbol{x}) + \boldsymbol{x}$.
(d) (Commutative) $\boldsymbol{x} + \boldsymbol{x}' = \boldsymbol{x}' + \boldsymbol{x}$.

Therefore, $(\mathbb{R}^N, +)$ is an "abelian group." As usual, we write $\boldsymbol{x} - \boldsymbol{x}'$ to denote $\boldsymbol{x} + (-\boldsymbol{x}')$.

We now define the product of an element of $\mathbb{R}^N$ by a real number. Given $\boldsymbol{x} = (x_1, x_2, \ldots, x_N) \in \mathbb{R}^N$ and $\alpha \in \mathbb{R}$, we set

$$\alpha \boldsymbol{x} = (\alpha x_1, \alpha x_2, \ldots, \alpha x_N).$$

The following properties hold:

(a) $\alpha(\beta \boldsymbol{x}) = (\alpha\beta)\boldsymbol{x}$.
(b) $(\alpha + \beta)\boldsymbol{x} = (\alpha \boldsymbol{x}) + (\beta \boldsymbol{x})$.
(c) $\alpha(\boldsymbol{x} + \boldsymbol{x}') = (\alpha \boldsymbol{x}) + (\alpha \boldsymbol{x}')$.
(d) $1\boldsymbol{x} = \boldsymbol{x}$.

With the preceding operations, $\mathbb{R}^N$ is a "vector space," and we will call its elements "vectors." In this environment, the real numbers will be called "scalars."

It would be useful to introduce here the "scalar product" of two vectors. Given $\boldsymbol{x}$ and $\boldsymbol{x}'$ as previously, we define the real number

$$\boldsymbol{x} \cdot \boldsymbol{x}' = \sum_{k=1}^{N} x_k x_k' = x_1 x_1' + x_2 x_2' + \cdots + x_N x_N'.$$

The scalar product is also denoted by a variety of symbols, for example,

$$\langle \boldsymbol{x} | \boldsymbol{x}' \rangle, \quad \langle \boldsymbol{x}, \boldsymbol{x}' \rangle, \quad (\boldsymbol{x} | \boldsymbol{x}'), \quad (\boldsymbol{x}, \boldsymbol{x}').$$

The following properties hold:

(a) $\boldsymbol{x} \cdot \boldsymbol{x} \geq 0$.
(b) $\boldsymbol{x} \cdot \boldsymbol{x} = 0 \Leftrightarrow \boldsymbol{x} = \boldsymbol{0}$.
(c) $(\boldsymbol{x} + \boldsymbol{x}') \cdot \boldsymbol{x}'' = (\boldsymbol{x} \cdot \boldsymbol{x}'') + (\boldsymbol{x}' \cdot \boldsymbol{x}'')$.
(d) $(\alpha \boldsymbol{x}) \cdot \boldsymbol{x}' = \alpha(\boldsymbol{x} \cdot \boldsymbol{x}')$.
(e) $\boldsymbol{x} \cdot \boldsymbol{x}' = \boldsymbol{x}' \cdot \boldsymbol{x}$.

If $\boldsymbol{x} \cdot \boldsymbol{x}' = 0$, we say that the two vectors $\boldsymbol{x}$ and $\boldsymbol{x}'$ are "orthogonal."

Let us finally define, but only in the three-dimensional space $\mathbb{R}^3$, the "cross product" of two vectors. Given $\boldsymbol{x} = (x_1, x_2, x_3)$ and $\boldsymbol{x}' = (x_1', x_2', x_3')$, we set

$$\boldsymbol{x} \times \boldsymbol{x}' = (x_2 x_3' - x_3 x_2', \ x_3 x_1' - x_1 x_3', \ x_1 x_2' - x_2 x_1').$$

It can be verified that $\boldsymbol{x} \times \boldsymbol{x}'$ is orthogonal to both $\boldsymbol{x}$ and $\boldsymbol{x}'$:

$$(\boldsymbol{x} \times \boldsymbol{x}') \cdot \boldsymbol{x} = 0, \qquad (\boldsymbol{x} \times \boldsymbol{x}') \cdot \boldsymbol{x}' = 0.$$

Moreover, when $\boldsymbol{x} \times \boldsymbol{x}' \neq \boldsymbol{0}$, since

$$\det \begin{pmatrix} x_1 & x_2 & x_3 \\ x_1' & x_2' & x_3' \\ x_2 x_3' - x_3 x_2' & x_3 x_1' - x_1 x_3' & x_1 x_2' - x_2 x_1' \end{pmatrix} > 0,$$

the direction of $\boldsymbol{x} \times \boldsymbol{x}'$ is provided by the so-called "right-hand rule." This means that the triple $(\boldsymbol{x}, \boldsymbol{x}', \boldsymbol{x} \times \boldsymbol{x}')$ has the same orientation as $(\boldsymbol{e}_1, \boldsymbol{e}_2, \boldsymbol{e}_3)$, the canonical basis

$$\boldsymbol{e}_1 = (1, 0, 0), \quad \boldsymbol{e}_2 = (0, 1, 0), \quad \boldsymbol{e}_3 = (0, 0, 1).$$

What follows are some properties of the cross product:

(a) $(\boldsymbol{x} + \boldsymbol{x}') \times \boldsymbol{x}'' = (\boldsymbol{x} \times \boldsymbol{x}'') + (\boldsymbol{x}' \times \boldsymbol{x}'')$;
(b) $(\alpha \boldsymbol{x}) \times \boldsymbol{x}' = \alpha (\boldsymbol{x} \times \boldsymbol{x}')$;
(c) $\boldsymbol{x} \times \boldsymbol{x}' = -\boldsymbol{x}' \times \boldsymbol{x}$.

One can also prove the **Jacobi identity**

$$\boldsymbol{x} \times (\boldsymbol{x}' \times \boldsymbol{x}'') + \boldsymbol{x}' \times (\boldsymbol{x}'' \times \boldsymbol{x}) + \boldsymbol{x}'' \times (\boldsymbol{x} \times \boldsymbol{x}') = \boldsymbol{0}.$$

Finally, note that

$$\boldsymbol{e}_1 \times \boldsymbol{e}_2 = \boldsymbol{e}_3, \quad \boldsymbol{e}_2 \times \boldsymbol{e}_3 = \boldsymbol{e}_1, \quad \boldsymbol{e}_3 \times \boldsymbol{e}_1 = \boldsymbol{e}_2.$$

1.4.1   Euclidean Norm and Distance

Starting from the scalar product, we can define the "Euclidean norm" of a vector $\boldsymbol{x} = (x_1, x_2, \ldots, x_N)$ as

$$\|\boldsymbol{x}\| = \sqrt{\boldsymbol{x} \cdot \boldsymbol{x}} = \sqrt{\sum_{k=1}^{N} x_k^2}.$$

The following properties hold:

(a) $\|\boldsymbol{x}\| \geq 0$.
(b) $\|\boldsymbol{x}\| = 0 \Leftrightarrow \boldsymbol{x} = \boldsymbol{0}$.
(c) $\|\alpha\boldsymbol{x}\| = |\alpha|\,\|\boldsymbol{x}\|$.
(d) $\|\boldsymbol{x} + \boldsymbol{x}'\| \leq \|\boldsymbol{x}\| + \|\boldsymbol{x}'\|$.

To prove the subadditivity property d), we need the following **Schwarz Inequality**.

Theorem 1.13 *For any two vectors $\boldsymbol{x}$, $\boldsymbol{x}'$, we have that*

$$|\boldsymbol{x} \cdot \boldsymbol{x}'| \leq \|\boldsymbol{x}\|\,\|\boldsymbol{x}'\|\,.$$

Equality holds if and only if $\boldsymbol{x}$ and $\boldsymbol{x}'$ are linearly dependent.

Proof The inequality surely holds if $\boldsymbol{x}' = \boldsymbol{0}$, since in that case $\boldsymbol{x} \cdot \boldsymbol{x}' = 0$ and $\|\boldsymbol{x}'\| = 0$. Assume, then, $\boldsymbol{x}' \neq \boldsymbol{0}$. For any $\alpha \in \mathbb{R}$ we have

$$0 \leq \|\boldsymbol{x} - \alpha\boldsymbol{x}'\|^2 = (\boldsymbol{x} - \alpha\boldsymbol{x}') \cdot (\boldsymbol{x} - \alpha\boldsymbol{x}') = \|\boldsymbol{x}\|^2 - 2\alpha\boldsymbol{x} \cdot \boldsymbol{x}' + \alpha^2\|\boldsymbol{x}'\|^2\,.$$

Taking $\alpha = \frac{1}{|\boldsymbol{x}'|2}\boldsymbol{x} \cdot \boldsymbol{x}'$, we obtain

$$0 \leq \|\boldsymbol{x}\|^2 - 2\frac{1}{\|\boldsymbol{x}'\|^2}(\boldsymbol{x} \cdot \boldsymbol{x}')^2 + \frac{1}{\|\boldsymbol{x}'\|^4}(\boldsymbol{x} \cdot \boldsymbol{x}')^2\|\boldsymbol{x}'\|^2 = \|\boldsymbol{x}\|^2 - \frac{1}{\|\boldsymbol{x}'\|^2}(\boldsymbol{x} \cdot \boldsymbol{x}')^2\,,$$

whence the inequality we wanted to prove.

Concerning the second part of the statement, we can assume that both vectors $\boldsymbol{x}$ and $\boldsymbol{x}'$ are different from $\boldsymbol{0}$. If $\boldsymbol{x}$ is equal to $\alpha\boldsymbol{x}'$, for some $\alpha \in \mathbb{R}$, then

$$|(\alpha\boldsymbol{x}') \cdot \boldsymbol{x}'| = |\alpha|\,|\boldsymbol{x}' \cdot \boldsymbol{x}'| = |\alpha|\,\|\boldsymbol{x}'\|^2 = \|\alpha\boldsymbol{x}'\|\,\|\boldsymbol{x}'\|\,.$$

Conversely, if $\boldsymbol{x} \cdot \boldsymbol{x}' = \|\boldsymbol{x}\|\,\|\boldsymbol{x}'\|$, then

$$\left\|\frac{\boldsymbol{x}}{\|\boldsymbol{x}\|} - \frac{\boldsymbol{x}'}{\|\boldsymbol{x}'\|}\right\|^2 = 1 - 2\frac{\boldsymbol{x} \cdot \boldsymbol{x}'}{\|\boldsymbol{x}\|\|\boldsymbol{x}'\|} + 1 = 0\,,$$

hence $\frac{\boldsymbol{x}}{\|\boldsymbol{x}\|} = \frac{\boldsymbol{x}'}{\|\boldsymbol{x}'\|}$. Because the two vectors have the same direction, they are linearly dependent. Similarly, if $\boldsymbol{x} \cdot \boldsymbol{x}' = -\|\boldsymbol{x}\|\,\|\boldsymbol{x}'\|$, then we can see that $\frac{\boldsymbol{x}}{\|\boldsymbol{x}\|} = -\frac{\boldsymbol{x}'}{\|\boldsymbol{x}'\|}$. ∎

Let us now prove property (d) of the norm. Using the Schwarz inequality,

$$
\begin{aligned}
\|\boldsymbol{x} + \boldsymbol{x}'\|^2 &= (\boldsymbol{x} + \boldsymbol{x}') \cdot (\boldsymbol{x} + \boldsymbol{x}') \\
&= \|\boldsymbol{x}\|^2 + 2\boldsymbol{x} \cdot \boldsymbol{x}' + \|\boldsymbol{x}'\|^2 \\
&\leq \|\boldsymbol{x}\|^2 + 2\|\boldsymbol{x}\| \, \|\boldsymbol{x}'\| + \|\boldsymbol{x}'\|^2 = (\|\boldsymbol{x}\| + \|\boldsymbol{x}'\|)^2 \,,
\end{aligned}
$$

whence the inequality we were looking for.

At this point, we cannot avoid taking a look at the **parallelogram identity**

$$
\|\boldsymbol{x} + \boldsymbol{x}'\|^2 + \|\boldsymbol{x} - \boldsymbol{x}'\|^2 = 2(\|\boldsymbol{x}\|^2 + \|\boldsymbol{x}'\|^2) \,,
$$

which in simple words states that, in any parallelogram, the sum of the squares of its two diagonals is equal to the sum of the squares of its four sides.

Let us define now, using the norm, the "Euclidean distance" between two elements $\boldsymbol{x} = (x_1, x_2, \ldots, x_N)$ and $\boldsymbol{x}' = (x_1', x_2', \ldots, x_N')$ of $\mathbb{R}^N$ as

$$
d(\boldsymbol{x}, \boldsymbol{x}') = \|\boldsymbol{x} - \boldsymbol{x}'\| = \sqrt{\sum_{k=1}^{N} (x_k - x_k')^2} \,.
$$

The following properties hold:

(a) $d(\boldsymbol{x}, \boldsymbol{x}') \geq 0$.
(b) $d(\boldsymbol{x}, \boldsymbol{x}') = 0 \Leftrightarrow \boldsymbol{x} = \boldsymbol{x}'$.
(c) $d(\boldsymbol{x}, \boldsymbol{x}') = d(\boldsymbol{x}', \boldsymbol{x})$.
(d) $d(\boldsymbol{x}, \boldsymbol{x}'') \leq d(\boldsymbol{x}, \boldsymbol{x}') + d(\boldsymbol{x}', \boldsymbol{x}'')$.

The last property is called the "triangle inequality." Let us prove it as follows:

$$
\begin{aligned}
d(\boldsymbol{x}, \boldsymbol{x}'') &= \|\boldsymbol{x} - \boldsymbol{x}''\| \\
&= \|(\boldsymbol{x} - \boldsymbol{x}') + (\boldsymbol{x}' - \boldsymbol{x}'')\| \\
&\leq \|\boldsymbol{x} - \boldsymbol{x}'\| + \|\boldsymbol{x}' - \boldsymbol{x}''\| = d(\boldsymbol{x}, \boldsymbol{x}') + d(\boldsymbol{x}', \boldsymbol{x}'') \,.
\end{aligned}
$$

In general, a real vector space V is said to be a "normed vector space" if there is a function $\| \cdot \| : V \to \mathbb{R}$, a "norm," with the following properties. For any x, x' in V, and $\alpha \in \mathbb{R}$

(a) $\|x\| \geq 0$.
(b) $\|x\| = 0 \Leftrightarrow x = 0$.
(c) $\|\alpha x\| = |\alpha| \, \|x\|$.
(d) $\|x + x'\| \leq \|x\| + \|x'\|$.

As we have seen, $\mathbb{R}^N$ is a normed vector space, with its Euclidean norm. Note however that different norms of a vector $\boldsymbol{x} = (x_1, x_2, \ldots, x_N)$ could also be defined on $\mathbb{R}^N$, for example,

$$\|\boldsymbol{x}\|_* = \sum_{k=1}^{N} |x_k|, \quad \text{or} \quad \|\boldsymbol{x}\|_{**} = \max\{|x_k| : k = 1, 2, \ldots, N\}.$$

1.5 Metric Spaces

For any nonempty set E, a function $d : E \times E \to \mathbb{R}$ is said to be a "distance" (on E) if it satisfies the following properties:

(a) $d(x, x') \geq 0$.
(b) $d(x, x') = 0 \Leftrightarrow x = x'$.
(c) $d(x, x') = d(x', x)$.
(d) $d(x, x'') \leq d(x, x') + d(x', x'')$ (the triangle inequality).

The set E, provided with the distance d, is said to be a "metric space." Its elements will often be referred to as "points."

We have seen that $\mathbb{R}^N$, provided with the Euclidean distance, is a metric space (in what follows, when speaking of $\mathbb{R}^N$ as a metric space, if not explicitly mentioned, we will always assume the given distance to be the Euclidean one). In the case $N = 1$, we have the usual distance on $\mathbb{R}$, i.e., $d(\alpha, \beta) = |\alpha - \beta|$.

More generally, any normed vector space is a metric space with the distance $d(x, x') = \|x - x'\|$. The fact that this is indeed a distance can be verified using the same approach as for the Euclidean distance.

It is also possible to consider different distances on the same set. For instance, taking two elements $\boldsymbol{x} = (x_1, x_2, \ldots, x_N)$ and $\boldsymbol{x}' = (x'_1, x'_2, \ldots, x'_N)$ of $\mathbb{R}^N$, the function

$$d_*(\boldsymbol{x}, \boldsymbol{x}') = \|\boldsymbol{x} - \boldsymbol{x}'\|_* = \sum_{k=1}^{N} |x_k - x'_k|$$

is also a distance on $\mathbb{R}^N$. The same is true of the function

$$d_{**}(\boldsymbol{x}, \boldsymbol{x}') = \|\boldsymbol{x} - \boldsymbol{x}'\|_{**} = \max\{|x_k - x'_k| : k = 1, 2, \ldots, N\},$$

and even

$$\hat{d}(\boldsymbol{x}, \boldsymbol{x}') = \begin{cases} 0 & \text{if } \boldsymbol{x} = \boldsymbol{x}', \\ 1 & \text{if } \boldsymbol{x} \neq \boldsymbol{x}'; \end{cases}$$

this is also a distance, however strange it might seem.

Now, let E be any metric space. Given a point $x_0 \in E$ and a real number $\rho > 0$, we define the "open ball" centered at x_0 with radius ρ as

$$B(x_0, \rho) = \{x \in E : d(x, x_0) < \rho\}.$$

Similarly, we define the "closed ball"

$$\overline{B}(x_0, \rho) = \{x \in E : d(x, x_0) \leq \rho\}$$

and the "sphere"

$$S(x_0, \rho) = \{x \in E : d(x, x_0) = \rho\}.$$

In $\mathbb{R}$, every interval $]a, b[$ is an open ball, and every interval $[a, b]$ is a closed ball; indeed,

$$]a, b[= B\left(\frac{a+b}{2}, \frac{b-a}{2}\right), \quad [a, b] = \overline{B}\left(\frac{a+b}{2}, \frac{b-a}{2}\right).$$

On the other hand, a sphere in $\mathbb{R}$ is a set having just two points.

In $\mathbb{R}^2$ (with the Euclidean distance), a ball is a disk; an open ball does not contain the external circle, but a closed ball does. A sphere is just the circle.

If in $\mathbb{R}^2$ we consider the distance d_* defined earlier, an open ball will be a square whose sides have an inclination of $45°$, having x_0 as its central point. A sphere will be the perimeter of such a square.

On the other hand, if we consider the distance d_{**}, a ball will still be a square, but with sides parallel to the cartesian axes. We will often denote a closed ball related to this distance by $\overline{B}[\boldsymbol{x}_0, \rho]$; if $\boldsymbol{x}_0 = (x_1^0, \ldots, x_N^0)$, then

$$\overline{B}[\boldsymbol{x}_0, \rho] = [x_1^0 - \rho, \, x_1^0 + \rho] \times \cdots \times [x_N^0 - \rho, \, x_N^0 + \rho].$$

A somewhat strange situation arises if we consider the distance $\hat{d}$ (on any set E). We have

$$B(x_0, \rho) = \begin{cases} \{x_0\} & \text{if } \rho \leq 1, \\ E & \text{if } \rho > 1, \end{cases} \qquad \overline{B}(x_0, \rho) = \begin{cases} \{x_0\} & \text{if } \rho < 1, \\ E & \text{if } \rho \geq 1, \end{cases}$$

hence

$$S(x_0, \rho) = \begin{cases} E \setminus \{x_0\} & \text{if } \rho = 1, \\ \varnothing & \text{if } \rho \neq 1. \end{cases}$$

We will now introduce a series of definitions that will be crucial for understanding the theory we want to develop.

Definition 1.14 A set $U \subseteq E$ is said to be a "neighborhood" of a point x_0 if there exists a $\rho > 0$ such that $B(x_0, \rho) \subseteq U$; in that case, the point x_0 is said to be an "internal point" of U. The set of all internal points of U is called the "interior" of U and is denoted by $\mathring{U}$. Clearly, we always have the inclusion $\mathring{U} \subseteq U$. It is said that U is an "open" set if it coincides with its interior, i.e., if $\mathring{U} = U$.

Here is an example of an open set.

Theorem 1.15 *An open ball is an open set.*

Proof Let $U = B(x_0, \rho)$ be an open ball, and take any point $x_1 \in U$. We want to prove that $x_1 \in \mathring{U}$, i.e., x_1 is an interior point of U. Choose $r > 0$ such that $r \leq \rho - d(x_0, x_1)$. If we show that $B(x_1, r) \subseteq U$, our proof will be completed.

For any $x \in B(x_1, r)$ we have

$$d(x, x_0) \leq d(x, x_1) + d(x_1, x_0) < r + d(x_1, x_0) \leq \rho,$$

so that $x \in B(x_0, \rho)$. We have thus shown that $B(x_1, r) \subseteq B(x_0, \rho)$. ∎

Examples Let us analyze three particular examples. In the first one, the set U coincides with E; in the second one, U is the empty set; in the third one, it is made of a single point.

1. Every point of E is internal to E since every ball is by definition contained in E, the universal set. Hence, the interior of E coincides with E, i.e., $\mathring{E} = E$. This means that E is an open set.
2. The empty set $\emptyset$ cannot have internal points. Hence, its interior, having no elements, is the empty set, i.e., $\mathring{\emptyset} = \emptyset$, meaning $\emptyset$ is an open set.
3. In general, the set $U = \{x_0\}$, made up of a single point, is not an open set (e.g., in $\mathbb{R}^N$ with the Euclidean distance), but it could be an open set in certain situations, i.e., when x_0 is an "isolated" point of E. This could happen, for instance, if $E = \mathbb{N}$, with the usual distance inherited from $\mathbb{R}$, or when considering the distance $\hat{d}$.

Theorem 1.16 *The interior of any set U is an open set.*

Proof If $\mathring{U} = \emptyset$, this is surely true. Assume, then, that $\mathring{U}$ is nonempty, and take any point $x_1 \in \mathring{U}$. Then there exists a $\rho > 0$ such that $B(x_1, \rho) \subseteq U$. If we show that $B(x_1, \rho) \subseteq \mathring{U}$, our proof will be completed, since we will have proved that every point x_1 of $\mathring{U}$ is an internal point of $\mathring{U}$.

To prove that $B(x_1, \rho) \subseteq \mathring{U}$, let x be an element of $B(x_1, \rho)$. Since $B(x_1, \rho)$ is an open set, there exists an $r > 0$ such that $B(x, r) \subseteq B(x_1, \rho)$. Then, $B(x, r) \subseteq U$, showing that x belongs to $\mathring{U}$. The proof is complete. ∎

The following implication holds:

$$U_1 \subseteq U_2 \quad \Rightarrow \quad \mathring{U}_1 \subseteq \mathring{U}_2 \, .$$

As a consequence, we see that $\mathring{U}$ is the greatest open set contained in U; indeed, if A is an open set and $A \subseteq U$, then $A \subseteq \mathring{U}$.

Definition 1.17 A point x_0 is said to be an "adherent point" of a set U if for every $\rho > 0$ we have that $B(x_0, \rho) \cap U \neq \emptyset$. The set of all adherent points of U is said to be the "closure" of U and is denoted by $\overline{U}$. Clearly, we always have the inclusion $U \subseteq \overline{U}$. It is said that U is a "closed" set if it coincides with its closure, i.e., if $U = \overline{U}$.

Here is an example of a closed set.

Theorem 1.18 *A closed ball is a closed set.*

Proof Let $U = \overline{B}(x_0, \rho)$ be a closed ball. To prove that $\overline{U} \subseteq U$, we will equivalently show that $CU \subseteq C\overline{U}$. This is surely true if $U = E$, i.e., if $CU = \emptyset$. Thus, assume now that CU is nonempty. Take any point $x_1 \in CU$, i.e., such that $d(x_1, x_0) > \rho$. We want to prove that $x_1 \in C\overline{U}$, i.e., that x_1 is not an adherent point of U. Choose $r > 0$ such that $r \leq d(x_0, x_1) - \rho$. If we show that $B(x_1, r) \cap U = \emptyset$, our proof will be completed.

Assume by contradiction that $B(x_1, r) \cap \overline{B}(x_0, \rho) \neq \emptyset$, and take an $x \in B(x_1, r) \cap \overline{B}(x_0, \rho)$. Then

$$d(x_0, x_1) \leq d(x_0, x) + d(x, x_1) < \rho + r \leq \rho + (d(x_0, x_1) - \rho) = d(x_0, x_1) \, ,$$

which is clearly impossible.   ∎

Examples Let us consider again the aforementioned three examples: $U = E$, $U = \emptyset$, and $U = \{x_0\}$.

1. Since E is the universal set, every adherent point of E necessarily belongs to E. Hence, the closure of E coincides with E, i.e., $\overline{E} = E$. This means that E is a closed set.
2. The empty set $\emptyset$ has no adherent points. Indeed, taking any point x_0 in E, for every $\rho > 0$ we have that $B(x_0, \rho) \cap \emptyset = \emptyset$. Hence, the closure of $\emptyset$, having no elements at all, is empty, i.e., $\overline{\emptyset} = \emptyset$, meaning $\emptyset$ is a closed set.
3. The set $U = \{x_0\}$, made up of a single point, is always a closed set. Indeed, if we take any $x_1 \notin U$, choosing $\rho > 0$ such that $\rho < d(x_0, x_1)$, we have that $B(x_1, \rho) \cap U = \emptyset$, thereby demonstrating that x_1 is not an adherent point of U.

Theorem 1.19 *The closure of any set U is a closed set.*

Proof Let $V = \overline{U}$. If $V = E$, this surely is a closed set. Let us then assume that $V \neq E$. We need to show that any adherent point of V belongs to V. By contradiction, assume that there exists some x_1 in $\overline{V}$ that does not belong to V. Since $x_1 \notin \overline{U}$, there is a $\rho > 0$ such that $B(x_1, \rho) \cap U = \emptyset$. On the other hand, since $x_1 \in \overline{V}$, we have that $B(x_1, \rho) \cap V \neq \emptyset$. Take an $x \in B(x_1, \rho) \cap V$. Then, since $B(x_1, \rho)$ is an open set, there exists $r > 0$ such that $B(x, r) \subseteq B(x_1, \rho)$. Since $x \in V = \overline{U}$, we have $B(x, r) \cap U \neq \emptyset$ and, hence, also $B(x_1, \rho) \cap U \neq \emptyset$, a contradiction. ∎

The following implication holds:

$$U_1 \subseteq U_2 \quad \Rightarrow \quad \overline{U}_1 \subseteq \overline{U}_2.$$

As a consequence, we see that $\overline{U}$ is the smallest closed set containing U: If C is a closed set and $C \supseteq U$, then $C \supseteq \overline{U}$.

We will now try to understand the relationships between the notions of interior and closure of a set and those between open and closed sets.

Theorem 1.20 *The following identities hold:*

$$\overline{CU} = C\mathring{U}, \qquad (\mathring{CU}) = C\overline{U}.$$

Proof Let us prove the first one. First of all notice that

$$\overline{CU} = \emptyset \;\Leftrightarrow\; CU = \emptyset \;\Leftrightarrow\; U = E \;\Leftrightarrow\; \mathring{U} = E \;\Leftrightarrow\; C\mathring{U} = \emptyset.$$

Assume now that $\overline{CU} \neq \emptyset$. Then

$$
\begin{aligned}
x \in \overline{CU} \quad &\Leftrightarrow \quad \forall \rho > 0 \quad B(x, \rho) \cap CU \neq \emptyset \\
&\Leftrightarrow \quad \forall \rho > 0 \quad B(x, \rho) \not\subseteq U \\
&\Leftrightarrow \quad x \notin \mathring{U} \\
&\Leftrightarrow \quad x \in C\mathring{U}.
\end{aligned}
$$

This proves the first identity. Now let $V = CU$. Then

$$\mathring{V} = C(C\mathring{V}) = C(\overline{CV}) = C\overline{U},$$

thereby also proving the second identity. ∎

As a consequence of the preceding theorem,

$$\overline{U} = \mathcal{C}(\mathcal{C}\mathring{U}), \qquad \mathring{U} = \mathcal{C}(\overline{\mathcal{C}U}).$$

Moreover, we have the following corollary.

Corollary 1.21 *A set is open (closed) if and only if its complementary is closed (open).*

Proof If U is open, then $U = \mathring{U}$, hence $\overline{\mathcal{C}U} = \mathcal{C}\mathring{U} = \mathcal{C}U$, so that $\mathcal{C}U$ is closed. On the other hand, if U is closed, then $U = \overline{U}$, hence $(\mathcal{C}\mathring{U}) = \mathcal{C}\overline{U} = \mathcal{C}U$, so that $\mathcal{C}U$ is open. ∎

It is possible to prove that the union and the intersection of two open (closed) sets is an open (closed) set. The same holds true for an arbitrary *finite* number of them.

However, if one considers an infinite number of open sets, it can be proved that their union is still an open set, whereas their intersection could not be. For example, in $\mathbb{R}$, taking the open sets

$$A_n = \left] -\frac{1}{n+1}, \frac{1}{n+1} \right[,$$

with $n \in \mathbb{N}$, their intersection is $\{0\}$, which is not an open set.

Analogously, if one considers an infinite number of closed sets, it can be proved that their intersection is still a closed set, whereas their union could not be such. For example, in $\mathbb{R}$, taking the closed sets

$$C_n = \left[-1 + \frac{1}{n+1}, 1 - \frac{1}{n+1} \right],$$

with $n \in \mathbb{N}$, their union is $]-1, 1[$, which is not a closed set.

Definition 1.22 The "boundary" of a set U, denoted by ∂U, is defined as the difference between its closure and its interior, i.e.,

$$\partial U = \overline{U} \setminus \mathring{U}.$$

We should be careful not to put too much trust in our intuition, naturally developed in an Euclidean world. For example, it is true in $\mathbb{R}^N$ that

$$\overline{B(\boldsymbol{x}_0, \rho)} = \overline{B}(\boldsymbol{x}_0, \rho), \qquad \partial B(\boldsymbol{x}_0, \rho) = S(\boldsymbol{x}_0, \rho).$$

However, these identities are not valid in any metric space E. For instance, if we take the previously defined distance $\hat{d}$, then $B(x_0, 1) = \{x_0\}$, which is a closed set, and $\overline{B}(x_0, 1) = E$, so that $\overline{B(x_0, 1)} \neq \overline{B}(x_0, 1)$. Moreover, $\partial B(x_0, 1) = \emptyset$, whereas $S(x_0, 1) = E \setminus \{x_0\}$, so that $\partial B(x_0, 1) \neq S(x_0, 1)$.

As a curious example, in $\mathbb{R}$, taking $U = \mathbb{Q}$, we have

$$\overline{\mathbb{Q}} = \mathbb{R}, \quad \mathring{\mathbb{Q}} = \emptyset, \quad \partial \mathbb{Q} = \mathbb{R}.$$

Continuity

2

In this chapter we introduce one of the most important concepts in mathematical analysis: the "continuity" of a function. This topic will be treated in the general framework of metric spaces.

2.1   Continuous Functions

Intuitively, a function f is "continuous" if the value $f(x)$ varies gradually when x varies in the domain, in other words, if we encounter no sudden variations in the values of the function. In order to make this intuitive idea rigorous enough, it will be convenient to focus our attention at a point x_0 of the domain and to clarify what we mean by

$$f \text{ is "continuous" at } x_0 .$$

We will proceed gradually.

First Attempt *We will say that f is "continuous" at x_0 when the following statement holds:*

$$\text{If } x \text{ is near } x_0, \text{ then } f(x) \text{ is near } f(x_0).$$

We immediately observe that, although the idea of continuity is already quite well formulated, the preceding proposition is not an acceptable definition, because the word "near," which appears twice, does not have a precise meaning. However, first of all, to measure how close x is to x_0 and how close $f(x)$ is to $f(x_0)$, we need to introduce distances. More precisely, we will have to assume that the domain and the codomain of the function are metric spaces.

Let, then, E and F be two metric spaces, with their distances d_E and d_F, respectively. Let x_0 be a point in E and $f : E \to F$ be our function. Let us make a second attempt at a definition.

Second Attempt *We will say that f is "continuous" at x_0 when the following statement holds:*

$$\text{If } d_E(x, x_0) \text{ is small, then } d_F(f(x), f(x_0)) \text{ is small.}$$

We immediately realize that the problem encountered in the first attempt at a definition has not been solved at all with this second attempt, since now the word "small," which appears twice, has no precise meaning.

We then ask ourselves: *How small* do we want the distance $d_F(f(x), f(x_0))$ to be? What we have in mind is that this distance can be made as small as we want (provided that the distance $d(x, x_0)$ is small enough, of course). To be able to measure it, we will then introduce a positive real number, which we call ε, and we will require $d_F(f(x), f(x_0))$ to be smaller than ε when $d(x, x_0)$ is sufficiently small. The arbitrariness of this positive number ε will allow us to take it as small as we like.

Third Attempt *We will say that f is "continuous" at x_0 when the following statement holds true: Taking any number $\varepsilon > 0$,*

$$\text{if } d_E(x, x_0) \text{ is small, then } d_F(f(x), f(x_0)) < \varepsilon .$$

Now the word "small" appears only once, whereas the distance $d_F(f(x), f(x_0))$ is simply controlled by the number ε. Hence, at least the second part of the proposition now has a precise meaning. We could then try to do the same for the distance $d(x, x_0)$, introducing a new positive number, which we call δ, so we can control it.

Fourth Attempt (the Good One!) *We will say that f is "continuous" at x_0 when the following statement holds true: Taking any number $\varepsilon > 0$, it is possible to find a number $\delta > 0$ for which,*

$$\text{if } d_E(x, x_0) < \delta, \text{ then } d_F(f(x), f(x_0)) < \varepsilon .$$

This last proposition, unlike the previous ones, contains no inaccurate words. The distances $d_E(x, x_0)$ and $d_F(f(x), f(x_0))$ are now simply controlled by the two positive numbers δ and ε, respectively. Let us rewrite it in a formal way.

Definition 2.1 We will say that f is "continuous" at x_0 if, for any positive number ε, there exists a positive number δ such that, if x is any element in the domain E whose distance from x_0 is less than δ, then the distance of $f(x)$ from $f(x_0)$ is less than ε. In symbols:

$$\forall \varepsilon > 0 \quad \exists \delta > 0 : \forall x \in E \quad d_E(x, x_0) < \delta \implies d_F(f(x), f(x_0)) < \varepsilon .$$

Rather often, in this formulation, "$\forall x \in E$" will be tacitly understood.

Let us note that one or both of the inequalities

$$d_E(x, x_0) < \delta, \qquad d_F(f(x), f(x_0)) < \varepsilon$$

may be replaced, respectively, by

$$d_E(x, x_0) \leq \delta, \qquad d_F(f(x), f(x_0)) \leq \varepsilon$$

without changing the definition at all. This is due to the fact that, on the one hand, ε is *any* positive number and, on the other hand, that if the implication holds for some positive number δ, then it holds a fortiori taking instead of that δ any smaller positive number.

Reading again the definition of continuity we see that f is continuous at x_0 if and only if

$$\forall \varepsilon > 0 \quad \exists \delta > 0 : \quad f(B(x_0, \delta)) \subseteq B(f(x_0), \varepsilon).$$

Moreover, it is equivalent if we take a closed ball instead of an open ball, one or both of them. Also equivalently we can say that f is continuous at x_0 if and only if

for every neighborhood V of $f(x_0)$

there exists a neighborhood U of x_0

such that $f(U) \subseteq V$.

In what follows, we will often denote the distances in E and F simply by d. We are confident that this will not create confusion.

When the function f happens to be continuous at all points x_0 of its domain E, we will say that "f is continuous on E" or simply "f is continuous."

Let us provide a few examples.

Example 1 The constant function: For some $\bar{c} \in F$ we have that $f(x) = \bar{c}$, for every $x \in E$. Since $d_F(f(x), f(x_0)) = d_F(\bar{c}, \bar{c}) = 0$ for every $x \in E$, such a function is clearly continuous (every choice of $\delta > 0$ is fine).

Example 2 Let x_0 be an "isolated point" of E, meaning there exists a $\rho > 0$ such that there are no points of E whose distance from x_0 is less than ρ, except x_0 itself. We can then see that, in this case, any function $f : E \to F$ is continuous at x_0. Indeed, for any $\varepsilon > 0$, taking $\delta = \rho$, we will have $B(x_0, \delta) = \{x_0\}$, so $f(B(x_0, \delta)) = \{f(x_0)\} \subseteq B(f(x_0), \varepsilon)$.

Example 3 Let $E = \mathbb{R}^N$ and $F = \mathbb{R}^N$. For a fixed number $\alpha \in \mathbb{R}$, let us consider the function $f : \mathbb{R}^N \to \mathbb{R}^N$ defined as $f(\boldsymbol{x}) = \alpha \boldsymbol{x}$. This is a continuous function. Indeed, if $\alpha = 0$, then we have a constant function with value $\boldsymbol{0}$, and we know already that such a function is continuous. Assume, in contrast, that $\alpha \neq 0$. Then, once $\varepsilon > 0$ has been fixed, since

$$\|f(\boldsymbol{x}) - f(\boldsymbol{x}_0)\| = \|\alpha\boldsymbol{x} - \alpha\boldsymbol{x}_0\| = \|\alpha(\boldsymbol{x} - \boldsymbol{x}_0)\| = |\alpha|\,\|\boldsymbol{x} - \boldsymbol{x}_0\|,$$

it is sufficient to take $\delta = \frac{\varepsilon}{|\alpha|}$ to verify that

$$\|\boldsymbol{x} - \boldsymbol{x}_0\| < \delta \;\Rightarrow\; \|f(\boldsymbol{x}) - f(\boldsymbol{x}_0)\| < \varepsilon\,.$$

Example 4 Let $E = \mathbb{R}^N$ and $F = \mathbb{R}$. Let us show that the function $f : \mathbb{R}^N \to \mathbb{R}$ defined as $f(\boldsymbol{x}) = \|\boldsymbol{x}\|$ is continuous on $\mathbb{R}^N$. This fact will be a simple consequence of the inequality

$$\Big|\|\boldsymbol{x}\| - \|\boldsymbol{x}'\|\Big| \le \|\boldsymbol{x} - \boldsymbol{x}'\|\,,$$

which we will now prove. We have

$$\|\boldsymbol{x}\| = \|(\boldsymbol{x} - \boldsymbol{x}') + \boldsymbol{x}'\| \le \|\boldsymbol{x} - \boldsymbol{x}'\| + \|\boldsymbol{x}'\|\,,$$

$$\|\boldsymbol{x}'\| = \|(\boldsymbol{x}' - \boldsymbol{x}) + \boldsymbol{x}\| \le \|\boldsymbol{x}' - \boldsymbol{x}\| + \|\boldsymbol{x}\|\,.$$

Since $\|\boldsymbol{x} - \boldsymbol{x}'\| = \|\boldsymbol{x}' - \boldsymbol{x}\|$, we have

$$\|\boldsymbol{x}\| - \|\boldsymbol{x}'\| \le \|\boldsymbol{x} - \boldsymbol{x}'\| \quad\text{and}\quad \|\boldsymbol{x}'\| - \|\boldsymbol{x}\| \le \|\boldsymbol{x} - \boldsymbol{x}'\|\,,$$

whence the inequality we wanted to prove. Now, considering any point $\boldsymbol{x}_0 \in \mathbb{R}^N$, once $\varepsilon > 0$ has been fixed, it is sufficient to take $\delta = \varepsilon$ to verify that

$$\|\boldsymbol{x} - \boldsymbol{x}_0\| < \delta \;\Rightarrow\; \Big|\|\boldsymbol{x}\| - \|\boldsymbol{x}_0\|\Big| < \varepsilon\,.$$

Example 5 Let E be any metric space, and $y_0 \in E$ be fixed. The function $f : E \to \mathbb{R}$ defined as $f(x) = d(x, y_0)$ is continuous. The proof of this fact is similar to the earlier one, since we can show that, for any $x_0 \in E$,

$$|d(x, y_0) - d(x_0, y_0)| \le d(x, x_0)\,.$$

Example 6 Let $E = \mathbb{R}$ and $F = \mathbb{R}$. Consider the "sign function" $f : \mathbb{R} \to \mathbb{R}$ defined as

$$f(x) = \begin{cases} -1 & \text{if } x < 0 \\ 0 & \text{if } x = 0 \\ 1 & \text{if } x > 0\,. \end{cases}$$

We can show that this function is continuous at all points except at $x_0 = 0$. Indeed, if $x_0 \ne 0$, then it will be sufficient to take $\delta < |x_0|$, so as to have f constant on the interval $]x_0 - \delta, x_0 + \delta[$ and, hence, continuous at x_0. To see that f is not continuous at 0, let us fix an $\varepsilon \in]0, 1[$; for any choice of $\delta > 0$, it is possible to find an $x \in] - \delta, \delta[$ such that $|f(x)| = 1$, hence $|f(x) - f(0)| > \varepsilon$.

Example 7 The "Dirichlet function" $\mathcal{D} : \mathbb{R} \to \mathbb{R}$ is defined as

$$\mathcal{D}(x) = \begin{cases} 1 & \text{if } x \in \mathbb{Q}, \\ 0 & \text{if } x \notin \mathbb{Q}. \end{cases}$$

It can be seen that, for any $x_0 \in \mathbb{R}$, this function is not continuous at x_0. Indeed, fixing $\varepsilon \in]0, 1[$, since both $\mathbb{Q}$ and $\mathbb{R} \setminus \mathbb{Q}$ are dense in $\mathbb{R}$, for every x_0 and any choice of $\delta > 0$ there will surely be a rational number x' and an irrational number x'' in $]x_0 - \delta, x_0 + \delta[$; hence, based on x_0 being rational or irrational, we will have that either $|\mathcal{D}(x'') - \mathcal{D}(x_0)| > \varepsilon$ or $|\mathcal{D}(x') - \mathcal{D}(x_0)| > \varepsilon$.

Let us study the behavior of continuity with respect to the sum of two functions and to the product with a constant. In the following theorem, we assume that F is a normed vector space.

Theorem 2.2 *Let F be a normed vector space and α a real constant. If $f, g : E \to F$ are continuous at x_0, then the same is true of $f \pm g$ and αf.*

Proof Let $\varepsilon > 0$ be fixed. By the continuity of f and g there exist $\delta_1 > 0$ and $\delta_2 > 0$ such that

$$d(x, x_0) < \delta_1 \implies \|f(x) - f(x_0)\| < \varepsilon,$$
$$d(x, x_0) < \delta_2 \implies \|g(x) - g(x_0)\| < \varepsilon.$$

Hence, taking $\delta = \min\{\delta_1, \delta_2\}$, we have that, if $d(x, x_0) < \delta$, then

$$\|(f \pm g)(x) - (f \pm g)(x_0)\| \le \|f(x) - f(x_0)\| + \|g(x) - g(x_0)\| < 2\varepsilon$$

and

$$\|(\alpha f)(x) - (\alpha f)(x_0)\| \le |\alpha| \, \|f(x) - f(x_0)\| < |\alpha| \, \varepsilon.$$

By the arbitrariness of ε, the statement is proved. ∎

Remark 2.3 The conclusion of the preceding proof is correct since the $\varepsilon > 0$ in the definition of continuity is arbitrary. Indeed, even if, for some constant $c > 0$, one proves that

$$\forall \varepsilon > 0 \quad \exists \delta > 0 : \quad \forall x \in E \quad d_E(x, x_0) < \delta \implies d_F(f(x), f(x_0)) < c\varepsilon,$$

this is sufficient to conclude that f is continuous at x_0. This observation will often be used in what follows.

We now state some properties of continuous functions with codomain $F = \mathbb{R}$.

Theorem 2.4 *If $f, g : E \to \mathbb{R}$ are continuous at x_0, the same is true of $f \cdot g$.*

Proof Let $\varepsilon > 0$ be fixed. It is not restrictive to assume $\varepsilon \le 1$, since we could always define $\varepsilon' = \min\{\varepsilon, 1\}$ and proceed with ε' instead of ε. By the continuity of f and g there exist $\delta_1 > 0$ and $\delta_2 > 0$ such that

$$d(x, x_0) < \delta_1 \implies |f(x) - f(x_0)| < \varepsilon ,$$
$$d(x, x_0) < \delta_2 \implies |g(x) - g(x_0)| < \varepsilon .$$

Here we note that, since $\varepsilon \le 1$, if $|f(x) - f(x_0)| < \varepsilon$, then $|f(x)| < |f(x_0)| + 1$. Hence, taking $\delta = \min\{\delta_1, \delta_2\}$, we have

$$d(x, x_0) < \delta \implies |(f \cdot g)(x) - (f \cdot g)(x_0)| =$$
$$= |f(x)g(x) - f(x)g(x_0) + f(x)g(x_0) - f(x_0)g(x_0)|$$
$$\le |f(x)| \cdot |g(x) - g(x_0)| + |g(x_0)| \cdot |f(x) - f(x_0)|$$
$$\le (|f(x_0)| + 1) \cdot |g(x) - g(x_0)| + |g(x_0)| \cdot |f(x) - f(x_0)|$$
$$< (|f(x_0)| + |g(x_0)| + 1) \, \varepsilon .$$

By the arbitrariness of ε, this proves that $f \cdot g$ is continuous at x_0. ∎

We now state the property of **sign permanence**.

Theorem 2.5 *If $g : E \to \mathbb{R}$ is continuous at x_0 and $g(x_0) > 0$, then there exists a neighborhood U of x_0 such that*

$$x \in U \implies g(x) > 0 .$$

Proof Let us fix $\varepsilon = g(x_0)$. By continuity, there exists $\delta > 0$ such that

$$d(x, x_0) < \delta \implies g(x_0) - \varepsilon < g(x) < g(x_0) + \varepsilon \implies 0 < g(x) < 2g(x_0) .$$

Then $U = B(x_0, \delta)$ is the neighborhood we are looking for. ∎

Clearly enough, if $g(x_0) < 0$ were true, then there would exist a neighborhood U of x_0 such that

$$x \in U \implies g(x) < 0 .$$

Theorem 2.6 *If $f, g : E \to \mathbb{R}$ are continuous at x_0 and $g(x_0) \neq 0$, then also $\frac{f}{g}$ is continuous at x_0.*

Proof Notice that, by the property of sign permanence, there exists a neighborhood U of x_0 such that the quotient $\frac{f(x)}{g(x)}$ is defined at least for all $x \in U$. Since $\frac{f}{g} = f \cdot \frac{1}{g}$, it will suffice to prove that $\frac{1}{g}$ is continuous at x_0. Let us fix $\varepsilon > 0$; we may assume without loss of generality that $\varepsilon < \frac{|g(x_0)|}{2}$. By the continuity of g there exists a $\delta > 0$ such that

$$d(x, x_0) < \delta \;\Rightarrow\; |g(x) - g(x_0)| < \varepsilon .$$

Since $\varepsilon < \frac{|g(x_0)|}{2}$, then also

$$d(x, x_0) < \delta \;\Rightarrow\; |g(x)| > |g(x_0)| - \varepsilon > \frac{|g(x_0)|}{2} .$$

As a consequence,

$$d(x, x_0) < \delta \;\Rightarrow\; \left| \frac{1}{g}(x) - \frac{1}{g}(x_0) \right| = \frac{|g(x_0) - g(x)|}{|g(x)|\,|g(x_0)|} < \frac{2}{|g(x_0)|^2}\,\varepsilon .$$

By the arbitrariness of ε, this proves that $\frac{1}{g}$ is continuous at x_0. ∎

We know that all constant functions are continuous, as is the function $f : \mathbb{R} \to \mathbb{R}$ defined as $f(x) = x$ (see Example 3 presented earlier, with $\alpha = 1$). By the previous theorems, *all polynomial functions are continuous*, as are all rational functions, defined by the quotient of two polynomials. More precisely, these latter are continuous at all points where they are defined, i.e., where the denominator is not equal to zero.

Let us now examine the behavior of a composition of continuous functions. In the following theorem, E, F, and G are three metric spaces.

Theorem 2.7 *Let $f : E \to F$ be continuous at x_0 and $g : F \to G$ be continuous at $f(x_0)$; then $g \circ f$ is continuous at x_0.*

Proof Let W be a fixed neighborhood of $[g \circ f](x_0) = g(f(x_0))$. By the continuity of g at $f(x_0)$ there exists a neighborhood V of $f(x_0)$ such that $g(V) \subseteq W$. Then, by the continuity of f at x_0, there exists a neighborhood U of x_0 such that $f(U) \subseteq V$. Hence, $[g \circ f](U) \subseteq W$, thereby proving the statement. ∎

Let us now define, for every $k = 1, 2, \ldots, N,$, the "kth projection" $p_k : \mathbb{R}^N \to \mathbb{R}$ as

$$p_k(x_1, x_2, \ldots, x_N) = x_k .$$

Theorem 2.8 *The functions p_k are continuous.*

Proof We consider a point $\boldsymbol{x}_0 = (x_1^0, x_2^0, \ldots, x_N^0) \in \mathbb{R}^N$ and fix an $\varepsilon > 0$. Notice that, for every $\boldsymbol{x} = (x_1, x_2, \ldots, x_N) \in \mathbb{R}^N$,

$$|x_k - x_k^0| \le \sqrt{\sum_{j=1}^{N}(x_j - x_j^0)^2} = d(\boldsymbol{x}, \boldsymbol{x}_0) \,;$$

hence, taking $\delta = \varepsilon$, we have that

$$d(\boldsymbol{x}, \boldsymbol{x}_0) < \delta \;\Rightarrow\; |p_k(\boldsymbol{x}) - p_k(\boldsymbol{x}_0)| = |x_k - x_k^0| < \varepsilon \,.$$

This proves that p_k is continuous at $\boldsymbol{x}_0$.   ∎

Let us consider now a function $f : E \to \mathbb{R}^M$ for some integer $M \ge 1$. We can define the "components" of f as $f_k = p_k \circ f : E \to \mathbb{R}$, with $k = 1, 2, \ldots, M$, so that

$$f(x) = (f_1(x), f_2(x), \ldots, f_M(x)) \,.$$

Theorem 2.9 *The function f is continuous at x_0 if and only if all its components are as well.*

Proof If f is continuous at x_0, then its components are also continuous since they are composed of two continuous functions. To prove the contrary, let us assume that all the components of f are continuous at x_0. Fixing $\varepsilon > 0$, for every $k = 1, 2, \ldots, M$ there is a $\delta_k > 0$ such that

$$d(x, x_0) < \delta_k \;\Rightarrow\; |f_k(x) - f_k(x_0)| < \varepsilon \,.$$

Setting $\delta = \min\{\delta_1, \delta_2, \ldots, \delta_M\}$, we have

$$d(x, x_0) < \delta \;\Rightarrow\; d(f(x), f(x_0)) = \sqrt{\sum_{j=1}^{M}(f_j(x) - f_j(x_0))^2} < \sqrt{M}\varepsilon \,.$$

By the arbitrariness of ε, the proof is complete.   ∎

Theorem 2.10 *Every linear function $\ell : \mathbb{R}^N \to \mathbb{R}^M$ is continuous.*

Proof We first observe that, since the projections p_k are linear functions, the components $\ell_k = p_k \circ \ell$ of the linear function ℓ are linear as well. Let $[e_1, e_2, \ldots, e_N]$

be the canonical basis of $\mathbb{R}^N$, i.e.,

$$e_1 = (1, 0, 0, \ldots, 0),$$
$$e_2 = (0, 1, 0, \ldots, 0),$$
$$\vdots$$
$$e_N = (0, 0, 0, \ldots, 1).$$

Every vector $\boldsymbol{x} = (x_1, x_2, \ldots, x_N) \in \mathbb{R}^N$ can be expressed as

$$\boldsymbol{x} = x_1 e_1 + x_2 e_2 + \cdots + x_N e_N = p_1(\boldsymbol{x}) e_1 + p_2(\boldsymbol{x}) e_2 + \cdots + p_N(\boldsymbol{x}) e_N.$$

Hence, for every $k \in \{1, 2, \ldots, M\}$,

$$\ell_k(\boldsymbol{x}) = p_1(\boldsymbol{x}) \ell_k(e_1) + p_2(\boldsymbol{x}) \ell_k(e_2) + \cdots + p_N(\boldsymbol{x}) \ell_k(e_N),$$

showing that ℓ_k is a linear combination of the projections $p_1, p_2, \ldots, p_N$. Since those functions are continuous, we have proved that ℓ_k is continuous for every $k \in \{1, 2, \ldots, M\}$. Therefore, since all its components are continuous, the function ℓ is continuous as well. ∎

We conclude this section with a characterization of continuity involving the counterimages of open and closed sets in arbitrary metric spaces.

Theorem 2.11 *The following propositions are equivalent:*

(i) $f : E \to F$ *is continuous.*
(ii) If A is open in F, then $f^{-1}(A)$ is open in E.
(iii) If C is closed in F, then $f^{-1}(C)$ is closed in E.

Proof Let us show that (i) implies (ii). Let $f : E \to F$ be continuous, and let A be an open set in F. Taking $x_0 \in f^{-1}(A)$, we have $f(x_0) \in A$. Since A is open, there exists a $\rho > 0$ for which $B(f(x_0), \rho) \subseteq A$. Since f is continuous at x_0, taking $\varepsilon = \rho$ in the definition, there exists a $\delta > 0$ such that $f(B(x_0, \delta)) \subseteq B(f(x_0), \rho)$. Then $B(x_0, \delta) \subseteq f^{-1}(B(f(x_0), \rho)) \subseteq f^{-1}(A)$, so that x_0 is in the interior of $f^{-1}(A)$. We have thus proved that every $x_0 \in f^{-1}(A)$ is in the interior of $f^{-1}(A)$, so that $f^{-1}(A)$ is open.

Let us prove now that (ii) implies (i). We consider a point $x_0 \in E$, fix $\varepsilon > 0$, and set $A = B(f(x_0), \varepsilon)$, which is an open set in F. If (ii) holds, then $f^{-1}(A)$ is an open set in E containing x_0. Hence, there exists a $\delta > 0$ such that $B(x_0, \delta) \subseteq f^{-1}(A)$, meaning $f(B(x_0, \delta)) \subseteq A = B(f(x_0), \varepsilon)$. The continuity of f at x_0 is thus proved.

We now show that (ii) implies (iii). Let C be a closed set in F, and let $A = \mathcal{C}C$, the complementary set of C. The set A is open in F so that, if (ii) holds, then $f^{-1}(A)$ is open in E. But $f^{-1}(A) = f^{-1}(\mathcal{C}C) = \mathcal{C}f^{-1}(C)$, so $f^{-1}(C)$ is closed.

In a very similar way one proves that (iii) implies (ii), concluding the proof of the theorem.   ∎

2.2   Intervals and Continuity

Here is a fundamental property of continuous functions defined on intervals.

Theorem 2.12 (Bolzano Theorem) *If* $f : [a, b] \to \mathbb{R}$ *is a continuous function such that*

$$\text{either} \quad f(a) < 0 < f(b), \quad \text{or} \quad f(a) > 0 > f(b),$$

then there exists a $c \in]a, b[$ *such that* $f(c) = 0$.

Proof We treat the case $f(a) < 0 < f(b)$, since the other one is completely analogous. We set $I_0 = [a, b]$ and consider the midpoint $\frac{a+b}{2}$ of the interval I_0. If f is equal to zero at that point, then we have found the point c we were looking for. Otherwise, either $f(\frac{a+b}{2}) < 0$ or $f(\frac{a+b}{2}) > 0$. If $f(\frac{a+b}{2}) < 0$, then we call I_1 the interval $[\frac{a+b}{2}, b]$; if $f(\frac{a+b}{2}) > 0$, then instead we refer to I_1 as the interval $[a, \frac{a+b}{2}]$. Taking now the midpoint of I_1 and following the same reasoning, we can define an interval I_2 and, by recurrence, a sequence of intervals $I_n = [a_n, b_n]$ such that

$$I_0 \supseteq I_1 \supseteq I_2 \supseteq I_3 \supseteq \dots$$

and $f(a_n) < 0 < f(b_n)$ for every n. By Cantor's Theorem 1.9, there exists a $c \in \mathbb{R}$ belonging to all these intervals. Let us prove that $f(c) = 0$. By contradiction, assume $f(c) \neq 0$. If $f(c) < 0$, by the property of sign permanence, there is a $\delta > 0$ such that $f(x) < 0$ for every $x \in]c - \delta, c + \delta[$. Now, since $b_n - c \leq b_n - a_n$ and $b_n - a_n = \frac{b-a}{2^n} < \frac{b-a}{n}$ for $n \geq 1$, taking $n > \frac{b-a}{\delta}$, we have $b_n \in]c - \delta, c + \delta[$. But then we should have $f(b_n) < 0$, in contradiction to the above inequality. A similar line of reasoning rules out the case $f(c) > 0$.   ∎

As a consequence of the foregoing theorem, we deduce that a continuous function "transforms intervals into intervals."

Corollary 2.13 *Let E be a subset of $\mathbb{R}$ and $f : E \to \mathbb{R}$ be a continuous function. If $I \subseteq E$ is an interval, then $f(I)$ is also an interval.*

Proof Excluding the trivial cases where I and $f(I)$ are made of a single element, let us take $\alpha, \beta \in f(I)$, with $\alpha < \beta$, and let γ be such that $\alpha < \gamma < \beta$. We want to see that $\gamma \in f(I)$. Let $g : E \to \mathbb{R}$ be the function defined by

$$g(x) = f(x) - \gamma.$$

We can find a, b in I such that $f(a) = \alpha$ and $f(b) = \beta$. Since I is an interval, the function g is defined on $[a, b]$ (or $[b, a]$ in the case $b < a$), and it is continuous there. Moreover, $g(a) < 0 < g(b)$, and, hence, by the foregoing theorem, there is a $c \in \,]a, b[$ such that $g(c) = 0$, i.e., $f(c) = \gamma$. ∎

2.3 Monotone Functions

Let E be a subset of $\mathbb{R}$. We will say that a function $f : E \to \mathbb{R}$ is:

"Increasing" if $[\, x_1 < x_2 \;\Rightarrow\; f(x_1) \leq f(x_2) \,]$.
"Decreasing" if $[\, x_1 < x_2 \;\Rightarrow\; f(x_1) \geq f(x_2) \,]$.
"Strictly increasing" if $[\, x_1 < x_2 \;\Rightarrow\; f(x_1) < f(x_2) \,]$.
"Strictly decreasing" if $[\, x_1 < x_2 \;\Rightarrow\; f(x_1) > f(x_2) \,]$.

We will say that f is "monotone" if it is either increasing or decreasing and "strictly monotone" if it is either strictly increasing or strictly decreasing.

Example The function $f : [0, +\infty[\,\to\, \mathbb{R}$ defined as $f(x) = x^n$ is strictly increasing. The case $n = 2$ was established in Lemma 1.6. The general case can be easily proved by induction.

Let us now show how one can characterize the continuity of invertible functions defined on an interval.

Theorem 2.14 *Let I and J be two intervals, and let $f : I \to J$ be an invertible function. Then*

$$f \text{ is continuous} \quad \Leftrightarrow \quad f \text{ is strictly monotone}.$$

In that case, $f^{-1} : J \to I$ is also strictly monotone and continuous.

Proof Assume f to be continuous and, by contradiction, that it is not strictly monotone. Then there exist $x_1 < x_2 < x_3$ in I such that either

$$f(x_1) < f(x_2) \quad \text{and} \quad f(x_2) > f(x_3)$$

or

$$f(x_1) > f(x_2) \quad \text{and} \quad f(x_2) < f(x_3).$$

(Equalities are not allowed since f is injective.) Let us consider the first case, the other being analogous. Choosing $\gamma \in \mathbb{R}$ such that $f(x_1) < \gamma < f(x_2)$ and $f(x_2) > \gamma > f(x_3)$, by Corollary 2.13 there exist $a \in]x_1, x_2[$ and $b \in]x_2, x_3[$ such that $f(a) = \gamma = f(b)$, in contradiction to the injectivity of f.

Assume now that f is strictly monotone, e.g., strictly increasing, the other case being analogous. Once we have fixed some $x_0 \in I$, we want to prove that f is continuous at x_0. Let us consider three distinct cases.

Case 1.   Assume that x_0 is not an endpoint of interval I and, consequently, $y_0 = f(x_0)$ is not an endpoint of J. Let $\varepsilon > 0$ be fixed; we can assume without loss of generality that $[y_0 - \varepsilon, y_0 + \varepsilon] \subseteq J$. Set $x_1 = f^{-1}(y_0 - \varepsilon)$ and $x_2 = f^{-1}(y_0 + \varepsilon)$, and notice that $x_1 < x_0 < x_2$. Since $f(x_1) = f(x_0) - \varepsilon$ and $f(x_2) = f(x_0) + \varepsilon$, taking $\delta = \min\{x_0 - x_1, x_2 - x_0\}$ we have

$$d(x, x_0) < \delta \;\; \Rightarrow \;\; x_1 < x < x_2$$
$$\Rightarrow \;\; f(x_1) < f(x) < f(x_2)$$
$$\Rightarrow \;\; d(f(x), f(x_0)) < \varepsilon,$$

showing that f is continuous at x_0.

Case 2.   Now let $x_0 = \min I$, hence also $y_0 = \min J$. Let $\varepsilon > 0$ be fixed; we can assume without loss of generality that $[y_0, y_0 + \varepsilon] \subseteq J$. Set, as previously, $x_2 = f^{-1}(y_0 + \varepsilon)$. Since $f(x_2) = f(x_0) + \varepsilon$, taking $\delta = x_2 - x_0$, we have

$$x_0 \le x < x_2 \Rightarrow f(x_0) \le f(x) < f(x_2) \Rightarrow d(f(x), f(x_0)) < \varepsilon,$$

demonstrating that f is continuous at x_0.

Case 3.   If $x_0 = \max I$, then the argument is similar to that in Case 2.

Finally, we observe that

$$f \;\; \text{strictly increasing} \;\; \Rightarrow \;\; f^{-1} \;\; \text{strictly increasing},$$
$$f \;\; \text{strictly decreasing} \;\; \Rightarrow \;\; f^{-1} \;\; \text{strictly decreasing}.$$

Therefore, if f is strictly monotone, then so is f^{-1}. Hence, since $f^{-1} : J \to I$ is invertible and strictly monotone, as proved earlier, it is necessarily continuous. ∎

2.4   The Exponential Function

Let us denote by $\mathbb{R}_+$ the set of positive real numbers, i.e.,

$$\mathbb{R}_+ =]0, +\infty[= \{x \in \mathbb{R} : x > 0\}.$$

The following theorem will be proved in Chap. 5.

Theorem 2.15 *Given $a > 0$, there exists a unique continuous function $f_a : \mathbb{R} \to \mathbb{R}_+$ such that*

(i) $f_a(x_1 + x_2) = f_a(x_1) f_a(x_2)$*, for every x_1, x_2 in $\mathbb{R}$.*
(ii) $f_a(1) = a$*.*

Moreover, if $a \neq 1$, then this function f_a is invertible.

The function f_a is called "exponential to base a" and is denoted by $\exp_a$. If $a \neq 1$, then the inverse function $f_a^{-1} : \mathbb{R}_+ \to \mathbb{R}$ is called the "logarithm to base a" and is denoted by $\log_a$. By Theorem 2.14, it is a continuous function. We can write, for $x \in \mathbb{R}$ and $y \in \mathbb{R}_+$,

$$\exp_a(x) = y \quad \Leftrightarrow \quad x = \log_a(y).$$

From the properties

(i) $\exp_a(x_1 + x_2) = \exp_a(x_1) \exp_a(x_2)$,
(ii) $\exp_a(1) = a$

we directly deduce the corresponding properties of the logarithm

(j) $\log_a(y_1 y_2) = \log_a(y_1) + \log_a(y_2)$,
(jj) $\log_a(a) = 1$.

Since the constant function $f(x) = 1$ verifies (i) and (ii), with $a = 1$, by the uniqueness of that function we deduce that $f = \exp_1$, i.e.,

$$\exp_1(x) = 1, \quad \text{for every } x \in \mathbb{R}.$$

Let us now deduce from (i) and (ii) some general properties of the exponential function. First of all, we observe that, since $\exp_a(1) = \exp_a(1 + 0) = \exp_a(1) \exp_a(0)$, it must be that

$$\exp_a(0) = 1.$$

Let us now prove that, for every $x \in \mathbb{R}$ and every $n \in \mathbb{N}$,

$$\exp_a(nx) = (\exp_a(x))^n .$$

We argue by induction. If $n = 0$, then we see that

$$\exp_a(0x) = 1 , \quad (\exp_a(x))^0 = 1 ,$$

hence the identity surely holds. Assume now that the formula is true for some $n \in \mathbb{N}$. Then

$$\begin{aligned}
\exp_a((n+1)x) &= \exp_a(nx + x) = \exp_a(nx) \exp_a(x) \\
&= (\exp_a(x))^n \exp_a(x) = (\exp_a(x))^{n+1} ,
\end{aligned}$$

hence it is true also for $n + 1$. The proof of the formula is thus completed.

Taking $x = 1$, we see that

$$\exp_a(n) = a^n$$

for every $n \in \mathbb{N}$. This fact motivates a new notation: For every $x \in \mathbb{R}$, we will often write a^x instead of $\exp_a(x)$.

Taking $n \in \mathbb{N} \setminus \{0\}$ and writing

$$a = \exp_a(1) = \exp_a\left(n\frac{1}{n}\right) = \left(\exp_a\left(\frac{1}{n}\right)\right)^n ,$$

we see that $\exp_a\left(\frac{1}{n}\right)$ is that number $u \in \mathbb{R}_+$ that solves the equation $u^n = a$. Such a number u is called the "nth root of a," and it is denoted by $u = \sqrt[n]{a}$, i.e.,

$$\exp_a\left(\frac{1}{n}\right) = \sqrt[n]{a} .$$

Hence, if $m \in \mathbb{N}$ and $n \in \mathbb{N} \setminus \{0\}$,

$$\exp_a\left(\frac{m}{n}\right) = \exp_a\left(m\frac{1}{n}\right) = \left(\exp_a\left(\frac{1}{n}\right)\right)^m = (\sqrt[n]{a})^m .$$

On the other hand, writing

$$1 = \exp_a(0) = \exp_a(x - x) = \exp_a(x) \exp_a(-x) ,$$

we see that

$$\exp_a(-x) = \frac{1}{\exp_a(x)} , \quad \text{for every } x \in \mathbb{R} .$$

In particular, if $m \in \mathbb{N}$ and $n \in \mathbb{N} \setminus \{0\}$,

$$\exp_a \left(\frac{-m}{n} \right) = \frac{1}{\exp_a \left(\frac{m}{n} \right)} = \frac{1}{(\sqrt[n]{a})^m} = (\sqrt[n]{a})^{-m}.$$

Let us check for a moment that, for every $m \in \mathbb{Z}$ and $n \in \mathbb{N} \setminus \{0\}$, we have

$$\left(\sqrt[n]{a} \right)^m = \sqrt[n]{a^m}.$$

Indeed, if $b = \sqrt[n]{a}$, then $a^m = (b^n)^m = b^{nm} = (b^m)^n$, whence $b^m = \sqrt[n]{a^m}$. We can thus conclude that

$$\exp_a \left(\frac{m}{n} \right) = \sqrt[n]{a^m}, \qquad \text{for every } \frac{m}{n} \in \mathbb{Q}.$$

If $a \neq 1$, then the exponential function $\exp_a : \mathbb{R} \to \mathbb{R}_+$ is continuous and invertible and, hence, strictly monotone. Since $\exp_a(0) = 1$ and $\exp_a(1) = a$,

$$\exp_a \text{ is :} \begin{cases} \text{striclty increasing} & \text{if } a > 1; \\ \text{striclty decreasing} & \text{if } 0 < a < 1. \end{cases}$$

(See Fig. 2.1) Let us also emphasize the following three important formulas:

$$(ab)^x = a^x b^x, \qquad \left(\frac{1}{a} \right)^x = \frac{1}{a^x} = a^{-x}, \qquad (a^y)^x = a^{yx}.$$

The first one follows from the fact that the function $f(x) = a^x b^x$ verifies the property $i)$ and $f(1) = ab$, hence $f = \exp_{ab}$. Analogously, for the second one, we take $f(x) = \frac{1}{a^x}$. For the third one, simply take $f(x) = a^{yx}$.

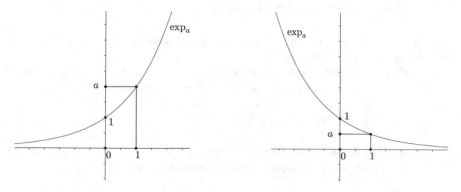

Fig. 2.1 The function $\exp_a$, with $a > 1$ and $a < 1$, respectively

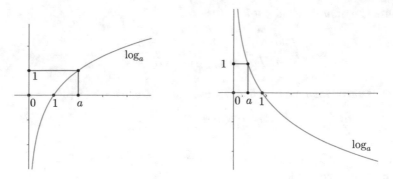

Fig. 2.2 The function $\log_a$, with $a > 1$ and $a < 1$, respectively

If $a \neq 1$, we have that

$$\log_a \text{ is } \begin{cases} \text{strictly increasing} & \text{if } a > 1, \\ \text{strictly decreasing} & \text{if } 0 < a < 1. \end{cases}$$

(See Fig. 2.2) The following are two important formulas for the logarithm:

$$\log_a(x^y) = y \log_a(x), \quad \log_b(x) = \frac{\log_a(x)}{\log_a(b)}.$$

To prove the first one, set $u = \log_a(x^y)$ and $v = \log_a(x)$. Then $a^u = x^y$ and $a^v = x$, and hence $a^u = (a^v)^y = a^{vy}$. By the injectivity of $\exp_a$, it must be that $u = vy$, and the first formula is proved. To prove the second formula, set $u = \log_b(x)$, $v = \log_a(x)$, and $w = \log_a(b)$. Then $b^u = x$, $a^v = x$, and $a^w = b$, whence $a^v = (a^w)^u = a^{wu}$. By injectivity, it must be that $v = wu$, and the second formula is also proved.

2.5   The Trigonometric Functions

We will now introduce the *trigonometric functions* following a path similar to that traced previously for the exponential function.

Given a real number $T > 0$, a function $F : \mathbb{R} \to \Omega$, where Ω is any possible set, is said to be "periodic with period T," or "T-periodic" short, if

$$F(x + T) = F(x), \quad \text{for every } x \in \mathbb{R}.$$

Clearly, if T is a period for the function F, then $2T, 3T, \ldots$ are also periods for F. We will say that T is the "minimal period" if there are no smaller periods.

Let us introduce the set

$$S^1 = \{z \in \mathbb{C} : |z| = 1\},$$

i.e., the circle centered at the origin, with radius 1, in the complex field $\mathbb{C}$.
The following theorem will be proved in Chap. 5.

Theorem 2.16 *Given $T > 0$, there exists a unique function $h_T : \mathbb{R} \to S^1$, continuous and periodic with minimal period T, such that*

(i) $h_T(x_1 + x_2) = h_T(x_1)h_T(x_2)$, for every x_1, x_2 in $\mathbb{R}$.
(ii) $h_T\left(\frac{T}{4}\right) = i$.

The function h_T is called the "circular function to base T." Since S^1 is indeed a subset of $\mathbb{R}^2$, the function h_T has two components, which will be denoted by $\cos_T$ and $\sin_T$; they will be called "cosine to base T" and "sine to base T," respectively. We can then write, for every $x \in \mathbb{R}$,

$$h_T(x) = (\cos_T(x), \sin_T(x)), \quad \text{or} \quad h_T(x) = \cos_T(x) + i \, \sin_T(x).$$

These functions are T-periodic, and from the properties of the circular function we have that

(a) $(\cos_T(x))^2 + (\sin_T(x))^2 = 1$.
(b) $\cos_T(x_1 + x_2) = \cos_T(x_1)\cos_T(x_2) - \sin_T(x_1)\sin_T(x_2)$.
(c) $\sin_T(x_1 + x_2) = \sin_T(x_1)\cos_T(x_2) + \cos_T(x_1)\sin_T(x_2)$.
(d) $\cos_T\left(\frac{T}{4}\right) = 0, \quad \sin_T\left(\frac{T}{4}\right) = 1$.

Let us now focus our attention on the interval $[0, T[$. Writing

$$i = h_T\left(\frac{T}{4}\right) = h_T\left(0 + \frac{T}{4}\right) = h_T(0)h_T\left(\frac{T}{4}\right) = h_T(0)i,$$

we see that $h_T(0) = 1$. Moreover,

$$h_T\left(\frac{T}{2}\right) = h_T\left(\frac{T}{4} + \frac{T}{4}\right) = h_T\left(\frac{T}{4}\right)h_T\left(\frac{T}{4}\right) = i^2 = -1,$$

whereas

$$h_T\left(\frac{3T}{4}\right) = h_T\left(\frac{T}{2} + \frac{T}{4}\right) = h_T\left(\frac{T}{2}\right)h_T\left(\frac{T}{4}\right) = (-1)i = -i.$$

Summing up,

$$\begin{aligned}
\cos_T (0) &= 1, & \sin_T (0) &= 0, \\
\cos_T \left(\tfrac{T}{4}\right) &= 0, & \sin_T \left(\tfrac{T}{4}\right) &= 1, \\
\cos_T \left(\tfrac{T}{2}\right) &= -1, & \sin_T \left(\tfrac{T}{2}\right) &= 0, \\
\cos_T \left(\tfrac{3T}{4}\right) &= 0, & \sin_T \left(\tfrac{3T}{4}\right) &= -1.
\end{aligned}$$

Now, from

$$1 = h_T(0) = h_T(x - x) = h_T(x)h_T(-x)$$

we have $h_T(-x) = h_T(x)^{-1} = h_T(x)^*$, being $|h_T(x)| = 1$. Hence,

$$\cos_T (-x) = \cos_T (x), \quad \sin_T (-x) = -\sin_T (x),$$

showing that $\cos_T$ is an even function, whereas $\sin_T$ is odd.

Let us prove now that $\widetilde{h}_T : [0, T[\to S^1$, the restriction of h_T to the interval $[0, T[$, is bijective. First, injectivity: Take $\alpha < \beta$ in $[0, T[$. By contradiction, if $h_T(\alpha) = h_T(\beta)$, then

$$h_T(\beta - \alpha) = h_T(\beta)h_T(-\alpha) = \frac{h_T(\beta)}{h_T(\alpha)} = 1,$$

and hence

$$h_T(x + (\beta - \alpha)) = h_T(x)h_T(\beta - \alpha) = h_T(x), \quad \text{for every } x \in \mathbb{R},$$

so that $\beta - \alpha$ would be a period for h_T smaller than T, while we know that T is the minimal period. Then $h_T(\alpha) \neq h_T(\beta)$, proving that $\widetilde{h}_T$ is injective.

We now prove that

$$\cos_T (x) \begin{cases} > 0 & \text{if } 0 < x < \frac{T}{4} \\ < 0 & \text{if } \frac{T}{4} < x < \frac{3T}{4} \\ > 0 & \text{if } \frac{3T}{4} < x < T \end{cases} \qquad \sin_T (x) \begin{cases} > 0 & \text{if } 0 < x < \frac{T}{2} \\ < 0 & \text{if } \frac{T}{2} < x < T . \end{cases}$$

(See Fig. 2.3) For example, if $x \in]0, \frac{T}{2}[$, then surely it cannot be that $\sin_T (x) = 0$, otherwise $h_T(x)$ would coincide with either $h_T(0)$ or $h_T(\frac{T}{2})$, contradicting the already proved injectivity. Then, by continuity (as a consequence of Bolzano's Theorem 2.12), $\sin_T$ must be either always positive or always negative on $]0, \frac{T}{2}[$. Since $\sin_T \left(\frac{T}{4}\right) = 1$, it needs to be always positive on that interval.

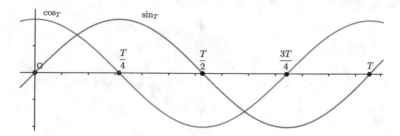

Fig. 2.3 The trigonometric functions $\cos_T$ and $\sin_T$

Now, to conclude, let us prove that $\tilde{h}_T : [0, T[\to S^1$ is surjective. Take a point $P = (X_1, X_2)$ in S^1. Notice that $X_1 \in [-1, 1]$. The two cases $X_1 = -1$ and $X_1 = 1$ imply $X_2 = 0$, and we already know that $h_T(\frac{T}{2}) = (-1, 0)$ and $h_T(0) = (1, 0)$. Assume, then, that $X_1 \in]-1, 1[$. Since $\cos_T(0) = 1$, $\cos_T(\frac{T}{2}) = -1$, and $\cos_T$ is continuous, by Bolzano's Theorem 2.12 there is a $\bar{x} \in]0, \frac{T}{2}[$ such that $\cos_T(\bar{x}) = X_1$. Then

$$| \sin_T(\bar{x})| = \sqrt{1 - (\cos_T(\bar{x}))^2} = \sqrt{1 - X_1^2} = |X_2| \,.$$

We have two possibilities: Either $\sin_T(\bar{x}) = X_2$, in which case $h_T(\bar{x}) = P$, or $\sin_T(\bar{x}) = -X_2$, and

$$h_T(T - \bar{x}) = h_T(-\bar{x}) = h_T(\bar{x})^* = (X_1, -X_2)^* = (X_1, X_2) = P \,.$$

Since $T - \bar{x} \in]\frac{T}{2}, T[$, we have proved that $\tilde{h}_T$ is surjective.

2.6   Other Examples of Continuous Functions

We define the "tangent to base T" as

$$\tan_T(x) = \frac{\sin_T(x)}{\cos_T(x)} \,.$$

Its natural domain is the set $\{x \in \mathbb{R} : x \neq \frac{T}{4} + k\frac{T}{2}, \ k \in \mathbb{Z}\}$. It is a continuous function, and it is periodic, with minimal period $\frac{T}{2}$ (See Fig. 2.4).

Let us also define the "hyperbolic functions"

$$\cosh_a(x) = \frac{a^x + a^{-x}}{2} \,, \qquad \sinh_a(x) = \frac{a^x - a^{-x}}{2} \,,$$

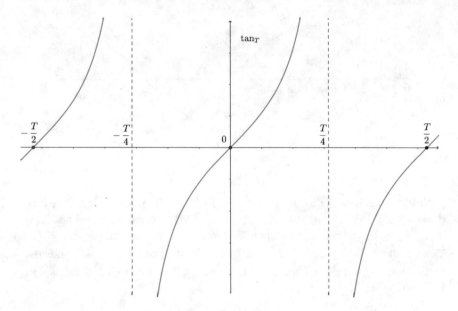

Fig. 2.4 The trigonometric function $\tan_T$

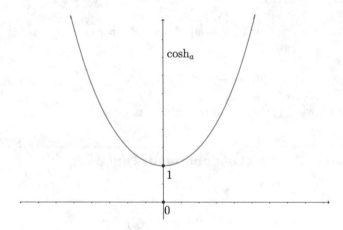

Fig. 2.5 The hyperbolic function $\cosh_a$

where $a > 0$ is fixed (Figs. 2.5 and 2.6). The "hyperbolic cosine" and "hyperbolic sine" to base a are continuous functions, and it can be verified that they satisfy the following identities:

(a) $(\cosh_a(x))^2 - (\sinh_a(x))^2 = 1.$
(b) $\cosh_a(x_1 + x_2) = \cosh_a(x_1)\cosh_a(x_2) + \sinh_a(x_1)\sinh_a(x_2).$
(c) $\sinh_a(x_1 + x_2) = \sinh_a(x_1)\cosh_a(x_2) + \cosh_a(x_1)\sinh_a(x_2).$

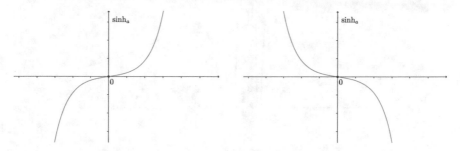

Fig. 2.6 The hyperbolic function $\sinh_a$ when $a > 1$ and $a < 1$, respectively

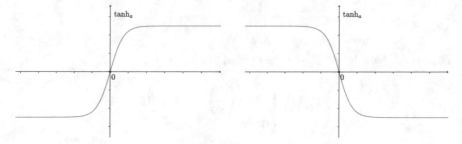

Fig. 2.7 The hyperbolic function $\tanh_a$ when $a > 1$ and $a < 1$, respectively

The striking analogies with the trigonometric functions can be explained recalling the similar properties of the exponential and circular functions. They will be further investigated in Sect. 8.4.2.

We can now define the "hyperbolic tangent" to base a as

$$\tanh_a(x) = \frac{\sinh_a(x)}{\cosh_a(x)}.$$

Its domain is the whole real line $\mathbb{R}$, and it is continuous (Fig. 2.7).

Let us now consider two examples of functions and examine their continuity.

Example 1 Let $f : \mathbb{R} \to \mathbb{R}$ be defined as

$$f(x) = \begin{cases} \sin_T\left(\dfrac{1}{x}\right) & \text{if } x \neq 0, \\ 0 & \text{if } x = 0. \end{cases}$$

If $x_0 \neq 0$, then the function f is continuous at x_0, since it is the composition of continuous functions. In contrast, if $x_0 = 0$, then it is not continuous at x_0 since in every neighborhood of 0 there are values of x for which $f(x) = 1$, whereas $f(0) = 0$.

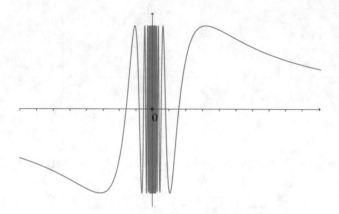

Example 2 Now let $f : \mathbb{R} \to \mathbb{R}$ be defined by

$$f(x) = \begin{cases} x \sin_T\left(\dfrac{1}{x}\right) & \text{if } x \neq 0, \\ 0 & \text{if } x = 0. \end{cases}$$

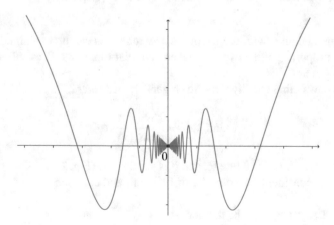

This function is continuous on the whole $\mathbb{R}$. Indeed, if $x_0 \neq 0$, then the situation is similar to the previously described one. If now $x_0 = 0$, then it is useful to observe that

$$|f(x)| \leq |x|, \quad \text{for every } x \in \mathbb{R}.$$

Thus, once $\varepsilon > 0$ is fixed, it is sufficient to choose $\delta = \varepsilon$ to have that

$$|x - 0| < \delta \implies |f(x) - f(0)| < \varepsilon.$$

Limits

We will now introduce another fundamental concept that, however, is strongly related to continuity. It is the notion of the "limit" of a function, a local notion, as we will see. As in Chap. 2, the theory will be developed within the framework of metric spaces.

3.1 The Notion of Limit

Our general setting involves two metric spaces, E and F, a point x_0 of E, and a function

$$f : E \to F \quad \text{or} \quad f : E \setminus \{x_0\} \to F,$$

not necessarily defined in x_0.

Definition 3.1 If there exists $l \in F$ such that the function $\tilde{f} : E \to F$, defined by

$$\tilde{f}(x) = \begin{cases} f(x) & \text{if } x \neq x_0, \\ l & \text{if } x = x_0, \end{cases}$$

is continuous at x_0, then l is said to be a "limit of f at x_0," or also a "limit of $f(x)$ as x tends to x_0," and we can write

$$l = \lim_{x \to x_0} f(x).$$

In other terms, l is a limit of f at x_0 if and only if

$$\forall \varepsilon > 0 \quad \exists \delta > 0 : \forall x \in E \quad 0 < d_E(x, x_0) < \delta \implies d_F(f(x), l) < \varepsilon,$$

© The Author(s), under exclusive license to Springer Nature Switzerland AG 2023
A. Fonda, *A Modern Introduction to Mathematical Analysis*,
https://doi.org/10.1007/978-3-031-23713-3_3

or, equivalently,

$\forall V$ *neighborhood of* l $\exists U$ *neighborhood of* x_0 : $f(U \setminus \{x_0\}) \subseteq V$.

Sometimes we can also write "$f(x) \to l$ as $x \to x_0$."

We know that if x_0 happens to be an isolated point, then the function $\tilde{f}$ defined earlier will be continuous at x_0 *for every* $l \in F$. Therefore, the notion of limit is of no interest at all in this case. This is why we will always assume that x_0 is *not* an isolated point, in which case we say that x_0 is a "cluster point" of E: Every neighborhood of x_0 contains some point of E that differs from x_0 itself.

Note that if x_0 is a cluster point of E, then every neighborhood U_0 of x_0 contains infinitely many points of E. Indeed, once we have found $x_1 \neq x_0$ in U_0, it is possible to choose a neighborhood U_1 of x_0 that does not contain x_1. Then we can find $x_2 \neq x_0$ in U_1, and so on.

From now on we will assume that x_0 is a cluster point of E. This assumption also enables us to prove the following proposition.

Proposition 3.2 *If a limit of* f *at* x_0 *exists, then it is unique.*

Proof Assume by contradiction that there are two limits, l and l', which are distinct. Let us take $\varepsilon = \frac{1}{2} d(l, l')$. Then there exists a $\delta > 0$ such that

$$0 < d(x, x_0) < \delta \implies d(f(x), l) < \varepsilon,$$

and there exists a $\delta' > 0$ such that

$$0 < d(x, x_0) < \delta' \implies d(f(x), l') < \varepsilon.$$

Let $x \neq x_0$ be such that $d(x, x_0) < \delta$ and $d(x, x_0) < \delta'$ (such an x exists because x_0 is a cluster point). Then

$$d(l', l) \leq d(l, f(x)) + d(f(x), l') < 2\varepsilon = d(l', l),$$

a contradiction.   ∎

The following relationship will surely be useful.

Proposition 3.3 *The equivalence*

$$\lim_{x \to x_0} f(x) = l \quad \Leftrightarrow \quad \lim_{x \to x_0} d(f(x), l) = 0$$

always holds true.

Proof The function $F(x) = d(f(x), l)$ has real values, and the distance in $\mathbb{R}$ is $d(\alpha, \beta) = |\alpha - \beta|$. The conclusion easily follows from the definitions. ∎

The following theorem underlines the strong relationship between the concepts of limit and continuity.

Proposition 3.4 *For any function* $f : E \to F$,

$$f \text{ is continuous at } x_0 \quad \Leftrightarrow \quad \lim_{x \to x_0} f(x) = f(x_0).$$

Proof In this case, the function $\tilde{f}$ coincides with f. ∎

3.2 Some Properties of Limits

Let us start with those properties of limits that are directly inherited from those of continuous functions. In the following statements, the functions f and g are defined on E or $E \setminus \{x_0\}$, indifferently, and x_0 is a cluster point of E.

Theorem 3.5 *Let F be a normed vector space, $f, g : E \setminus \{x_0\} \to F$ two functions such that*

$$l_1 = \lim_{x \to x_0} f(x), \quad l_2 = \lim_{x \to x_0} g(x),$$

and $\alpha \in \mathbb{R}$. Then

$$\lim_{x \to x_0} [f(x) \pm g(x)] = l_1 \pm l_2, \quad \lim_{x \to x_0} [\alpha f](x) = \alpha l_1.$$

Assume now that $F = \mathbb{R}^M$ for some integer $M \geq 1$. For any function $f : E \setminus \{x_0\} \to \mathbb{R}^M$ we can consider its components $f_k : E \setminus \{x_0\} \to \mathbb{R}$, with $k = 1, 2, \ldots, M$, and we can write

$$f(x) = (f_1(x), f_2(x), \ldots, f_M(x)).$$

Theorem 3.6 *The limit $\lim_{x \to x_0} f(x) = l \in \mathbb{R}^M$ exists if and only if all the limits $\lim_{x \to x_0} f_k(x) = l_k \in \mathbb{R}$ exist, with $k = 1, 2, \ldots, M$. In that case, $l = (l_1, l_2, \ldots, l_M)$, i.e.,*

$$\lim_{x \to x_0} f(x) = \left(\lim_{x \to x_0} f_1(x), \lim_{x \to x_0} f_2(x), \ldots, \lim_{x \to x_0} f_M(x) \right).$$

Proof This is a direct consequence of Theorem 2.9 for continuous functions. ∎

We now assume that $F = \mathbb{R}$ and state the property of **sign permanence**.

Theorem 3.7 *Let* $g : E \setminus \{x_0\} \to \mathbb{R}$ *be such that*

$$\lim_{x \to x_0} g(x) > 0.$$

Then there exists a neighborhood U of x_0 such that

$$x \in U \setminus \{x_0\} \;\Rightarrow\; g(x) > 0.$$

Similarly, if

$$\lim_{x \to x_0} g(x) < 0,$$

then there exists a neighborhood U of x_0 such that

$$x \in U \setminus \{x_0\} \;\Rightarrow\; g(x) < 0.$$

As an immediate consequence, we have the following corollary.

Corollary 3.8 *If there is a neighborhood U of x_0 such that $g(x) \leq 0$ for every $x \in U \setminus \{x_0\}$, then, if the limit exists,*

$$\lim_{x \to x_0} g(x) \leq 0.$$

Similarly, if $g(x) \geq 0$ for every $x \in U \setminus \{x_0\}$, then, if the limit exists,

$$\lim_{x \to x_0} g(x) \geq 0.$$

Still assuming that $F = \mathbb{R}$, we now consider the product and the quotient of two functions.

Theorem 3.9 *Let* $f, g : E \setminus \{x_0\} \to \mathbb{R}$ *be such that*

$$l_1 = \lim_{x \to x_0} f(x), \quad l_2 = \lim_{x \to x_0} g(x).$$

Then

$$\lim_{x \to x_0} [f(x)g(x)] = l_1 l_2.$$

Moreover, if $l_2 \neq 0$, then

$$\lim_{x \to x_0} \frac{f(x)}{g(x)} = \frac{l_1}{l_2}.$$

Now let E be any metric space and $F = \mathbb{R}$. The following theorem has a strange name, indeed.

Theorem 3.10 (Squeeze Theorem)   *Let $\mathcal{F}_1, \mathcal{F}_2 : E \setminus \{x_0\} \to \mathbb{R}$ be such that*

$$\lim_{x \to x_0} \mathcal{F}_1(x) = \lim_{x \to x_0} \mathcal{F}_2(x) = l\,.$$

If $f : E \setminus \{x_0\} \to \mathbb{R}$ has the property that

$$\mathcal{F}_1(x) \le f(x) \le \mathcal{F}_2(x) \quad \text{for every } x \in E \setminus \{x_0\}\,,$$

then

$$\lim_{x \to x_0} f(x) = l\,.$$

Proof Once $\varepsilon > 0$ has been fixed, there exist $\delta_1 > 0$ and $\delta_2 > 0$ such that

$$0 < d(x, x_0) < \delta_1 \quad \Rightarrow \quad l - \varepsilon < \mathcal{F}_1(x) < l + \varepsilon\,,$$
$$0 < d(x, x_0) < \delta_2 \quad \Rightarrow \quad l - \varepsilon < \mathcal{F}_2(x) < l + \varepsilon\,.$$

Taking $\delta = \min\{\delta_1, \delta_2\}$, we have

$$0 < d(x, x_0) < \delta \quad \Rightarrow \quad l - \varepsilon < \mathcal{F}_1(x) \le f(x) \le \mathcal{F}_2(x) < l + \varepsilon\,,$$

thereby completing the proof. ∎

As a consequence, we have the following corollary.

Corollary 3.11   *Let $f, g : E \setminus \{x_0\} \to \mathbb{R}$ be such that*

$$\lim_{x \to x_0} f(x) = 0\,,$$

and there is a constant $C > 0$ such that

$$|g(x)| \le C\,, \quad \text{for every } x \in E \setminus \{x_0\}\,.$$

Then

$$\lim_{x \to x_0} f(x)g(x) = 0\,.$$

Proof After noticing that

$$-C|f(x)| \le f(x)g(x) \le C|f(x)|,$$

and recalling that, by Proposition 3.3,

$$\lim_{x \to x_0} f(x) = 0 \quad \Leftrightarrow \quad \lim_{x \to x_0} |f(x)| = 0,$$

the result follows from the Squeeze Theorem 3.10, taking $\mathcal{F}_1(x) = -C|f(x)|$ and $\mathcal{F}_2(x) = C|f(x)|$. ∎

Remark 3.12 Returning to the statements of the preceding theorem and corollary, we realize that "for every $x \in E \setminus \{x_0\}$" could be weakened to "for every $x \neq x_0$ in a neighborhood of x_0." This is due to the fact that the notion of limit relates only to the local behavior of the function near x_0. This observation holds in general when dealing with limits and will often be used in what follows.

3.3   Change of Variables in the Limit

We now return to the general setting in metric spaces and examine the composition of two functions, f and g. We have two interesting situations. In the first one, some continuity of g is needed.

Theorem 3.13 *Let $f : E \to F$, or $f : E \setminus \{x_0\} \to F$, be such that*

$$\lim_{x \to x_0} f(x) = l.$$

If $g : F \to G$ is continuous at l, then

$$\lim_{x \to x_0} g(f(x)) = g(l),$$

i.e.,

$$\lim_{x \to x_0} g(f(x)) = g\left(\lim_{x \to x_0} f(x) \right).$$

Proof Recalling the definition of limit, we know that the function $\tilde{f} : E \to \mathbb{R}$ is continuous at x_0, whereas g is continuous at $l = \tilde{f}(x_0)$. Hence, $g \circ \tilde{f}$ is continuous at x_0, so that, recalling that $f(x) = \tilde{f}(x)$ when $x \neq x_0$,

$$\lim_{x \to x_0} g(f(x)) = \lim_{x \to x_0} g(\tilde{f}(x)) = g(\tilde{f}(x_0)) = g(l),$$

as we wanted to prove. ∎

The following theorem gives us the **change of variables formula**. The function g does not have to be continuous at the limit point of f, and indeed it could even not be defined there.

Theorem 3.14 *Let* $f : E \to F$, *or* $f : E \setminus \{x_0\} \to F$, *be such that*

$$\lim_{x \to x_0} f(x) = l.$$

Assume, moreover, that $f(x) \neq l$ *for every* $x \neq x_0$ *in a neighborhood of* x_0. *Let* $g : F \to G$, *or* $g : F \setminus \{l\} \to G$, *be such that*

$$\lim_{y \to l} g(y) = L.$$

Then

$$\lim_{x \to x_0} g(f(x)) = L,$$

i.e.,

$$\lim_{x \to x_0} g(f(x)) = \lim_{\substack{y \to \lim_{x \to x_0} f(x)}} g(y). \tag{3.1}$$

In the preceding formula, we say that the "change of variables $y = f(x)$" has been performed in the limit.

Proof We first observe that, in view of the assumptions, $g \circ f$ is defined on $U \setminus \{x_0\}$ for some neighborhood U of x_0. Moreover, l is a cluster point of F. Recalling again the definition of limit, we know that the function $\tilde{f} : E \to F$ is continuous at x_0, with $\tilde{f}(x_0) = l$. Similarly, let us introduce the function $\tilde{g} : F \to G$, defined as

$$\tilde{g}(y) = \begin{cases} g(y) & \text{if } y \neq l, \\ L & \text{if } y = l. \end{cases}$$

This function is continuous at l, so the composition $\tilde{g} \circ \tilde{f}$ is continuous at x_0. For every $x \in U \setminus \{x_0\}$, since $f(x) \neq l$, we have

$$g(f(x)) = \tilde{g}(f(x)) = \tilde{g}(\tilde{f}(x)),$$

and hence

$$\lim_{x \to x_0} g(f(x)) = \lim_{x \to x_0} \tilde{g}(\tilde{f}(x)) = \tilde{g}(\tilde{f}(x_0)) = L,$$

thereby proving the result. ∎

3.4   On the Limit of Restrictions

We have studied some properties of limits in a context where E and F are metric spaces, x_0 is a cluster point of E, and either $f : E \to F$ or $f : E \setminus \{x_0\} \to F$. We now note that all the aforementioned considerations still hold if we assume that the domain of f is a subset of E, say, $D \subseteq E$, provided that x_0 is a "cluster point of" D. By this we mean that every neighborhood of x_0 contains some point of D that differs from x_0 itself. Notice that x_0 might not be an element of D.

Now let $f : E \setminus \{x_0\} \to F$, and let $\widehat{E} \subseteq E$. We can consider the restriction of f to $\widehat{E} \setminus \{x_0\}$, i.e., the function $\hat{f} : \widehat{E} \setminus \{x_0\} \to F$ such that $\hat{f}(x) = f(x)$ for every $x \in \widehat{E} \setminus \{x_0\}$.

Theorem 3.15 *If the limit of f at x_0 exists and x_0 is a cluster point of $\widehat{E}$, then the limit of $\hat{f}$ at x_0 also exists, and it has the same value:*

$$\lim_{x \to x_0} \hat{f}(x) = \lim_{x \to x_0} f(x).$$

Proof The proof follows directly from the definition of $\hat{f}$.   ∎

The previous theorem is often used to establish the nonexistence of the limit of f at some point x_0, trying to find two restrictions along which the two limits differ.

Example 1 The function $f : \mathbb{R}^2 \setminus \{(0,0)\} \to \mathbb{R}$, defined by

$$f(x, y) = \frac{xy}{x^2 + y^2},$$

has no limit as $(x, y) \to (0, 0)$, since the restrictions of f to the lines $\widehat{E}_1 = \{(x, y) : x = 0\}$ and $\widehat{E}_2 = \{(x, y) : x = y\}$ have different limits.

Example 2 More surprising is the case of the function

$$f(x, y) = \frac{x^2 y}{x^4 + y^2}.$$

It can be seen that all its restrictions to the lines passing through $(0, 0)$ have limits equal to 0. Indeed, this is easily seen for $\widehat{E}_1 = \{(x, y) : x = 0\}$ and $\widehat{E}_2 = \{(x, y) : y = 0\}$, whereas for any $m \neq 0$,

$$\lim_{x \to 0} f(x, mx) = \lim_{x \to 0} \frac{mx^3}{x^4 + m^2 x^2} = \lim_{x \to 0} \frac{mx}{x^2 + m^2} = 0.$$

However, the restriction to the parabola $\{(x, y) : y = x^2\}$ is constantly equal to $\frac{1}{2}$, thereby leading to a different limit. Hence, the function f has no limit at $(0, 0)$.

Example 3 Quite unlike the two preceding examples, let us now prove that

$$\lim_{(x,y)\to(0,0)} \frac{x^2 y^2}{x^2 + y^2} = 0.$$

Let $\varepsilon > 0$ be fixed. After having verified that

$$\frac{x^2 y^2}{x^2 + y^2} \leq \frac{1}{2}(x^2 + y^2),$$

it is natural to take $\delta = \sqrt{2\varepsilon}$, so that

$$d((x, y), (0, 0)) < \delta \implies \left| \frac{x^2 y^2}{x^2 + y^2} - 0 \right| < \varepsilon.$$

Now let E be a subset of $\mathbb{R}$ and F be a general metric space. Given $f : E \to F$ or $f : E \setminus \{x_0\} \to F$, we can consider the two restrictions $\hat{f}_1$ and $\hat{f}_2$ to the sets $\widehat{E}_1 = E \cap]-\infty, x_0[$ and $\widehat{E}_2 = E \cap]x_0 + \infty[$, respectively. If x_0 is a cluster point of $\widehat{E}_1$, then we call the "left limit" of f at x_0, whenever it exists, the limit of $\hat{f}_1(x)$ when x approaches x_0 (in $\widehat{E}_1$), and we denote it by

$$\lim_{x \to x_0^-} f(x).$$

Analogously, if x_0 is a cluster point of $\widehat{E}_2$, we call the "right limit" of f at x_0, whenever it exists, the limit of $\hat{f}_2(x)$ when x approaches x_0 (in $\widehat{E}_2$), and we denote it by

$$\lim_{x \to x_0^+} f(x).$$

Theorem 3.16 *If x_0 is a cluster point of both $E \cap]-\infty, x_0[$ and $E \cap]x_0 + \infty[$, then the limit of f at x_0 exists if and only if both the left limit and the right limit exist, and they are equal to each other.*

Proof We already know that if the limit of f at x_0 exists, then all restrictions of f must have the same limit at x_0. Conversely, let us assume that the left limit and the right limit exist and that $l \in F$ is their common value. Let $\varepsilon > 0$ be fixed. Then there exist $\delta_1 > 0$ and $\delta_2 > 0$ such that if $x \in E$,

$$x_0 - \delta_1 < x < x_0 \implies d(f(x), l) < \varepsilon$$

and

$$x_0 < x < x_0 + \delta_2 \implies d(f(x), l) < \varepsilon.$$

Defining $\delta = \min\{\delta_1, \delta_2\}$, we then have that if $x \in E \setminus \{x_0\}$,

$$x_0 - \delta < x < x_0 + \delta \quad \Rightarrow \quad d(f(x), l) < \varepsilon,$$

showing that the limit of f at x_0 exists and is equal to l.   ∎

Example The "sign function" $f : \mathbb{R} \to \mathbb{R}$, defined as

$$f(x) = \begin{cases} 1 & \text{if } x > 0, \\ 0 & \text{if } x = 0, \\ -1 & \text{if } x < 0, \end{cases}$$

has no limit at $x_0 = 0$ since $\lim\limits_{x \to 0^-} f(x) = -1$ and $\lim\limits_{x \to 0^+} f(x) = 1$.

3.5   The Extended Real Line

Let us consider the function $\varphi : \mathbb{R} \to]-1, 1[$ defined as

$$\varphi(x) = \frac{x}{1 + |x|}.$$

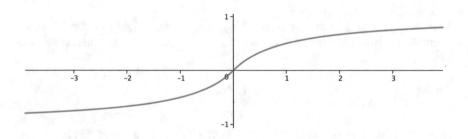

This is an invertible function, with inverse $\varphi^{-1} :]-1, 1[\to \mathbb{R}$ given by

$$\varphi^{-1}(y) = \frac{y}{1 - |y|}.$$

We can now define a new distance on $\mathbb{R}$ as

$$\tilde{d}(x, x') = |\varphi(x) - \varphi(x')|.$$

It can indeed be verified that it satisfies the four properties characterizing a distance. Let us denote by $\tilde{B}(x_0, \rho)$ the open ball for this new distance centered at x_0, with radius $\rho > 0$, i.e.,

$$\widetilde{B}(x_0, \rho) = \{x \in \mathbb{R} : |\varphi(x) - \varphi(x_0)| < \rho\}.$$

We claim that the neighborhoods of any point $x_0 \in \mathbb{R}$ remain the same as those provided by the usual distance on $\mathbb{R}$. Indeed, since φ is continuous at x_0, for every $\rho_1 > 0$, there exists a $\rho_2 > 0$ such that

$$|x - x_0| < \rho_2 \quad \Rightarrow \quad |\varphi(x) - \varphi(x_0)| < \rho_1,$$

i.e.,

$$]x_0 - \rho_2, x_0 + \rho_2[\subseteq \widetilde{B}(x_0, \rho_1).$$

Conversely, since φ^{-1} is continuous at $y_0 = \varphi(x_0) \in]-1, 1[$, for every $\rho_1 > 0$ there exists a $\rho_2 > 0$ such that

$$|y - y_0| < \rho_2 \quad \Rightarrow \quad y \in]-1, 1[\text{ and } |\varphi^{-1}(y) - \varphi^{-1}(y_0)| < \rho_1.$$

In particular, taking $y = \varphi(x)$,

$$|\varphi(x) - \varphi(x_0)| < \rho_2 \quad \Rightarrow \quad \varphi(x) \in]-1, 1[\text{ and } |x - x_0| < \rho_1,$$

i.e.,

$$\widetilde{B}(x_0, \rho_2) \subseteq]x_0 - \rho_1, x_0 + \rho_1[.$$

We have thus proved our claim.

Let us now introduce a new set, $\widetilde{\mathbb{R}}$, defined by adding to $\mathbb{R}$ two new elements denoted by $-\infty$ and $+\infty$, i.e.,

$$\widetilde{\mathbb{R}} = \mathbb{R} \cup \{-\infty, +\infty\}.$$

The set $\widetilde{\mathbb{R}}$ is totally ordered, maintaining the usual order on the reals while setting

$$-\infty < x < +\infty \quad \text{for every } x \in \mathbb{R}.$$

Let us define the function $\tilde{\varphi} : \widetilde{\mathbb{R}} \to [-1, 1]$ as

$$\tilde{\varphi}(x) = \begin{cases} -1 & \text{if } x = -\infty, \\ \varphi(x) & \text{if } x \in \mathbb{R}, \\ 1 & \text{if } x = +\infty. \end{cases}$$

It is invertible, with inverse $\tilde{\varphi}^{-1} : [-1, 1] \to \widetilde{\mathbb{R}}$ given by

$$\tilde{\varphi}^{-1}(y) = \begin{cases} -\infty & \text{if } y = -1, \\ \varphi^{-1}(y) & \text{if } y \in]-1, 1[, \\ +\infty & \text{if } y = 1. \end{cases}$$

We now define, for every $x, x' \in \widetilde{\mathbb{R}}$,

$$\tilde{d}(x, x') = |\tilde{\varphi}(x) - \tilde{\varphi}(x')|.$$

It is readily verified that $\tilde{d}$ is a distance on $\widetilde{\mathbb{R}}$, so that $\widetilde{\mathbb{R}}$ is now a metric space. Let us see, for example, what a ball centered at $+\infty$ looks like:

$$B(+\infty, \rho) = \{x \in \widetilde{\mathbb{R}} : |\tilde{\varphi}(x) - 1| < \rho\} = \{x \in \widetilde{\mathbb{R}} : \tilde{\varphi}(x) > 1 - \rho\},$$

hence

$$B(+\infty, \rho) = \begin{cases} \widetilde{\mathbb{R}} & \text{if } \rho > 2, \\ ]-\infty, +\infty] & \text{if } \rho = 2, \\ ]\varphi^{-1}(1 - \rho), +\infty] & \text{if } \rho < 2, \end{cases}$$

where we have used the notation

$$]a, +\infty] = \{x \in \widetilde{\mathbb{R}} : x > a\} =]a, +\infty[\cup \{+\infty\}.$$

We can thus state that a neighborhood of $+\infty$ is a set that contains, besides $+\infty$ itself, an interval of the type $]\alpha, +\infty[$ for some $\alpha \in \mathbb{R}$.

Analogously, a neighborhood of $-\infty$ is a set containing $-\infty$ and an interval of the type $]-\infty, \beta[$ for some $\beta \in \mathbb{R}$.

Let us see how the definition of limit translates in some cases where the new elements $-\infty$ and $+\infty$ appear.

To start with, let $f : E \to F$ be a function with $E \subseteq \mathbb{R}$, whose codomain F is any metric space. Considering E a subset of $\widetilde{\mathbb{R}}$, we have that $+\infty$ is a cluster point of E if and only if E is not bounded from above. In that case,

$$\lim_{x \to +\infty} f(x) = l \in F \Leftrightarrow \quad \forall V \text{ neighbourhood of } l$$
$$\exists U \text{ neighbourhood of } +\infty : f(U \cap E) \subseteq V$$
$$\Leftrightarrow \quad \forall \varepsilon > 0 \quad \exists \alpha \in \mathbb{R} : \quad x > \alpha \Rightarrow d(f(x), l) < \varepsilon.$$

Similarly, if E is not bounded from below,

$$\lim_{x \to -\infty} f(x) = l \in F \quad \Leftrightarrow \quad \forall \varepsilon > 0 \quad \exists \beta \in \mathbb{R} : \quad x < \beta \Rightarrow d(f(x), l) < \varepsilon.$$

Notice that

$$\lim_{x \to -\infty} f(x) = l \quad \Leftrightarrow \quad \lim_{x \to +\infty} f(-x) = l.$$

Let us now consider a function $f : E \to \mathbb{R}$, or $f : E \setminus \{x_0\} \to \mathbb{R}$, where E is any metric space and x_0 is a cluster point of E. If we consider the codomain $F = \mathbb{R}$ a subset of $\widetilde{\mathbb{R}}$, then

$$\lim_{x \to x_0} f(x) = +\infty \Leftrightarrow \quad \forall V \text{ neighbourhood of } +\infty$$
$$\exists U \text{ neighbourhood of } x_0 : \ f(U \setminus \{x_0\}) \subseteq V$$
$$\Leftrightarrow \quad \forall \alpha \in \mathbb{R} \quad \exists \delta > 0 : \quad 0 < d(x, x_0) < \delta \ \Rightarrow \ f(x) > \alpha.$$

Similarly,

$$\lim_{x \to x_0} f(x) = -\infty \quad \Leftrightarrow \quad \forall \beta \in \mathbb{R} \quad \exists \delta > 0 : \quad 0 < d(x, x_0) < \delta \ \Rightarrow \ f(x) < \beta.$$

Notice that

$$\lim_{x \to x_0} f(x) = -\infty \quad \Leftrightarrow \quad \lim_{x \to x_0} [-f(x)] = +\infty.$$

The foregoing situations can be combined together. For example, if $E \subseteq \mathbb{R}$ is not bounded from above and $F = \mathbb{R}$ is considered a subset of $\widetilde{\mathbb{R}}$, then

$$\lim_{x \to +\infty} f(x) = +\infty \Leftrightarrow \quad \forall V \text{ neighborhood of } +\infty$$
$$\exists U \text{ neighborhood of } +\infty : \quad f(U \cap E) \subseteq V$$
$$\Leftrightarrow \quad \forall \alpha \in \mathbb{R} \quad \exists \alpha' \in \mathbb{R} : \quad x > \alpha' \ \Rightarrow \ f(x) > \alpha,$$

and

$$\lim_{x \to +\infty} f(x) = -\infty \quad \Leftrightarrow \quad \forall \beta \in \mathbb{R} \quad \exists \alpha' \in \mathbb{R} : \quad x > \alpha' \ \Rightarrow \ f(x) < \beta.$$

On the other hand, if $E \subseteq \mathbb{R}$ is not bounded from below,

$$\lim_{x \to -\infty} f(x) = +\infty \quad \Leftrightarrow \quad \forall \alpha \in \mathbb{R} \quad \exists \beta' \in \mathbb{R} : \quad x < \beta' \ \Rightarrow \ f(x) > \alpha,$$

and

$$\lim_{x \to -\infty} f(x) = -\infty \quad \Leftrightarrow \quad \forall \beta \in \mathbb{R} \quad \exists \beta' \in \mathbb{R} : \quad x < \beta' \ \Rightarrow \ f(x) < \beta.$$

An important particular situation is encountered when dealing with a sequence $(a_n)_n$ in a metric space F. We are thus given a function $f : \mathbb{N} \to F$ defined as

$f(n) = a_n$. Considering $\mathbb{N}$ a subset of $\widetilde{\mathbb{R}}$, it is readily seen that the only cluster point of $\mathbb{N}$ is $+\infty$, and, adapting the definition of limit to this case, we can write

$$\lim_{n\to+\infty} a_n = l \in F \quad \Leftrightarrow \quad \forall \varepsilon > 0 \ \exists \bar{n} \in \mathbb{N}: \quad n \geq \bar{n} \ \Rightarrow \ d(a_n, l) < \varepsilon.$$

As a particular case we may have $F = \mathbb{R}$, considered as a subset of $\widetilde{\mathbb{R}}$, and we thus recover the preceding definitions when $l = -\infty$ or $l = +\infty$.

The limit of a sequence will often be denoted simply by $\lim_n a_n$, tacitly implying that $n \to +\infty$.

3.6   Some Operations with $-\infty$ and $+\infty$

When the limits are $-\infty$ or $+\infty$, the normal operations with limits cannot be used. We will provide here a few useful rules for some of these cases. In what follows, all the functions will be defined either on the whole metric space E or on $E \setminus \{x_0\}$, and x_0 will always be assumed to be a cluster point of E. Let us start with the sum of two functions.

Theorem 3.17 *If*

$$\lim_{x\to x_0} f(x) = +\infty$$

and there exists a $\gamma \in \mathbb{R}$ such that

$$g(x) \geq \gamma, \quad \text{for every } x \neq x_0 \text{ in a neighborhood of } x_0,$$

then

$$\lim_{x\to x_0} [f(x) + g(x)] = +\infty.$$

Proof Let $\alpha \in \mathbb{R}$ be fixed. Defining $\tilde{\alpha} = \alpha - \gamma$, there exists a $\delta > 0$ such that

$$0 < d(x, x_0) < \delta \ \Rightarrow \ f(x) > \tilde{\alpha}.$$

Hence,

$$0 < d(x, x_0) < \delta \ \Rightarrow \ f(x) + g(x) > \tilde{\alpha} + \gamma = \alpha,$$

thereby proving the result.   ∎

Corollary 3.18 *If*

$$\lim_{x \to x_0} f(x) = +\infty \quad and \quad \lim_{x \to x_0} g(x) = l \in \mathbb{R} \ (or \, l = +\infty),$$

then

$$\lim_{x \to x_0} [f(x) + g(x)] = +\infty.$$

Proof If the limit of g is some $l \in \mathbb{R}$, then there exists a $\delta > 0$ such that

$$0 < d(x, x_0) < \delta \implies g(x) \geq l - 1.$$

On the other hand, if the limit of g is $+\infty$, then we can find a $\delta > 0$ such that

$$0 < d(x, x_0) < \delta \implies g(x) \geq 0.$$

In any case, the previous theorem can be applied to obtain the conclusion. ∎

As a mnemonic rule, we will briefly write

$$(+\infty) + l = +\infty \quad \text{if } l \text{ is a real number;}$$
$$(+\infty) + (+\infty) = +\infty.$$

In perfect analogy, we can state a theorem, with a related corollary, in the case where the limit of f is $-\infty$. As a mnemonic rule, we will then write

$$(-\infty) + l = -\infty \quad \text{if } l \text{ is a real number;}$$
$$(-\infty) + (-\infty) = -\infty.$$

Regarding the product of two functions, we have the following theorem.

Theorem 3.19 *If*

$$\lim_{x \to x_0} f(x) = +\infty$$

and there exists a $\gamma > 0$ such that

$$g(x) \geq \gamma, \quad \textit{for every } x \neq x_0 \textit{ in a neighbourhood of } x_0,$$

then

$$\lim_{x \to x_0} [f(x)g(x)] = +\infty.$$

Proof Let $\alpha \in \mathbb{R}$ be fixed. We may assume with no loss of generality that $\alpha > 0$. Setting $\tilde{\alpha} = \frac{\alpha}{\gamma}$, there exists a $\delta > 0$ such that

$$0 < d(x, x_0) < \delta \implies f(x) > \tilde{\alpha}.$$

Hence,

$$0 < d(x, x_0) < \delta \implies f(x)g(x) > \tilde{\alpha}\gamma = \alpha,$$

thereby proving the statement.   ∎

Corollary 3.20 *If*

$$\lim_{x \to x_0} f(x) = +\infty \quad and \quad \lim_{x \to x_0} g(x) = l > 0 \ \ (or \ l = +\infty),$$

then

$$\lim_{x \to x_0} [f(x)g(x)] = +\infty.$$

Proof If the limit of g is a real number $l > 0$, then there exists a $\delta > 0$ such that

$$0 < d(x, x_0) < \delta \implies g(x) \geq \frac{l}{2}.$$

On the other hand, if the limit of g is $+\infty$, then there is a $\delta > 0$ such that

$$0 < d(x, x_0) < \delta \implies g(x) \geq 1.$$

In any case, the previous theorem provides the conclusion.   ∎

In the same spirit, we will briefly write

$$(+\infty) \cdot l = +\infty \quad \text{if } l > 0 \text{ is a real number};$$
$$(+\infty) \cdot (+\infty) = +\infty,$$

with all the following variants:

$$(+\infty) \cdot l = -\infty \quad \text{if } l < 0 \text{ is a real number};$$
$$(-\infty) \cdot l = -\infty \quad \text{if } l > 0 \text{ is a real number};$$

$$(-\infty) \cdot l = +\infty \quad \text{if } l < 0 \text{ is a real number};$$

$$(+\infty) \cdot (-\infty) = -\infty;$$

$$(-\infty) \cdot (-\infty) = +\infty.$$

Let us now analyze the reciprocal of a function. We have two theorems.

Theorem 3.21 *If*

$$\lim_{x \to x_0} |f(x)| = +\infty,$$

then

$$\lim_{x \to x_0} \frac{1}{f(x)} = 0.$$

Proof Let $\varepsilon > 0$ be fixed. Setting $\alpha = \frac{1}{\varepsilon}$, there exists a $\delta > 0$ such that

$$0 < d(x, x_0) < \delta \implies |f(x)| > \alpha.$$

Hence,

$$0 < d(x, x_0) < \delta \implies \left| \frac{1}{f(x)} - 0 \right| = \frac{1}{|f(x)|} < \frac{1}{\alpha} = \varepsilon,$$

thereby proving the claim. ∎

Theorem 3.22 *If*

$$\lim_{x \to x_0} f(x) = 0$$

and

$$f(x) > 0 \quad \textit{for every } x \neq x_0 \textit{ in a neighborhood of } x_0,$$

then

$$\lim_{x \to x_0} \frac{1}{f(x)} = +\infty.$$

However, if

$$f(x) < 0 \quad \textit{for every } x \neq x_0 \textit{ in a neighborhood of } x_0,$$

then

$$\lim_{x \to x_0} \frac{1}{f(x)} = -\infty.$$

Proof We treat the first case, the second one being similar. Let $\alpha \in \mathbb{R}$ be fixed; we can assume without loss of generality that $\alpha > 0$. Setting $\varepsilon = \frac{1}{\alpha}$, there exists a $\delta > 0$ such that

$$0 < d(x, x_0) < \delta \implies 0 < f(x) < \varepsilon.$$

Then

$$0 < d(x, x_0) < \delta \implies \frac{1}{f(x)} > \frac{1}{\varepsilon} = \alpha,$$

and the proof is completed. ∎

Finally, we present two useful variants of the Squeeze Theorem 3.10 in the case where the limit is $+\infty$, where only one comparison function will be needed.

Theorem 3.23 *Let $\mathcal{F}_1$ be such that*

$$\lim_{x \to x_0} \mathcal{F}_1(x) = +\infty.$$

If

$$f(x) \geq \mathcal{F}_1(x) \quad \text{for every } x \neq x_0 \text{ in a neighborhood of } x_0,$$

then

$$\lim_{x \to x_0} f(x) = +\infty.$$

Proof Setting $g(x) = f(x) - \mathcal{F}_1(x)$, we have $g(x) \geq 0$ for every x in a neighborhood of x_0 and $f(x) = \mathcal{F}_1(x) + g(x)$. The result then follows directly from Theorem 3.17. ∎

In the case where the limit is $-\infty$, we have the following analogous result.

Theorem 3.24 *Let $\mathcal{F}_2$ be such that*

$$\lim_{x \to x_0} \mathcal{F}_2(x) = -\infty.$$

If

$$f(x) \le \mathcal{F}_2(x), \quad \text{for every } x \ne x_0 \text{ in a neighbourhood of } x_0,$$

then

$$\lim_{x \to x_0} f(x) = -\infty.$$

We will now deal with some elementary situations when x approaches either $+\infty$ or $-\infty$.

1. Let us first consider the function

$$f(x) = x^n,$$

where n is an integer. It can be verified by induction that for every $n \ge 1$,

$$x \ge 1 \quad \Rightarrow \quad x^n \ge x.$$

Since clearly $\lim\limits_{x \to +\infty} x = +\infty$, as a consequence of the preceding theorems we have

$$\lim_{x \to +\infty} x^n = \begin{cases} +\infty & \text{if } n \ge 1, \\ 1 & \text{if } n = 0, \\ 0 & \text{if } n \le -1. \end{cases}$$

If we then take into account that

$$(-x)^n = x^n \text{ if } n \text{ is even}, \qquad (-x)^n = -x^n \text{ if } n \text{ is odd},$$

we also conclude that

$$\lim_{x \to -\infty} x^n = \begin{cases} +\infty & \text{if } n \ge 1 \text{ is even}, \\ -\infty & \text{if } n \ge 1 \text{ is odd}, \\ 1 & \text{if } n = 0, \\ 0 & \text{if } n \le -1. \end{cases}$$

2. Let us consider the polynomial function

$$f(x) = a_n x^n + a_{n-1} x^{n-1} + \cdots + a_2 x^2 + a_1 x + a_0,$$

where $n \geq 1$ and $a_n \neq 0$. Writing

$$f(x) = x^n \left(a_n + \frac{a_{n-1}}{x} + \cdots + \frac{a_2}{x^{n-2}} + \frac{a_1}{x^{n-1}} + \frac{a_0}{x^n} \right)$$

and using the fact that

$$\lim_{x \to +\infty} \left(a_n + \frac{a_{n-1}}{x} + \cdots + \frac{a_2}{x^{n-2}} + \frac{a_1}{x^{n-1}} + \frac{a_0}{x^n} \right) = a_n,$$

we see that

$$\lim_{x \to +\infty} f(x) = \begin{cases} +\infty & \text{if } a_n > 0, \\ -\infty & \text{if } a_n < 0, \end{cases}$$

whereas

$$\lim_{x \to -\infty} f(x) = \begin{cases} +\infty & \text{if either } [n \text{ is even and } a_n > 0] \text{ or } [n \text{ is odd and } a_n < 0], \\ -\infty & \text{if either } [n \text{ is even and } a_n < 0] \text{ or } [n \text{ is odd and } a_n > 0]. \end{cases}$$

3. Consider now the rational function

$$f(x) = \frac{a_n x^n + a_{n-1} x^{n-1} + \cdots + a_2 x^2 + a_1 x + a_0}{b_m x^m + b_{m-1} x^{m-1} + \cdots + b_2 x^2 + b_1 x + b_0},$$

where $n, m \geq 1$ and $a_n \neq 0$, $b_m \neq 0$. As previously, writing

$$f(x) = x^{n-m} \frac{a_n + \frac{a_{n-1}}{x} + \cdots + \frac{a_2}{x^{n-2}} + \frac{a_1}{x^{n-1}} + \frac{a_0}{x^n}}{b_m + \frac{b_{m-1}}{x} + \cdots + \frac{b_2}{x^{m-2}} + \frac{b_1}{x^{m-1}} + \frac{b_0}{x^m}},$$

we can conclude that

$$\lim_{x \to +\infty} f(x) = \lim_{x \to +\infty} \frac{a_n}{b_m} x^{n-m} = \begin{cases} +\infty & \text{if } n > m \text{ and } a_n, b_m \text{ have the same sign,} \\ -\infty & \text{if } n > m \text{ and } a_n, b_m \text{ have opposite signs,} \\ \dfrac{a_n}{b_m} & \text{if } n = m, \\ 0 & \text{if } n < m. \end{cases}$$

In a similar way, once it is observed that

$$\lim_{x \to -\infty} f(x) = \lim_{x \to -\infty} \frac{a_n}{b_m} x^{n-m},$$

the limit can be computed in all the different cases.

3.7   Limits of Monotone Functions

We will now see how the monotonicity of a function makes it possible to establish the existence of left or right limits.

Let E be a subset of $\mathbb{R}$, and let x_0 be a cluster point of $E \cap]x_0, +\infty[$.

Theorem 3.25 *If $f : E \cap]x_0, +\infty[\to \mathbb{R}$ is increasing, then*

$$\lim_{x \to x_0^+} f(x) = \inf f(E \cap]x_0, +\infty[).$$

On the other hand, if f is decreasing, then

$$\lim_{x \to x_0^+} f(x) = \sup f(E \cap]x_0, +\infty[).$$

Proof We prove only the first statement, since the proof of the second one is analogous. Set $\bar{\iota} = \inf f(E \cap]x_0, +\infty[)$. If it happens that $\bar{\iota} \in \mathbb{R}$, then we fix an $\varepsilon > 0$. By the properties of the infimum, there exists a $\bar{y} \in f(E \cap]x_0, +\infty[)$ such that $\bar{y} < \bar{\iota} + \varepsilon$. Then, taking $\bar{x} \in E \cap]x_0, +\infty[$ satisfying $f(\bar{x}) = \bar{y}$ and using the fact that f is increasing, we have

$$x_0 < x < \bar{x} \;\Rightarrow\; \bar{\iota} \leq f(x) \leq f(\bar{x}) < \bar{\iota} + \varepsilon,$$

thereby completing the proof in this case.

If it happens that $\bar{\iota} = -\infty$, then we fix a $\beta \in \mathbb{R}$. Then, since $f(E \cap]x_0, +\infty[)$ is unbounded from below, there exists a $\bar{x} \in E \cap]x_0, +\infty[$ satisfying $f(\bar{x}) < \beta$. Using the fact that f is increasing, we have

$$x_0 < x < \bar{x} \;\Rightarrow\; f(x) \leq f(\bar{x}) < \beta,$$

thereby proving that $\lim_{x \to x_0^+} f(x) = -\infty$.   ∎

Notice that the previous statement also includes the case where $x_0 = -\infty$, provided that E is unbounded from below.

We now state the analogous result by assuming that x_0 is a cluster point of $E \cap]-\infty, x_0[$.

Theorem 3.26 *If $f : E \cap]-\infty, x_0[\to \mathbb{R}$ is increasing, then*

$$\lim_{x \to x_0^-} f(x) = \sup f(E \cap]-\infty, x_0[).$$

On the other hand, if f is decreasing, then

$$\lim_{x \to x_0^-} f(x) = \inf f(E \cap] - \infty, x_0[).$$

Proof Defining $g = -f$, we are led back to the previous theorem, and the conclusion rapidly follows. ∎

The previous statement also includes the case where $x_0 = +\infty$, provided that E is unbounded from above. This happens, e.g., for real sequences, leading to the following corollary.

Corollary 3.27 *Every monotone sequence of real numbers has a limit.*

Proof If $(a_n)_n$ is increasing, then $\lim_n a_n = \sup\{a_n : n \in \mathbb{N}\}$, and this limit can be either a real number or $+\infty$. Similarly, if $(a_n)_n$ is decreasing, then the limit will be either a real number or $-\infty$. ∎

As an example, consider the sequence $(a_n)_n$ defined for $n \geq 1$ by the formula

$$a_n = \left(1 + \frac{1}{n}\right)^n.$$

Let us prove that it is increasing:

$$\frac{a_{n+1}}{a_n} = \frac{\left(1 + \frac{1}{n+1}\right)^{n+1}}{\left(1 + \frac{1}{n}\right)^n}$$

$$= \left(\frac{n+2}{n+1}\right)^{n+1} \left(\frac{n}{n+1}\right)^{n+1} \frac{n+1}{n}$$

$$= \left(\frac{n^2 + 2n}{(n+1)^2}\right)^{n+1} \frac{n+1}{n}$$

$$= \left(1 + \frac{-1}{(n+1)^2}\right)^{n+1} \frac{n+1}{n},$$

so that, by the Bernoulli inequality,

$$\frac{a_{n+1}}{a_n} \geq \left(1 + (n+1)\frac{-1}{(n+1)^2}\right) \frac{n+1}{n} = 1.$$

We have thus shown that $a_n \leq a_{n+1}$ for every $n \geq 1$; hence, $(a_n)_n$ is increasing.

Let us now consider the sequence $(b_n)_n$ defined for $n \geq 1$ by

$$b_n = \left(1 + \frac{1}{n}\right)^{n+1}.$$

Let us prove that $(b_n)_n$ is decreasing:

$$
\begin{aligned}
\frac{b_n}{b_{n+1}} &= \frac{\left(1 + \frac{1}{n}\right)^{n+1}}{\left(1 + \frac{1}{n+1}\right)^{n+2}} \\
&= \frac{n}{n+1}\left(\frac{n+1}{n}\right)^{n+2}\left(\frac{n+1}{n+2}\right)^{n+2} \\
&= \frac{n}{n+1}\left(\frac{(n+1)^2}{n^2+2n}\right)^{n+2} \\
&= \frac{n}{n+1}\left(1 + \frac{1}{n^2+2n}\right)^{n+2} \\
&\geq \frac{n}{n+1}\left(1 + (n+2)\frac{1}{n^2+2n}\right) = 1,
\end{aligned}
$$

showing that $b_n \geq b_{n+1}$ for every $n \geq 1$ (once again we have used the Bernoulli inequality). Since $0 < a_n < b_n$ for every $n \geq 1$, both the sequences $(a_n)_n$ and $(b_n)_n$ have a finite positive limit. Additionally, since

$$\frac{\lim_n b_n}{\lim_n a_n} = \lim_n \frac{b_n}{a_n} = \lim_n \left(1 + \frac{1}{n}\right) = 1,$$

we can conclude that $\lim_n a_n = \lim_n b_n$. This is a real number, and it is called either "Euler's number" or "Napier's constant"; it is denoted by the letter e. We can thus write

$$e = \lim_n \left(1 + \frac{1}{n}\right)^n.$$

It can be proved that it is an irrational number:

$$e = 2.71828\ldots$$

3.8 Limits for Exponentials and Logarithms

First of all, we want to prove that, when x varies in $\mathbb{R}$,

$$\lim_{x \to +\infty} \left(1 + \frac{1}{x}\right)^x = e. \tag{3.2}$$

For every $x \geq 0$, let $n(x)$ be a (unique) natural number such that

$$n(x) \leq x < n(x) + 1,$$

i.e., the "integer part" of x. Then, for $x \geq 1$,

$$\left(1 + \frac{1}{n(x) + 1}\right)^{n(x)} < \left(1 + \frac{1}{x}\right)^{n(x)} \leq$$

$$\leq \left(1 + \frac{1}{x}\right)^x <$$

$$< \left(1 + \frac{1}{x}\right)^{n(x)+1} \leq \left(1 + \frac{1}{n(x)}\right)^{n(x)+1}.$$

Since $\lim_{x \to +\infty} n(x) = +\infty$, we have

$$\lim_{x \to +\infty} \left(1 + \frac{1}{n(x)}\right)^{n(x)+1} = \lim_{n} \left(1 + \frac{1}{n}\right)^{n+1}$$

$$= \lim_{n} \left(1 + \frac{1}{n}\right)^n \left(1 + \cdot \frac{1}{n}\right)$$

$$= e \cdot 1 = e$$

and

$$\lim_{x \to +\infty} \left(1 + \frac{1}{n(x) + 1}\right)^{n(x)} = \lim_{n} \left(1 + \frac{1}{n}\right)^{n-1}$$

$$= \lim_{n} \left(1 + \frac{1}{n}\right)^n \left(1 + \frac{1}{n}\right)^{-1}$$

$$= e \cdot 1 = e.$$

By the Squeeze Theorem 3.10, identity (3.2) follows.

We now prove that also

$$\lim_{x \to -\infty} \left(1 + \frac{1}{x}\right)^x = e. \tag{3.3}$$

Indeed, using the formula $\lim_{x \to -\infty} f(x) = \lim_{x \to +\infty} f(-x)$, we have

$$\lim_{x \to -\infty} \left(1 + \frac{1}{x}\right)^x = \lim_{x \to +\infty} \left(1 - \frac{1}{x}\right)^{-x} = \lim_{x \to +\infty} \left(1 + \frac{1}{x-1}\right)^x$$

$$= \lim_{y \to +\infty} \left(1 + \frac{1}{y}\right)^{y+1} = \lim_{y \to +\infty} \left(1 + \frac{1}{y}\right)^y \left(1 + \frac{1}{y}\right) = e \cdot 1 = e.$$

We are now in a position to establish the following important result involving the exponential and the logarithm functions.

Theorem 3.28 *We have*

$$\lim_{x \to 0} \frac{\log_a(1+x)}{x} = \log_a(e), \qquad \lim_{x \to 0} \frac{a^x - 1}{x} = \frac{1}{\log_a(e)}.$$

Proof By (3.2) and the continuity of $\log_a$, we have

$$\lim_{x \to 0+} \frac{\log_a(1+x)}{x} = \lim_{y \to +\infty} y \log_a \left(1 + \frac{1}{y}\right) = \lim_{y \to +\infty} \log_a \left(1 + \frac{1}{y}\right)^y = \log_a(e),$$

and by (3.3) the same is true for the left limit. Moreover,

$$\lim_{x \to 0} \frac{a^x - 1}{x} = \lim_{y \to 0} \frac{y}{\log_a(1+y)} = \frac{1}{\lim_{y \to 0} \dfrac{\log_a(1+y)}{y}} = \frac{1}{\log_a(e)},$$

thereby completing the proof. ∎

Notice that the choice $a = e$ considerably simplifies the preceding formulas: We have

$$\lim_{x \to 0} \frac{\log_e(1+x)}{x} = 1, \qquad \lim_{x \to 0} \frac{e^x - 1}{x} = 1.$$

That is why we will almost always choose as the base of the exponential and the logarithm the Euler number e, which is the "natural base." We will write $\exp(x)$ (or even $\exp x$) instead of $\exp_e(x)$ and $\ln(x)$ (or even $\ln x$) instead of $\log_e(x)$. The following formulas may be useful:

$$a^x = e^{x \ln a}, \qquad \log_a(x) = \frac{\ln x}{\ln a}.$$

Also, for the hyperbolic functions the base e will always be preferred, and we will write $\cosh(x)$ (or even $\cosh x$) instead of $\cosh_e(x)$ and $\sinh(x)$ (or even $\sinh x$) instead of $\sinh_e(x)$.

The following identities hold:

$$\lim_{x \to 0} \frac{\sinh x}{x} = 1, \qquad \lim_{x \to 0} \frac{\cosh x - 1}{x^2} = \frac{1}{2}, \qquad \lim_{x \to 0} \frac{\tanh x}{x} = 1.$$

Let us prove, for example, the first one:

$$\lim_{x \to 0} \frac{e^x - e^{-x}}{2x} = \frac{1}{2} \left(\lim_{x \to 0} \frac{e^x - 1}{x} + \lim_{x \to 0} \frac{e^{-x} - 1}{-x} \right)$$

$$= \frac{1}{2} \left(1 + \lim_{y \to 0} \frac{e^y - 1}{y} \right) = \frac{1}{2}(1 + 1) = 1.$$

Let us now concentrate on the behavior of the exponential and the logarithm at $+\infty$. Using the properties of monotonicity and surjectivity of $\exp_a : \mathbb{R} \to {]}0, +\infty{[}$, we see that

$$\lim_{x \to +\infty} a^x = \begin{cases} +\infty & \text{if } a > 1, \\ 0 & \text{if } 0 < a < 1, \end{cases}$$

while

$$\lim_{x \to +\infty} \log_a(x) = \begin{cases} +\infty & \text{if } a > 1, \\ -\infty & \text{if } 0 < a < 1. \end{cases}$$

Writing $x^\alpha = \exp(\alpha \ln x)$, we see that

$$\lim_{x \to +\infty} x^\alpha = \begin{cases} +\infty & \text{if } \alpha > 0, \\ 1 & \text{if } \alpha = 0, \\ 0 & \text{if } \alpha < 0. \end{cases}$$

In the following theorem, we compare the growth of e^x, x^α, and $\ln x$ at $+\infty$.

Theorem 3.29 *For every $\alpha > 0$ we have*

$$\lim_{x \to +\infty} \frac{e^x}{x^\alpha} = +\infty, \qquad \lim_{x \to +\infty} \frac{\ln x}{x^\alpha} = 0.$$

Proof Let us start proving that, if $a > 1$,

$$\lim_n \frac{a^n}{n} = +\infty .$$

Indeed, writing $a = 1 + b$, with $b > 0$, we see that, if $n \geq 2$,

$$a^n = (1+b)^n = 1 + nb + \frac{n(n-1)}{2} b^2 + \cdots + b^n > \frac{n(n-1)}{2} b^2 .$$

Hence, for every $n \geq 2$,

$$\frac{a^n}{n} > \frac{n-1}{2} b^2 ,$$

whence the result, by Theorem 3.23. Let us now show that for every integer $k \geq 1$,

$$\lim_n \frac{a^n}{n^k} = +\infty .$$

Indeed, writing

$$\frac{a^n}{n^k} = \left(\frac{a^{n/k}}{n} \right)^k = \left(\frac{(\sqrt[k]{a})^n}{n} \right)^k ,$$

we can use the fact that $\lim_n \dfrac{(\sqrt[k]{a})^n}{n} = +\infty$ in order to arrive at the conclusion.
We now assume that $x \geq 1$. Let $n(x)$ and $n(\alpha)$ be natural numbers such that

$$n(x) \leq x < n(x) + 1 , \qquad n(\alpha) \leq \alpha < n(\alpha) + 1 .$$

Setting $k = n(\alpha) + 1$, we have

$$\frac{e^x}{x^\alpha} \geq \frac{e^x}{x^{n(\alpha)+1}} = \frac{e^x}{x^k} \geq \frac{e^x}{(n(x)+1)^k} \geq \frac{e^{n(x)}}{(n(x)+1)^k} .$$

Moreover,

$$\lim_{x \to +\infty} \frac{e^{n(x)}}{(n(x)+1)^k} = \lim_n \frac{e^n}{(n+1)^k} = \frac{1}{e} \lim_n \frac{e^{n+1}}{(n+1)^k} = \frac{1}{e} \lim_m \frac{e^m}{m^k} = +\infty ,$$

and the first identity follows.

Now, by the change of variables "$y = \ln x$," we obtain

$$\lim_{x \to +\infty} \frac{\ln x}{x^\alpha} = \lim_{y \to +\infty} \frac{y}{(e^y)^\alpha} = \lim_{y \to +\infty} \left(\frac{y^{1/\alpha}}{e^y} \right)^\alpha = \lim_{y \to +\infty} \left(\frac{e^y}{y^{1/\alpha}} \right)^{-\alpha} = 0,$$

thereby also proving the second identity.   ∎

We have thus seen that the exponential function e^x grows at $+\infty$ faster than any power x^α. We now show that the factorial grows still faster.

Theorem 3.30 *For every $a \in \mathbb{R}$,*

$$\lim_n \frac{a^n}{n!} = 0.$$

Proof If $|a| < 1$, then $\lim_n a^n = 0$, whence the result. Let us now assume $|a| \geq 1$ and prove by induction that for every $n \geq n(|a|)$,

$$|a|^{n-n(|a|)} \leq n!.$$

Indeed, this is surely true for $n = n(|a|)$. On the other hand, if the inequality is true for some $n \geq n(|a|)$ then, since $|a| < n + 1$,

$$|a|^{n+1-n(|a|)} = |a|^{n-n(|a|)}|a| \leq n!\,|a| \leq n!\,(n + 1) = (n + 1)!,$$

so that the inequality is also true for $n + 1$.

Now, for $n \geq n(|a|) + 1$,

$$\frac{|a|^n}{n!} = \frac{|a|^{n-1-n(|a|)}|a|^{1+n(|a|)}}{n!} \leq \frac{(n - 1)!\,|a|^{1+n(|a|)}}{n!} = \frac{|a|^{1+n(|a|)}}{n},$$

and the result follows.   ∎

3.9   Liminf and Limsup

Let $(a_n)_n$ be a sequence of real numbers. For every couple of natural numbers n, ℓ we define

$$\alpha_{n,\ell} = \min\{a_n, a_{n+1}, \ldots, a_{n+\ell}\}, \qquad \beta_{n,\ell} = \max\{a_n, a_{n+1}, \ldots, a_{n+\ell}\}.$$

If we keep n fixed, the sequence $(\alpha_{n,\ell})_\ell$ is decreasing, and the sequence $(\beta_{n,\ell})_\ell$ is increasing, so the following limits exist:

$$\overline{\alpha}_n = \lim_\ell \alpha_{n,\ell} = \inf\{a_n, a_{n+1}, \dots\}, \qquad \overline{\beta}_n = \lim_\ell \beta_{n,\ell} = \sup\{a_n, a_{n+1}, \dots\}.$$

Notice that $\overline{\alpha}_n$ could be equal to $-\infty$, and $\overline{\beta}_n$ could be equal to $+\infty$. Moreover,

$$\overline{\alpha}_n \le a_n \le \overline{\beta}_n \quad \text{for every } n.$$

Now the sequence $(\overline{\alpha}_n)_n$ either is constantly equal to $-\infty$ or has real values and is increasing; similarly, the sequence $(\overline{\beta}_n)_n$ either is constantly equal to $+\infty$ or has real values and is decreasing. We can then define the "lower limit" and the "upper limit" of $(a_n)_n$ as

$$\liminf_n a_n = \lim_n \overline{\alpha}_n, \qquad \limsup_n a_n = \lim_n \overline{\beta}_n.$$

Let us see how the lower limit can be characterized. We have three cases:

$$\liminf_n a_n = \ell \in \mathbb{R} \iff \begin{cases} (i) \ \forall \varepsilon > 0 \quad \exists \bar{n} \in \mathbb{N}: \quad n \ge \bar{n} \implies a_n > \ell - \varepsilon \\ (ii) \ \forall \varepsilon > 0, \ a_n < \ell + \varepsilon \ \text{for infinite values of } n. \end{cases}$$

$$\liminf_n a_n = -\infty \iff \forall \beta \in \mathbb{R}, \ a_n < \beta \ \text{for infinite values of } n.$$

$$\liminf_n a_n = +\infty \iff \forall \alpha \in \mathbb{R} \quad \exists \bar{n} \in \mathbb{N}: \quad n \ge \bar{n} \implies a_n > \alpha.$$

Notice that this last case is equivalent to $\lim_n a_n = +\infty$.

Analogously, for the upper limit we also have three cases:

$$\limsup_n a_n = \ell \in \mathbb{R} \iff \begin{cases} (i) \ \forall \varepsilon > 0 \quad \exists \bar{n} \in \mathbb{N}: \quad n \ge \bar{n} \implies a_n < \ell + \varepsilon \\ (ii) \ \forall \varepsilon > 0, \ a_n > \ell - \varepsilon \ \text{for infinite values of } n. \end{cases}$$

$$\limsup_n a_n = +\infty \iff \forall \alpha \in \mathbb{R}, \ a_n > \alpha \ \text{for infinite values of } n.$$

$$\limsup_n a_n = -\infty \iff \forall \beta \in \mathbb{R} \quad \exists \bar{n} \in \mathbb{N}: \quad n \ge \bar{n} \implies a_n < \beta.$$

This last case is equivalent to $\lim_n a_n = -\infty$.

The advantage of considering the lower and upper limits is that they always exist, while the limit, as we know, could not exist. The following theorem explains this situation better.

Theorem 3.31 *The sequence $(a_n)_n$ has a limit (possibly equal to either $-\infty$ or $+\infty$) if and only if $\liminf\limits_{n} a_n = \limsup\limits_{n} a_n$; in that case, this value coincides with* $\lim\limits_{n} a_n$.

Proof It is a direct consequence of the foregoing characterizations. We avoid the details for brevity's sake. ∎

The following property will be useful.

Proposition 3.32 *Let $(a_n)_n$ be a sequence of positive real numbers. Then*

$$\liminf_{n} \frac{a_{n+1}}{a_n} \leq \liminf_{n} \sqrt[n]{a_n} \leq \limsup_{n} \sqrt[n]{a_n} \leq \limsup_{n} \frac{a_{n+1}}{a_n} \,.$$

Proof Let us prove the last inequality. Let $\ell = \limsup\limits_{n} \frac{a_{n+1}}{a_n}$. If $\ell = +\infty$, then there is nothing to be proved. Thus, assume $\ell < +\infty$, and notice that surely $\ell \geq 0$. Let $\varepsilon > 0$ be fixed. Then there exists a $\bar{n} \in \mathbb{N}$ such that

$$n \geq \bar{n} \quad \Rightarrow \quad \frac{a_{n+1}}{a_n} < \ell + \frac{\varepsilon}{2}\,.$$

As a consequence,

$$a_{\bar{n}+1} < \left(\ell + \frac{\varepsilon}{2}\right)a_{\bar{n}}\,,$$

$$a_{\bar{n}+2} < \left(\ell + \frac{\varepsilon}{2}\right)a_{\bar{n}+1} < \left(\ell + \frac{\varepsilon}{2}\right)^2 a_{\bar{n}}\,,$$

$$a_{\bar{n}+3} < \left(\ell + \frac{\varepsilon}{2}\right)a_{\bar{n}+2} < \left(\ell + \frac{\varepsilon}{2}\right)^3 a_{\bar{n}}\,,$$

$$\dots$$

and thus it can be proved by induction that

$$n \geq \bar{n}+1 \quad \Rightarrow \quad a_n < \left(\ell + \frac{\varepsilon}{2}\right)^n \frac{a_{\bar{n}}}{(\ell + \frac{\varepsilon}{2})^{\bar{n}}}\,.$$

Since for every $\alpha > 0$ we have $\lim_{n} \sqrt[n]{\alpha} = 1$, there exists a $\tilde{n} \geq \bar{n}+1$ such that

$$n \geq \tilde{n} \quad \Rightarrow \quad \sqrt[n]{a_n} < \left(\ell + \frac{\varepsilon}{2}\right)\sqrt[n]{\frac{a_{\bar{n}}}{(\ell + \frac{\varepsilon}{2})^{\bar{n}}}} < \ell + \varepsilon\,.$$

We have thus proved that $\limsup\limits_{n} \sqrt[n]{a_n} \leq \ell$.

The first inequality can be proved similarly. ∎

Compactness and Completeness

4

In this chapter we discover some more subtle properties of the set of real numbers. This investigation will emphasize two important concepts, which will then be analyzed in the general setting of metric spaces: compactness and completeness.

4.1 Some Preliminaries on Sequences

Let U be a subset of a metric space E. Let us recall that a point x_0 is an "adherent point" of U if for every $\rho > 0$ one has that $B(x_0, \rho) \cap U \neq \emptyset$. On the other hand, x_0 is a "cluster point" for U if for every $\rho > 0$ one has that $B(x_0, \rho) \cap U$ contains infinitely many elements of U.

We can characterize the notion of "adherent point" by making use of sequences.

Proposition 4.1 *An element x of E is an adherent point of U if and only if there exists a sequence $(a_n)_n$ in U such that $\lim_n a_n = x$.*

Proof If x is an adherent point of U, then for every $n \in \mathbb{N}$ the intersection $B\left(x, \frac{1}{n+1}\right) \cap U$ is nonempty, and we can select one of its elements, calling it a_n. In such a way, we have constructed a sequence $(a_n)_n$ in U, and it is now a simple task to verify that $\lim_n a_n = x$.

Assume now that there exists a sequence $(a_n)_n$ in U such that $\lim_n a_n = x$. Then, for any $\rho > 0$, there exists a $\bar{n} \in \mathbb{N}$ such that

$$n \geq \bar{n} \;\Rightarrow\; a_n \in B(x, \rho).$$

Hence, $B(x, \rho) \cap U$ is nonempty, proving that x is an adherent point of U. ∎

Let us now consider two metric spaces E and F and a function $f : E \to F$. We want to characterize the continuity of f at a point $x_0 \in E$ by the use of sequences.

Proposition 4.2 *The function f is continuous at x_0 if and only if, for any sequence $(a_n)_n$ in E,*

$$\lim_n a_n = x_0 \quad \Rightarrow \quad \lim_n f(a_n) = f(x_0).$$

Proof Assume that f is continuous at x_0, and let $(a_n)_n$ be a sequence in E such that $\lim_n a_n = x_0$. By Theorem 3.13 on the limit of the composition of functions,

$$\lim_n f(a_n) = f(\lim_n a_n) = f(x_0),$$

so that one of the two implications is proved.

Let us now assume that f is not continuous at x_0. Then there is an $\varepsilon > 0$ such that, for every $\delta > 0$, there exists an $x \in E$ such that $d(x, x_0) < \delta$ and $d(f(x), f(x_0)) \geq \varepsilon$. Taking $\delta = \frac{1}{n+1}$, for every $n \in \mathbb{N}$ there exists an a_n in E such that $d(a_n, x_0) < \frac{1}{n+1}$ and $d(f(a_n), f(x_0)) \geq \varepsilon$. Then $\lim_n a_n = x_0$, but surely it cannot be that $\lim_n f(a_n) = f(x_0)$. The proof is thus completed. ■

As an immediate corollary, we have the following characterization of the limit, assuming x_0 to be a cluster point.

Proposition 4.3 *We have that $\lim\limits_{x \to x_0} f(x) = l$ if and only if, for any sequence $(a_n)_n$ in $E \setminus \{x_0\}$,*

$$\lim_n a_n = x_0 \quad \Rightarrow \quad \lim_n f(a_n) = l.$$

Given any sequence $(a_n)_n$, we define a "subsequence" by selecting a strictly increasing sequence of indices $(n_k)_k$ and considering the composition

$$k \mapsto n_k \mapsto a_{n_k}.$$

We will denote by $(a_{n_k})_k$ such a subsequence.

Notice that, since the indices n_k are in $\mathbb{N}$ and $n_{k+1} > n_k$, it must be that $n_{k+1} \geq n_k + 1$. As a consequence, one proves by induction that $n_k \geq k$, for every k, whence

$$\lim_k n_k = +\infty.$$

Proposition 4.4 *If a sequence has a limit, then all its subsequences must have the same limit.*

Proof Indeed, by the Change of Variables Formula (3.1),

$$\lim_{k \to +\infty} a_{n_k} = \lim_{n \to \lim_{k \to +\infty} n_k} a_n = \lim_{n \to +\infty} a_n,$$

thereby proving the result. ∎

Theorem 4.5 *Any sequence of real numbers has a monotone subsequence.*

Proof Let $(a_n)_n$ be a sequence in $\mathbb{R}$. We say that $\bar{n}$ is a "lookout point" for the sequence if $a_{\bar{n}} \geq a_n$ for every $n \geq \bar{n}$. Now we distinguish three cases.

Case 1. There are infinitely many lookout points; let us order them in a strictly increasing sequence of indices $(n_k)_k$. Then the subsequence $(a_{n_k})_k$ is decreasing.

Case 2. There are only finitely many lookout points. Let N be the largest one, and choose $n_0 > N$. Then, since n_0 is not a lookout point, there exists $n_1 > n_0$ such that $a_{n_1} > a_{n_0}$. By induction, we construct a strictly increasing sequence of indices $(n_k)_k$ in this way: Once n_k has been defined, since it is not a lookout point, there exists a $n_{k+1} > n_k$ such that $a_{n_{k+1}} > a_{n_k}$. The subsequence $(a_{n_k})_k$ thus constructed is strictly increasing.

Case 3. There are no lookout points. In this case, choose n_0 arbitrarily, and proceed as in Case 2. ∎

4.2 Compact Sets

Here is a fundamental property of closed and bounded intervals.

Theorem 4.6 (Bolzano–Weierstrass Theorem) *Every sequence $(a_n)_n$ in $[a, b]$ has a subsequence $(a_{n_k})_k$ having a limit in $[a, b]$.*

Proof By Theorem 4.5, there is a monotone subsequence $(a_{n_k})_k$, which by Corollary 3.27 has a limit $\lim_k a_{n_k} = l$. Since $a \leq a_{n_k} \leq b$ for every k, by Corollary 3.8 it must be that $a \leq l \leq b$, thereby proving the result. ∎

In a metric space E, we will say that a subset U is "compact" if every sequence $(a_n)_n$ in U has a subsequence $(a_{n_k})_k$ having a limit in U.

Bolzano–Weierstrass Theorem 4.6 thus states that if $E = \mathbb{R}$, then the intervals of the type $U = [a, b]$ are compact sets. In what follows, a subset of a metric space will be said to be "bounded" whenever it is contained in a ball.

Theorem 4.7 *Every compact subset of E is closed and bounded.*

Proof Assume that $U \subseteq E$ is compact. Taking $x \in \overline{U}$, by Proposition 4.1, there is a sequence $(a_n)_n$ in U such that $\lim_n a_n = x$. Since U is compact, there exists a subsequence $(a_{n_k})_k$ having a limit in U. But, since it is a subsequence, $\lim_k a_{n_k} = x$, and hence $x \in U$. We have thus shown that every adherent point of U belongs to U; hence, U is closed.

Now fix some $x_0 \in U$ arbitrarily. We will show that if $n \in \mathbb{N}$ is sufficiently large, then $U \subseteq B(x_0, n)$. By contradiction, if this is false, then we can build a sequence $(a_n)_n$ in U such that $d(a_n, x_0) \geq n$ for every $n \in \mathbb{N}$. Since U is compact, there exists a subsequence $(a_{n_k})_k$ having a limit $\bar{x} \in U$. Using the triangle inequality,

$$|d(a_{n_k}, x_0) - d(\bar{x}, x_0)| \leq d(a_{n_k}, \bar{x}),$$

whence $\lim_k d(a_{n_k}, x_0) = d(\bar{x}, x_0)$, whereas it should be

$$\lim_k d(a_{n_k}, x_0) = +\infty,$$

a contradiction. Therefore, U must be bounded. ∎

Let us focus our attention now on the compact subsets of $\mathbb{R}^N$, with $N \geq 1$.

Theorem 4.8 *A subset of $\mathbb{R}^N$ is compact if and only if it is closed and bounded.*

Proof We already know that every compact set is closed and bounded. Assume now that U is a closed and bounded subset of $\mathbb{R}^N$. For simplicity, we will assume that $N = 2$. Then U is contained in a rectangle $I = [a, b] \times [c, d]$. Let $(a_n)_n$ be a sequence in U. Then $a_n = (a_n^1, a_n^2)$, with $a_n^1 \in [a, b]$ and $a_n^2 \in [c, d]$. By the Bolzano–Weierstrass Theorem 4.6, the sequence $(a_n^1)_n$ has a subsequence $(a_{n_k}^1)_k$ having a limit $l_1 \in [a, b]$. Let us now consider the sequence $(a_{n_k}^2)_k$, with the same indices n_k as the one we just found; it is a subsequence of $(a_n^2)_n$. By Bolzano–Weierstrass Theorem 4.6, the sequence $(a_{n_k}^2)_k$ has a subsequence $(a_{n_{k_j}}^2)_j$ having a limit $l_2 \in [c, d]$. By Theorem 3.6,

$$\lim_j a_{n_{k_j}} = (\lim_j a_{n_{k_j}}^1, \lim_j a_{n_{k_j}}^2) = (l_1, l_2).$$

By Proposition 4.1, $l = (l_1, l_2)$ is an adherent point of U. Since U is closed, l is necessarily an element of U. ∎

The following property of compact sets will be useful.

Theorem 4.9 *Let $U \subseteq \mathbb{R}^N$ be a compact set. If $(A_i)_{i \in \mathcal{I}}$ is a family (not necessarily a countable family) of open sets such that*

$$U \subseteq \bigcup_{i \in \mathcal{I}} A_i,$$

then there exists a finite subfamily $(A^1, \ldots, A^n)$ of $(A_i)_{i \in \mathcal{I}}$ such that

$$U \subseteq A^1 \cup \cdots \cup A^n.$$

Proof For simplicity, we assume $N = 2$. Let us first prove the statement in the case where U is a closed rectangle, and let us denote it by $R_0 = [a_0, b_0] \times [c_0, d_0]$. By contradiction, assume that there is an open covering $(A_i)_{i \in \mathcal{I}}$ of R_0 without finite subcoverings. We split the rectangle R_0 into four smaller equal closed rectangles, connecting the midpoints of its sides. Among these four rectangles, there is at least one for which there is no finite subfamily of $(A_i)_{i \in \mathcal{I}}$ covering it. Let us call it R_1. We now proceed recursively and construct in this way a sequence of closed rectangles $R_k = [a_k, b_k] \times [c_k, d_k]$ such that

$$R_0 \supseteq R_1 \supseteq R_2 \supseteq \cdots \supseteq R_k \supseteq R_{k+1}, \supseteq \cdots,$$

for each of which there is no finite subfamily of $(A_i)_{i \in \mathcal{I}}$ covering it. By the Cantor Theorem 1.9, there exist $\bar{x}$ belonging to all intervals $[a_k, b_k]$ and $\bar{y}$ belonging to all intervals $[c_k, d_k]$, so that $(\bar{x}, \bar{y}) \in R_k$ for every $k \in \mathbb{N}$. Since $(\bar{x}, \bar{y})$ belongs to U, there is at least one A_i containing it. This set A_i is open, and the dimensions of R_k tend to zero as k tends to $+\infty$. Then, for k sufficiently large, the rectangle R_k will be entirely contained in A_i. But this is a contradiction, since there is no finite subfamily of $(A_i)_{i \in \mathcal{I}}$ covering R_k.

Now let U be any closed and bounded subset of $\mathbb{R}^2$. Then U is contained in a rectangle $[a, b] \times [c, d]$. If $(A_i)_{i \in \mathcal{I}}$ is an open covering of U, then

$$[a, b] \times [c, d] \subseteq \left(\bigcup_{i \in \mathcal{I}} A_i \right) \cup (\mathbb{R}^2 \setminus U).$$

Since $\mathbb{R}^2 \setminus U$ is open, we now have an open covering of $[a, b] \times [c, d]$, and by the first part of the proof, there is a finite subfamily $(A^1, \ldots, A^n)$ of $(A_i)_{i \in \mathcal{I}}$ such that

$$[a, b] \times [c, d] \subseteq (A^1 \cup \cdots \cup A^n) \cup (\mathbb{R}^2 \setminus U).$$

Consequently, $U \subseteq A^1 \cup \cdots \cup A^n$, and the proof is thus completed. ∎

Note The preceding theorem indeed holds in any metric space, and it can be shown that the stated property is necessary and sufficient for the compactness of a set U.

4.3 Compactness and Continuity

In what follows, we will say that a function $f : A \to \mathbb{R}$ is "bounded from above" (or "bounded from below" or "bounded") if that is its image $f(A)$. We will say that "f has a maximum" (or "f has a minimum") if $f(A)$ does. In the case where

f has a maximum or a minimum, we will call "maximum point" any $\bar{x}$ for which $f(\bar{x}) = \max f(A)$ and "minimum point" any $\bar{x}$ for which $f(\bar{x}) = \min f(A)$.

Theorem 4.10 (Weierstrass Theorem) *If U is a compact set and $f : U \to \mathbb{R}$ is a continuous function, then f has a maximum and a minimum.*

Proof Let $s = \sup f(U)$. We will prove that there is a maximum point, i.e., a $\bar{x} \in U$, such that $f(\bar{x}) = s$.

We first note that there is a sequence $(y_n)_n$ in $f(U)$ such that $\lim_n y_n = s$. Indeed, if $s \in \mathbb{R}$, then for every $n \geq 1$ we can find a $y_n \in f(U)$ such that $s - \frac{1}{n} < y_n \leq s$; and if $s = +\infty$, then for every n there is a $y_n \in f(U)$ such that $y_n > n$. In both cases, we have $\lim_n y_n = s$.

Correspondingly, we can find a sequence $(x_n)_n$ in U such that $f(x_n) = y_n$. Since U is compact, there exists a subsequence $(x_{n_k})_k$ having a limit $\bar{x} \in U$. Because $\lim_n y_n = s$ and $y_{n_k} = f(x_{n_k})$, the subsequence $(y_{n_k})_k$ also has the same limit s. Then, by the continuity of f,

$$f(\bar{x}) = f(\lim_k x_{n_k}) = \lim_k f(x_{n_k}) = \lim_k y_{n_k} = s .$$

The theorem is thus proved in what concerns the existence of the maximum. To deal with the minimum, either one proceeds analogously or one considers the continuous function $g = -f$ and uses the fact that g has a maximum. ∎

The following theorem holds for a general metric space F.

Theorem 4.11 *If U is a compact set and $f : U \to F$ is a continuous function, then $f(U)$ is a compact set.*

Proof Let $(y_n)_n$ be a sequence in $f(U)$. We can then find a sequence $(x_n)_n$ in U such that $f(x_n) = y_n$ for every $n \in \mathbb{N}$. Since U is compact, there exists a subsequence $(x_{n_k})_k$ that has a limit $\bar{x} \in U$. Recalling that $y_{n_k} = f(x_{n_k})$ and that f is continuous,

$$\lim_k f(x_{n_k}) = f(\lim_k x_{n_k}) = f(\bar{x}) .$$

Therefore, the subsequence $(y_{n_k})_k$ has a limit, precisely $f(\bar{x})$, in $f(U)$. ∎

We now introduce the concept of "uniform continuity." First, recall the meaning of $f : E \to F$ is "continuous." This means that f is continuous at every point $x_0 \in E$, i.e.,

$$\forall x_0 \in E \quad \forall \varepsilon > 0 \quad \exists \delta > 0 : \quad \forall x \in E \quad d(x, x_0) < \delta \implies d(f(x), f(x_0)) < \varepsilon .$$

Notice that, in general, the choice of δ depends on both ε and x_0. We will say that f is "uniformly continuous" whenever such a δ does not depend on x_0, i.e.,

$$\forall \varepsilon > 0 \ \exists \delta > 0 : \ \forall x_0 \in E \ \forall x \in E \quad d(x, x_0) < \delta \ \Rightarrow \ d(f(x), f(x_0)) < \varepsilon .$$

The following theorem states that continuity implies uniform continuity when the domain is a compact set.

Theorem 4.12 (Heine Theorem) *If U is a compact set and $f : U \to F$ is a continuous function, then f is uniformly continuous.*

Proof By contradiction, assume that f is not uniformly continuous, i.e.,

$$\exists \varepsilon > 0 : \ \forall \delta > 0 \ \exists x_0 \in E \ \exists x \in E : \ d(x, x_0) < \delta \ \text{ and } \ d(f(x), f(x_0)) \geq \varepsilon .$$

Let us fix such an $\varepsilon > 0$, and choose $\delta = \frac{1}{n+1}$, with $n \in \mathbb{N}$. Correspondingly, there are x_n^0 and x_n in U such that

$$d(x_n, x_n^0) < \frac{1}{n+1} \quad \text{and} \quad d(f(x_n), f(x_n^0)) \geq \varepsilon .$$

We thus have two sequences, $(x_n)_n$ and $(x_n^0)_n$, in U. Since U is compact, there exists a subsequence $(x_{n_k})_k$ having a limit $\bar{x} \in U$. Let us now consider the subsequence $(x_{n_k}^0)_k$, with the same indices n_k as the one we just found. Since $d(x_{n_k}, x_{n_k}^0)$ tends to zero, this subsequence $(x_{n_k}^0)_k$ has the same limit $\bar{x}$. By the continuity of f,

$$\lim_k f(x_{n_k}) = f(\bar{x}) \quad \text{and} \quad \lim_k f(x_{n_k}^0) = f(\bar{x}) ,$$

implying that

$$\lim_k d(f(x_{n_k}), f(x_{n_k}^0)) = 0 ,$$

in contradiction to the fact that $d(f(x_{n_k}), f(x_{n_k}^0)) \geq \varepsilon > 0$, for every $k \in \mathbb{N}$. ∎

4.4 Complete Metric Spaces

We will now introduce the concept of "completeness" for a metric space E. To this end, we first need to introduce a special class of sequences. We will say that $(a_n)_n$ is a "Cauchy sequence" in E if

$$\forall \varepsilon > 0 \ \exists \bar{n} : \ [m \geq \bar{n} \text{ and } n \geq \bar{n}] \ \Rightarrow \ d(a_m, a_n) < \varepsilon .$$

The metric space E will be said to be "complete" if every Cauchy sequence has a limit in E.

It is readily seen that if $(a_n)_n$ has a limit $l \in E$, then it is a Cauchy sequence. Indeed, for any fixed $\varepsilon > 0$, taking m and n large enough, we have

$$d(a_m, a_n) \le d(a_m, l) + d(l, a_n) < 2\varepsilon .$$

In contrast, a Cauchy sequence in E might not have a limit in the space E. As an example, take $\mathbb{Q}$ with the usual distance and the sequence $a_n = \left(1 + \frac{1}{n}\right)^n$ whose limit is $e \notin \mathbb{Q}$. Indeed, $\mathbb{Q}$ is not complete, whereas $\mathbb{R}$ is, as we will now prove.

Theorem 4.13 $\mathbb{R}$ *is complete.*

Proof Let $(a_n)_n$ be a Cauchy sequence in $\mathbb{R}$. By definition (taking $\varepsilon = 1$), there exists a $\bar{n}_1$ such that, for every $m \ge \bar{n}_1$ and $n \ge \bar{n}_1$, we have $d(a_n, a_m) < 1$. Taking $m = \bar{n}_1$ and setting $a = a_{\bar{n}_1} - 1$, $b = a_{\bar{n}_1} + 1$, we thus see that the sequence $(a_n)_{n \ge \bar{n}_1}$ is contained in the interval $[a, b]$. By Bolzano–Weierstrass Theorem 4.6, there exists a subsequence $(a_{n_k})_k$ having a limit $l \in [a, b]$. We now want to prove that

$$\lim_n a_n = l .$$

Let $\varepsilon > 0$ be fixed. Since $(a_n)_n$ is a Cauchy sequence,

$$\exists \bar{n} : \quad m \ge \bar{n} \text{ and } n \ge \bar{n} \; \Rightarrow \; d(a_m, a_n) < \varepsilon .$$

Moreover, since $\lim_k a_{n_k} = l$ and $\lim_k n_k = +\infty$,

$$\exists \bar{k} : \quad k \ge \bar{k} \; \Rightarrow \; d(a_{n_k}, l) < \varepsilon \text{ and } n_k \ge \bar{n} .$$

Then for every $n \ge \bar{n}$,

$$d(a_n, l) \le d(a_n, a_{n_{\bar{k}}}) + d(a_{n_{\bar{k}}}, l) < \varepsilon + \varepsilon = 2\varepsilon ,$$

thereby completing the proof. ∎

We now extend the previous theorem to higher dimensions.

Theorem 4.14 $\mathbb{R}^N$ *is complete.*

Proof For simplicity, we assume $N = 2$. Let $(a_n)_n$ be a Cauchy sequence in $\mathbb{R}^2$. We write each vector $a_n \in \mathbb{R}^2$ in its coordinates

$$a_n = (a_{n,1}, a_{n,2}) .$$

Since

$$|a_{m,1} - a_{n,1}| \le \|a_n - a_m\|, \quad |a_{m,2} - a_{n,2}| \le \|a_n - a_m\|,$$

we see that both $(a_{n,1})_n$ and $(a_{n,2})_n$ are Cauchy sequences in $\mathbb{R}$. Hence, since $\mathbb{R}$ is complete, each of them has a limit

$$\lim_n a_{n,1} = l_1 \in \mathbb{R}, \quad \lim_n a_{n,2} = l_2 \in \mathbb{R}.$$

Then

$$\lim_n a_n = (\lim_n a_{n,1}, \ \lim_n a_{n,2}) = (l_1, l_2),$$

which is an element of $\mathbb{R}^2$. ∎

A normed vector space that is complete with respect to the distance given by its norm is said to be a "Banach space." We have thus proved that $\mathbb{R}^N$ is a Banach space.

The following theorem provides us the **Cauchy criterion** for functions.

Theorem 4.15 *Let F be a complete metric space. Then a function $f : E \to F$ has a limit $\lim\limits_{x \to x_0} f(x)$ in F if and only if the following property holds:*

$$\forall \varepsilon > 0 \ \exists \delta > 0 : [\, 0 < d(x, x_0) < \delta \ \text{and} \ 0 < d(x', x_0) < \delta \,] \Rightarrow d(f(x), f(x')) < \varepsilon .$$

Proof Assume that $\lim_{x \to x_0} f(x) = l \in F$. The conclusion then follows from the definition of limit and the triangle inequality

$$d(f(x), f(x')) \le d(f(x), l) + d(f(x'), l) .$$

On the other hand, if the property stated in the theorem holds, take a sequence $(a_n)_n$ such that $\lim_n a_n = x_0$. Then we see that $(f(a_n))_n$ is a Cauchy sequence, so it has a limit $l \in F$. To see that this limit does not depend on the sequence, let $(a'_n)_n$ be another sequence such that $\lim_n a'_n = x_0$. By the foregoing property, for every $\varepsilon > 0$ it will be $d(f(a_n), f(a'_n)) < \varepsilon$ for n sufficiently large, from what follows that $(f(a'_n))_n$ has the same limit as $(f(a_n))_n$. Having thus proved that $\lim_n f(a_n) = l$ for any sequence $(a_n)_n$ such that $\lim_n a_n = x_0$, the conclusion follows from Proposition 4.3. ∎

4.5   Completeness and Continuity

The following theorem provides a useful extension property for uniformly continuous functions. We say that a set is "dense" in E if its closure coincides with E.

Theorem 4.16 *Let $\widehat{E}$ be a dense subset of E, and let F be a complete metric space. If $\hat{f} : \widehat{E} \to F$ is uniformly continuous, then there exists a unique continuous function $f : E \to F$ whose restriction to $\widehat{E}$ coincides with $\hat{f}$.*

Proof Taking $x \in E$, there exists a sequence $(x_n)_n$ in $\widehat{E}$ such that $\lim_n x_n = x$. Since $\hat{f}$ is uniformly continuous and $(x_n)_n$ is a Cauchy sequence, it follows that $(\hat{f}(x_n))_n$ is also a Cauchy sequence. Hence, since F is complete, it has a limit $y \in F$. We define $f(x) = \lim_n \hat{f}(x_n) = y$.

Let us verify that this is a good definition. If $(\tilde{x}_n)_n$ is another sequence in $\widehat{E}$ such that $\lim_n \tilde{x}_n = x$, then $\lim_n d(x_n, \tilde{x}_n) = 0$, and since $\hat{f}$ is uniformly continuous, then also $\lim_n d(\hat{f}(x_n), \hat{f}(\tilde{x}_n)) = 0$. Hence, $(\hat{f}(\tilde{x}_n))_n$ necessarily has the same limit y of $(\hat{f}(x_n))_n$, and the definition is consistent.

Clearly, the function f thus defined extends $\hat{f}$ since, if $x \in U$, we can take the sequence $(x_n)_n$ as being constantly equal to x. Let us now prove that f is (uniformly) continuous. Once $\varepsilon > 0$ has been fixed, let $\delta > 0$ be such that, taking $u, v \in \widehat{E}$,

$$d(u, v) \leq 2\delta \quad \Rightarrow \quad d(\hat{f}(u), \hat{f}(v)) \leq \frac{\varepsilon}{3}\,.$$

If x, y are two points in E such that $d(x, y) \leq \delta$, then we can take two sequences $(x_n)_n$ and $(y_n)_n$ in $\widehat{E}$ such that $\lim_n x_n = x$ and $\lim_n y_n = y$. Then, since $\lim_n \hat{f}(x_n) = f(x)$ and $\lim_n \hat{f}(y_n) = f(y)$, for all sufficiently large n it will be that $d(x_n, y_n) \leq 2\delta$ and

$$d(f(x), f(y)) \leq d(f(x), \hat{f}(x_n)) + d(\hat{f}(x_n), \hat{f}(y_n)) + d(\hat{f}(y_n), f(y))$$

$$\leq \frac{\varepsilon}{3} + \frac{\varepsilon}{3} + \frac{\varepsilon}{3} = \varepsilon\,,$$

which proves that f is uniformly continuous.

To conclude the proof, let $\tilde{f} : E \to F$ be any continuous function extending $\hat{f}$. Then, for every $x \in E$, taking a sequence $(x_n)_n$ in $\widehat{E}$ such that $\lim_n x_n = x$,

$$\tilde{f}(x) = \lim_n \tilde{f}(x_n) = \lim_n \hat{f}(x_n) = f(x)\,.$$

We have thus proved that f is the only possible continuous extension of $\hat{f}$ to E. ∎

4.6   Spaces of Continuous Functions

Let E and F be two metric spaces. We consider a sequence of functions $f_n : E \to F$, and we want to examine, whenever it exists, the limit

$$\lim_n f_n(x) .$$

Clearly enough, this limit could exist for some $x \in E$ and not exist at all for others. So assume that, for some subset $U \subseteq E$, there is a function $f : U \to F$ for which

$$\lim_n f_n(x) = f(x) , \quad \text{for every } x \in U .$$

In this case we will say that the sequence $(f_n)_n$ "converges pointwise" to f on U; it thus happens that

$$\forall x \in U \quad \forall \varepsilon > 0 \quad \exists \bar{n} \in \mathbb{N} : \quad n \geq \bar{n} \implies d(f_n(x), f(x)) < \varepsilon .$$

If the preceding choice of $\bar{n}$ does not depend on $x \in U$, we will say that the sequence $(f_n)_n$ "converges uniformly" to f on U; in this case,

$$\forall \varepsilon > 0 \quad \exists \bar{n} \in \mathbb{N} : \quad \forall x \in U \quad n \geq \bar{n} \implies d(f_n(x), f(x)) < \varepsilon ,$$

i.e., equivalently,

$$\lim_n \left[\sup\{d(f_n(x), f(x)) : x \in U\} \right] = 0 .$$

Let us provide an example of a sequence $(f_n)_n$ which converges pointwise, but not uniformly. Let $f_n : [0, 1] \to \mathbb{R}$ be defined for $n \geq 1$ as

$$f_n(x) = \begin{cases} nx & \text{if } 0 \leq x \leq \frac{1}{n} , \\ 2 - nx & \text{if } \frac{1}{n} \leq x \leq \frac{2}{n} , \\ 0 & \text{if } \frac{2}{n} \leq x \leq 1 . \end{cases}$$

It is easily seen that $\lim_n f_n(x) = 0$ for every $x \in [0, 1]$, but the convergence is not uniform, since $f_n(\frac{1}{n}) = 1$ for every $n \geq 1$.

The uniform convergence has good behavior with respect to continuity, as the following theorem states.

Theorem 4.17 *If each function $f_n : E \to F$ is continuous on $U \subseteq E$ and $(f_n)_n$ converges uniformly to f on U, then f is also continuous on U.*

Proof Consider an arbitrary point x_0 of U; we will show that f is continuous at x_0. Let $\varepsilon > 0$ be fixed. Then there exists a $\bar{n} \in \mathbb{N}$ such that

$$\forall x \in U \quad n \geq \bar{n} \implies d(f_n(x), f(x)) < \frac{1}{3}\varepsilon .$$

Since

$$d(f(x), f(x_0)) \leq d(f(x), f_n(x)) + d(f_n(x), f_n(x_0)) + d(f_n(x_0), f(x_0)) ,$$

taking $n = \bar{n}$, we have that

$$d(f(x), f(x_0)) < \frac{2}{3}\varepsilon + d(f_{\bar{n}}(x), f_{\bar{n}}(x_0)) .$$

By the continuity of $f_{\bar{n}}$ at x_0, we can find a $\delta > 0$ such that

$$d(x, x_0) < \delta \quad \implies \quad d(f_{\bar{n}}(x), f_{\bar{n}}(x_0)) < \frac{1}{3}\varepsilon .$$

Hence,

$$d(x, x_0) < \delta \quad \implies \quad d(f(x), f(x_0)) < \frac{2}{3}\varepsilon + \frac{1}{3}\varepsilon = \varepsilon ,$$

thereby proving that f is continuous at x_0. ∎

Note When $(f_n)_n$ is a sequence of continuous functions that converges uniformly on U to some function, for every $x_0 \in U$ we can write

$$\lim_n \left(\lim_{x \to x_0} f_n(x) \right) = \lim_{x \to x_0} \left(\lim_n f_n(x) \right) .$$

In what follows, we say that a function $f : E \to F$ is "bounded" if its image $f(E)$ is bounded, and we denote by $\mathcal{B}(E, F)$ the set of all those functions. We define, in $\mathcal{B}(E, F)$,

$$d_\infty(f, g) = \sup\{d(f(x), g(x)) : x \in E\} .$$

It is readily verified that this is a distance; hence, $\mathcal{B}(E, F)$ will now be treated as a metric space. Taking a sequence $(f_n)_n$ and a function f in the space $\mathcal{B}(E, F)$, we have

$$\lim_n f_n = f \quad \Leftrightarrow \quad (f_n)_n \text{ converges uniformly to } f \text{ on } E .$$

Let us now investigate some further properties of this space of functions.

Theorem 4.18 *If F is complete, then $\mathcal{B}(E, F)$ is also complete.*

Proof Let $(f_n)_n$ be a Cauchy sequence in $\mathcal{B}(E, F)$. Since, for every $x \in E$,

$$d(f_m(x), f_n(x)) \le d_\infty(f_m, f_n),$$

we have that $(f_n(x))_n$ is a Cauchy sequence in F for each $x \in E$. Since F is complete, the sequence $(f_n(x))_n$ has a limit in F; we will denote it by $f(x)$. In this way, we have indeed defined a function $f : E \to F$, and we will now prove that $\lim_n f_n = f$ in $\mathcal{B}(E, F)$. Let $\varepsilon > 0$ be fixed. Since $(f_n)_n$ is a Cauchy sequence, there exists a $\bar{n} \in \mathbb{N}$ such that

$$[m \ge \bar{n} \text{ and } n \ge \bar{n}] \Rightarrow d_\infty(f_n, f_m) < \varepsilon$$
$$\Rightarrow d(f_n(x), f_m(x)) < \varepsilon, \text{ for every } x \in E.$$

By the continuity of the distance, we have

$$\lim_m d(f_n(x), f_m(x)) = d(f_n(x), f(x)),$$

whence

$$n \ge \bar{n} \quad \Rightarrow \quad d(f_n(x), f(x)) \le \varepsilon, \text{ for every } x \in E.$$

Hence, f belongs to $\mathcal{B}(E, F)$, and

$$n \ge \bar{n} \quad \Rightarrow \quad d_\infty(f_n, f) \le \varepsilon.$$

The statement is thus proved.   ∎

If $f \in \mathcal{B}(E, F)$ and F is a normed vector space, we can define

$$\|f\|_\infty = \sup\{\|f(x)\| : x \in E\}.$$

One easily verifies that this is indeed a norm on $\mathcal{B}(E, F)$ and that

$$d_\infty(f, g) = \|f - g\|_\infty.$$

As an immediate consequence of the preceding theorem, we have the following corollary.

Corollary 4.19 *If F is a Banach space, then $\mathcal{B}(E, F)$ is also a Banach space.*

Proof Since F is complete with respect to the distance induced by its norm, $\mathcal{B}(E, F)$ is also complete, by Theorem 4.18.   ∎

Let us denote by $C(E, F)$ the set of continuous functions $f : E \to F$. We are now interested in considering the space $C(E, F) \cap B(E, F)$, made up of bounded and continuous functions.

Theorem 4.20 *The set $C(E, F) \cap B(E, F)$ is closed in $B(E, F)$.*

Proof Let $f \in B(E, F)$ be an adherent point of $C(E, F) \cap B(E, F)$. By Proposition 4.1, there exists a sequence $(f_n)_n$ in $C(E, F) \cap B(E, F)$ such that $\lim_n f_n = f$. By Theorem 4.17, f is continuous, since it is the uniform limit of continuous functions, hence $f \in C(E, F) \cap B(E, F)$. We have thus proved that the closure of $C(E, F) \cap B(E, F)$ coincides with $C(E, F) \cap B(E, F)$ itself. ∎

The set $C(E, F) \cap B(E, F)$ inherits the distance d_∞ from $B(E, F)$ and, when F is a normed vector space, also its norm $\| \cdot \|_\infty$. The following corollaries will be useful.

Corollary 4.21 *If F is complete, then $C(E, F) \cap B(E, F)$ is also complete. Hence, if F is a Banach space, then $C(E, F) \cap B(E, F)$ is also a Banach space.*

Proof Any Cauchy sequence in $C(E, F) \cap B(E, F)$ has a limit in $B(E, F)$, since $B(E, F)$ is complete. Since $C(E, F) \cap B(E, F)$ is closed in $B(E, F)$, this limit belongs to $C(E, F) \cap B(E, F)$. Hence, $C(E, F) \cap B(E, F)$ is complete. ∎

Corollary 4.22 *If E is compact and F is complete, then $C(E, F)$ is also complete. Hence, if E is compact and F is a Banach space, then $C(E, F)$ is also a Banach space.*

Proof Since E is compact, every $f \in C(E, F)$ is bounded, hence $C(E, F)$ coincides with $C(E, F) \cap B(E, F)$. The conclusion follows by the previous corollary. ∎

Exponential and Circular Functions

5

The aim of this chapter is to provide a unified construction of the exponential and the trigonometric functions using geometrical arguments in the complex plane. The basis of this construction will lie in the proof of the following statement.

Theorem 5.1 *Let ζ be a nonzero complex number, with $\Re\zeta \geq 0$ and $\Im\zeta \geq 0$, and let τ be a positive real number. There exists a unique continuous function $f : \mathbb{R} \to \mathbb{C} \setminus \{0\}$ with the following properties:*

(a) $f(0) = 1,\ f(\tau) = \zeta$.
(b) $f(x_1 + x_2) = f(x_1)f(x_2)$, *for every* $x_1, x_2 \in \mathbb{R}$.
(c) $\Re f(x) \geq 0$ *and* $\Im f(x) \geq 0$, *for every* $x \in [0, \tau]$.

Its proof will be developed in the next section. We will first learn how to measure the length of an arc of the unit circle. This will also lead to the definition of the number π. Then the function f will be defined on some dense subset of $\mathbb{R}$ in the most natural way, indeed the only possible one. Finally, it will be extended on the whole real line.

So let us move forward this plan.

5.1 The Construction

It is not restrictive to assume $\tau = 1$. Indeed, if we denote by $f_\tau : \mathbb{R} \to \mathbb{C} \setminus \{0\}$ the function we are looking for, once we have found $f_1 : \mathbb{R} \to \mathbb{C} \setminus \{0\}$, it is sufficient to define

$$f_\tau(x) = f_1\left(\frac{x}{\tau}\right),$$

so that all the requirements are satisfied.

© The Author(s), under exclusive license to Springer Nature Switzerland AG 2023
A. Fonda, *A Modern Introduction to Mathematical Analysis*,
https://doi.org/10.1007/978-3-031-23713-3_5

We will denote the first quadrant of the complex plane by

$$Q_1 = \{z \in \mathbb{C} : \mathfrak{R}(x) \geq 0, \ \mathfrak{I}(z) \geq 0\}.$$

The proof of Theorem 5.1 is divided into the following three subsections, as explained earlier.

5.1.1   Preliminaries for the Proof

We provide here a rigorous definition of the *argument* of the complex number ζ, passing through some sequences that have a simple geometric interpretation in the complex plane.

Let $(\sigma_n)_n$ be the sequence of complex numbers

$$\sigma_n = x_n + iy_n \in Q_1,$$

such that

$$\sigma_0 = \zeta, \quad \text{and} \quad \sigma_{n+1}^2 = \sigma_n, \quad \text{for every } n \in \mathbb{N}.$$

More explicitly, $\sigma_{n+1} = (x_{n+1}, y_{n+1})$, with

$$x_{n+1} = \sqrt{\frac{x_n + \sqrt{x_n^2 + y_n^2}}{2}}, \qquad y_{n+1} = \frac{y_n}{\sqrt{2\left(x_n + \sqrt{x_n^2 + y_n^2}\right)}}.$$

Notice that $x_n > 0$ for every $n \geq 1$. It is easily seen, by induction, that

$$\sigma_n^{2^n} = \zeta, \quad \text{for every } n \in \mathbb{N}. \tag{5.1}$$

Now let $(\tilde{\sigma}_n)_n$ be defined as

$$\tilde{\sigma}_n = \frac{\sigma_n}{|\sigma_n|}.$$

These are unit vectors, i.e., $|\tilde{\sigma}_n| = 1$, and, setting

$$\tilde{\zeta} = \frac{\zeta}{|\zeta|},$$

it is easily seen that

$$\tilde{\sigma}_0 = \tilde{\zeta}, \quad \text{and} \quad \tilde{\sigma}_{n+1}^2 = \tilde{\sigma}_n, \quad \text{for every } n \in \mathbb{N}.$$

The two sequences $(\sigma_n)_n$ and $(\tilde{\sigma}_n)_n$ are drawn in Fig. 5.1 in the case $\Im(\zeta) \neq 0$ (the following pictures will also be drawn for this case). On the other hand, if $\Im(\zeta) = 0$, then $\tilde{\sigma}_n = 1$ for every $n \in \mathbb{N}$.

Let us now define the two sequences $(\ell_n)_n$ and $(L_n)_n$ as

$$\ell_n = |\tilde{\sigma}_n - 1|, \qquad L_n = \frac{2\ell_n}{\sqrt{4 - \ell_n^2}}.$$

They represent the length of the segments depicted in Fig. 5.2.

Observe now that, for every m,

$$|\tilde{\sigma}_n^{m+1} - \tilde{\sigma}_n^m| = |\tilde{\sigma}_n^m(\tilde{\sigma}_n - 1)| = |\tilde{\sigma}_n|^m |\tilde{\sigma}_n - 1| = |\tilde{\sigma}_n - 1| = \ell_n,$$

so that the points

$$1, \tilde{\sigma}_n, \tilde{\sigma}_n^2, \tilde{\sigma}_n^3, \tilde{\sigma}_n^4, \ldots$$

all lie on the unit circle $S^1 = \{z \in \mathbb{C} : |z| = 1\}$, and ℓ_n is the distance from each one of them to the next one. This fact can be visualized in Fig. 5.3.

Fig. 5.1 The definition of σ_n and $\tilde{\sigma}_n$

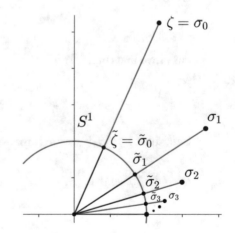

Fig. 5.2 The definition of ℓ_n and L_n

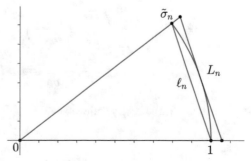

Fig. 5.3 The equidistant
points

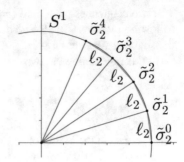

Fig. 5.4 A geometric view
of ℓ_{n+1} in terms of ℓ_n

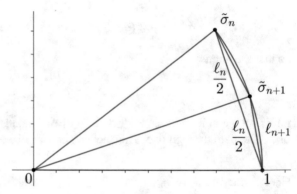

Let us prove that (Fig. 5.4)

$$\ell_{n+1} = \sqrt{2 - \sqrt{4 - \ell_n^2}}.\tag{5.2}$$

Indeed, since $1 = |\tilde{\sigma}_n|^2 = \tilde{\sigma}_n \tilde{\sigma}_n^*$, for every $n \in \mathbb{N}$, we have

$$\ell_n^2 = (\tilde{\sigma}_n - 1)(\tilde{\sigma}_n - 1)^* = (\tilde{\sigma}_n - 1)(\tilde{\sigma}_n^* - 1) = (\tilde{\sigma}_n - 1)\left(\frac{1}{\tilde{\sigma}_n} - 1\right) = 2 - \tilde{\sigma}_n - \frac{1}{\tilde{\sigma}_n},$$

hence

$$(2 - \ell_{n+1}^2)^2 = \left(\tilde{\sigma}_{n+1} + \frac{1}{\tilde{\sigma}_{n+1}}\right)^2 = 2 + \tilde{\sigma}_{n+1}^2 + \frac{1}{\tilde{\sigma}_{n+1}^2}$$

$$= 2 + \tilde{\sigma}_n + \frac{1}{\tilde{\sigma}_n} = 4 - \left(2 - \tilde{\sigma}_n - \frac{1}{\tilde{\sigma}_n}\right) = 4 - \ell_n^2,$$

and formula (5.2) directly follows.

It can now be noted that, since $\tilde{\sigma}_0$ belongs to $S^1 \cap \mathcal{Q}_1$, it must be that $\ell_0 = |\tilde{\sigma}_0 - 1| \leq \sqrt{2}$ and, by formula (5.2),

$$0 < \ell_n \leq \sqrt{2}, \quad \text{for every } n \in \mathbb{N}. \tag{5.3}$$

We finally define the two sequences $(a_n)_n$ and $(b_n)_n$ as

$$a_n = 2^n \ell_n, \qquad b_n = 2^n L_n.$$

Note that if $\Im(\zeta) = 0$, then $\ell_n = 0$, hence $a_n = b_n = 0$ for every $n \in \mathbb{N}$. Let us now concentrate on the case $\Im(\zeta) > 0$. In this case,

$$\frac{b_n}{a_n} = \frac{2}{\sqrt{4 - \ell_n^2}} > 1,$$

hence $a_n < b_n$, for every $n \in \mathbb{N}$. Let us see that the sequence $(a_n)_n$ is strictly increasing; by (5.2),

$$\frac{a_{n+1}}{a_n} = 2\frac{\ell_{n+1}}{\ell_n} = 2\frac{\sqrt{2 - \sqrt{4 - \ell_n^2}}}{\ell_n} = \frac{2}{\sqrt{2 + \sqrt{4 - \ell_n^2}}} > \frac{2}{\sqrt{2 + 2}} = 1,$$

hence $a_n < a_{n+1}$, for every $n \in \mathbb{N}$. On the other hand, let us prove that the sequence $(b_n)_n$ is strictly decreasing; by (5.2) again,

$$\begin{aligned}
\frac{b_n}{b_{n+1}} &= \frac{1}{2}\frac{\ell_n}{\sqrt{4 - \ell_n^2}}\frac{\sqrt{4 - \ell_{n+1}^2}}{\ell_{n+1}} \\
&= \frac{1}{2}\frac{\ell_n}{\sqrt{4 - \ell_n^2}}\frac{\sqrt{2 + \sqrt{4 - \ell_n^2}}}{\sqrt{2 - \sqrt{4 - \ell_n^2}}} \\
&= \frac{1}{2}\frac{2 + \sqrt{4 - \ell_n^2}}{\sqrt{4 - \ell_n^2}} \\
&= \frac{1}{2}\left(\frac{2}{\sqrt{4 - \ell_n^2}} + 1\right) \\
&> \frac{1}{2}(1 + 1) = 1,
\end{aligned}$$

hence $b_n > b_{n+1}$ for every $n \in \mathbb{N}$.

Thus, the sequences $(a_n)_n$ and $(b_n)_n$ are monotone, so they both have a finite limit. Since, then,

$$\lim_n \ell_n = \lim_n \frac{a_n}{2^n} = 0 \, ,$$

we have

$$\frac{\lim_n b_n}{\lim_n a_n} = \lim_n \frac{b_n}{a_n} = \lim_n \frac{2}{\sqrt{4 - \ell_n^2}} = 1 \, ,$$

so we can conclude that the two sequences do indeed have the same limit. We call this real number *argument of* ζ and denote it by $\mathrm{Arg}(\zeta)$. We can thus write

$$\mathrm{Arg}(\zeta) = \lim_n 2^n |\tilde{\sigma}_n - 1| \, .$$

In this way, we have rigorously defined the "length" of the arc on the unitary circle S^1 starting from $(1, 0)$ and arriving at $\tilde{\zeta} = \zeta/|\zeta|$, moving in counterclockwise direction.

It may surprise the reader that such an intuitive notion has required so much work! However, the precise definition of the length of a curve will only be given later on in this book and requires some deeper analytical tools (Chap. 11).

We are now ready to introduce an important number in mathematics, the number π, pronounced "*pie*," defined as

$$\pi = 2\mathrm{Arg}(i) = 3.14159\ldots$$

The importance of this number π will emerge later on. It measures twice the length of the arc on the unitary circle S^1 starting from $(1, 0)$ and arriving at $(0, 1)$, moving in a counterclockwise direction, so half the length of S^1 itself. It can be proved that it is an irrational number.

In the case where $\Im(\zeta) = 0$, i.e., when ζ is a positive real number, we set

$$\mathrm{Arg}(\zeta) = 0 \, .$$

In what follows, we will require the inequality

$$|\tilde{\sigma}_n - 1| \le \frac{1}{2^n} \mathrm{Arg}(\zeta) \, , \quad \text{for every } n \in \mathbb{N} \, , \tag{5.4}$$

which is a direct consequence of the fact that $(a_n)_n$ is increasing.

5.1.2 Definition on a Dense Set

We first define the function f on the set

$$E = \left\{ \frac{m}{2^n} : m \in \mathbb{Z}, n \in \mathbb{N} \right\},$$

which is a dense subset of $\mathbb{R}$. We will see that if we want a function $f : E \to \mathbb{C} \setminus \{0\}$ to satisfy the conditions (a), (b), and (c) of the statement, then its definition is uniquely determined.

Thus, assume that (a), (b), and (c) hold for some function $f : E \to \mathbb{C} \setminus \{0\}$. Since $f(1) = \zeta = \sigma_0$, by (b),

$$\sigma_0 = f(1) = f\left(\frac{1}{2} + \frac{1}{2}\right) = f\left(\frac{1}{2}\right) f\left(\frac{1}{2}\right) = \left[f\left(\frac{1}{2}\right)\right]^2,$$

and since $\Re f\left(\frac{1}{2}\right) \geq 0$, $\Im f\left(\frac{1}{2}\right) \geq 0$, it must be that $f\left(\frac{1}{2}\right) = \sigma_1$. Similarly, since

$$\sigma_1 = f\left(\frac{1}{2}\right) = f\left(\frac{1}{4} + \frac{1}{4}\right) = f\left(\frac{1}{4}\right) f\left(\frac{1}{4}\right) = \left[f\left(\frac{1}{4}\right)\right]^2,$$

we see that $f\left(\frac{1}{4}\right) = \sigma_2$. Iterating this process, we see that we must set

$$f\left(\frac{1}{2^n}\right) = \sigma_n, \quad \text{for every } n \in \mathbb{N}.$$

Moreover,

$$f\left(\frac{m}{2^n}\right) = f\left(m \frac{1}{2^n}\right) = \left[f\left(\frac{1}{2^n}\right)\right]^m.$$

This shows that if (a), (b), and (c) hold, then the definition of f on the set E must be

$$f\left(\frac{m}{2^n}\right) = \sigma_n^m. \tag{5.5}$$

This is a good definition since

$$\frac{m}{2^n} = \frac{m'}{2^{n'}} \quad \Rightarrow \quad \sigma_n^m = \sigma_{n'}^{m'}.$$

Indeed, if, for instance, $n' \geq n$, then we see by (5.1) that $\sigma_n = \sigma_{n'}^{2^{n'-n}}$, hence

$$\sigma_n^m = (\sigma_{n'}^{2^{n'-n}})^m = \sigma_{n'}^{m2^{n'-n}} = \sigma_{n'}^{m'}.$$

Let us now prove that the function $f : E \to \mathbb{C} \setminus \{0\}$ defined by (5.5) satisfies the properties (a), (b), and (c). Notice that $f(0) = 1$ and $f(1) = \zeta$, so that property (a) holds. Let us prove that

$$f(x_1 + x_2) = f(x_1) f(x_2), \quad \text{for every } x_1, x_2 \in E, \tag{5.6}$$

which is property (b) on the domain E. Taking $x_1 = \dfrac{k}{2^n}$ and $x_2 = \dfrac{m}{2^n}$ (we can now choose the same denominator), we have

$$f\left(\frac{k}{2^n} + \frac{m}{2^n}\right) = f\left(\frac{k+m}{2^n}\right) = \sigma_n^{k+m} = \sigma_n^k \sigma_n^m = f\left(\frac{k}{2^n}\right) f\left(\frac{m}{2^n}\right).$$

Finally, with the aim of verifying property (c), we claim that

$$1, \tilde{\sigma}_n, \tilde{\sigma}_n^2, \ldots \tilde{\sigma}_n^{2^n} \text{ belong to } \mathcal{Q}_1, \quad \text{for every } n \in \mathbb{N}.$$

This is surely true if $n = 0$ or 1. If $n = 2$, then we have that $1, \tilde{\sigma}_2$ and $\tilde{\sigma}_2^2 = \tilde{\sigma}_1$ surely belong to $\mathcal{Q}_1$, as well as $\tilde{\sigma}_2^4 = \tilde{\zeta}$. Concerning $\tilde{\sigma}_2^3$, we notice that

$$|\tilde{\sigma}_2^3| = 1 \quad \text{and} \quad |\tilde{\sigma}_2^3 - \tilde{\sigma}_2^2| = |\tilde{\sigma}_2^3 - \tilde{\sigma}_2^4| = \ell_2.$$

In principle, two points satisfy these properties, one in the first quadrant and one in the third. However, since $\ell_2 < \sqrt{2}$, it must be that $\tilde{\sigma}_2^3$ belongs to $\mathcal{Q}_1$. By induction, the same argument can be used to prove the claim for every $n \in \mathbb{N}$.

Thus, we have constructed a function $f : E \to \mathbb{C} \setminus \{0\}$ that verifies the properties (a), (b), and (c) on its domain. And this is the only possible function with these properties.

5.1.3   Extension to the Whole Real Line

To extend the function $f : E \to \mathbb{C} \setminus \{0\}$ defined by (5.5) to the whole real line $\mathbb{R}$, we will apply Theorem 4.16. To this end, we first need to verify that f is uniformly continuous on any bounded subset of its domain E. We fix a real number $R > 0$ and consider the restriction of f to $E \cap [-R, R]$.

We define the two functions $g : E \to]0, +\infty[$ and $h : E \to S^1$ by

$$g(x) = |f(x)|, \qquad h(x) = \frac{f(x)}{|f(x)|},$$

and we remark that

$$g(x_1 + x_2) = g(x_1)g(x_2) \quad \text{for every } x_1, x_2 \in E \tag{5.7}$$

and

$$h(x_1 + x_2) = h(x_1)h(x_2) \quad \text{for every } x_1, x_2 \in E. \tag{5.8}$$

Let us first concentrate on the function g and prove that it is uniformly continuous on $E \cap [-R, R]$. We need some preliminary considerations.

It is easily seen that if $|\zeta| = 1$, then g is constant. Assume now that $|\zeta| > 1$. In this case, it can be seen that $|\sigma_n| > 1$ for every $n \in \mathbb{N}$, so also $|\sigma_n^m| > 1$ for every $n \in \mathbb{N}$ and $m \geq 1$, i.e.,

$$g(x) > 1, \quad \text{for every } x \in E \cap]0, +\infty[.$$

Consequently,

$$x_1 < x_2 \quad \Rightarrow \quad g(x_2) = g(x_1)g(x_2 - x_1) > g(x_1),$$

proving that g is strictly increasing. Let us now show that

$$\lim_{x \to 0^+} g(x) = 1.$$

Fix an $\epsilon > 0$. Let $\bar{n} \in \mathbb{N}$ be such that $\bar{n} \geq (|\zeta| - 1)/\epsilon$. Then, for every $n \geq \bar{n}$, using the Bernoulli inequality,

$$\left[g\left(\frac{1}{2^n} \right) \right]^{2^n} = |\zeta| \leq 1 + n\epsilon \leq 1 + 2^n \epsilon \leq (1 + \epsilon)^{2^n},$$

so that

$$1 < g\left(\frac{1}{2^n} \right) \leq 1 + \epsilon.$$

Since g is increasing, this proves the claim.

We are now ready to prove that g is uniformly continuous on $E \cap [-R, R]$. Let us fix $\varepsilon > 0$. By the above considerations, there exists $\delta > 0$ such that

$$0 < x < \delta \quad \Rightarrow \quad 1 < g(x) < 1 + \frac{\varepsilon}{g(R)}.$$

Then, taking $x_1, x_2 \in E \cap [-R, R]$ such that $x_1 < x_2$ and $x_2 - x_1 \le \delta$, we have

$$0 < g(x_2) - g(x_1) = g(x_1)(g(x_2 - x_1) - 1) < g(R)\frac{\varepsilon}{g(R)} = \varepsilon.$$

This proves that g is uniformly continuous on $E \cap [-R, R]$, with values in the interval $[g(-R), g(R)]$.

If $|\zeta| < 1$, then the proof is similar (but g is strictly decreasing in this case). We now concentrate on the function h. Notice that

$$h\left(\frac{m}{2^n}\right) = \frac{\sigma_n^m}{|\sigma_n^m|} = \tilde{\sigma}_n^m.$$

Take $x_1 = \dfrac{k}{2^n}$ and $x_2 = \dfrac{m}{2^n}$, with $k < m$. Then

$$h\left(\frac{m}{2^n}\right) - h\left(\frac{k}{2^n}\right) = \tilde{\sigma}_n^m - \tilde{\sigma}_n^k = \tilde{\sigma}_n^k(\tilde{\sigma}_n^{m-k} - 1)$$

$$= \tilde{\sigma}_n^k(\tilde{\sigma}_n - 1)(1 + \tilde{\sigma}_n + \tilde{\sigma}_n^2 + \cdots + \tilde{\sigma}_n^{m-k-1}),$$

and hence, by (5.4),

$$\left|h\left(\frac{m}{2^n}\right) - h\left(\frac{k}{2^n}\right)\right| \le |\tilde{\sigma}_n - 1|(m - k) \le \mathrm{Arg}(\zeta)\left|\frac{m}{2^n} - \frac{k}{2^n}\right|.$$

This proves that h is uniformly continuous on the whole domain E.

The restriction of the function f on $E \cap [-R, R]$ is uniformly continuous, since, when $x_1, x_2 \in E \cap [-R, R]$,

$$
\begin{aligned}
|f(x_2) - f(x_1)| &= |g(x_2)h(x_2) - g(x_1)h(x_1)| \\
&\le |g(x_2) - g(x_1)|\,|h(x_2)| + |h(x_2) - h(x_1)|\,|g(x_1)| \\
&\le |g(x_2) - g(x_1)| + |h(x_2) - h(x_1)|\,\max\{g(-R), g(R)\},
\end{aligned}
$$

and both g and h are uniformly continuous. This restriction of the function f takes its values in the compact set

$$F = \{z \in \mathbb{C} : r_1 \le |z| \le r_2\},$$

where $r_1 = g(-R)$ and $r_2 = g(R)$, or vice versa. Since F is a closed subset of the complete metric space $\mathbb{C}$, it is a complete metric space itself (with Euclidean distance). Hence, by Theorem 4.16, f can be extended in a unique way to a continuous function on $[-R, R]$, with values in the same set F. Since this can be done for an arbitrary $R > 0$, we have thus defined a continuous extension of f on the whole real axis $\mathbb{R}$, with nonzero values. And this is the only possible continuous extension. We will still denote by f this new function. We now verify properties (a), (b), and (c) for this function.

Recalling that $\tau = 1$, property (a) was already verified earlier since $f(0) = 1$ and $f(1) = \zeta$.

Concerning property (b), take x_1, x_2 in $\mathbb{R}$, and let $(x_{1,n})_n$ and $(x_{2,n})_n$ be two sequences in E such that $\lim_n x_{1,n} = x_1$ and $\lim_n x_{2,n} = x_2$. Then, from (5.6), by continuity,

$$f(x_1 + x_2) = f(\lim_n(x_{1,n} + x_{2,n})) = \lim_n f(x_{1,n} + x_{2,n}) = \lim_n f(x_{1,n}) f(x_{2,n})$$

$$= \lim_n f(x_{1,n}) \lim_n f(x_{2,n}) = f(x_1) f(x_2).$$

Finally, to verify property (c), take $x \in [0, 1]$, and let $(x_n)_n$ be a sequence in $E \cap [0, 1]$ such that $\lim_n x_n = x$. Since $f(x_n)$ belongs to $\mathcal{Q}_1$, which is a closed set, then, by continuity,

$$f(x) = \lim_n f(x_n) \in \mathcal{Q}_1.$$

We have thus constructed a continuous function $f : \mathbb{R} \to \mathbb{C} \setminus \{0\}$ that verifies the properties (a), (b), and (c). And this is the only possible one. Then the proof of Theorem 5.1 is complete.

5.2 Exponential and Circular Functions

In this section we define the exponential and circular functions, providing a proof for the previously stated Theorems 2.15 and 2.16.

Proof of Theorem 2.15 If $\tau = 1$ and ζ is a positive real number, say, $\zeta = a > 0$, then the function h is constantly equal to 1 and f coincides with $g : \mathbb{R} \to]0, +\infty[$. This is the *exponential with base a*, i.e., the function $f_a : \mathbb{R} \to \mathbb{R}_+$ whose existence was stated in Theorem 2.15. Indeed, property (i) follows from (5.7), and $g(1) = a$. We need to prove that if $a \neq 1$, then $g : \mathbb{R} \to]0, +\infty[$ is invertible.

As seen previously, when $a > 1$, the function g is strictly increasing on E, and so also on $\mathbb{R}$, whereas if $a < 1$, then it is strictly decreasing. Hence, if $a \neq 1$, then the function g is injective; let us show that it is also surjective.

We have shown, after the statement of Theorem 2.15, that $g(n) = a^n$ for every $n \in \mathbb{N}$. Assume, for instance, $a > 1$; then $\lim_n a^n = +\infty$, hence, by monotonicity, also

$$\lim_{x \to +\infty} g(x) = +\infty.$$

On the other hand, since $g(x)g(-x) = g(x - x) = g(0) = 1$,

$$\lim_{x \to -\infty} g(x) = \lim_{x \to +\infty} g(-x) = \lim_{x \to +\infty} \frac{1}{g(x)} = 0.$$

We can then conclude, by Bolzano's Theorem 2.12, that the image of g is the whole interval $]0, +\infty[$.

We have thus proved that if $a > 1$, then $g : \mathbb{R} \to]0, +\infty[$ is invertible. The same conclusion holds when $0 < a < 1$, and the proof is analogous. The proof of Theorem 2.15 is thus complete. ∎

Proof of Theorem 2.16 Now let $\zeta = i$ and $\tau > 0$ be arbitrary. Then the function g is constantly equal to 1, so f coincides with $h : \mathbb{R} \to S^1$. Notice that, since $h(\tau) = i$,

$$h(2\tau) = h(\tau + \tau) = h(\tau)^2 = i^2 = -1,$$

$$h(3\tau) = h(2\tau + \tau) = h(2\tau)h(\tau) = -i,$$

$$h(4\tau) = h(3\tau + \tau) = h(3\tau)h(\tau) = 1,$$

and then

$$h(x + 4\tau) = h(x)h(4\tau) = h(x), \quad \text{for every } x \in \mathbb{R},$$

showing that h is a periodic function, with period $T = 4\tau$. We would like to prove that T is indeed the *minimal* period of h.

Since h is continuous and nonconstant, its minimal period is T/k for some integer $k \geq 1$. Assume by contradiction that $k \geq 2$. Then

$$\frac{T}{2k} = \frac{2\tau}{k} \leq \tau$$

and

$$1 = h\left(\frac{T}{k}\right) = \left[h\left(\frac{T}{2k}\right)\right]^2,$$

and since $h(T/2k) \in \mathcal{Q}_1$, it must be that $h(T/2k) = 1$. Then we will have

$$1 = h\left(\frac{T}{2k}\right) = \left[h\left(\frac{T}{4k}\right)\right]^2,$$

and since $h(T/4k) \in \mathcal{Q}_1$, it must be that $h(T/4k) = 1$, too. Proceeding in this way, we see that it must be that

$$h\left(\frac{T}{2^j k}\right) = 1, \quad \text{for every } j \in \mathbb{N}.$$

Hence, also

$$h\left(\frac{mT}{2^j k}\right) = 1, \quad \text{for every } j \in \mathbb{N} \text{ and } m \in \mathbb{Z}.$$

Since the set $\{mT/2^j k : j \in \mathbb{N}, m \in \mathbb{Z}\}$ is dense in $\mathbb{R}$ and h is continuous, this would imply that h is constantly equal to 1, a contradiction, since $h(\tau) = i$.

We have thus proved that

$$T = 4\tau \text{ is the minimal period of } h,$$

and henceforth we will write h_T instead of h. Since (i) follows from (5.8) and $h(\tau) = i$, the proof of Theorem 2.16 is thus now complete.   ■

5.3   Limits for Trigonometric Functions

In the following theorem, the number π enters the picture.

Theorem 5.2 *We have*

$$\lim_{x \to 0^+} \frac{h_T(x) - 1}{x} = \frac{2\pi}{T} i.$$

Proof Since $T = 4\tau$ and

$$h_{4\tau}(x) = h_4\left(\frac{x}{\tau}\right),$$

it will be equivalent to proving that

$$\lim_{x \to 0^+} \frac{h_4(x) - 1}{x} = \frac{\pi}{2} i.$$

First of all, we show that this is true when $x = \frac{1}{2^n}$, i.e., that

$$\lim_n 2^n (\sigma_n - 1) = \frac{\pi}{2} i. \tag{5.9}$$

(Recall that σ_n and $\tilde{\sigma}_n$ coincide in this case.) We already know that

$$\lim_n |2^n(\sigma_n - 1)| = \text{Arg}(i) = \frac{\pi}{2} = \left|\frac{\pi}{2}i\right| ;$$

hence, since $\Im(2^n(\sigma_n - 1)) > 0$, it will be sufficient to show that

$$\lim_n 2^n \Re(\sigma_n - 1) = 0. \tag{5.10}$$

Since $\sigma_n^* = \sigma_n^{-1}$, we see that

$$\Re(\sigma_n - 1) = \frac{(\sigma_n - 1) + (\sigma_n - 1)^*}{2} = \frac{\sigma_n + \sigma_n^{-1} - 2}{2} = \frac{\sigma_n^2 + 1 - 2\sigma_n}{2\sigma_n} = \frac{(\sigma_n - 1)^2}{2\sigma_n}.$$

Recalling (5.4) with $\zeta = i$, since $\tilde{\sigma}_n = \sigma_n$ and $\text{Arg}(i) = \frac{\pi}{2}$, we have that

$$2^{n+1}|\sigma_n - 1| < \pi. \tag{5.11}$$

Hence,

$$|2^n \Re(\sigma_n - 1)| \leq 2^n \frac{|\sigma_n - 1|^2}{|2\sigma_n|} \leq \frac{\pi^2}{2^{n+3}},$$

thereby proving (5.10) and, hence, (5.9).

We now prove the stated limit when x varies in E, i.e., when $x = \frac{m}{2^n} > 0$. In such a case,

$$\left|\frac{h_4(x) - 1}{x} - \frac{\pi}{2}i\right| = \left|\frac{\sigma_n^m - 1}{\frac{m}{2^n}} - \frac{\pi}{2}i\right|$$

$$= \left|2^n(\sigma_n - 1)\frac{1 + \sigma_n + \sigma_n^2 + \cdots + \sigma_n^{m-1}}{m} - \frac{\pi}{2}i\right|$$

$$\leq 2^n|\sigma_n - 1|\left|\frac{1 + \sigma_n + \sigma_n^2 + \cdots + \sigma_n^{m-1}}{m} - 1\right| + \left|2^n(\sigma_n - 1) - \frac{\pi}{2}i\right|$$

$$\leq 2^n|\sigma_n - 1|\frac{|\sigma_n - 1| + |\sigma_n^2 - 1| + \cdots + |\sigma_n^{m-1} - 1|}{m} + \left|2^n(\sigma_n - 1) - \frac{\pi}{2}i\right|.$$

By (5.11), for $k = 1, 2, \ldots, m - 1$ we have

$$|\sigma_n^k - 1| = \left|\sum_{j=1}^k (\sigma_n^j - \sigma_n^{j-1})\right| \leq \sum_{j=1}^k |\sigma_n^j - \sigma_n^{j-1}| = k|\sigma_n - 1| \leq k\frac{\pi}{2^{n+1}}.$$

Using the formula

$$1 + 2 + 3 + \cdots + (m-1) = \frac{(m-1)m}{2},$$

we obtain

$$\frac{|\sigma_n - 1| + |\sigma_n^2 - 1| + \cdots + |\sigma_n^{m-1} - 1|}{m}$$

$$\leq \frac{1}{m}\left[\frac{\pi}{2^{n+1}} + 2\frac{\pi}{2^{n+1}} + \cdots + (m-1)\frac{\pi}{2^{n+1}}\right]$$

$$= \frac{1}{m}\frac{(m-1)m}{2}\frac{\pi}{2^{n+1}} < \frac{\pi}{4}\frac{m}{2^n}.$$

In conclusion, if $x = \frac{m}{2^n} > 0$, then

$$\left|\frac{h_4(x) - 1}{x} - \frac{\pi}{2}i\right| \leq \frac{\pi}{2}\frac{\pi}{4}\frac{m}{2^n} + \left|2^n(\sigma_n - 1) - \frac{\pi}{2}i\right|.$$

As $x = \frac{m}{2^n}$ tends to 0, necessarily n tends to $+\infty$, and the result follows by (5.9).

We finally look for the limit as $x \to 0^+$, without further restrictions on x, and assume by contradiction that either such a limit does not exist or that it is not equal to $\frac{\pi}{2}i$. Then there is $\varepsilon > 0$ and a strictly decreasing sequence $(x_n)_n$, with $x_n \to 0^+$, such that, for every n,

$$\left|\frac{h_4(x_n) - 1}{x_n} - \frac{\pi}{2}i\right| > \varepsilon.$$

By the continuity of the function $\frac{h_4(x)-1}{x}$ and the density of E in $\mathbb{R}$, for every sufficiently large n one can find a positive number $x'_n \in E$ such that

$$|x_n - x'_n| \leq \frac{1}{n} \quad \text{and} \quad \left|\frac{h_4(x'_n) - 1}{x'_n} - \frac{\pi}{2}i\right| > \varepsilon,$$

contradicting the previous part of the proof.   ∎

As a consequence of the preceding theorem, we have the following corollary.

Corollary 5.3 *We have*

$$\lim_{x \to 0} \frac{\sin_T(x)}{x} = \frac{2\pi}{T}.$$

Proof Writing $h_T(x) = \cos_T(x) + i \sin_T(x)$, we have

$$\frac{h_T(x) - 1}{x} = \frac{\cos_T(x) - 1}{x} + i \frac{\sin_T(x)}{x}.$$

Hence, by Theorem 5.2,

$$\lim_{x \to 0^+} \frac{\cos_T(x) - 1}{x} = 0, \qquad \lim_{x \to 0^+} \frac{\sin_T(x)}{x} = \frac{2\pi}{T}.$$

Now, we have shown, after the statement of Theorem 2.16, that $\sin_T$ is an odd function; hence,

$$\lim_{x \to 0^-} \frac{\sin_T(x)}{x} = \lim_{x \to 0^+} \frac{\sin_T(-x)}{-x} = \lim_{x \to 0^+} \frac{\sin_T(x)}{x} = \frac{2\pi}{T},$$

and the proof is completed. ∎

Notice that the choice $T = 2\pi$ simplifies the preceding formula. This is why we will always choose as the base of the trigonometric functions the number $T = 2\pi$. We will write $\cos(x)$, $\sin(x)$, $\tan(x)$ (or simply $\cos x$, $\sin x$, $\tan x$) instead of $\cos_{2\pi}(x)$, $\sin_{2\pi}(x)$, $\tan_{2\pi}(x)$. Hence,

$$\lim_{x \to 0} \frac{\sin x}{x} = 1.$$

The knowledge of this limit now allows us to prove that

$$\lim_{x \to 0} \frac{\tan x}{x} = 1.$$

Indeed, we have

$$\lim_{x \to 0} \frac{\tan x}{x} = \lim_{x \to 0} \frac{\sin x}{x} \cos x = \lim_{x \to 0} \frac{\sin x}{x} \lim_{x \to 0} \cos x = 1 \cdot \cos(0) = 1.$$

Moreover, we can also prove that

$$\lim_{x \to 0} \frac{\cos x - 1}{x^2} = -\frac{1}{2}.$$

Indeed, we have

$$\lim_{x \to 0} \frac{\cos x - 1}{x^2} = \lim_{x \to 0} \frac{\cos^2 x - 1}{x^2 (\cos x + 1)}$$

$$= -\lim_{x \to 0} \frac{\sin^2 x}{x^2} \lim_{x \to 0} \frac{1}{\cos x + 1} = -1 \cdot \frac{1}{2} = -\frac{1}{2}.$$

It will be useful to keep in mind these remarkable limits for further applications.

Part II

Differential and Integral Calculus in $\mathbb{R}$

The Derivative

We start by introducing the concept of "derivative" of a function defined on a subset of $\mathbb{R}$, taking its values in $\mathbb{R}$.

Let $\mathcal{O}$, a subset of $\mathbb{R}$, be the domain of the function $f : \mathcal{O} \to \mathbb{R}$, and consider as fixed a point $x_0 \in \mathcal{O}$. For every $x \in \mathcal{O} \setminus \{x_0\}$, we may write the "difference quotient"

$$\frac{f(x) - f(x_0)}{x - x_0} \, ;$$

it is precisely the slope of the line passing through the points $(x_0, f(x_0))$ and $(x, f(x))$.

Henceforth, x_0 will be assumed to be a cluster point of $\mathcal{O}$.

Definition 6.1 The limit

$$\lim_{x \to x_0} \frac{f(x) - f(x_0)}{x - x_0} \, ,$$

whenever it exists, is called the "derivative" of f at x_0, and is denoted by one of the following symbols:

$$f'(x_0), \quad Df(x_0), \quad \frac{df}{dx}(x_0) \, .$$

We say that f is "differentiable" at x_0 when the derivative exists and is a real number (hence, not equal to $+\infty$ or $-\infty$). In such a case, the line passing through the point $(x_0, f(x_0))$ and having $f'(x_0)$ as its slope, whose equation is

$$y = f(x_0) + f'(x_0)(x - x_0) \, ,$$

is called the "tangent line" to the graph of f at the point $(x_0, f(x_0))$.

© The Author(s), under exclusive license to Springer Nature Switzerland AG 2023
A. Fonda, *A Modern Introduction to Mathematical Analysis*,
https://doi.org/10.1007/978-3-031-23713-3_6

Note that, in some cases, the derivative of f at x_0 could only be a left limit or a right limit. Typically this situation arises when $\mathcal{O}$ is an interval and x_0 coincides with an endpoint.

It is sometimes useful to write, equivalently,

$$f'(x_0) = \lim_{x \to x_0} \frac{f(x) - f(x_0)}{x - x_0} = \lim_{h \to 0} \frac{f(x_0 + h) - f(x_0)}{h}.$$

Example 1 Let $f : \mathbb{R} \to \mathbb{R}$ be defined as $f(x) = mx + q$. Then

$$f'(x_0) = \lim_{x \to x_0} \frac{(mx + q) - (mx_0 + q)}{x - x_0} = m.$$

The tangent line in this case coincides with the graph of the function itself. The particular case $m = 0$ tells us that the derivative of a constant function is always equal to 0.

Example 2 Let $f(x) = x^n$; then

$$f'(x_0) = \lim_{x \to x_0} \frac{x^n - x_0^n}{x - x_0} = \lim_{x \to x_0} \left(\sum_{k=0}^{n-1} x^k x_0^{n-1-k} \right) = n x_0^{n-1}.$$

Let us prove the same formula using a different approach:

$$f'(x_0) = \lim_{h \to 0} \frac{(x_0 + h)^n - x_0^n}{h} = \lim_{h \to 0} \frac{1}{h} \left(\sum_{k=0}^{n} \binom{n}{k} x_0^{n-k} h^k - x_0^n \right)$$

$$= \lim_{h \to 0} \frac{1}{h} \left(\sum_{k=1}^{n} \binom{n}{k} x_0^{n-k} h^k \right) = \lim_{h \to 0} \left(\sum_{k=1}^{n} \binom{n}{k} x_0^{n-k} h^{k-1} \right) = n x_0^{n-1}.$$

Example 3 Now let $f(x) = e^x$; then

$$f'(x_0) = \lim_{h \to 0} \frac{e^{x_0+h} - e^{x_0}}{h} = \lim_{h \to 0} e^{x_0} \frac{e^h - 1}{h} = e^{x_0}.$$

Example 4 Choosing $f(x) = \cos x$, we have

$$f'(x_0) = \lim_{h \to 0} \frac{\cos(x_0 + h) - \cos(x_0)}{h}$$

$$= \lim_{h \to 0} \frac{\cos(x_0) \cos(h) - \sin(x_0) \sin(h) - \cos(x_0)}{h}$$

$$= -\cos(x_0) \lim_{h \to 0} h \frac{1 - \cos(h)}{h^2} - \sin(x_0) \lim_{h \to 0} \frac{\sin(h)}{h}$$

$$= -\sin(x_0).$$

Example 5 On the other hand, if $g(x) = \sin x$, then

$$g'(x_0) = \lim_{h \to 0} \frac{\sin(x_0 + h) - \sin(x_0)}{h}$$

$$= \lim_{h \to 0} \frac{\sin(x_0) \cos(h) + \cos(x_0) \sin(h) - \sin(x_0)}{h}$$

$$= -\sin(x_0) \lim_{h \to 0} h \frac{1 - \cos(h)}{h^2} + \cos(x_0) \lim_{h \to 0} \frac{\sin(h)}{h}$$

$$= \cos(x_0).$$

The following theorem provides us a characterization of differentiability.

Theorem 6.2 *The function f is differentiable at x_0 if and only if there exists a real number ℓ for which one can write*

$$f(x) = f(x_0) + \ell(x - x_0) + r(x), \tag{6.1}$$

where r is a function such that

$$\lim_{x \to x_0} \frac{r(x)}{x - x_0} = 0. \tag{6.2}$$

In that case, we have $\ell = f'(x_0)$.

Proof Assume that f is differentiable at x_0. Then

$$\lim_{x \to x_0} \frac{f(x) - f(x_0) - f'(x_0)(x - x_0)}{x - x_0} = 0.$$

Hence, setting $r(x) = f(x) - f(x_0) - f'(x_0)(x - x_0)$, the desired properties (6.1) and (6.2) are readily verified, taking $\ell = f'(x_0)$.
  Conversely, assume that (6.1) and (6.2) hold. Then

$$\lim_{x \to x_0} \frac{f(x) - f(x_0) - \ell(x - x_0)}{x - x_0} = 0,$$

and hence

$$\lim_{x \to x_0} \frac{f(x) - f(x_0)}{x - x_0} = \lim_{x \to x_0} \left(\frac{f(x) - f(x_0) - \ell(x - x_0)}{x - x_0} + \ell \right) = \ell \,,$$

showing that f is differentiable at x_0.   ∎

We now prove that differentiability implies continuity.

Theorem 6.3 *If f is differentiable at x_0, then f is continuous at x_0.*

Proof Since f is differentiable at x_0, we have that

$$\lim_{x \to x_0} f(x) = \lim_{x \to x_0} \left[f(x_0) + \frac{f(x) - f(x_0)}{x - x_0}(x - x_0) \right]$$
$$= f(x_0) + f'(x_0) \cdot 0 = f(x_0) \,,$$

which means that f is continuous at x_0.   ∎

6.1   Some Differentiation Rules

Let us review some rules for the computation of the derivative.

Theorem 6.4 *If $f, g : \mathcal{O} \to \mathbb{R}$ are differentiable at x_0, then so is $f + g$, and*

$$(f + g)'(x_0) = f'(x_0) + g'(x_0) \,.$$

Proof We compute

$$\lim_{x \to x_0} \frac{(f + g)(x) - (f + g)(x_0)}{x - x_0} = \lim_{x \to x_0} \left[\frac{f(x) - f(x_0)}{x - x_0} + \frac{g(x) - g(x_0)}{x - x_0} \right]$$
$$= \lim_{x \to x_0} \frac{f(x) - f(x_0)}{x - x_0} + \lim_{x \to x_0} \frac{g(x) - g(x_0)}{x - x_0}$$
$$= f'(x_0) + g'(x_0) \,,$$

and the formula is proved.   ∎

Theorem 6.5 *If $f, g : \mathcal{O} \to \mathbb{R}$ are differentiable at x_0, then so is $f \cdot g$, and*

$$(f \cdot g)'(x_0) = f'(x_0)g(x_0) + f(x_0)g'(x_0) \,.$$

Proof We can write

$$\lim_{x\to x_0} \frac{(f\cdot g)(x) - (f\cdot g)(x_0)}{x - x_0} = \lim_{x\to x_0}\left[\frac{f(x) - f(x_0)}{x - x_0}g(x_0) + f(x)\frac{g(x) - g(x_0)}{x - x_0}\right]$$

$$= \lim_{x\to x_0} \frac{f(x) - f(x_0)}{x - x_0}g(x_0) + \lim_{x\to x_0} f(x) \lim_{x\to x_0}\frac{g(x) - g(x_0)}{x - x_0},$$

and the conclusion follows, recalling that $\lim_{x\to x_0} f(x) = f(x_0)$, since f is continuous at x_0. ■

The particular case where g is constant with a value $\alpha \in \mathbb{R}$ gives us the formula

$$(\alpha f)'(x_0) = \alpha f'(x_0).$$

Moreover, writing $f - g = f + (-1)g$, we have that

$$(f - g)'(x_0) = f'(x_0) - g'(x_0).$$

Theorem 6.6 *If $f, g : \mathcal{O} \to \mathbb{R}$ are differentiable at x_0 and $g(x_0) \neq 0$, then so is $\frac{f}{g}$, and*

$$\left(\frac{f}{g}\right)'(x_0) = \frac{f'(x_0)g(x_0) - f(x_0)g'(x_0)}{[g(x_0)]^2}.$$

Proof Since $\frac{f}{g} = f \cdot \frac{1}{g}$, it will be useful to first show that $\frac{1}{g}$ is differentiable at x_0. Indeed, we have

$$\lim_{x\to x_0} \frac{\frac{1}{g}(x) - \frac{1}{g}(x_0)}{x - x_0} = \lim_{x\to x_0} \frac{g(x_0) - g(x)}{(x - x_0)g(x)g(x_0)} = -\frac{g'(x_0)}{[g(x_0)]^2}.$$

Then

$$\left(\frac{f}{g}\right)'(x_0) = f'(x_0)\frac{1}{g}(x_0) + f(x_0)\left(\frac{1}{g}\right)'(x_0) = \frac{f'(x_0)}{g(x_0)} - f(x_0)\frac{g'(x_0)}{[g(x_0)]^2},$$

whence the conclusion. ■

Example 1 Let us take into consideration the tangent function

$$\tan x = \frac{\sin x}{\cos x}.$$

Choosing $f(x) = \sin x$ and $g(x) = \cos x$, we have[1]

$$D \tan x_0 = \frac{f'(x_0)g(x_0) - f(x_0)g'(x_0)}{[g(x_0)]^2} = \frac{\cos^2 x_0 + \sin^2 x_0}{\cos^2 x_0} = \frac{1}{\cos^2 x_0}.$$

Example 2 We now compute the derivative of the hyperbolic functions. Let

$$\cosh(x) = \frac{e^x + e^{-x}}{2} = \frac{1}{2}\left(e^x + \frac{1}{e^x}\right).$$

Then

$$D\cosh(x_0) = \frac{1}{2}\left(e^{x_0} - \frac{e^{x_0}}{[e^{x_0}]^2}\right) = \frac{e^{x_0} - e^{-x_0}}{2} = \sinh(x_0).$$

Similarly, writing

$$\sinh(x) = \frac{e^x - e^{-x}}{2} = \frac{1}{2}\left(e^x - \frac{1}{e^x}\right),$$

we have

$$D\sinh(x_0) = \frac{1}{2}\left(e^{x_0} + \frac{e^{x_0}}{[e^{x_0}]^2}\right) = \frac{e^{x_0} + e^{-x_0}}{2} = \cosh(x_0).$$

Moreover,

$$D\tanh(x_0) = \frac{\cosh(x_0)\cosh(x_0) - \sinh(x_0)\sinh(x_0)}{\cosh^2(x_0)} = \frac{1}{\cosh^2(x_0)}.$$

Example 3 All the polynomial functions

$$F(x) = a_n x^n + a_{n-1}x^{n-1} + \cdots + a_2 x^2 + a_1 x + a_0$$

are differentiable, with derivative

$$F'(x_0) = na_n x_0^{n-1} + (n-1)a_{n-1}x_0^{n-2} + \cdots + 2a_2 x_0 + a_1.$$

---

[1] Here and in what follows we will often write $\cos^2 x$ and $\sin^2 x$ instead of $(\cos x)^2$ and $(\sin x)^2$, respectively.

Hence, all rational functions of the type

$$F(x) = \frac{p(x)}{q(x)},$$

with $p(x)$ and $q(x)$ polynomials, are also differentiable at all points x_0 where $q(x_0) \neq 0$.

Let us now see how to compute the derivative of the composition of two functions. Subsequently in this chapter, we will always consider only nondegenerate intervals, i.e., those not reduced to a single point.

Theorem 6.7 *If $f : \mathcal{O} \to \mathbb{R}$ is differentiable at x_0, and $g : J \to \mathbb{R}$ is differentiable at $f(x_0)$, where J is an interval containing $f(\mathcal{O})$, then $g \circ f$ is differentiable at x_0, and*

$$(g \circ f)'(x_0) = g'(f(x_0)) f'(x_0).$$

Proof Setting $y_0 = f(x_0)$, let $R : J \to \mathbb{R}$ be the auxiliary function defined as

$$R(y) = \begin{cases} \dfrac{g(y) - g(y_0)}{y - y_0} & \text{if } y \neq y_0, \\ g'(y_0) & \text{if } y = y_0. \end{cases}$$

We observe that the function R is continuous at y_0 and

$$g(y) - g(y_0) = R(y)(y - y_0) \qquad \text{for every } y \in J.$$

Hence, if $x \neq x_0$,

$$\frac{g(f(x)) - g(f(x_0))}{x - x_0} = R(f(x)) \frac{f(x) - f(x_0)}{x - x_0}.$$

Since f is continuous at x_0 and R is continuous at $y_0 = f(x_0)$, the function $R \circ f$ is continuous at x_0, hence

$$\lim_{x \to x_0} \frac{g(f(x)) - g(f(x_0))}{x - x_0} = \lim_{x \to x_0} R(f(x)) \lim_{x \to x_0} \frac{f(x) - f(x_0)}{x - x_0}$$

$$= R(f(x_0)) f'(x_0) = g'(f(x_0)) f'(x_0),$$

which is what we wanted to prove. ∎

Example 1 Let $h : \mathbb{R} \to \mathbb{R}$ be defined as $h(x) = \cos(e^x)$. Then $h = g \circ f$, with $f(x) = e^x$ and $g(y) = \cos y$. For any $x_0 \in \mathbb{R}$, we have that $f'(x_0) = e^{x_0}$, and if $y_0 = f(x_0)$, then $g'(y_0) = -\sin y_0$. Therefore,

$$h'(x_0) = g'(f(x_0))f'(x_0) = -\sin(e^{x_0})\, e^{x_0}\,.$$

Example 2 Now let $h : \mathbb{R} \to \mathbb{R}$ be defined as $h(x) = e^{\cos x}$. Then $h = g \circ f$, with $f(x) = \cos x$ and $g(y) = e^y$. For any $x_0 \in \mathbb{R}$, we have that $f'(x_0) = -\sin x_0$, and if $y_0 = f(x_0)$, then $g'(y_0) = e^{y_0}$. Therefore,

$$h'(x_0) = g'(f(x_0))f'(x_0) = e^{\cos x_0}\, (-\sin x_0)\,.$$

We will now show how to compute the derivative of the inverse of an invertible function.

Theorem 6.8 *Let I, J be two intervals, and $f : I \to J$ be a continuous invertible function. If f is differentiable at x_0 and $f'(x_0) \neq 0$, then f^{-1} is differentiable at $y_0 = f(x_0)$, and*

$$(f^{-1})'(y_0) = \frac{1}{f'(x_0)}\,.$$

Proof We first observe that, by Theorem 2.14, the function $f^{-1} : J \to I$ is continuous. Then, by the change of variable formula,

$$\lim_{y \to y_0} \frac{f^{-1}(y) - f^{-1}(y_0)}{y - y_0} = \lim_{\substack{x \to \lim f^{-1}(y) \\ y \to y_0}} \frac{x - x_0}{f(x) - f(x_0)}$$

$$= \lim_{x \to f^{-1}(y_0)} \frac{x - x_0}{f(x) - f(x_0)}$$

$$= \lim_{x \to x_0} \frac{1}{\frac{f(x) - f(x_0)}{x - x_0}} = \frac{1}{f'(x_0)}\,,$$

thereby proving the result. ∎

Example 1 If $f(x) = e^x$, then $f^{-1}(y) = \ln y$, and, for any $y_0 > 0$, writing $y_0 = e^{x_0}$, we have that

$$D \ln(y_0) = (f^{-1})'(y_0) = \frac{1}{f'(x_0)} = \frac{1}{e^{x_0}} = \frac{1}{y_0}\,.$$

Example 2 Let α be any real number, and let $h :]0, +\infty[\to \mathbb{R}$ be defined by $h(x) = x^\alpha$. Since

$$x^\alpha = e^{\alpha \ln x},$$

we can write $h = g \circ f$, with $f(x) = \alpha \ln x$ and $g(y) = e^y$. Then

$$h'(x_0) = g'(f(x_0)) f'(x_0) = e^{\alpha \ln x_0} \alpha \frac{1}{x_0} = x_0^\alpha \alpha \frac{1}{x_0} = \alpha x_0^{\alpha-1}.$$

We thus see that the same formula we had found for an exponent $n \in \mathbb{N}$ also holds for any exponent $\alpha \in \mathbb{R}$.

6.2   The Derivative Function

We will now assume that the function $f : I \to \mathbb{R}$ is defined over an interval $I \subseteq \mathbb{R}$. We say that "f is differentiable" if it is differentiable at every point of I. In this case, we can associate to every $x \in I$ the real number $f'(x)$, thereby defining a function $f' : I \to \mathbb{R}$, which is called the "derivative function" or simply "derivative" of f. Looking back at our previous examples, we can summarize the derivatives we have found in the following table:

$f(x)$	$f'(x)$
x^α	$\alpha x^{\alpha-1}$
e^x	e^x
$\ln x$	$\dfrac{1}{x}$
$\cos x$	$-\sin x$
$\sin x$	$\cos x$
$\tan x$	$\dfrac{1}{\cos^2 x}$
$\cosh x$	$\sinh x$
$\sinh x$	$\cosh x$
$\tanh x$	$\dfrac{1}{\cosh^2 x}$
$\ldots$	$\ldots$

Some care must be taken concerning the domains, of course.

It might be interesting at this point to see whether the derivative function f' has a derivative at some point x_0 of I. If it does, then we call $(f')'(x_0)$ the "second derivative" of f at x_0 and denote it by one of the following symbols:

$$f''(x_0), \quad D^2 f(x_0), \quad \frac{d^2 f}{dx^2}(x_0).$$

It is now possible to proceed by induction and define the nth derivative of f at x_0, using the notation

$$f^{(n)}(x_0), \quad D^n f(x_0), \quad \frac{d^n f}{dx^n}(x_0),$$

by setting $f^{(n)}(x_0) = (f^{(n-1)})'(x_0)$.

We say that $f : I \to \mathbb{R}$ is "n times differentiable" if it is so at every point of I. If, moreover, the nth derivative $f^{(n)} : I \to \mathbb{R}$ is continuous, we say that f is of class $\mathcal{C}^n$. The set of those functions is denoted by $\mathcal{C}^n(I, \mathbb{R})$ or sometimes by $\mathcal{C}^n(I)$. In this setting, $\mathcal{C}^0(I, \mathbb{R})$ is just $\mathcal{C}(I, \mathbb{R})$, the set of continuous functions.

If f is of class $\mathcal{C}^n$ for every $n \in \mathbb{N}$, we say that f is "infinitely differentiable." The set of those functions is denoted by $\mathcal{C}^\infty(I, \mathbb{R})$ or sometimes by $\mathcal{C}^\infty(I)$. For example, the exponential function $f(x) = e^x$ belongs to this set, since

$$D^n e^x = e^x, \quad \text{for every } n \geq 1.$$

It can be verified that all the functions in the preceding table are infinitely differentiable on their domains.

6.3   Remarkable Properties of the Derivative

We say that $x_0 \in \mathcal{O}$ is a "local maximum point" for the function $f : \mathcal{O} \to \mathbb{R}$ if there exists a neighborhood U of x_0 for which $f(U)$ has a maximum and $f(x_0) = \max f(U)$. Equivalently, if

$$\exists \rho > 0: \quad x \in B(x_0, \rho) \cap \mathcal{O} \implies f(x) \leq f(x_0).$$

A similar definition holds for "local minimum point."

We will now compute the derivative of a function f at the local maximum or minimum points, provided that they are not at the endpoints of the domain, the interval I.

Theorem 6.9 (Fermat Theorem—I) *Let x_0 be an internal point of I, and assume $f : I \to \mathbb{R}$ to be differentiable at x_0. If, moreover, x_0 is a local maximum or minimum point for f, then $f'(x_0) = 0$.*

Proof If x_0 is an internal local maximum point for f, there exists a $\rho > 0$ such that $]x_0 - \rho, x_0 + \rho[\subseteq I$ and

$$\frac{f(x) - f(x_0)}{x - x_0} \quad \begin{cases} \geq 0 & \text{if } x_0 - \rho < x < x_0, \\ \leq 0 & \text{if } x_0 < x < x_0 + \rho. \end{cases}$$

Since f is differentiable at x_0, the limit of the difference quotient exists, and it coincides with the left and right limits, i.e.,

$$f'(x_0) = \lim_{x \to x_0^-} \frac{f(x) - f(x_0)}{x - x_0} = \lim_{x \to x_0^+} \frac{f(x) - f(x_0)}{x - x_0}.$$

By the foregoing inequalities, as a consequence of sign permanence,

$$\lim_{x \to x_0^-} \frac{f(x) - f(x_0)}{x - x_0} \geq 0 \geq \lim_{x \to x_0^+} \frac{f(x) - f(x_0)}{x - x_0}.$$

Then it must be that $f'(x_0) = 0$. In the case of a local minimum point, one proceeds similarly. ∎

It is natural that the derivative, as any limit, provides us with *local* information on the behavior of a function. However, the following theorems will open the door to the study of the *global* properties of the graph of a function.

Theorem 6.10 (Rolle Theorem)   *If* $f : [a, b] \to \mathbb{R}$ *is a continuous function, differentiable on* $]a, b[$, *and*

$$f(a) = f(b),$$

then there exists a point $\xi \in]a, b[$ *such that* $f'(\xi) = 0$.

Proof If the function is constant, then its derivative is equal to zero at every point, and the conclusion trivially follows. Assume, then, that f is not constant. Then there exists a $\bar{x} \in]a, b[$ such that

$$\text{either} \quad f(\bar{x}) < f(a) = f(b), \quad \text{or} \quad f(\bar{x}) > f(a) = f(b).$$

Let us consider the first case. By Weierstrass' Theorem 4.10, f has a minimum in $[a, b]$, and in this case any minimum point cannot be an endpoint of $[a, b]$; hence, it must be in $]a, b[$. Let $\xi \in]a, b[$ be such a point. By Fermat's Theorem 6.9, it must be that $f'(\xi) = 0$.

The situation is analogous in the second case. By Weierstrass' Theorem 4.10, f has a maximum in $[a, b]$, and in this case any maximum point must be in $]a, b[$. By Fermat's Theorem 6.9, if $\xi \in]a, b[$ is such a point, then $f'(\xi) = 0$. ∎

What follows is a generalization of the preceding theorem; it is also known as the **mean value theorem**.

Theorem 6.11 (Lagrange Theorem) *If $f : [a, b] \to \mathbb{R}$ is a continuous function, differentiable on $]a, b[$, then there exists a point $\xi \in]a, b[$ such that*

$$f'(\xi) = \frac{f(b) - f(a)}{b - a}.$$

Proof We define the function $g : [a, b] \to \mathbb{R}$ as

$$g(x) = f(x) - \left[f(a) + \frac{f(b) - f(a)}{b - a}(x - a) \right].$$

Clearly g is continuous on $[a, b]$, differentiable on $]a, b[$, and such that

$$g(a) = 0 = g(b).$$

By Rolle's Theorem 6.10, there exists a point $\xi \in]a, b[$ where

$$g'(\xi) = f'(\xi) - \frac{f(b) - f(a)}{b - a} = 0,$$

whence the conclusion.   ∎

Corollary 6.12 *Let I be an interval and $f : I \to \mathbb{R}$ a continuous function, differentiable on $\mathring{I}$. The following propositions hold:*

(a) If $f'(x) \geq 0$ for every $x \in \mathring{I}$, then f is increasing.
(b) If $f'(x) > 0$ for every $x \in \mathring{I}$, then f is strictly increasing.
(c) If $f'(x) \leq 0$ for every $x \in \mathring{I}$, then f is decreasing.
(d) If $f'(x) < 0$ for every $x \in \mathring{I}$, then f is strictly decreasing.
(e) If $f'(x) = 0$ for every $x \in \mathring{I}$, then f is constant.

Proof To prove (a), let $x_1 < x_2$ in I. By Lagrange's Theorem 6.11, there exists a $\xi \in]x_1, x_2[$ such that

$$f'(\xi) = \frac{f(x_2) - f(x_1)}{x_2 - x_1}.$$

Hence, since $f'(\xi) \geq 0$, it must be that $f(x_1) \leq f(x_2)$. This proves that f is increasing. All the other propositions follow similarly.   ∎

Remark 6.13 Note that if f is increasing, then every difference quotient for f is greater than or equal to zero, and therefore $f'(x) \geq 0$ for every $x \in \mathring{I}$. Hence, in (a), and the same also in (c) and (e), the implication can be reversed. But this is not the case for (b) and (d); indeed, if f is strictly increasing, it is not true in general that $f'(x) > 0$ for every $x \in \mathring{I}$. The derivative could be equal to zero somewhere, as the example $f(x) = x^3$ shows.

6.4 Inverses of Trigonometric and Hyperbolic Functions

Recalling the sign properties of the trigonometric functions and that $D \cos x = -\sin x$ and $D \sin x = \cos x$, we have the following properties:

$$\cos \text{ is } \begin{cases} \text{strictly decreasing on } [0, \pi] \,, \\ \text{strictly increasing on } [\pi, 2\pi] \,, \end{cases}$$

$$\sin \text{ is } \begin{cases} \text{strictly increasing on } \left[-\dfrac{\pi}{2}, \dfrac{\pi}{2} \right], \\ \text{strictly decreasing on } \left[\dfrac{\pi}{2}, \dfrac{3\pi}{2} \right]. \end{cases}$$

Let us consider the two functions $F : [0, \pi] \to [-1, 1]$ and $G : [-\frac{\pi}{2}, \frac{\pi}{2}] \to [-1, 1]$ defined by $F(x) = \cos x$ and $G(x) = \sin x$. They are strictly monotone, hence injective. Moreover, because they are continuous, their image is an interval. Since $F(\pi) = -1 = G(-\frac{\pi}{2})$ and $F(0) = 1 = G(\frac{\pi}{2})$, both images coincide with $[-1, 1]$. Therefore, the two functions thus defined are bijective. We will call the functions $F^{-1} : [-1, 1] \to [0, \pi]$ and $G^{-1} : [-1, 1] \to [-\frac{\pi}{2}, \frac{\pi}{2}]$ "arccosine" and "arcsine," respectively, and we will write

$$F^{-1}(y) = \arccos y \,, \qquad G^{-1}(y) = \arcsin y \,.$$

The first one is strictly decreasing, whereas the second one is strictly increasing. Let us compute their derivatives. Setting $y = F(x)$, for $x \in \,]0, \pi[$ we have

$$(F^{-1})'(y) = \frac{1}{F'(x)} = -\frac{1}{\sin x} = -\frac{1}{\sqrt{1 - \cos^2 x}} = -\frac{1}{\sqrt{1 - y^2}} \,,$$

while setting $y = G(x)$, for $x \in \,] - \frac{\pi}{2}, \frac{\pi}{2}[$ we have

$$(G^{-1})'(y) = \frac{1}{G'(x)} = \frac{1}{\cos x} = \frac{1}{\sqrt{1 - \sin^2 x}} = \frac{1}{\sqrt{1 - y^2}} \,.$$

Note that $\arccos + \arcsin$, having a derivative always equal to zero, is constant. Since its value at 0 is $\frac{\pi}{2}$, we have that

$$\arccos y + \arcsin y = \frac{\pi}{2} \qquad \text{for every } y \in [-1, 1] \,.$$

Let us consider now the function $H :] - \frac{\pi}{2}, \frac{\pi}{2}[\rightarrow \mathbb{R}$ defined as $H(x) = \tan x$. Considerations similar to those given previously show that it is invertible. We will call the function $H^{-1} : \mathbb{R} \rightarrow] - \frac{\pi}{2}, \frac{\pi}{2}[$ "arctangent," and we will write

$$H^{-1}(y) = \arctan y.$$

It is strictly increasing, with

$$\lim_{y \to -\infty} \arctan y = -\frac{\pi}{2}, \qquad \lim_{y \to +\infty} \arctan y = \frac{\pi}{2}.$$

Let us compute its derivative. Setting $y = H(x)$, for $x \in] - \frac{\pi}{2}, \frac{\pi}{2}[$ we have

$$(H^{-1})'(y) = \frac{1}{H'(x)} = \cos^2 x = \frac{1}{1 + \tan^2 x} = \frac{1}{1 + y^2}.$$

Let us now switch to the hyperbolic functions. The hyperbolic sine $\sinh : \mathbb{R} \rightarrow \mathbb{R}$ is strictly increasing and invertible. The inverse function can be written explicitly as

$$\sinh^{-1}(y) = \ln(y + \sqrt{y^2 + 1}).$$

The derivative of this function can be computed either directly or using the formula for the inverse function. If $y = \sinh(x)$, then

$$D \sinh^{-1}(y) = \frac{1}{D \sinh(x)} = \frac{1}{\cosh(x)} = \frac{1}{\sqrt{1 + \sinh^2(x)}} = \frac{1}{\sqrt{1 + y^2}}.$$

The hyperbolic cosine $\cosh : \mathbb{R} \rightarrow \mathbb{R}$ is neither injective (being even) nor surjective (since $\cosh x \geq 1$ for every $x \in \mathbb{R}$). On the other hand, the function $F : [0, +\infty[\rightarrow [1, +\infty[$, defined as $F(x) = \cosh x$, is strictly increasing and invertible. Its inverse function $F^{-1} : [1, +\infty[\rightarrow [0, +\infty[$ can be written explicitly as

$$F^{-1}(y) = \ln(y + \sqrt{y^2 - 1}).$$

It is often denoted, by an abuse of notation, by $\cosh^{-1}$. Let us compute its derivative. If $y = \cosh(x)$, with $x > 0$, then

$$D \cosh^{-1}(y) = \frac{1}{D \cosh(x)} = \frac{1}{\sinh(x)} = \frac{1}{\sqrt{\cosh^2(x) - 1}} = \frac{1}{\sqrt{y^2 - 1}}.$$

The function $\tanh : \mathbb{R} \rightarrow \mathbb{R}$ is strictly increasing, but it is not surjective, since $-1 < \tanh x < 1$ for every $x \in \mathbb{R}$. On the other hand, the function $H : \mathbb{R} \rightarrow] - 1, 1[$, defined as $H(x) = \tanh x$, is invertible, and its inverse $H^{-1} :]-1, 1[\to \mathbb{R}$ is given by

$$H^{-1}(y) = \frac{1}{2} \ln \left(\frac{1+y}{1-y} \right).$$

It is often denoted, by an abuse of notation, by $\tanh^{-1}$. Let us compute its derivative. If $y = \tanh(x)$, then

$$D \tanh^{-1}(y) = \frac{1}{D \tanh(x)} = \cosh^2(x) = \frac{1}{1 - \tanh^2(x)} = \frac{1}{1 - y^2}.$$

We can now return to the table of derivatives and enrich it with some of those found earlier.

$f(x)$	$f'(x)$	$f(x)$	$f'(x)$
x^α	$\alpha x^{\alpha-1}$	$\arccos x$	$-\dfrac{1}{\sqrt{1-x^2}}$
e^x	e^x	$\arcsin x$	$\dfrac{1}{\sqrt{1-x^2}}$
$\ln x$	$\dfrac{1}{x}$		
$\cos x$	$-\sin x$	$\arctan x$	$\dfrac{1}{1+x^2}$
$\sin x$	$\cos x$	$\cosh^{-1} x$	$\dfrac{1}{\sqrt{x^2-1}}$
$\tan x$	$\dfrac{1}{\cos^2 x}$	$\sinh^{-1} x$	$\dfrac{1}{\sqrt{x^2+1}}$
$\cosh x$	$\sinh x$		
$\sinh x$	$\cosh x$	$\tanh^{-1} x$	$\dfrac{1}{1-x^2}$
$\tanh x$	$\dfrac{1}{\cosh^2 x}$	$\cdots$	$\cdots$

6.5   Convexity and Concavity

As usual, in what follows, $I \subseteq \mathbb{R}$ will denote a nondegenerate interval.

We will say that a function $f : I \to \mathbb{R}$ is "convex" if, taking arbitrarily three points $x_1 < x_2 < x_3$ in I, the following inequality holds:

$$(a) \qquad \frac{f(x_2) - f(x_1)}{x_2 - x_1} \leq \frac{f(x_3) - f(x_2)}{x_3 - x_2}.$$

Let us show that inequality (a) is equivalent to the following ones:

$$(b) \qquad \frac{f(x_2) - f(x_1)}{x_2 - x_1} \leq \frac{f(x_3) - f(x_1)}{x_3 - x_1},$$

$$(c) \qquad \frac{f(x_3) - f(x_1)}{x_3 - x_1} \leq \frac{f(x_3) - f(x_2)}{x_3 - x_2}.$$

Indeed,

$$\frac{f(x_2) - f(x_1)}{x_2 - x_1} \leq \frac{f(x_3) - f(x_2)}{x_3 - x_2}$$

$$\Leftrightarrow (f(x_2) - f(x_1))(x_3 - x_2) \leq (f(x_3) - f(x_2))(x_2 - x_1)$$

$$\Leftrightarrow (f(x_2) - f(x_1))(x_3 - x_1 + x_1 - x_2) \leq (f(x_3) - f(x_1) + f(x_1) - f(x_2))(x_2 - x_1)$$

$$\Leftrightarrow (f(x_2) - f(x_1))(x_3 - x_1) \leq (f(x_3) - f(x_1))(x_2 - x_1)$$

$$\Leftrightarrow \frac{f(x_2) - f(x_1)}{x_2 - x_1} \leq \frac{f(x_3) - f(x_1)}{x_3 - x_1},$$

proving that $(a) \Leftrightarrow (b)$. The proof of the equivalence $(a) \Leftrightarrow (c)$ is analogous.

We may now observe that $f : I \to \mathbb{R}$ is convex if and only if, for every $x_0 \in I$, the difference quotient function $F : I \setminus \{x_0\} \to \mathbb{R}$, defined by

$$F(x) = \frac{f(x) - f(x_0)}{x - x_0},$$

is increasing. Indeed, taking x, x' in $I \setminus \{x_0\}$ such that $x < x'$, we can see that $F(x) \leq F(x')$ in all three possible cases: $x < x' < x_0$, or $x < x_0 < x'$, or $x_0 < x < x'$.

The following characterization of a convex differentiable function will now be of no surprise.

Theorem 6.14 *If $f : I \to \mathbb{R}$ is continuous, differentiable on $\mathring{I}$, then*

$$f \text{ is convex} \qquad \Leftrightarrow \qquad f' \text{ is increasing on } \mathring{I}.$$

Proof Assume that f is convex. Let $\alpha < \beta$ be two points in $\mathring{I}$. If $\alpha < x < \beta$, then, by (b), we have

$$\frac{f(x) - f(\alpha)}{x - \alpha} \leq \frac{f(\beta) - f(\alpha)}{\beta - \alpha},$$

whence, since f is differentiable at α,

$$f'(\alpha) = \lim_{x \to \alpha^+} \frac{f(x) - f(\alpha)}{x - \alpha} \leq \frac{f(\beta) - f(\alpha)}{\beta - \alpha} .$$

Analogously, by (c), we have

$$\frac{f(\beta) - f(\alpha)}{\beta - \alpha} \leq \frac{f(\beta) - f(x)}{\beta - x} .$$

whence, since f is differentiable at β,

$$f'(\beta) = \lim_{x \to \beta^-} \frac{f(\beta) - f(x)}{\beta - x} \geq \frac{f(\beta) - f(\alpha)}{\beta - \alpha} .$$

Then $f'(\alpha) \leq f'(\beta)$, showing that f' is increasing on $\mathring{I}$.

Conversely, assume f' to be increasing on $\mathring{I}$. Taking $x_1 < x_2 < x_3$ arbitrarily in I, by Lagrange's Theorem 6.11,

$$\exists\, \xi_1 \in]x_1, x_2[\colon\ f'(\xi_1) = \frac{f(x_2) - f(x_1)}{x_2 - x_1}$$

and

$$\exists\, \xi_2 \in]x_2, x_3[\colon\ f'(\xi_2) = \frac{f(x_3) - f(x_2)}{x_3 - x_2} .$$

Notice that $\xi_1 < \xi_2$. Since f' is increasing on $\mathring{I}$, it must be that $f'(\xi_1) \leq f'(\xi_2)$, thereby yielding inequality (a). ∎

We will say that f is "strictly convex" if, taking arbitrarily three points $x_1 < x_2 < x_3$ in I, we have that

$$(a')\qquad \frac{f(x_2) - f(x_1)}{x_2 - x_1} < \frac{f(x_3) - f(x_2)}{x_3 - x_2} .$$

Equivalently,

$$(b')\qquad \frac{f(x_2) - f(x_1)}{x_2 - x_1} < \frac{f(x_3) - f(x_1)}{x_3 - x_1}$$

or

$$(c')\qquad \frac{f(x_3) - f(x_1)}{x_3 - x_1} < \frac{f(x_3) - f(x_2)}{x_3 - x_2} .$$

The following characterization also holds true in this case.

Theorem 6.15 *If* $f : I \to \mathbb{R}$ *is continuous, differentiable on* $\overset{\circ}{I}$, *then*

$$f \text{ is strictly convex} \quad \Leftrightarrow \quad f' \text{ is strictly increasing on } \overset{\circ}{I} .$$

Proof We need to slightly modify the proof of the previous theorem. Assume that f is strictly convex, and let $\alpha < \beta$ be two points in $\overset{\circ}{I}$. If $\alpha < x < \frac{1}{2}(\alpha + \beta)$, by (b') we have

$$\frac{f(x) - f(\alpha)}{x - \alpha} < \frac{f(\frac{\alpha+\beta}{2}) - f(\alpha)}{\frac{\alpha+\beta}{2} - \alpha} < \frac{f(\beta) - f(\alpha)}{\beta - \alpha} ,$$

whence

$$f'(\alpha) = \lim_{x \to \alpha^+} \frac{f(x) - f(\alpha)}{x - \alpha} \le \frac{f(\frac{\alpha+\beta}{2}) - f(\alpha)}{\frac{\alpha+\beta}{2} - \alpha} < \frac{f(\beta) - f(\alpha)}{\beta - \alpha} .$$

Analogously, if $\frac{1}{2}(\alpha + \beta) < x < \beta$, then, by (c'), we have

$$\frac{f(\beta) - f(\alpha)}{\beta - \alpha} < \frac{f(\beta) - f(\frac{\alpha+\beta}{2})}{\beta - \frac{\alpha+\beta}{2}} < \frac{f(\beta) - f(x)}{\beta - x} ,$$

whence

$$f'(\beta) = \lim_{x \to \beta^-} \frac{f(\beta) - f(x)}{\beta - x} \ge \frac{f(\beta) - f(\frac{\alpha+\beta}{2})}{\beta - \frac{\alpha+\beta}{2}} > \frac{f(\beta) - f(\alpha)}{\beta - \alpha} .$$

Then $f'(\alpha) < f'(\beta)$, thereby proving that f' is strictly increasing on $\overset{\circ}{I}$.

Conversely, assume f' to be strictly increasing on $\overset{\circ}{I}$. Taking $x_1 < x_2 < x_3$ in I, by Lagrange's Theorem 6.11, exactly as in the proof of the previous theorem, we obtain inequality (a'). ∎

We will say that f is "concave" if the function $(-f)$ is convex or, equivalently, the opposite inequality in (a) (or in (b) or in (c)) holds. Analogously, we will say that f is "strictly concave" if the function $(-f)$ is strictly convex or, equivalently, the opposite inequality in (a') (or in (b') or in (c')) holds. Clearly enough, analogous theorems can be written characterizing either the concavity or the strict concavity of f when f is differentiable and f' is either decreasing or strictly decreasing, respectively.

We can now state the following corollary, which is widely applied in practice.

Corollary 6.16 *Let I be an interval and $f : I \rightarrow \mathbb{R}$ be a continuous function, twice differentiable on $\overset{\circ}{I}$. The following propositions hold:*

(a) *If $f''(x) \geq 0$ for every $x \in \overset{\circ}{I}$, then f is convex.*
(b) *If $f''(x) > 0$ for every $x \in \overset{\circ}{I}$, then f is strictly convex.*
(c) *If $f''(x) \leq 0$ for every $x \in \overset{\circ}{I}$, then f is concave.*
(d) *If $f''(x) < 0$ for every $x \in \overset{\circ}{I}$, then f is strictly concave.*

Proof Let us prove (a). Since $f''(x) \geq 0$ for every $x \in \overset{\circ}{I}$, by Corollary 6.12, the function $f' : \overset{\circ}{I} \rightarrow \mathbb{R}$ is increasing. Hence, by Theorem 6.14, the function $f : I \rightarrow \mathbb{R}$ is convex. The other properties follow similarly.　∎

Recalling Remark 6.13, we can observe that in (a) and (c) the implications can be reversed. If f is convex, then $f''(x) \geq 0$ for every $x \in \overset{\circ}{I}$, and similarly, if f is concave, then $f''(x) \leq 0$ for every $x \in \overset{\circ}{I}$. But this is not permitted either in (b), as the example $f(x) = x^4$ shows, or in (d).

Example 1 The exponential function $f(x) = e^x$ is strictly convex since

$$f''(x) = e^x > 0, \quad \text{for every } x \in \mathbb{R}.$$

Its inverse $\ln(x)$, the natural logarithm, is strictly concave.

Example 2 Since $D^2 \cos x = -\cos x$ and $D^2 \sin x = -\sin x$, recalling the sign properties of these functions, we have that

$$\cos \text{ is} \begin{cases} \text{strictly concave on } \left[-\dfrac{\pi}{2}, \dfrac{\pi}{2}\right], \\[2ex] \text{strictly convex on } \left[\dfrac{\pi}{2}, \dfrac{3\pi}{2}\right], \end{cases}$$

$$\sin \text{ is} \begin{cases} \text{strictly concave on } [0, \pi], \\ \text{strictly convex on } [\pi, 2\pi]. \end{cases}$$

The points separating an interval where the function is convex from an interval where it is concave are called "inflexion points." For the cosine function, the set of inflexion points is $\{\frac{\pi}{2} + k\pi : k \in \mathbb{Z}\}$, whereas for the sine function it is $\{k\pi : k \in \mathbb{Z}\}$.

A similar analysis can be made on all the other elementary functions introduced till now.

The following property of a differentiable convex function can be useful. It states, roughly speaking, that its graph always lies above any of its tangents.

Theorem 6.17 *If* $f : I \to \mathbb{R}$ *is convex and it is differentiable at some point* $x_0 \in I$, *then*

$$f(x) \geq f'(x_0)(x - x_0) + f(x_0), \quad \text{for every } x \in I.$$

Proof The inequality surely holds if $x = x_0$. Thus, let us assume $x \neq x_0$.

If $x > x_0$, taking $h > 0$ such that $x_0 < x_0 + h < x$, then, by the convexity of f, we have that

$$\frac{f(x) - f(x_0)}{x - x_0} \geq \frac{f(x_0 + h) - f(x_0)}{h}.$$

Taking the limit as $h \to 0$ we find

$$\frac{f(x) - f(x_0)}{x - x_0} \geq f'(x_0),$$

thereby leading to the inequality we want to prove.

On the other hand, if $x < x_0$, taking $h < 0$ such that $x < x_0 + h < x_0$, then, by the convexity of f, we have that

$$\frac{f(x_0) - f(x)}{x_0 - x} \leq \frac{f(x_0) - f(x_0 + h)}{-h},$$

i.e.,

$$\frac{f(x) - f(x_0)}{x - x_0} \leq \frac{f(x_0 + h) - f(x_0)}{h}.$$

Taking the limit as $h \to 0$, the conclusion follows as well. ∎

6.6 L'Hôpital's Rules

We first need to prove the following generalization of the Lagrange Theorem 6.11.

Theorem 6.18 (Cauchy Theorem) *If* $f, g : [a, b] \to \mathbb{R}$ *are two continuous functions, differentiable on* $]a, b[$, *with* $g'(x) \neq 0$ *for every* $x \in]a, b[$, *then there exists a point* $\xi \in]a, b[$ *such that*

$$\frac{f'(\xi)}{g'(\xi)} = \frac{f(b) - f(a)}{g(b) - g(a)}.$$

Proof We define the function $h : [a, b] \to \mathbb{R}$ as

$$h(x) = (g(b) - g(a))f(x) - (f(b) - f(a))g(x).$$

It is continuous on $[a, b]$, differentiable on $]a, b[$, and such that $h(a) = h(b)$. Then Rolle's Theorem 6.10 guarantees the existence of a point $\xi \in]a, b[$ where $h'(\xi) = 0$, whence the conclusion.   ∎

Notice that Lagrange's Theorem 6.11 can now be seen as a corollary of Cauchy's theorem by taking $g(x) = x$.

In the remainder of the book, it will be convenient to adopt the following notation. Whenever a is greater than b, the symbol $[a, b]$ indicates the interval $[b, a]$, and $]a, b[$ indicates $]b, a[$. Note that the statement of Cauchy's Theorem 6.18 remains valid in this case as well.

The following result is known as "L'Hôpital's rule in the indeterminate case $\frac{0}{0}$."

Theorem 6.19 (L'Hôpital Theorem—I) *Let I be an interval containing a point x_0, and let $f, g : I \setminus \{x_0\} \to \mathbb{R}$ be two differentiable functions, with $g'(x) \neq 0$ for every $x \in I \setminus \{x_0\}$, such that*

$$\lim_{x \to x_0} f(x) = \lim_{x \to x_0} g(x) = 0.$$

If the limit

$$\lim_{x \to x_0} \frac{f'(x)}{g'(x)}$$

exists, then the limit

$$\lim_{x \to x_0} \frac{f(x)}{g(x)}$$

also exists, and the two coincide.

Proof Set $l = \lim\limits_{x \to x_0} \dfrac{f'(x)}{g'(x)}$ (allowing the possibility that $l = +\infty$ or $-\infty$). Let us extend the two functions at the point x_0 by setting $f(x_0) = g(x_0) = 0$; in this way, f and g will be continuous on the whole interval I. By Cauchy's Theorem 6.18, for every $x \neq x_0$ in I there is a point $\xi_x \in]x_0, x[$ (depending on x) such that

$$\frac{f'(\xi_x)}{g'(\xi_x)} = \frac{f(x) - f(x_0)}{g(x) - g(x_0)} = \frac{f(x)}{g(x)}.$$

Notice that $\lim_{x \to x_0} \xi_x = x_0$. Then, using the change of variables formula (3.1),

$$\lim_{x \to x_0} \frac{f(x)}{g(x)} = \lim_{x \to x_0} \frac{f'(\xi_x)}{g'(\xi_x)} = \lim_{\substack{y \to \lim \xi_x \\ x \to x_0}} \frac{f'(y)}{g'(y)} = \lim_{y \to x_0} \frac{f'(y)}{g'(y)} = l \,,$$

and the proof is complete. ∎

Note that the preceding theorem does not exclude the possibility of x_0 being an endpoint of the interval I, in which case we are dealing with left or right limits.

We also observe that the conclusion of the statement is written as an implication: If the limit of the quotient of the two derivatives exists, then the limit of the quotient of the two functions exists. The opposite implication is not true, as we can see in the following example. Let $x_0 = 0$,

$$f(x) = x^2 \sin\left(\frac{1}{x}\right), \qquad g(x) = x \,.$$

Then $\lim_{x \to 0} f(x) = \lim_{x \to 0} g(x) = 0$,

$$\lim_{x \to 0} \frac{f(x)}{g(x)} = \lim_{x \to 0} x \sin\left(\frac{1}{x}\right) = 0 \,,$$

while

$$\frac{f'(x)}{g'(x)} = 2x \sin\left(\frac{1}{x}\right) - \cos\left(\frac{1}{x}\right),$$

so that the limit $\lim_{x \to 0} \frac{f'(x)}{g'(x)}$ does not exist.

As an example of the application of L'Hôpital's rule, let $I = \mathbb{R}$, $x_0 = 0$, $f(x) = \sin x - x$, and $g(x) = x^3$. Then, from

$$\lim_{x \to 0} \frac{f'(x)}{g'(x)} = \lim_{x \to 0} \frac{\cos x - 1}{3x^2} = -\frac{1}{6} \,,$$

we deduce that

$$\lim_{x \to 0} \frac{\sin x - x}{x^3} = -\frac{1}{6} \,.$$

The following corollary can be useful in determining whether a function is differentiable at some point x_0.

Corollary 6.20 *Let I be an interval containing a point x_0, and let $f : I \to \mathbb{R}$ be a continuous function, differentiable at all $x \neq x_0$. If the limit*

$$l = \lim_{x \to x_0} f'(x)$$

exists, then the derivative of f at x_0 exists, and it coincides with l.

Proof Let $F(x) = f(x) - f(x_0)$ and $G(x) = x - x_0$. We have that $G'(x) \neq 0$ for every $x \neq x_0$,

$$\lim_{x \to x_0} F(x) = \lim_{x \to x_0} G(x) = 0,$$

and

$$\lim_{x \to x_0} \frac{F'(x)}{G'(x)} = \lim_{x \to x_0} f'(x) = l.$$

By L'Hôpital's rule we have that

$$\lim_{x \to x_0} \frac{f(x) - f(x_0)}{x - x_0} = \lim_{x \to x_0} \frac{F(x)}{G(x)} = l,$$

i.e., $f'(x_0) = l$. ∎

L'Hôpital's rule can be extended to cases where $x_0 = +\infty$ or $-\infty$. Let us analyze here the first case; the other one is analogous.

Theorem 6.21 (L'Hôpital Theorem—II) *Let I be an interval, unbounded from above, and let $f, g : I \to \mathbb{R}$ be two differentiable functions, with $g'(x) \neq 0$ for every $x \in I$, such that*

$$\lim_{x \to +\infty} f(x) = \lim_{x \to +\infty} g(x) = 0.$$

If the limit

$$\lim_{x \to +\infty} \frac{f'(x)}{g'(x)}$$

exists, then the limit

$$\lim_{x \to +\infty} \frac{f(x)}{g(x)}$$

also exists, and the two coincide.

Proof Set $l = \lim\limits_{x \to +\infty} \dfrac{f'(x)}{g'(x)}$. Defining the two functions $F(x) = f(x^{-1})$ and $G(x) = g(x^{-1})$, we see that $G'(x) \neq 0$ for every x, and

$$\lim_{x \to 0^+} F(x) = \lim_{x \to 0^+} G(x) = 0.$$

Moreover,

$$\lim_{x \to 0^+} \frac{F'(x)}{G'(x)} = \lim_{x \to 0^+} \frac{f'(x^{-1})(-x^{-2})}{g'(x^{-1})(-x^{-2})} = \lim_{x \to 0^+} \frac{f'(x^{-1})}{g'(x^{-1})} = \lim_{y \to +\infty} \frac{f'(y)}{g'(y)} = l.$$

Then, by Theorem 6.19, $\lim\limits_{x \to 0^+} \dfrac{F(x)}{G(x)} = l$, and hence

$$\lim_{x \to +\infty} \frac{f(x)}{g(x)} = \lim_{u \to 0^+} \frac{f(u^{-1})}{g(u^{-1})} = \lim_{u \to 0^+} \frac{F(u)}{G(u)} = l,$$

thereby proving the result.   ∎

We will now state what is called "L'Hôpital's rule in the indeterminate case $\frac{\infty}{\infty}$."
In the following theorem, ∞ can be either $+\infty$ or $-\infty$.

Theorem 6.22 (L'Hôpital Theorem—III) *Let I be an interval containing a point x_0, and let $f, g : I \setminus \{x_0\} \to \mathbb{R}$ be two differentiable functions, with $g'(x) \neq 0$ for every $x \in I \setminus \{x_0\}$, such that*

$$\lim_{x \to x_0} f(x) = \lim_{x \to x_0} g(x) = \infty.$$

If the limit

$$\lim_{x \to x_0} \frac{f'(x)}{g'(x)}$$

exists, then the limit

$$\lim_{x \to x_0} \frac{f(x)}{g(x)}$$

also exists, and the two coincide.

Proof Set $l = \lim\limits_{x \to x_0} \dfrac{f'(x)}{g'(x)}$. We first assume that $l \in \mathbb{R}$, and that x_0 is not the right endpoint of I. Let $\varepsilon > 0$ be fixed. Then there exists a $\delta_1 > 0$ such that

$$x_0 < x < x_0 + \delta_1 \;\Rightarrow\; \left| \frac{f'(x)}{g'(x)} - l \right| \le \frac{\varepsilon}{2}.$$

By Cauchy's Theorem 6.18, for every $x \in \,]x_0, x_0 + \delta_1[$ there is a $\xi_x \in \,]x, x_0 + \delta_1[$ such that

$$\frac{f'(\xi_x)}{g'(\xi_x)} = \frac{f(x_0 + \delta_1) - f(x)}{g(x_0 + \delta_1) - g(x)},$$

and hence

$$x_0 < x < x_0 + \delta_1 \;\Rightarrow\; \left| \frac{f(x_0 + \delta_1) - f(x)}{g(x_0 + \delta_1) - g(x)} - l \right| \le \frac{\varepsilon}{2}.$$

We can moreover assume that δ_1 was chosen small enough so that

$$x_0 < x < x_0 + \delta_1 \;\Rightarrow\; f(x) \ne 0 \ \text{ and } \ g(x) \ne 0.$$

Let us write

$$\frac{f(x_0 + \delta_1) - f(x)}{g(x_0 + \delta_1) - g(x)} = \psi(x) \frac{f(x)}{g(x)}$$

and observe that

$$\lim_{x \to x_0} \psi(x) = \lim_{x \to x_0} \frac{1 - f(x_0 + \delta_1)/f(x)}{1 - g(x_0 + \delta_1)/g(x)} = 1.$$

In particular,

$$\lim_{x \to x_0} \frac{1}{\psi(x)} \left(l - \frac{\varepsilon}{2} \right) = l - \frac{\varepsilon}{2}, \qquad \lim_{x \to x_0} \frac{1}{\psi(x)} \left(l + \frac{\varepsilon}{2} \right) = l + \frac{\varepsilon}{2},$$

so that there exists a $\delta \in \,]0, \delta_1[$ such that, if $x_0 < x < x_0 + \delta$, then $\psi(x) > 0$ and

$$\frac{1}{\psi(x)} \left(l - \frac{\varepsilon}{2} \right) \ge l - \varepsilon, \qquad \frac{1}{\psi(x)} \left(l + \frac{\varepsilon}{2} \right) \le l + \varepsilon.$$

Therefore, if $x_0 < x < x_0 + \delta$, then

$$l - \varepsilon \le \frac{1}{\psi(x)} \left(l - \frac{\varepsilon}{2} \right) \le \frac{1}{\psi(x)} \frac{f(x_0 + \delta_1) - f(x)}{g(x_0 + \delta_1) - g(x)} \le \frac{1}{\psi(x)} \left(l + \frac{\varepsilon}{2} \right) \le l + \varepsilon,$$

and hence

$$\left| \frac{f(x)}{g(x)} - l \right| \leq \varepsilon .$$

We have thus proved that

$$\lim_{x \to x_0^+} \frac{f(x)}{g(x)} = l .$$

In a perfectly analogous way one proves that, if x_0 is not the left endpoint of I, then

$$\lim_{x \to x_0^-} \frac{f(x)}{g(x)} = l ,$$

so that the theorem is proved in the case where $l \in \mathbb{R}$.

Assume now that $l = +\infty$ and that x_0 is not the right endpoint of I. Let $\alpha > 0$ be fixed. Then there exists a $\delta_1 > 0$ such that

$$x_0 < x < x_0 + \delta_1 \;\Rightarrow\; \frac{f'(x)}{g'(x)} \geq 2\alpha .$$

Proceeding as previously, we have that

$$x_0 < x < x_0 + \delta_1 \;\Rightarrow\; \frac{f(x_0 + \delta_1) - f(x)}{g(x_0 + \delta_1) - g(x)} \geq 2\alpha .$$

We can, moreover, assume that δ_1 has been chosen small enough so that

$$x_0 < x < x_0 + \delta_1 \;\Rightarrow\; f(x) \neq 0 \;\text{ and }\; g(x) \neq 0 .$$

Let $\psi(x)$ be defined as previously. There exists a $\delta \in \,]0, \delta_1[$ such that

$$x_0 < x < x_0 + \delta \;\Rightarrow\; 0 < \psi(x) < 2 .$$

Therefore, if $x_0 < x < x_0 + \delta$, then

$$\frac{1}{\psi(x)} \frac{f(x_0 + \delta_1) - f(x)}{g(x_0 + \delta_1) - g(x)} \geq \frac{1}{\psi(x)} 2\alpha \geq \alpha ,$$

and hence

$$\frac{f(x)}{g(x)} \geq \alpha .$$

We have thus proved that

$$\lim_{x \to x_0^+} \frac{f(x)}{g(x)} = +\infty \,.$$

In a perfectly analogous way one proves that, if x_0 is not the left endpoint of I, then

$$\lim_{x \to x_0^-} \frac{f(x)}{g(x)} = +\infty \,,$$

so that the theorem is proved also in the case $l = +\infty$. Finally, the case $l = -\infty$ can be ruled out observing that a change of sign in one of the two functions leads back to the previously proved case. ∎

Example We want to compute

$$\lim_{x \to 0^+} x \ln x \,.$$

Setting $f(x) = \ln x$ and $g(x) = 1/x$, we see that $\lim_{x \to 0^+} f(x) = -\infty$ and $\lim_{x \to 0^+} g(x) = +\infty$. Moreover,

$$\lim_{x \to 0^+} \frac{f'(x)}{g'(x)} = \lim_{x \to 0^+} \frac{1/x}{-1/x^2} = \lim_{x \to 0^+} (-x) = 0 \,.$$

Hence, also

$$\lim_{x \to 0^+} x \ln x = \lim_{x \to 0^+} \frac{f(x)}{g(x)} = 0 \,.$$

Even in the indeterminate case $\frac{\infty}{\infty}$ we can extend L'Hôpital's rule to cases where $x_0 = +\infty$ or $-\infty$. Let us see, e.g., the first case.

Theorem 6.23 (L'Hôpital Theorem—IV) *Let I be an interval, unbounded from above, and let $f, g : I \to \mathbb{R}$ be two differentiable functions, with $g'(x) \neq 0$ for every $x \in I$, such that*

$$\lim_{x \to +\infty} f(x) = \lim_{x \to +\infty} g(x) = \infty \,.$$

If the limit

$$\lim_{x \to +\infty} \frac{f'(x)}{g'(x)}$$

exists, then the limit

$$\lim_{x \to +\infty} \frac{f(x)}{g(x)}$$

also exists, and the two coincide.

The proof is analogous to that of Theorem 6.21, so we omit it for brevity's sake.

6.7   Taylor Formula

The following theorem provides us the so-called "Taylor formula with Lagrange's form of the remainder."

Theorem 6.24 (Taylor Theorem—I) *Let $x \neq x_0$ be two points of an interval I and $f : I \to \mathbb{R}$ be $n + 1$ times differentiable. Then there exists a $\xi \in]x_0, x[$ such that*

$$f(x) = p_n(x) + r_n(x) ,$$

where

$$p_n(x) = f(x_0) + f'(x_0)(x - x_0) + \frac{1}{2!} f''(x_0)(x - x_0)^2 + \cdots + \frac{1}{n!} f^{(n)}(x_0)(x - x_0)^n$$

is the "nth-order Taylor polynomial associated with the function f at the point x_0," and

$$r_n(x) = \frac{1}{(n + 1)!} f^{(n+1)}(\xi)(x - x_0)^{n+1}$$

is the "Lagrange form of the remainder."

Proof We first observe that the polynomial p_n satisfies the following properties:

$$\begin{cases} p_n(x_0) = f(x_0) , \\ p_n'(x_0) = f'(x_0) , \\ p_n''(x_0) = f''(x_0) , \\ \vdots \\ p_n^{(n)}(x_0) = f^{(n)}(x_0) . \end{cases}$$

By Cauchy's Theorem 6.18, we can find a point $\xi_1 \in]x_0, x[$ such that

$$\frac{f(x) - p_n(x)}{(x - x_0)^{n+1}} = \frac{(f(x) - p_n(x)) - (f(x_0) - p_n(x_0))}{(x - x_0)^{n+1} - (x_0 - x_0)^{n+1}}$$

$$= \frac{f'(\xi_1) - p'_n(\xi_1)}{(n+1)(\xi_1 - x_0)^n} .$$

Again by Cauchy's Theorem 6.18, we can find a point $\xi_2 \in]x_0, \xi_1[$ such that

$$\frac{f'(\xi_1) - p'_n(\xi_1)}{(n+1)(\xi_1 - x_0)^n} = \frac{(f'(\xi_1) - p'_n(\xi_1)) - (f'(x_0) - p'_n(x_0))}{(n+1)(\xi_1 - x_0)^n - (n+1)(x_0 - x_0)^n}$$

$$= \frac{f''(\xi_2) - p''_n(\xi_2)}{(n+1)n(\xi_2 - x_0)^{n-1}} .$$

Proceeding by induction, we find $n + 1$ points $\xi_1, \xi_2, \ldots, \xi_{n+1}$ such that

$$\frac{f(x) - p_n(x)}{(x - x_0)^{n+1}} = \frac{f'(\xi_1) - p'_n(\xi_1)}{(n+1)(\xi_1 - x_0)^n}$$

$$= \frac{f''(\xi_2) - p''_n(\xi_2)}{(n+1)n(\xi_2 - x_0)^{n-1}}$$

$$\vdots$$

$$= \frac{f^{(n+1)}(\xi_{n+1}) - p_n^{(n+1)}(\xi_{n+1})}{(n+1)!(\xi_{n+1} - x_0)^0} .$$

If $x > x_0$, these points satisfy the inequalities

$$x_0 < \xi_{n+1} < \xi_n < \cdots < \xi_2 < \xi_1 < x ,$$

whereas if $x < x_0$, they are in the opposite order. Since the $(n + 1)$th derivative of an nth-order polynomial is constantly equal to zero, we have that $p_n^{(n+1)}(\xi_{n+1}) = 0$, and setting $\xi = \xi_{n+1}$ we conclude.   ∎

If $n = 0$, then the preceding Taylor Formula is simply

$$f(x) = f(x_0) + f'(\xi)(x - x_0), \quad \text{for some } \xi \in]x_0, x[,$$

which is the outcome of Lagrange's Theorem 6.11.

Note that the Taylor polynomial

$$p_n(x) = \sum_{k=0}^{n} \frac{f^{(k)}(x_0)}{k!}(x - x_0)^k \tag{6.3}$$

could have a degree smaller than n (here $f^{(0)}$ simply denotes f). For example, if f is a constant function, then the degree of $p_n(x)$ is equal to 0.

Examples Let us now determine the Taylor polynomial of some elementary functions, taking for simplicity $x_0 = 0$ (in which case it is sometimes called a "Maclaurin polynomial").

1. Let $f(x) = e^x$. Then

$$p_n(x) = 1 + x + \frac{x^2}{2!} + \frac{x^3}{3!} + \cdots + \frac{x^n}{n!} = \sum_{k=0}^{n} \frac{x^k}{k!}.$$

2. Let $f(x) = \cos x$. Then, if either $n = 2m$ or $n = 2m + 1$,

$$p_n(x) = 1 - \frac{x^2}{2!} + \frac{x^4}{4!} - \frac{x^6}{6!} + \cdots + (-1)^m \frac{x^{2m}}{(2m)!} = \sum_{k=0}^{m} (-1)^k \frac{x^{2k}}{(2k)!}.$$

3. Let $f(x) = \sin x$. Then, if either $n = 2m + 1$ or $n = 2m + 2$,

$$p_n(x) = x - \frac{x^3}{3!} + \frac{x^5}{5!} - \frac{x^7}{7!} + \cdots + (-1)^m \frac{x^{2m+1}}{(2m+1)!} = \sum_{k=0}^{m} (-1)^k \frac{x^{2k+1}}{(2k+1)!}.$$

4. Now let $f(x) = \frac{1}{1-x}$. It can be shown by induction that

$$f^{(n)}(x) = \frac{n!}{(1-x)^{n+1}}.$$

Then $f^{(n)}(0) = n!$, and hence

$$p_n(x) = 1 + x + x^2 + x^3 + \cdots + x^n.$$

5. We proceed similarly for the function $f(x) = \frac{1}{1+x}$ and find

$$p_n(x) = 1 - x + x^2 - x^3 + \cdots + (-1)^n x^n.$$

6. Consider now the function $f(x) = \ln(1 + x)$. Its derivative coincides with the previous function, and we easily obtain

$$p_n(x) = x - \frac{x^2}{2} + \frac{x^3}{3} - \frac{x^4}{4} + \cdots + (-1)^{n-1} \frac{x^n}{n}.$$

7. Another example where the Taylor polynomial has an explicit formula is given by the function $f(x) = \frac{1}{1+x^2}$. If either $n = 2m$ or $n = 2m + 1$, then we have

$$p_n(x) = 1 - x^2 + x^4 - x^6 + \cdots + (-1)^m x^{2m}.$$

8. At this point it is easy to deal with the function $f(x) = \arctan x$, whose derivative is the previous function. If either $n = 2m + 1$ or $n = 2m + 2$, then

$$p_n(x) = x - \frac{x^3}{3} + \frac{x^5}{5} - \frac{x^7}{7} + \cdots + (-1)^m \frac{x^{2m+1}}{2m + 1}.$$

In a similar way we can find the Taylor polynomials of the hyperbolic functions $\cosh x$, $\sinh x$, and that of $\tanh^{-1} x$. The following summary table may be useful:

$f(x)$	$p_n(x)$ at the point $x_0 = 0$
e^x	$1 + x + \dfrac{x^2}{2!} + \dfrac{x^3}{3!} + \cdots + \dfrac{x^n}{n!}$
$\ln(1 + x)$	$x - \dfrac{x^2}{2} + \dfrac{x^3}{3} - \dfrac{x^4}{4} + \cdots + \dfrac{(-1)^{n-1} x^n}{n}$
$\cos x$	$1 - \dfrac{x^2}{2!} + \dfrac{x^4}{4!} - \dfrac{x^6}{6!} + \cdots + (-1)^m \dfrac{x^{2m}}{(2m)!}$
$\sin x$	$x - \dfrac{x^3}{3!} + \dfrac{x^5}{5!} - \dfrac{x^7}{7!} + \cdots + (-1)^m \dfrac{x^{2m+1}}{(2m+1)!}$
$\arctan x$	$x - \dfrac{x^3}{3} + \dfrac{x^5}{5} - \dfrac{x^7}{7} + \cdots + (-1)^m \dfrac{x^{2m+1}}{2m+1}$
$\cosh x$	$1 + \dfrac{x^2}{2!} + \dfrac{x^4}{4!} + \dfrac{x^6}{6!} + \cdots + \dfrac{x^{2m}}{(2m)!}$
$\sinh x$	$x + \dfrac{x^3}{3!} + \dfrac{x^5}{5!} + \dfrac{x^7}{7!} + \cdots + \dfrac{x^{2m+1}}{(2m+1)!}$
$\tanh^{-1} x$	$x + \dfrac{x^3}{3} + \dfrac{x^5}{5} + \dfrac{x^7}{7} + \cdots + \dfrac{x^{2m+1}}{2m+1}$

On the other hand, there is no elementary expression for the Taylor polynomial of the functions $\tan x$ and $\tanh x$. We report here only the first few terms:

$$\tan x \quad x + \frac{x^3}{3} + \frac{2x^5}{15} + \frac{17x^7}{315} + \cdots$$

$$\tanh x \quad x - \frac{x^3}{3} + \frac{2x^5}{15} - \frac{17x^7}{315} + \cdots$$

6.8   Local Maxima and Minima

Assuming x_0 to be fixed, we would like to take some limit in the Taylor formula as x tends to x_0. Hence, for every $x \neq x_0$, to emphasize the fact that the point $\xi \in]x_0, x[$ in the Taylor formula depends on x, we will write $\xi = \xi_x$.

Whenever $f^{(n+1)}$ happens to be *bounded in a neighbourhood of* x_0, we see that

$$\lim_{x \to x_0} \frac{r_n(x)}{(x - x_0)^n} = \lim_{x \to x_0} \frac{1}{(n+1)!} f^{(n+1)}(\xi_x)(x - x_0) = 0 \, .$$

This relation is sometimes written using the following notation:

$$r_n(x) = o(|x - x_0|^n) \quad \text{if } x \to x_0 \, .$$

This is surely true if $f^{(n+1)}$ is *continuous at* x_0, in which case

$$\lim_{x \to x_0} \frac{r_n(x)}{(x - x_0)^{n+1}} = \lim_{x \to x_0} \frac{1}{(n+1)!} f^{(n+1)}(\xi_x) = \frac{1}{(n+1)!} f^{(n+1)}(x_0) \, .$$

Theorem 6.25 *Let $f \in C^2(I, \mathbb{R})$, and assume that x_0 is an internal point of I. If $f'(x_0) = 0$ and $f''(x_0) > 0$, then x_0 is a local minimum point for f. On the other hand, if $f'(x_0) = 0$ and $f''(x_0) < 0$, then x_0 is a local maximum point for f.*

Proof Let us prove the first statement; the second one is analogous. Using the Taylor formula with $n = 1$, we have that

$$\lim_{x \to x_0} \frac{f(x) - f(x_0)}{(x - x_0)^2} = \lim_{x \to x_0} \frac{f(x) - [f(x_0) + f'(x_0)(x - x_0)]}{(x - x_0)^2}$$

$$= \lim_{x \to x_0} \frac{r_1(x)}{(x - x_0)^2} = \frac{1}{2!} f''(x_0) > 0 \, .$$

Hence, using sign permanence, there is a neighborhood U of x_0 such that $f(x) > f(x_0)$ for every $x \in U \setminus \{x_0\}$. ∎

Whenever $f'(x_0) = 0$ and $f''(x_0) = 0$, we will need further information. Always assuming that x_0 is an internal point of I and that f is sufficiently regular, it can be seen that if $f'''(x_0) \neq 0$, then x_0 will be neither a local minimum nor a local maximum point for f. On the other hand, if $f'''(x_0) = 0$, then

$$\text{if } f'(x_0) = f''(x_0) = f'''(x_0) = 0 \text{ and } f''''(x_0) > 0,$$

$$\text{then } x_0 \text{ is a local minimum point,}$$

whereas

$$\text{if } f'(x_0) = f''(x_0) = f'''(x_0) = 0 \text{ and } f''''(x_0) < 0,$$

$$\text{then } x_0 \text{ is a local maximum point.}$$

This procedure can be continued, of course, but we avoid the details for brevity's sake.

6.9 Analyticity of Some Elementary Functions

Somewhat surprisingly, the Taylor polynomial at x_0 may be a good approximation of a function even at distant points x if the degree is taken large enough. An example follows, provided by the exponential function.

Theorem 6.26 *For every $x \in \mathbb{R}$, we have that*

$$e^x = \lim_n \left(1 + x + \frac{x^2}{2!} + \frac{x^3}{3!} + \cdots + \frac{x^n}{n!} \right).$$

Proof The formula clearly holds when $x = 0$. Assuming $x \neq 0$, by Taylor's Theorem 6.24 there exists a $\xi \in]0, x[$ such that $f(x) = p_n(x) + r_n(x)$, with

$$r_n(x) = e^\xi \frac{x^{n+1}}{(n+1)!}.$$

We want to prove that $\lim_n r_n(x) = 0$. Notice that for any $x \in \mathbb{R}$,

$$|r_n(x)| \leq e^{|x|} \frac{|x|^{n+1}}{(n+1)!}.$$

Since we proved in Theorem 3.30 that $\lim_n \frac{a^n}{n!} = 0$ for every $a \in \mathbb{R}$, the conclusion follows. ∎

Instead of

$$e^x = \lim_{n \to +\infty} \sum_{k=0}^{n} \frac{x^k}{k!},$$

we will briefly write

$$e^x = \sum_{k=0}^{\infty} \frac{x^k}{k!}.$$

This is the "Taylor series" associated with the exponential function at the point $x_0 = 0$.

A similar phenomenon holds for the cosine and sine functions.

Theorem 6.27 *For every $x \in \mathbb{R}$, we have that*

$$\cos x = \lim_m \left(1 - \frac{x^2}{2!} + \frac{x^4}{4!} - \frac{x^6}{6!} + \cdots + (-1)^m \frac{x^{2m}}{(2m)!} \right),$$

$$\sin x = \lim_m \left(x - \frac{x^3}{3!} + \frac{x^5}{5!} - \frac{x^7}{7!} + \cdots + (-1)^m \frac{x^{2m+1}}{(2m+1)!} \right).$$

Proof Following the lines of the previous proof and using the fact that $|\cos \xi| \leq 1$ and $|\sin \xi| \leq 1$ for every $\xi \in \mathbb{R}$, we see that

$$|r_n(x)| \leq \frac{|x|^{n+1}}{(n+1)!}.$$

Since $\lim_n \dfrac{a^n}{n!} = 0$ for every $a \in \mathbb{R}$, the conclusion follows. ∎

We will briefly write

$$\cos x = \sum_{k=0}^{\infty} (-1)^k \frac{x^{2k}}{(2k)!}, \qquad \sin x = \sum_{k=0}^{\infty} (-1)^k \frac{x^{2k+1}}{(2k+1)!}.$$

Similarly, one can prove that

$$\cosh x = \sum_{k=0}^{\infty} \frac{x^{2k}}{(2k)!}, \qquad \sinh x = \sum_{k=0}^{\infty} \frac{x^{2k+1}}{(2k+1)!}.$$

The functions $f \in C^{\infty}(I, \mathbb{R})$ for which $f(x) = \lim_n p_n(x)$ for every $x \in I$ are called "analytic" on I. This is not the case for every function. For instance, the function $f : \mathbb{R} \to \mathbb{R}$, defined as

$$f(x) = \begin{cases} e^{-1/x^2} & \text{if } x \neq 0, \\ 0 & \text{if } x = 0, \end{cases}$$

is infinitely differentiable, and $f^{(k)}(0) = 0$ for every $k \in \mathbb{N}$, hence $p_n(x)$ is identically equal to zero. The reader is invited to verify this.

The Integral

7

In this chapter, we denote by I a *compact* interval of the real line $\mathbb{R}$, i.e.,

$$I = [a, b], \quad \text{for some } a < b.$$

7.1 Riemann Sums

First, we choose in I some points

$$a = a_0 < a_1 < \cdots < a_{m-1} < a_m = b,$$

thereby obtaining a "partition" of I made by the intervals $[a_{j-1}, a_j]$, with $j = 1, \ldots, m$. Then, for each j, we choose a point

$$x_j \in [a_{j-1}, a_j].$$

A "tagged partition" of I is the set

$$\mathring{\mathcal{P}} = \left\{ (x_1, [a_0, a_1]), \ldots, (x_m, [a_{m-1}, a_m]) \right\}.$$

Examples Let $I = [0, 1]$. Here are some tagged partitions of I:

$$\mathring{\mathcal{P}} = \left\{ \left(\tfrac{1}{6}, [0, 1] \right) \right\}$$

$$\mathring{\mathcal{P}} = \left\{ \left(0, \left[0, \tfrac{1}{3} \right] \right), \left(\tfrac{1}{2}, \left[\tfrac{1}{3}, 1 \right] \right) \right\}$$

$$\mathring{P} = \left\{ \left(\tfrac{1}{3}, \left[0, \tfrac{1}{3} \right] \right), \left(\tfrac{1}{3}, \left[\tfrac{1}{3}, \tfrac{2}{3} \right] \right), \left(\tfrac{2}{3}, \left[\tfrac{2}{3}, 1 \right] \right) \right\}$$

$$\mathring{P} = \left\{ \left(\tfrac{1}{8}, \left[0, \tfrac{1}{4} \right] \right), \left(\tfrac{3}{8}, \left[\tfrac{1}{4}, \tfrac{1}{2} \right] \right), \left(\tfrac{5}{8}, \left[\tfrac{1}{2}, \tfrac{3}{4} \right] \right), \left(\tfrac{7}{8}, \left[\tfrac{3}{4}, 1 \right] \right) \right\}.$$

We now consider a function $f : I \to \mathbb{R}$. For each tagged partition $\mathring{P}$ as above we define the number

$$S(f, \mathring{P}) = \sum_{j=1}^{m} f(x_j)(a_j - a_{j-1}),$$

which is called the "Riemann sum" associated with f and $\mathring{P}$.

To better understand this definition, assume for simplicity that the function f is positive on I. Then to each tagged partition of I we associate the sum of the areas of the rectangles having base $[a_{j-1}, a_j]$ and height $[0, f(x_j)]$.

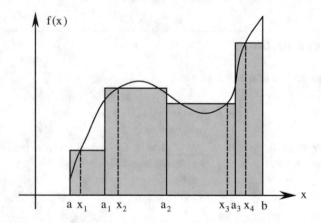

If f is not positive on I, the areas will be considered with a positive or negative sign depending on whether $f(x_j)$ is positive or negative, respectively. If $f(x_j) = 0$, the jth term of the sum will clearly be equal to zero.

Example Let $f : [0, 1] \to \mathbb{R}$ be defined as $f(x) = 4x^2 - 1$, and let

$$\mathring{P} = \left\{ \left(\tfrac{1}{8}, \left[0, \tfrac{1}{4} \right] \right), \left(\tfrac{1}{2}, \left[\tfrac{1}{4}, \tfrac{3}{4} \right] \right), \left(\tfrac{7}{8}, \left[\tfrac{3}{4}, 1 \right] \right) \right\}.$$

Then

$$S(f, \mathring{P}) = -\tfrac{15}{16} \cdot \tfrac{1}{4} + 0 \cdot \tfrac{1}{2} + \tfrac{33}{16} \cdot \tfrac{1}{4} = \tfrac{9}{32}.$$

7.2   δ-Fine Tagged Partitions

To measure how "fine" a tagged partition is, we will have to deal with a "gauge", i.e., a *positive* function $\delta : I \to \mathbb{R}$.

If δ is a gauge on I, we say that the tagged partition $\mathring{\mathcal{P}}$ is "δ-fine" if, for every $j = 1, \ldots, m$,

$$x_j - a_{j-1} \leq \delta(x_j) \quad \text{and} \quad a_j - x_j \leq \delta(x_j) ;$$

equivalently, we may write

$$[a_{j-1}, a_j] \subseteq [x_j - \delta(x_j), x_j + \delta(x_j)] .$$

We will now show that it is always possible to find a δ-fine tagged partition of the compact interval I, whatever the gauge δ.

Theorem 7.1 (Cousin Theorem) *For every gauge δ on $I = [a, b]$ there is a δ-fine tagged partition of I.*

Proof Set $I_0 = I$, and assume by contradiction that there exists a gauge $\delta : I_0 \to \mathbb{R}$ for which there are no δ-fine tagged partitions. Taking the midpoint of I_0, we divide it in two *closed* subintervals. At least one of these two subintervals will not have any δ-fine tagged partition (otherwise we could glue together the two δ-fine tagged partitions to get a δ-fine tagged partition of the original interval I_0). Let us choose it and denote it by I_1. We now iterate the same procedure, thereby constructing a sequence

$$I = I_0 \supseteq I_1 \supseteq I_2 \supseteq I_3 \supseteq \ldots$$

of closed subintervals, none of which has any δ-fine tagged partitions. By Cantor's Theorem 1.9, there is a point c belonging to all of these intervals. For n sufficiently large, I_n will be contained in $[c - \delta(c), c + \delta(c)]$. But then the set $\mathring{\mathcal{P}} = \{(c, I_n)\}$, whose only element is the couple (c, I_n), is a δ-fine tagged partition of I_n, a contradiction. ∎

Examples Let us provide some examples of δ-fine tagged partitions of the interval $I = [0, 1]$.

We start with a constant gauge: $\delta(x) = \frac{1}{5}$. Since the previous theorem does not give any information on how to find a δ-fine tagged partition, we will proceed by guessing. As a first guess, we choose the a_j equally spaced and the x_j as the midpoints of the intervals $[a_{j-1}, a_j]$, i.e.,

$$a_j = \frac{j}{m} , \quad x_j = \frac{a_{j-1} + a_j}{2} = \frac{2j - 1}{2m} .$$

For the corresponding tagged partition to be δ-fine, it must be that

$$x_j - a_{j-1} = \frac{1}{2m} \le \frac{1}{5} \quad \text{and} \quad a_j - x_j = \frac{1}{2m} \le \frac{1}{5}.$$

These inequalities are satisfied choosing $m \ge 3$. If $m = 3$, we have the δ-fine tagged partition

$$\overset{\circ}{\mathcal{P}} = \left\{ \left(\tfrac{1}{6}, \left[0, \tfrac{1}{3} \right] \right), \left(\tfrac{1}{2}, \left[\tfrac{1}{3}, \tfrac{2}{3} \right] \right), \left(\tfrac{5}{6}, \left[\tfrac{2}{3}, 1 \right] \right) \right\}.$$

If, instead of taking the points x_j in the middle of the respective intervals we would like to choose them, for example, at the left endpoint, i.e., $x_j = \frac{j-1}{m}$, then in order to have a δ-fine tagged partition we should ask that

$$x_j - a_{j-1} = 0 \le \frac{1}{5} \quad \text{and} \quad a_j - x_j = \frac{1}{m} \le \frac{1}{5}.$$

These inequalities are verified if $m \ge 5$. For instance, if $m = 5$, then we have the δ-fine tagged partition

$$\overset{\circ}{\mathcal{P}} = \left\{ \left(0, \left[0, \tfrac{1}{5} \right] \right), \left(\tfrac{1}{5}, \left[\tfrac{1}{5}, \tfrac{2}{5} \right] \right), \left(\tfrac{2}{5}, \left[\tfrac{2}{5}, \tfrac{3}{5} \right] \right), \left(\tfrac{3}{5}, \left[\tfrac{3}{5}, \tfrac{4}{5} \right] \right), \left(\tfrac{4}{5}, \left[\tfrac{4}{5}, 1 \right] \right) \right\}.$$

Notice that, with such a choice of a_j, if $m \ge 5$, then the points x_j can actually be taken arbitrarily in the corresponding intervals $[a_{j-1}, a_j]$, still yielding δ-fine tagged partitions.

The previous example shows how it is possible to construct δ-fine tagged partitions in the case of a gauge δ that is constant with value $\frac{1}{5}$. It is clear that a similar procedure can be used for a constant gauge with arbitrary positive value. Consider now the case where δ is a continuous function. Then Weierstrass' Theorem 4.10 says that $\delta(x)$ has a minimum positive value: let it be $\bar{\delta}$. Consider then the constant gauge with value $\bar{\delta}$, and construct a $\bar{\delta}$-fine tagged partition with the procedure we saw earlier. Clearly, such a tagged partition must be δ-fine as well. This argument shows how the case of a continuous gauge can be reduced to that of a constant gauge.

Consider now the noncontinuous gauge

$$\delta(x) = \begin{cases} \frac{1}{2} & \text{if } x = 0, \\ \frac{x}{2} & \text{if } x \in \,]0, 1]. \end{cases}$$

As previously, we proceed by guessing. Let us try, as earlier, taking the a_j equally distant and the x_j as the midpoints of the intervals $[a_{j-1}, a_j]$. This time, however, we are going to fail; indeed, we should have

$$x_1 = x_1 - a_0 \leq \delta(x_1) = \frac{x_1}{2},$$

which is clearly impossible if $x_1 > 0$. The only way to solve this problem is to choose $x_1 = 0$. We decide, then, for instance, to take the x_j to coincide with a_{j-1}, as was also done earlier. We thus find the δ-fine tagged partition

$$\overset{\circ}{\mathcal{P}} = \left\{ \left(0, \left[0, \tfrac{1}{2}\right]\right), \left(\tfrac{1}{2}, \left[\tfrac{1}{2}, \tfrac{3}{4}\right]\right), \left(\tfrac{3}{4}, \left[\tfrac{3}{4}, 1\right]\right) \right\}.$$

Notice that a more economic choice might have been

$$\overset{\circ}{\mathcal{P}} = \left\{ \left(0, \left[0, \tfrac{1}{2}\right]\right), \left(1, \left[\tfrac{1}{2}, 1\right]\right) \right\}.$$

The choice $x_1 = 0$ is, however, unavoidable.

Finally, once a point $c \in \,]0, 1[$ is fixed, let the gauge $\delta : [0, 1] \to \mathbb{R}$ be defined as

$$\delta(x) = \begin{cases} \dfrac{c - x}{2} & \text{if } x \in [0, c[\,, \\[2mm] \dfrac{1}{5} & \text{if } x = c\,, \\[2mm] \dfrac{x - c}{2} & \text{if } x \in \,]c, 1]\,. \end{cases}$$

Similar considerations to those made in the previous case lead to the conclusion that, in order to have a δ-fine tagged partition, it is necessary for one of the x_j to be equal to c. For example, if $c = \tfrac{1}{2}$, a possible choice is

$$\overset{\circ}{\mathcal{P}} = \left\{ \left(0, \left[0, \tfrac{1}{4}\right]\right), \left(\tfrac{1}{4}, \left[\tfrac{1}{4}, \tfrac{3}{8}\right]\right), \left(\tfrac{1}{2}, \left[\tfrac{3}{8}, \tfrac{5}{8}\right]\right), \left(\tfrac{3}{4}, \left[\tfrac{5}{8}, \tfrac{3}{4}\right]\right), \left(1, \left[\tfrac{3}{4}, 1\right]\right) \right\}.$$

7.3 Integrable Functions on a Compact Interval

We now want to define some kind of convergence of the Riemann sums when the tagged partitions become "finer and finer". The following definition is due to Jaroslav Kurzweil and Ralph Henstock.

Definition 7.2 A function $f : I \to \mathbb{R}$ is said to be "integrable" if there is a real number $\mathcal{J}$ with the following property: Given $\varepsilon > 0$, it is possible to find a gauge $\delta : I \to \mathbb{R}$ such that, for every δ-fine tagged partition $\overset{\circ}{\mathcal{P}}$ of I,

$$|S(f, \overset{\circ}{\mathcal{P}}) - \mathcal{J}| \leq \varepsilon.$$

We will also say that f is "integrable on I."

Let us prove that there is at most one $\mathcal{J} \in \mathbb{R}$ that verifies the conditions of the definition. If there were a second one, say, $\mathcal{J}'$, then, for every $\varepsilon > 0$, there would be two gauges δ and δ' on I associated respectively with $\mathcal{J}$ and $\mathcal{J}'$, satisfying the condition of the definition. Define the gauge

$$\delta''(x) = \min\{\delta(x), \delta'(x)\}.$$

Once a δ''-fine tagged partition $\overset{\circ}{\mathcal{P}}$ of I is chosen, we have that $\overset{\circ}{\mathcal{P}}$ is both δ-fine and δ'-fine, and hence

$$|\mathcal{J} - \mathcal{J}'| \leq |\mathcal{J} - S(f, \overset{\circ}{\mathcal{P}})| + |S(f, \overset{\circ}{\mathcal{P}}) - \mathcal{J}'| \leq 2\varepsilon.$$

Since this holds for every $\varepsilon > 0$, it necessarily must be that $\mathcal{J} = \mathcal{J}'$.

If $f : I \to \mathbb{R}$ is an integrable function, the only element $\mathcal{J} \in \mathbb{R}$ verifying the conditions of the definition is called the "integral" of f on I and is denoted by one of the following symbols:

$$\int_I f, \qquad \int_a^b f, \qquad \int_I f(x)\,dx, \qquad \int_a^b f(x)\,dx.$$

The presence of the letter x in the preceding notation has no independent importance. It could be replaced by any other letter t, u, α, ..., or by any other symbol, unless already used with another meaning. For reasons to be explained later on, we set, moreover,

$$\int_b^a f = -\int_a^b f, \quad \text{and} \quad \int_a^a f = 0.$$

Examples

1. As a first example, consider a constant function $f(x) = c$. In this case, for any tagged partition $\overset{\circ}{\mathcal{P}}$ of $[a, b]$,

$$S(f, \overset{\circ}{\mathcal{P}}) = \sum_{j=1}^m c(a_j - a_{j-1}) = c \sum_{j=1}^m (a_j - a_{j-1}) = c(b - a),$$

hence also

$$\int_a^b c\,dx = c(b - a).$$

Indeed, once we fix $\varepsilon > 0$, it is readily seen in this simple case that any gauge $\delta : [a, b] \to \mathbb{R}$ satisfies the condition of the definition, with $\mathcal{J} = c(b - a)$, since for every δ-fine tagged partition $\overset{\circ}{\mathcal{P}}$ of I we have $|S(f, \overset{\circ}{\mathcal{P}}) - \mathcal{J}| = 0 < \varepsilon$.

2. As a second example, consider the function $f(x) = x$. Then

$$S(f, \overset{\circ}{\mathcal{P}}) = \sum_{j=1}^{m} x_j(a_j - a_{j-1}) \, .$$

To find a candidate for the integral, let us consider a particular tagged partition where the x_j are the midpoints of the intervals $[a_{j-1}, a_j]$. In this particular case, we have

$$\sum_{j=1}^{m} x_j(a_j - a_{j-1}) = \sum_{j=1}^{m} \frac{a_{j-1} + a_j}{2}(a_j - a_{j-1}) = \frac{1}{2}\sum_{j=1}^{m}(a_j^2 - a_{j-1}^2) = \frac{1}{2}(b^2 - a^2).$$

We want to prove now that the function $f(x) = x$ is integrable on $[a, b]$ and that its integral is really $\frac{1}{2}(b^2 - a^2)$. Fix $\varepsilon > 0$. For any tagged partition $\overset{\circ}{\mathcal{P}}$ we have

$$\left| S(f, \overset{\circ}{\mathcal{P}}) - \frac{1}{2}(b^2 - a^2) \right| = \left| \sum_{j=1}^{m} x_j(a_j - a_{j-1}) - \sum_{j=1}^{m} \frac{a_{j-1} + a_j}{2}(a_j - a_{j-1}) \right|$$

$$\leq \sum_{j=1}^{m} \left| x_j - \frac{a_{j-1} + a_j}{2} \right| (a_j - a_{j-1})$$

$$\leq \sum_{j=1}^{m} \frac{a_j - a_{j-1}}{2}(a_j - a_{j-1}) \, .$$

If we choose the gauge δ to be constant with value $\frac{\varepsilon}{b-a}$, then for every δ-fine tagged partition $\overset{\circ}{\mathcal{P}}$ we have

$$\left| S(f, \overset{\circ}{\mathcal{P}}) - \frac{1}{2}(b^2 - a^2) \right| \leq \sum_{j=1}^{m} \frac{2\delta}{2}(a_j - a_{j-1}) = \frac{\varepsilon}{b - a}\sum_{j=1}^{m}(a_j - a_{j-1}) = \varepsilon \, .$$

The condition of the definition is thus verified with this choice of the gauge, and we have proved that

$$\int_{a}^{b} x \, dx = \frac{1}{2}(b^2 - a^2) \, .$$

7.4   Elementary Properties of the Integral

Let $f : I \to \mathbb{R}$ and $g : I \to \mathbb{R}$ be two real functions and $\alpha \in \mathbb{R}$ a constant. It is easy to verify that for every tagged partition $\mathring{\mathcal{P}}$ of I,

$$S(f + g, \mathring{\mathcal{P}}) = S(f, \mathring{\mathcal{P}}) + S(g, \mathring{\mathcal{P}})$$

and

$$S(\alpha f, \mathring{\mathcal{P}}) = \alpha S(f, \mathring{\mathcal{P}}) .$$

These linearity properties are inherited by the integral, as will be proved in the following two propositions.

Proposition 7.3 *If f and g are integrable on I, then $f + g$ is integrable on I and*

$$\int_I (f + g) = \int_I f + \int_I g .$$

Proof Set $\mathcal{J}_1 = \int_I f$ and $\mathcal{J}_2 = \int_I g$. Once $\varepsilon > 0$ is fixed, there are two gauges δ_1 and δ_2 on I such that, for every tagged partition $\mathring{\mathcal{P}}$ of I, if $\mathring{\mathcal{P}}$ is δ_1-fine, then

$$|S(f, \mathring{\mathcal{P}}) - \mathcal{J}_1| \le \frac{\varepsilon}{2} ,$$

whereas if $\mathring{\mathcal{P}}$ is δ_2-fine, then

$$|S(g, \mathring{\mathcal{P}}) - \mathcal{J}_2| \le \frac{\varepsilon}{2} .$$

Let us define the gauge $\delta : I \to \mathbb{R}$ as $\delta(x) = \min\{\delta_1(x), \delta_2(x)\}$. Let $\mathring{\mathcal{P}}$ be a δ-fine tagged partition of I. It is thus both δ_1-fine and δ_2-fine, hence

$$
\begin{aligned}
|S(f + g, \mathring{\mathcal{P}}) - (\mathcal{J}_1 + \mathcal{J}_2)| &= |S(f, \mathring{\mathcal{P}}) - \mathcal{J}_1 + S(g, \mathring{\mathcal{P}}) - \mathcal{J}_2| \\
&\le |S(f, \mathring{\mathcal{P}}) - \mathcal{J}_1| + |S(g, \mathring{\mathcal{P}}) - \mathcal{J}_2| \\
&\le \frac{\varepsilon}{2} + \frac{\varepsilon}{2} = \varepsilon .
\end{aligned}
$$

This completes the proof.   ∎

Proposition 7.4 *If f is integrable on I and $\alpha \in \mathbb{R}$, then αf is integrable on I and*

$$\int_I (\alpha f) = \alpha \int_I f .$$

Proof If $\alpha = 0$, then the identity is surely true. If $\alpha \neq 0$, then set $\mathcal{J} = \int_I f$ and fix $\varepsilon > 0$. There is a gauge δ on I such that

$$|S(f, \mathring{\mathcal{P}}) - \mathcal{J}| \leq \frac{\varepsilon}{|\alpha|}$$

for every δ-fine tagged partition $\mathring{\mathcal{P}}$ of I. Then, for every δ-fine tagged partition $\mathring{\mathcal{P}}$ of I, we have

$$|S(\alpha f, \mathring{\mathcal{P}}) - \alpha \mathcal{J}| = |\alpha S(f, \mathring{\mathcal{P}}) - \alpha \mathcal{J}| = |\alpha| \, |S(f, \mathring{\mathcal{P}}) - \mathcal{J}| \leq |\alpha| \frac{\varepsilon}{|\alpha|} = \varepsilon,$$

and the proof is thus completed. ∎

We have just proved that the set of integrable functions is a real vector space and that the integral is a linear function on it.

We now study the behavior of the integral with respect to the order relation in $\mathbb{R}$.

Proposition 7.5 *If f is integrable on I and $f(x) \geq 0$ for every $x \in I$, then*

$$\int_I f \geq 0.$$

Proof Fix $\varepsilon > 0$. There is a gauge δ on I such that

$$\left| S(f, \mathring{\mathcal{P}}) - \int_I f \right| \leq \varepsilon$$

for every δ-fine tagged partition $\mathring{\mathcal{P}}$ of I. Hence,

$$\int_I f \geq S(f, \mathring{\mathcal{P}}) - \varepsilon \geq -\varepsilon,$$

since clearly $S(f, \mathring{\mathcal{P}}) \geq 0$. Since this is true for every $\varepsilon > 0$, it must be that $\int_I f \geq 0$, thereby proving the result. ∎

Corollary 7.6 *If f and g are integrable on I and $f(x) \leq g(x)$ for every $x \in I$, then*

$$\int_I f \leq \int_I g.$$

Proof It is sufficient to apply the preceding proposition to the function $g - f$. ∎

Corollary 7.7 *If f and $|f|$ are integrable on I, then*

$$\left| \int_I f \right| \le \int_I |f| \, .$$

Proof Applying the preceding corollary to the inequalities

$$- |f| \le f \le |f| \, ,$$

we have

$$- \int_I |f| \le \int_I f \le \int_I |f| \, ,$$

whence the conclusion.   ∎

7.5   The Fundamental Theorem

The following theorem establishes an unexpected link between differential and integral calculus. It is called the **Fundamental Theorem** of differential and integral calculus.

Theorem 7.8 (Fundamental Theorem—I) *Let $F : [a, b] \to \mathbb{R}$ be a differentiable function, and let f be its derivative: $F'(x) = f(x)$ for every $x \in [a, b]$. Then f is integrable on $[a, b]$ and*

$$\int_a^b f = F(b) - F(a) \, .$$

Proof Let $\varepsilon > 0$ be fixed. We know that for every $x \in [a, b]$,

$$f(x) = F'(x) = \lim_{u \to x} \frac{F(u) - F(x)}{u - x} \, .$$

Then for every $x \in [a, b]$, there is a $\delta(x) > 0$ such that, for every $u \in [a, b]$,

$$0 < |u - x| \le \delta(x) \;\Rightarrow\; \left| \frac{F(u) - F(x)}{u - x} - f(x) \right| \le \frac{\varepsilon}{b - a} \, ,$$

i.e.,

$$|u - x| \le \delta(x) \;\Rightarrow\; |F(u) - F(x) - f(x)(u - x)| \le \frac{\varepsilon}{b - a} |u - x| \, .$$

We have thus defined a gauge $\delta : [a, b] \to \mathbb{R}$.

Consider now a δ-fine tagged partition of I,

$$\mathring{P} = \{(x_1, [a_0, a_1]), \ldots, (x_m, [a_{m-1}, a_m])\}.$$

Since, for every $j = 1, \ldots, m$,

$$|a_{j-1} - x_j| \le \delta(x_j) \quad \text{and} \quad |a_j - x_j| \le \delta(x_j),$$

we have that

$$\left| F(a_j) - F(a_{j-1}) - f(x_j)(a_j - a_{j-1}) \right|$$
$$= \left| [F(a_j) - F(x_j) - f(x_j)(a_j - x_j)] \right.$$
$$\left. + [F(x_j) - F(a_{j-1}) + f(x_j)(a_{j-1} - x_j)] \right|$$
$$\le \left| F(a_j) - F(x_j) - f(x_j)(a_j - x_j) \right|$$
$$+ \left| F(a_{j-1}) - F(x_j) - f(x_j)(a_{j-1} - x_j) \right|$$
$$\le \frac{\varepsilon}{b-a} |a_j - x_j| + \frac{\varepsilon}{b-a} |a_{j-1} - x_j|$$
$$= \frac{\varepsilon}{b-a}(a_j - x_j + x_j - a_{j-1}) = \frac{\varepsilon}{b-a}(a_j - a_{j-1}).$$

Hence,

$$|F(b) - F(a) - S(f, \mathring{P})| = \left| \sum_{j=1}^{m} [F(a_j) - F(a_{j-1})] - \sum_{j=1}^{m} f(x_j)(a_j - a_{j-1}) \right|$$
$$= \left| \sum_{j=1}^{m} [F(a_j) - F(a_{j-1}) - f(x_j)(a_j - a_{j-1})] \right|$$
$$\le \sum_{j=1}^{m} \left| F(a_j) - F(a_{j-1}) - f(x_j)(a_j - a_{j-1}) \right|$$
$$\le \sum_{j=1}^{m} \frac{\varepsilon}{b-a}(a_j - a_{j-1}) = \varepsilon,$$

and the theorem is proved.   ∎

7.6   Primitivable Functions

In this section and the following one, we denote by $\mathcal{I}$ any interval in $\mathbb{R}$ (not necessarily a compact interval).

A function $f : \mathcal{I} \to \mathbb{R}$ is said to be "primitivable" (or "primitivable on $\mathcal{I}$") if there is a differentiable function $F : \mathcal{I} \to \mathbb{R}$ such that $F'(x) = f(x)$ for every $x \in \mathcal{I}$. Such a function F is called a "primitive" of f.

The Fundamental Theorem establishes that all primitivable functions defined on a compact interval $I = [a, b]$ are integrable and that their integral is easily computable once a primitive is known. It can be reformulated as follows.

Theorem 7.9 (Fundamental Theorem—II)   *Let $f : [a, b] \to \mathbb{R}$ be a primitivable function, and let F be one of its primitives. Then f is integrable on $[a, b]$ and*

$$\int_a^b f = F(b) - F(a) \, .$$

It is sometimes useful to denote the difference $F(b) - F(a)$ by the symbols

$$[F]_a^b \, , \quad [F(x)]_{x=a}^{x=b}$$

or variants of these, for instance $[F(x)]_a^b$, when no ambiguities can arise.

Example Consider the function $f(x) = x^n$. It is easy to see that $F(x) = \frac{1}{n+1}x^{n+1}$ is a primitive. The Fundamental Theorem tells us that

$$\int_a^b x^n \, dx = \left[\frac{1}{n+1} x^{n+1} \right]_a^b = \frac{1}{n+1}(b^{n+1} - a^{n+1}) \, .$$

The fact that the difference $F(b) - F(a)$ does not depend on the chosen primitive is explained by the following proposition.

Proposition 7.10 *Let $f : \mathcal{I} \to \mathbb{R}$ be a primitivable function, and let F be one of its primitives. Then a function $G : \mathcal{I} \to \mathbb{R}$ is a primitive of f if and only if $F - G$ is a constant function on $\mathcal{I}$.*

Proof If $F - G$ is constant, then

$$G'(x) = (F - (F - G))'(x) = F'(x) - (F - G)'(x) = F'(x) = f(x)$$

for every $x \in \mathcal{I}$, and hence G is a primitive of f. On the other hand, if G is a primitive of f, then we have

$$(F - G)'(x) = F'(x) - G'(x) = f(x) - f(x) = 0$$

for every $x \in \mathcal{I}$. Consequently, $F - G$ is constant on $\mathcal{I}$.   ∎

Note that if $f : \mathcal{I} \rightarrow \mathbb{R}$ is a primitivable function, then it is also primitivable on every subinterval of $\mathcal{I}$. In particular, it is integrable on every interval $[a, x] \subseteq \mathcal{I}$, and therefore it is possible to define a function

$$x \mapsto \int_a^x f \,,$$

which we call the "integral function" of f and denote by one of the following symbols:

$$\int_a^{\cdot} f \,, \qquad \int_a^{\cdot} f(t) \, dt \,.$$

In this last notation it is convenient to use a letter other than x for the variable of f; for instance, here we have chosen the letter t. The Fundamental Theorem tells us that if F is a primitive of f, then

$$\int_a^x f = F(x) - F(a) \,, \qquad \text{for every } x \in [a, b] \,.$$

We thus see that $\int_a^x f$ differs from $F(x)$ by a constant, whence the following corollary.

Corollary 7.11 *Let $f : [a, b] \rightarrow \mathbb{R}$ be a primitivable function. Then the integral function $\int_a^{\cdot} f$ is one of its primitives; it is differentiable on $[a, b]$ and*

$$\left(\int_a^{\cdot} f \right)'(x) = f(x) \,, \qquad \text{for every } x \in [a, b] \,.$$

Notice that the choice of the point a in the definition of $\int_a^{\cdot} f$ is not at all mandatory. If $f : \mathcal{I} \rightarrow \mathbb{R}$ is primitivable, one could take any point $\omega \in \mathcal{I}$ and consider the function $\int_{\omega}^{\cdot} f$. The conventions made on the integral with exchanged endpoints are such that the previously stated corollary still holds with this new integral function. Indeed, if F is a primitive of f, even if $x < \omega$, then we have

$$\int_{\omega}^x f = - \int_x^{\omega} f = -(F(\omega) - F(x)) = F(x) - F(\omega) \,,$$

so that $\int_{\omega}^{\cdot} f$ is still a primitive of f. We can then write

$$\frac{d}{dx} \int_{\omega}^x f = f(x) \,, \qquad \text{or, equivalently,} \qquad \frac{d}{dx} \int_{\omega}^x f(t) \, dt = f(x) \,.$$

This formula can be generalized; if $\alpha, \beta \; : \; [a, b] \; \to \; \mathbb{R}$ are two differentiable functions, then

$$\frac{d}{dx} \int_{\alpha(x)}^{\beta(x)} f(t)\, dt = f(\beta(x))\beta'(x) - f(\alpha(x))\alpha'(x) .$$

Indeed, if F is a primitive of f, then the preceding formula is easily obtained by writing $\int_{\alpha(x)}^{\beta(x)} f(t)\, dt = F(\beta(x)) - F(\alpha(x))$ and differentiating.

We will denote the set of all primitives of f by one of the following symbols:

$$\int f , \qquad \int f(x)\, dx .$$

One should be careful with the notation $\int$ introduced for the primitives, which looks similar to that for the integral, even if the two concepts are completely different. Concerning the use of x, an observation analogous to the one made for the integral can be made here, as well: it can be replaced by any other letter or symbol, with due precaution. When applying the theory to practical problems, however, if F denotes a primitive of f instead of correctly writing

$$\int f = \{F + c : c \in \mathbb{R}\} ,$$

it is common to use improper expressions of the type

$$\int f(x)\, dx = F(x) + c ,$$

where $c \in \mathbb{R}$ stands for an arbitrary constant; we will adapt to this habit, too. Let us make a list of primitives of some elementary functions:

$$\int e^x\, dx = e^x + c ,$$

$$\int \sin x\, dx = -\cos x + c ,$$

$$\int \cos x\, dx = \sin x + c ,$$

$$\int x^\alpha\, dx = \frac{x^{\alpha+1}}{\alpha + 1} + c , \quad \text{with } \alpha \neq -1 ,$$

$$\int \frac{1}{x}\, dx = \ln |x| + c ,$$

$$\int \frac{1}{1+x^2}\, dx = \arctan x + c\,,$$

$$\int \frac{1}{\sqrt{1-x^2}}\, dx = \arcsin x + c\,.$$

Notice that the definition of primitivable function makes sense even in some cases where f is not necessarily defined on an interval, and indeed the preceding formulas should be interpreted on the natural domains of the considered functions. For example,

$$\int \frac{1}{x}\, dx = \begin{cases} \ln x + c & \text{if } x \in \,]0,+\infty[\,, \\ \ln(-x) + c & \text{if } x \in \,]-\infty,0[\,. \end{cases}$$

Example Using the Fundamental Theorem we find

$$\int_0^\pi \sin x\, dx = [-\cos x]_0^\pi = -\cos \pi + \cos 0 = 2\,.$$

Notice that the presence of the arbitrary constant c can sometimes lead to apparently different results. For example, we know that $\int \frac{1}{\sqrt{1-x^2}}\, dx = \arcsin x + c$, but it is readily verified that we also have

$$\int \frac{1}{\sqrt{1-x^2}}\, dx = -\arccos x + c\,.$$

This is explained by the fact that $\arcsin x = \frac{\pi}{2} - \arccos x$ for every $x \in [-1,1]$, hence the difference of arcsin and $-\arccos$ is constant. The same notation c for the arbitrary constant in the two formulas could sometimes be misleading!

From the known properties of derivatives we can easily prove the following two propositions.

Proposition 7.12 *Let f and g be primitivable on $\mathcal{I}$, and let F and G be two corresponding primitives. Then $f + g$ is primitivable on $\mathcal{I}$, and $F + G$ is one of its primitives; we will briefly write*[1]

$$\int (f+g) = \int f + \int g\,.$$

---

[1] Here and in what follows, we use in an intuitive way the algebraic operations involving sets. To be precise, the sum of two sets A and B is defined as

$$A + B = \{a + b : a \in A,\ b \in B\}.$$

Proposition 7.13 *Let f be primitivable on $\mathcal{I}$, and let F be one of its primitives. If $\alpha \in \mathbb{R}$ is any given constant, then αf is primitivable on $\mathcal{I}$, and αF is one of its primitives; we will briefly write*

$$\int (\alpha f) = \alpha \int f.$$

As a consequence of these propositions, we have that the set of primitivable functions on $\mathcal{I}$ is a real vector space.

We conclude this section by presenting an interesting class of integrable functions that are not primitivable. Let the function $f : [a, b] \to \mathbb{R}$ be such that the set

$$E = \{x \in [a, b] : f(x) \neq 0\}$$

is finite or countable (for instance, a function that is zero everywhere except at a point, or the Dirichlet function $\mathcal{D} : [a, b] \to \mathbb{R}$, defined by $\mathcal{D}(x) = 1$ if x is rational, and $\mathcal{D}(x) = 0$ if x is irrational).

Let us prove that such a function is integrable, with $\int_a^b f = 0$. Assume for definiteness that E is infinite (the case where E is finite can be treated in an analogous way). Since it is countable, we can write $E = \{e_n : n \in \mathbb{N}\}$. Once $\varepsilon > 0$ has been fixed, we construct a gauge δ on $[a, b]$ as follows. If $x \notin E$, then we set $\delta(x) = 1$; if, instead, for a certain n we have $x = e_n$, then we set

$$\delta(e_n) = \frac{\varepsilon}{2^{n+3} |f(e_n)|}.$$

Now let $\overset{\circ}{\mathcal{P}} = \{(x_1, [a_0, a_1]), \ldots, (x_m, [a_{m-1}, a_m])\}$ be a δ-fine tagged partition of $[a, b]$. By the way in which f is defined, the associated Riemann sum becomes

$$S(f, \overset{\circ}{\mathcal{P}}) = \sum_{\{1 \leq j \leq m \,:\, x_j \in E\}} f(x_j)(a_j - a_{j-1}).$$

Let $N = \max\{1 \leq j \leq m : x_j \in E\}$. Since $[a_{j-1}, a_j] \subseteq [x_j - \delta(x_j), x_j + \delta(x_j)]$, we have that $a_j - a_{j-1} \leq 2\delta(x_j)$, and if x_j is in E, it must be that $x_j = e_n$ for some $n \in \mathbb{N}$. To any such e_n can, however, correspond one or two points x_j, so that we will have

$$\left| \sum_{\{1 \leq j \leq m \,:\, x_j \in E\}} f(x_j)(a_j - a_{j-1}) \right| \leq 2 \sum_{n=0}^{N} |f(e_n)| 2\delta(e_n)$$

$$= 4 \sum_{n=0}^{N} \frac{\varepsilon}{2^{n+3}} = \frac{\varepsilon}{2} \sum_{n=0}^{N} \left(\frac{1}{2}\right)^n = \frac{\varepsilon}{2} \frac{1 - (\frac{1}{2})^{N+1}}{1 - \frac{1}{2}} < \varepsilon.$$

This shows that f is integrable on $[a, b]$ and that $\int_a^b f = 0$.

Let us see now that if E is nonempty, then f is not primitivable on $[a, b]$. Indeed, if it were, its integral function $\int_a^x f$ should be one of its primitives. But the foregoing procedure shows that $\int_a^x f = 0$ for every $x \in [a, b]$. Then f should be identically zero, being the derivative of a constant function, a contradiction.

7.7 Primitivation by Parts and by Substitution

We now present two methods frequently used for finding the primitives of certain functions. The first one is known as the method of "primitivation by parts."

Proposition 7.14 Let $F, G : \mathcal{I} \to \mathbb{R}$ be two differentiable functions, and let f, g be the corresponding derivatives. One has that fG is primitivable on $\mathcal{I}$ if and only if Fg is, in which case a primitive of fG is obtained subtracting from FG a primitive of Fg; we will briefly write

$$\int fG = FG - \int Fg .$$

Proof Since F and G are differentiable, then so is FG, and we have

$$(FG)' = fG + Fg .$$

Hence, $fG + Fg$ is primitivable on $\mathcal{I}$ with primitive FG, and the conclusion follows from Proposition 7.12. ∎

Example We would like to find a primitive of the function $h(x) = xe^x$. Define the following functions: $f(x) = e^x$, $G(x) = x$, and consequently $F(x) = e^x$, $g(x) = 1$. Applying the formula given by the foregoing proposition, we have

$$\int e^x x \, dx = e^x x - \int e^x \, dx = xe^x - e^x + c ,$$

where c stands, as usual, for an arbitrary constant.

As an immediate consequence of Proposition 7.14, we have the rule of "integration by parts":

$$\int_a^b fG = F(b)G(b) - F(a)G(a) - \int_a^b Fg .$$

Examples Applying the formula to the function $h(x) = xe^x$ of the previous example, we compute

$$\int_0^1 e^x x\, dx = e^1 \cdot 1 - e^0 \cdot 0 - \int_0^1 e^x\, dx = e - [e^x]_0^1 = e - (e^1 - e^0) = 1\,.$$

Note that we could have obtained the same result using the Fundamental Theorem; having already found earlier that a primitive of h is given by $H(x) = xe^x - e^x$, we have that

$$\int_0^1 e^x x\, dx = H(1) - H(0) = (e - e) - (0 - 1) = 1\,.$$

Let us consider some additional examples. Let $h(x) = \sin^2 x$. With the obvious choice of the functions f and G, we find

$$\int \sin^2 x\, dx = -\cos x \sin x + \int \cos^2 x\, dx$$

$$= -\cos x \sin x + \int (1 - \sin^2 x)\, dx$$

$$= x - \cos x \sin x - \int \sin^2 x\, dx\,,$$

from which we obtain

$$\int \sin^2 x\, dx = \frac{1}{2}(x - \cos x \sin x) + c\,.$$

Consider now the case of the function $h(x) = \ln x$, with $x > 0$. To apply the formula of primitivation by parts, we choose the functions $f(x) = 1$, $G(x) = \ln x$. In this way, we find

$$\int \ln x\, dx = x \ln x - \int x \frac{1}{x}\, dx = x \ln x - \int 1\, dx = x \ln x - x + c\,.$$

The second method we want to study is known as the method of "primitivation by substitution."

Proposition 7.15 *Let $\varphi : \mathcal{I} \to \mathbb{R}$ be a differentiable function and $f : \varphi(\mathcal{I}) \to \mathbb{R}$ be a primitivable function on the interval $\varphi(\mathcal{I})$, with primitive F. Then the function $(f \circ \varphi)\varphi'$ is primitivable on $\mathcal{I}$, and one of its primitives is given by $F \circ \varphi$. We will briefly write*

$$\int (f \circ \varphi)\varphi' = \left(\int f \right) \circ \varphi\,.$$

Proof The function $F \circ \varphi$ is differentiable on $\mathcal{I}$ and

$$(F \circ \varphi)' = (F' \circ \varphi)\varphi' = (f \circ \varphi)\varphi'.$$

It follows that $(f \circ \varphi)\varphi'$ is primitivable on $\mathcal{I}$, with primitive $F \circ \varphi$. ∎

As an example, we look for a primitive of the function $h(x) = xe^{x^2}$. Defining $\varphi(x) = x^2$, $f(t) = \frac{1}{2}e^t$ (it is advisable to use different letters to indicate the variables of φ and f), we have that $h = (f \circ \varphi)\varphi'$. Since a primitive of f is seen to be $F(t) = \frac{1}{2}e^t$, a primitive of h is $F \circ \varphi$, i.e.,

$$\int xe^{x^2} \, dx = F(\varphi(x)) + c = \frac{1}{2}e^{x^2} + c.$$

The formula of primitivation by substitution is often written in the form

$$\int f(\varphi(x))\varphi'(x) \, dx = \int f(t) \, dt \Big|_{t=\varphi(x)},$$

where, if F is a primitive of f, the term on the right-hand side should be read

$$\int f(t) \, dt \Big|_{t=\varphi(x)} = F(t) + c \Big|_{t=\varphi(x)} = F(\varphi(x)) + c.$$

Formally, there is a "change of variable" $t = \varphi(x)$, and the symbol dt joins the game to replace $\varphi'(x) \, dx$ (the Leibniz notation $\frac{dt}{dx} = \varphi'(x)$ is a useful mnemonic rule).

Example To find a primitive of the function $h(x) = \frac{\ln x}{x}$, we can choose $\varphi(x) = \ln x$ and apply the formula

$$\int \frac{\ln x}{x} \, dx = \int t \, dt \Big|_{t=\ln x} = \frac{1}{2}t^2 + c \Big|_{t=\ln x} = \frac{1}{2}(\ln x)^2 + c.$$

In this case, writing $t = \ln x$, we have that the symbol dt replaces $\frac{1}{x}dx$.

As a consequence of the preceding formulas, we have the rule of "integration by substitution":

$$\int_a^b f(\varphi(x))\varphi'(x) \, dx = \int_{\varphi(a)}^{\varphi(b)} f(t) \, dt.$$

Indeed, if F is a primitive of f on $\varphi(\mathcal{I})$, applying the Fundamental Theorem twice, we have

$$\int_a^b (f \circ \varphi)\varphi' = (F \circ \varphi)(b) - (F \circ \varphi)(a) = F(\varphi(b)) - F(\varphi(a)) = \int_{\varphi(a)}^{\varphi(b)} f.$$

Example Taking the function $h(x) = xe^{x^2}$ defined previously, we have

$$\int_0^2 xe^{x^2}\, dx = \int_0^4 \frac{1}{2}e^t\, dt = \frac{1}{2}[e^t]_0^4 = \frac{e^4 - 1}{2}.$$

Clearly, the same result is obtainable directly by the Fundamental Theorem once we know that a primitive of h is given by $H(x) = \frac{1}{2}e^{x^2}$. Indeed, we have

$$\int_0^2 xe^{x^2}\, dx = H(2) - H(0) = \frac{1}{2}e^4 - \frac{1}{2}e^0 = \frac{e^4 - 1}{2}.$$

When the function $\varphi : \mathcal{I} \to \varphi(\mathcal{I})$ is invertible, we can also write

$$\int f(t)\, dt = \int f(\varphi(x))\varphi'(x)\, dx \Big|_{x=\varphi^{-1}(t)},$$

with the corresponding formula for the integral:

$$\int_\alpha^\beta f(t)\, dt = \int_{\varphi^{-1}(\alpha)}^{\varphi^{-1}(\beta)} f(\varphi(x))\varphi'(x)\, dx.$$

Example Looking for a primitive of $f(t) = \sqrt{1 - t^2}$, with $t \in\,]-1, 1[$, we may consider the function $\varphi :\,]0, \pi[\to\,]-1, 1[$ defined as $\varphi(x) = \cos x$, so that

$$f(\varphi(x))\varphi'(x) = \sqrt{1 - \cos^2 x}\, (-\sin x) = -\sin^2 x,$$

since $\sin x > 0$ when $x \in\,]0, \pi[$. Therefore, we can write

$$\int \sqrt{1 - t^2}\, dt = -\int \sin^2 x\, dx \Big|_{x=\arccos t}$$

$$= -\frac{1}{2}(x - \cos x \sin x) + c \Big|_{x=\arccos t}$$

$$= -\frac{1}{2}\left(\arccos t - t\sqrt{1 - t^2}\right) + c.$$

7.8   The Taylor Formula with Integral Form Remainder

Here we have the "Taylor formula with integral form of the remainder."

Theorem 7.16 (Taylor Theorem—II) *Let* $x \neq x_0$ *be two points of an interval* I *and* $f : I \to \mathbb{R}$ *be* $n + 1$ *times differentiable. Then*

$$f(x) = p_n(x) + \frac{1}{n!} \int_{x_0}^{x} f^{(n+1)}(u)(x - u)^n \, du \,,$$

where $p_n(x)$ *is the nth-order Taylor polynomial associated with* f *at* x_0.

Proof Let us first prove by induction that if f is $n + 1$ times differentiable, then the function $g_n(u) = f^{(n+1)}(u)(x - u)^n$ is primitivable (here x is fixed). If $n = 0$, then we have that $g_0(u) = f'(u)$, hence the proposition is true. Assume now that the proposition is true for some $n \in \mathbb{N}$. Then, if f is $n + 2$ times differentiable,

$$D_u(f^{(n+1)}(u)(x - u)^{n+1}) = f^{(n+2)}(u)(x - u)^{n+1} - (n + 1)f^{(n+1)}(u)(x - u)^n \,,$$

i.e.,

$$g_{n+1}(u) = (n + 1)g_n(u) + D_u(f^{(n+1)}(u)(x - u)^{n+1}) \,.$$

Since we know that g_n is primitivable, the preceding formula tells us that g_{n+1} is too, since it is the sum of two primitivable functions. We have thus proved the assertion.

Let us now prove the formula by induction. If $n = 0$, then, by the Fundamental Theorem,

$$f(x) = f(x_0) + \int_{x_0}^{x} f'(u) \, du = p_0(x) + \frac{1}{0!} \int_{x_0}^{x} f^{(0+1)}(u)(x - u)^0 \, du \,,$$

hence the formula is true. Assume now that the formula holds true for some $n \in \mathbb{N}$, and let f be $n + 2$ times differentiable. Then

$$f(x) - p_{n+1}(x) = f(x) - \left(p_n(x) + \frac{1}{(n + 1)!} f^{(n+1)}(x_0)(x - x_0)^{n+1} \right)$$

$$= \frac{1}{n!} \int_{x_0}^{x} f^{(n+1)}(u)(x - u)^n \, du - \frac{1}{(n + 1)!} f^{(n+1)}(x_0)(x - x_0)^{n+1}$$

$$= \frac{1}{n!} \left(\int_{x_0}^{x} f^{(n+1)}(u)(x - u)^n \, du - \frac{1}{n + 1} f^{(n+1)}(x_0)(x - x_0)^{n+1} \right).$$

Integrating by parts (we know that g_n and g_{n+1} are primitivable),

$$\int_{x_0}^x f^{(n+1)}(u)(x-u)^n\, du =$$

$$= \left[\left(-\frac{(x-u)^{n+1}}{n+1} \right) f^{(n+1)}(u) \right]_{u=x_0}^{u=x} - \int_{x_0}^x \left(-\frac{(x-u)^{n+1}}{n+1} \right) f^{(n+2)}(u)\, du$$

$$= \frac{1}{n+1} f^{(n+1)}(x_0)(x-x_0)^{n+1} + \frac{1}{n+1} \int_{x_0}^x f^{(n+2)}(u)(x-u)^{n+1}\, du\,,$$

and substituting,

$$f(x) - p_{n+1}(x) = \frac{1}{n!} \left(\frac{1}{n+1} \int_{x_0}^x f^{(n+2)}(u)(x-u)^{n+1}\, du \right)$$

$$= \frac{1}{(n+1)!} \int_{x_0}^x f^{(n+2)}(u)(x-u)^{n+1}\, du\,.$$

Hence, the formula holds also for $n+1$, and the proof is complete. ∎

7.9   The Cauchy Criterion

We have already encountered the Cauchy criterion for sequences in complete metric spaces and for the limit of functions (Theorem 4.15). It is not surprising that a similar criterion holds also for integrability, which can be thought as a kind of "limit" of the Riemann sums.

Theorem 7.17 (Cauchy Criterion) *A function $f : I \to \mathbb{R}$ is integrable if and only if for every $\varepsilon > 0$ there is a gauge $\delta : I \to \mathbb{R}$ such that, taking two δ-fine tagged partitions $\mathring{\mathcal{P}}$, $\mathring{\mathcal{Q}}$ of I, we have*

$$|S(f, \mathring{\mathcal{P}}) - S(f, \mathring{\mathcal{Q}})| \le \varepsilon\,.$$

Proof Let us first prove the necessary condition. Let f be integrable on I, with integral $\mathcal{J}$, and fix $\varepsilon > 0$. Then there is a gauge δ on I such that, for every δ-fine tagged partition $\mathring{\mathcal{P}}$ of I, we have

$$|S(f, \mathring{\mathcal{P}}) - \mathcal{J}| \le \frac{\varepsilon}{2}\,.$$

If $\mathring{\mathcal{P}}$ and $\mathring{\mathcal{Q}}$ are two δ-fine tagged partitions, then we have

$$|S(f, \mathring{\mathcal{P}}) - S(f, \mathring{\mathcal{Q}})| \le |S(f, \mathring{\mathcal{P}}) - \mathcal{J}| + |\mathcal{J} - S(f, \mathring{\mathcal{Q}})| \le \frac{\varepsilon}{2} + \frac{\varepsilon}{2} = \varepsilon\,.$$

Let us now prove the sufficiency. Once the stated condition is assumed, let us choose $\varepsilon = 1$ so that we can find a gauge δ_1 on I such that

$$|S(f, \mathring{\mathcal{P}}) - S(f, \mathring{\mathcal{Q}})| \leq 1,$$

whenever $\mathring{\mathcal{P}}$ and $\mathring{\mathcal{Q}}$ are δ_1-fine tagged partitions of I. Taking $\varepsilon = 1/2$, we can find a gauge δ_2 on I that we can choose so that $\delta_2(x) \leq \delta_1(x)$ for every $x \in I$, such that

$$|S(f, \mathring{\mathcal{P}}) - S(f, \mathring{\mathcal{Q}})| \leq \frac{1}{2}$$

whenever $\mathring{\mathcal{P}}$ and $\mathring{\mathcal{Q}}$ are δ_2-fine tagged partitions of I. We can continue this way, choosing $\varepsilon = 1/k$, with k a positive integer, and find a sequence $(\delta_k)_k$ of gauges on I such that, for every $x \in I$,

$$\delta_1(x) \geq \delta_2(x) \geq \cdots \geq \delta_k(x) \geq \delta_{k+1}(x) \geq \ldots,$$

and such that

$$|S(f, \mathring{\mathcal{P}}) - S(f, \mathring{\mathcal{Q}})| \leq \frac{1}{k}$$

whenever $\mathring{\mathcal{P}}$ and $\mathring{\mathcal{Q}}$ are δ_k-fine tagged partitions of I.

Let us fix, for every k, a δ_k-fine tagged partition $\mathring{\mathcal{P}}_k$ of I. We want to show that $(S(f, \mathring{\mathcal{P}}_k))_k$ is a Cauchy sequence of real numbers. Let $\bar{\varepsilon} > 0$ be given. Let us choose a positive integer N such that $N\bar{\varepsilon} \geq 1$. If $k_1 \geq N$ and $k_2 \geq N$, assuming, for instance, $k_2 \geq k_1$, then we have

$$|S(f, \mathring{\mathcal{P}}_{k_1}) - S(f, \mathring{\mathcal{P}}_{k_2})| \leq \frac{1}{k_1} \leq \frac{1}{N} \leq \bar{\varepsilon}.$$

This proves that $(S(f, \mathring{\mathcal{P}}_k))_k$ is a Cauchy sequence; hence, it has a finite limit, which we denote by $\mathcal{J}$.

Now we show that $\mathcal{J}$ is just the integral of f on I. Fix $\varepsilon > 0$, let n be a positive integer such that $n\varepsilon \geq 1$, and consider the gauge $\delta = \delta_n$. For every δ-fine tagged partition $\mathring{\mathcal{P}}$ of I and for every $k \geq n$, we have

$$|S(f, \mathring{\mathcal{P}}) - S(f, \mathring{\mathcal{P}}_k)| \leq \frac{1}{n} \leq \varepsilon.$$

Letting k tend to $+\infty$, we have that $S(f, \mathring{\mathcal{P}}_k)$ tends to $\mathcal{J}$, and consequently

$$|S(f, \mathring{\mathcal{P}}) - \mathcal{J}| \leq \varepsilon.$$

The proof is thus completed. ∎

7.10   Integrability on Subintervals

In this section we will see that if a function is integrable on an interval $I = [a, b]$, then it is also integrable on any of its subintervals. In particular, it is possible to consider its integral function. Moreover, we will see that if a function is integrable on two contiguous intervals, then it is also integrable on their union.

In what follows, it will be useful to consider the so-called "tagged subpartitions" of the interval I. A tagged subpartition is a set of the type

$$\Xi = \{(\xi_j, [\alpha_j, \beta_j]) : j = 1, \ldots, m\},$$

where the intervals $[\alpha_j, \beta_j]$ are nonoverlapping, but not necessarily contiguous, and $\xi_j \in [\alpha_j, \beta_j]$ for every $j = 1, \ldots, m$. For a tagged subpartition Ξ, it is still meaningful to consider the associated Riemann sum

$$S(f, \Xi) = \sum_{j=1}^{m} f(\xi_j)(\beta_j - \alpha_j).$$

Moreover, given a gauge δ on I, the tagged subpartition Ξ is δ-fine if, for every j, we have

$$\xi_j - \alpha_j \leq \delta(\xi_j) \quad \text{and} \quad \beta_j - \xi_j \leq \delta(\xi_j).$$

Let us state the property of "additivity on subintervals."

Theorem 7.18 *Given three points $a < c < b$, the function $f : [a, b] \to \mathbb{R}$ is integrable on $[a, b]$ if and only if it is integrable on both $[a, c]$ and $[c, b]$. In this case,*

$$\int_a^b f = \int_a^c f + \int_c^b f.$$

Proof We denote by $f_1 : [a, c] \to \mathbb{R}$ and $f_2 : [c, b] \to \mathbb{R}$ the two restrictions of f to $[a, c]$ and $[c, b]$, respectively.

Let us first assume that f is integrable on $[a, b]$ and prove that f_1 is integrable on $[a, c]$ by the Cauchy criterion. Fix $\varepsilon > 0$; since f is integrable on $[a, b]$, this verifies the Cauchy condition, and hence there is a gauge $\delta : [a, b] \to \mathbb{R}$ such that

$$|S(f, \mathring{\mathcal{P}}) - S(f, \mathring{\mathcal{Q}})| \leq \varepsilon$$

for every two δ-fine tagged partitions $\mathring{\mathcal{P}}, \mathring{\mathcal{Q}}$ of $[a, b]$. The restrictions of δ to $[a, c]$ and $[c, b]$ are two gauges $\delta_1 : [a, c] \to \mathbb{R}$ and $\delta_2 : [c, b] \to \mathbb{R}$. Now let $\mathring{\mathcal{P}}_1$ and $\mathring{\mathcal{Q}}_1$ be two δ_1-fine tagged partitions of $[a, c]$. Let us fix a δ_2-fine tagged partition $\mathring{\mathcal{P}}_2$ of

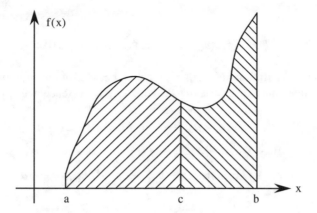

[c, b] and consider the tagged partition $\overset{\circ}{\mathcal{P}}$ of $[a, b]$ made by $\overset{\circ}{\mathcal{P}}_1 \cup \overset{\circ}{\mathcal{P}}_2$ and the tagged partition $\overset{\circ}{\mathcal{Q}}$ of $[a, b]$ made by $\overset{\circ}{\mathcal{Q}}_1 \cup \overset{\circ}{\mathcal{P}}_2$. It is clear that both $\overset{\circ}{\mathcal{P}}$ and $\overset{\circ}{\mathcal{Q}}$ are δ-fine. Moreover, we have

$$|S(f_1, \overset{\circ}{\mathcal{P}}_1) - S(f_1, \overset{\circ}{\mathcal{Q}}_1)| \; = |S(f, \overset{\circ}{\mathcal{P}}) - S(f, \overset{\circ}{\mathcal{Q}})| \leq \varepsilon ;$$

the Cauchy criterion thus applies, so that f is integrable on $[a, c]$. Analogously it can be proved that f is integrable on $[c, b]$.

Suppose now that f_1 is integrable on $[a, c]$ and f_2 on $[c, b]$. Let us then prove that f is integrable on $[a, b]$ with integral $\int_a^c f + \int_c^b f$. Once $\varepsilon > 0$ is fixed, there is a gauge δ_1 on $[a, c]$ and a gauge δ_2 on $[c, b]$ such that, for every δ_1-fine tagged partition $\overset{\circ}{\mathcal{P}}_1$ of $[a, c]$, we have

$$\left| S(f_1, \overset{\circ}{\mathcal{P}}_1) - \int_a^c f \right| \leq \frac{\varepsilon}{2} ,$$

and for every δ_2-fine tagged partition $\overset{\circ}{\mathcal{P}}_2$ of $[c, b]$, we have

$$\left| S(f_2, \overset{\circ}{\mathcal{P}}_2) - \int_c^b f \right| \leq \frac{\varepsilon}{2} .$$

We now define a gauge δ on $[a, b]$ as follows:

$$\delta(x) = \begin{cases} \min \left\{ \delta_1(x), \dfrac{c - x}{2} \right\} & \text{if } a \leq x < c , \\[2mm] \min \left\{ \delta_1(c), \delta_2(c) \right\} & \text{if } x = c , \\[2mm] \min \left\{ \delta_2(x), \dfrac{x - c}{2} \right\} & \text{if } c < x \leq b . \end{cases}$$

Let

$$\overset{\circ}{P} = \{(x_1, [a_0, a_1]), \ldots, (x_m, [a_{m-1}, a_m])\}$$

be a δ-fine tagged partition of $[a, b]$. Notice that, because of the particular choice of the gauge δ, there must be a certain $\bar{j}$ for which $x_{\bar{j}} = c$. Hence, we have

$$S(f, \overset{\circ}{P}) = f(x_1)(a_1 - a_0) + \cdots + f(x_{\bar{j}-1})(a_{\bar{j}-1} - a_{\bar{j}-2}) +$$
$$+ f(c)(c - a_{\bar{j}-1}) + f(c)(a_{\bar{j}} - c) +$$
$$+ f(x_{\bar{j}+1})(a_{\bar{j}+1} - a_{\bar{j}}) + \cdots + f(x_m)(a_m - a_{m-1}).$$

Let us set

$$\overset{\circ}{P}_1 = \{(x_1, [a_0, a_1]), \ldots, (x_{\bar{j}-1}, [a_{\bar{j}-2}, a_{\bar{j}-1}]), (c, [a_{\bar{j}-1}, c])\}$$

and

$$\overset{\circ}{P}_2 = \{(c, [c, a_{\bar{j}}]), (x_{\bar{j}+1}, [a_{\bar{j}}, a_{\bar{j}+1}]), \ldots, (x_m, [a_{m-1}, a_m])\}$$

(but in case $a_{\bar{j}-1}$ or $a_{\bar{j}}$ coincides with c, we will have to take away an element from one of the two). Then $\overset{\circ}{P}_1$ is a δ_1-fine tagged partition of $[a, c]$ and $\overset{\circ}{P}_2$ is a δ_2-fine tagged partition of $[c, b]$, and we have

$$S(f, \overset{\circ}{P}) = S(f_1, \overset{\circ}{P}_1) + S(f_2, \overset{\circ}{P}_2).$$

Consequently,

$$\left| S(f, \overset{\circ}{P}) - \left(\int_a^c f + \int_c^b f \right) \right| \leq \left| S(f_1, \overset{\circ}{P}_1) - \int_a^c f \right| + \left| S(f_2, \overset{\circ}{P}_2) - \int_c^b f \right|$$
$$\leq \frac{\varepsilon}{2} + \frac{\varepsilon}{2} = \varepsilon,$$

which completes the proof.   ∎

Example Consider the function $f : [0, 2] \to \mathbb{R}$ defined by

$$f(x) = \begin{cases} 3 & \text{if } x \in [0, 1], \\ 5 & \text{if } x \in {]}1, 2]. \end{cases}$$

Since f is constant on $[0, 1]$ with value 3, it is integrable there, and $\int_0^1 f = 3$. Moreover, on the interval $[1, 2]$ the function f differs from the constant 5 at only one point: We have that the function $g(x) = f(x) - 5$ is zero except for $x = 1$.

As we have shown at the end of Sect. 7.6, g is integrable on $[1, 2]$ with zero integral and so, since $f(x) = g(x) + 5$, even f is integrable and

$$\int_1^2 f(x)\,dx = \int_1^2 g(x)\,dx + \int_1^2 5\,dx = 0 + 5 = 5\,.$$

In conclusion,

$$\int_0^2 f(x)\,dx = \int_0^1 f(x)\,dx + \int_1^2 f(x)\,dx = 3 + 5 = 8\,.$$

It is easy to see from the theorem just proved that if a function is integrable on an interval I, it is still integrable on any subinterval of I. Moreover, we have the following corollary.

Corollary 7.19 *If $f : I \to \mathbb{R}$ is integrable, for any three arbitrarily chosen points u, v, w in I one has*

$$\int_u^w f = \int_u^v f + \int_v^w f\,.$$

Proof The case $u < v < w$ follows immediately from the previous theorem. The other possible cases are easily obtained using the conventions on the integrals with exchanged or equal endpoints.   ∎

7.11   *R*-Integrable and Continuous Functions

Let us introduce an important class of integrable functions. As usual, $I = [a, b]$.

Definition 7.20 We say that an integrable function $f : I \to \mathbb{R}$ is "*R*-integrable" (or "integrable according to Riemann") if among all possible gauges $\delta : I \to \mathbb{R}$ that verify the definition of integrability it is always possible to choose one that is constant on I.

We can immediately see, repeating the proofs, that the set of *R*-integrable functions is a vector subspace of the space of integrable functions. Moreover, the following Cauchy criterion holds for *R*-integrable functions whenever one considers only constant gauges.

Theorem 7.21 *A function $f : I \to \mathbb{R}$ is R-integrable if and only if for every $\varepsilon > 0$ there is a $\delta > 0$ (i.e., a constant gauge δ) such that, taking two δ-fine tagged* *partitions* $\mathring{\mathcal{P}}, \mathring{\mathcal{Q}}$ *of I, one has*

$$|S(f, \mathring{\mathcal{P}}) - S(f, \mathring{\mathcal{Q}})| \le \varepsilon .$$

We now want to establish the integrability of continuous functions. Indeed, in the following two theorems we will prove that they are both R-integrable and primitivable.

To simplify the expressions to come, we will denote by $\mu(K)$ the length of a bounded interval K. In particular,

$$\mu([a, b]) = b - a .$$

It will be useful, moreover, to set $\mu(\varnothing) = 0$. Here is the first theorem.

Theorem 7.22 *Every continuous function* $f : I \to \mathbb{R}$ *is R-integrable.*

Proof Fix $\varepsilon > 0$. Since f is continuous on a compact interval, by Heine's Theorem 4.12, it is uniformly continuous there, so that there is a $\delta > 0$ such that, for x and x' in I,

$$|x - x'| \le 2\delta \quad \Rightarrow \quad |f(x) - f(x')| \le \frac{\varepsilon}{b - a} .$$

We will verify the Cauchy criterion for the R-integrability by considering the constant gauge δ thus found. Let

$$\mathring{\mathcal{P}} = \{(x_1, [a_0, a_1]), \dots, (x_m, [a_{m-1}, a_m])\}$$

and

$$\mathring{\mathcal{Q}} = \{(\tilde{x}_1, [\tilde{a}_0, \tilde{a}_1]), \dots, (\tilde{x}_{\widetilde{m}}, [\tilde{a}_{\widetilde{m}-1}, \tilde{a}_{\widetilde{m}}])\}$$

be two δ-fine tagged partitions of I. Let us define the intervals (perhaps empty or reduced to a single point)

$$I_{j,k} = [a_{j-1}, a_j] \cap [\tilde{a}_{k-1}, \tilde{a}_k] .$$

Then we have

$$a_j - a_{j-1} = \sum_{k=1}^{\widetilde{m}} \mu(I_{j,k}), \qquad \tilde{a}_k - \tilde{a}_{k-1} = \sum_{j=1}^{m} \mu(I_{j,k}),$$

and, if $I_{j,k}$ is nonempty, $|x_j - \tilde{x}_k| \leq 2\delta$. Hence,

$$|S(f, \mathring{\mathcal{P}}) - S(f, \mathring{\mathcal{Q}})| = \left| \sum_{j=1}^{m} \sum_{k=1}^{\tilde{m}} f(x_j) \mu(I_{j,k}) - \sum_{k=1}^{\tilde{m}} \sum_{j=1}^{m} f(\tilde{x}_k) \mu(I_{j,k}) \right|$$

$$= \left| \sum_{j=1}^{m} \sum_{k=1}^{\tilde{m}} [f(x_j) - f(\tilde{x}_k)] \mu(I_{j,k}) \right|$$

$$\leq \sum_{j=1}^{m} \sum_{k=1}^{\tilde{m}} |f(x_j) - f(\tilde{x}_k)| \mu(I_{j,k})$$

$$\leq \sum_{j=1}^{m} \sum_{k=1}^{\tilde{m}} \frac{\varepsilon}{b-a} \mu(I_{j,k}) = \varepsilon.$$

Therefore, the Cauchy criterion applies, and the proof is completed.   ∎

Here is the second theorem.

Theorem 7.23 *Every continuous function $f : [a, b] \to \mathbb{R}$ is primitivable.*

Proof Since it is continuous, f is integrable on every subinterval of $[a, b]$, so we can consider its integral function $\int_a^\cdot f$. Let us prove that it is a primitive of f, i.e., that if a point x_0 is taken in $[a, b]$, the derivative of $\int_a^\cdot f$ at x_0 coincides with $f(x_0)$. We first consider the case where $x_0 \in \,]a, b[$. We want to prove that

$$\lim_{h \to 0} \frac{1}{h} \left(\int_a^{x_0+h} f - \int_a^{x_0} f \right) = f(x_0).$$

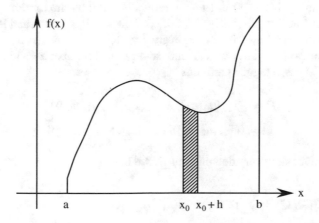

Equivalently, since

$$\frac{1}{h}\left(\int_a^{x_0+h} f - \int_a^{x_0} f\right) - f(x_0) = \frac{1}{h}\int_{x_0}^{x_0+h} (f(x) - f(x_0))\, dx\,,$$

we will show that

$$\lim_{h\to 0}\frac{1}{h}\int_{x_0}^{x_0+h} (f(x) - f(x_0))\, dx = 0\,.$$

Fix $\varepsilon > 0$. Since f is continuous in x_0, there is a $\delta > 0$ such that, for every $x \in [a, b]$ satisfying $|x - x_0| \le \delta$, one has $|f(x) - f(x_0)| \le \varepsilon$. Taking h such that $0 < |h| \le \delta$, we distinguish two cases. If $0 < h \le \delta$, then

$$\left|\frac{1}{h}\int_{x_0}^{x_0+h} (f(x) - f(x_0))\, dx\right| \le \frac{1}{h}\int_{x_0}^{x_0+h} |f(x) - f(x_0)|\, dx \le \frac{1}{h}\int_{x_0}^{x_0+h} \varepsilon\, dx = \varepsilon\,;$$

on the other hand, if $-\delta \le h < 0$, then we have

$$\left|\frac{1}{h}\int_{x_0}^{x_0+h} (f(x) - f(x_0))\, dx\right| \le \frac{1}{-h}\int_{x_0+h}^{x_0} |f(x) - f(x_0)|\, dx \le \frac{1}{-h}\int_{x_0+h}^{x_0} \varepsilon\, dx = \varepsilon\,,$$

and the proof is completed when $x_0 \in {]}a, b{[}$. In case $x_0 = a$ or $x_0 = b$, we proceed analogously, considering the right or the left derivative, respectively. ∎

Notice that it is not always possible to find an elementary expression for the primitive of a continuous function. As an example, the function $f(x) = \sin(x^2)$, being continuous, is primitivable, but there is no elementary formula defining any of its primitives. By "elementary formula" we mean an analytic formula where only polynomials, exponentials, logarithms, and trigonometric functions appear.

Let us now prove that the Dirichlet function $\mathcal{D}$ is not R-integrable on any interval $[a, b]$ (remember that $\mathcal{D}(x)$ is 1 on the rationals and 0 on the irrationals). We will show that the Cauchy criterion is not verified. Take $\delta > 0$ constant, and let $a = a_0 < a_1 < \cdots < a_m = b$ be such that, for every $j = 1, \dots, m$, one has $a_j - a_{j-1} \le \delta$. In every interval $[a_{j-1}, a_j]$ we can choose a rational number x_j and an irrational number $\tilde{x}_j$. The two tagged partitions

$$\mathring{\mathcal{P}} = \{(x_1, [a_0, a_1]), \dots, (x_m, [a_{m-1}, a_m])\}\,,$$

$$\mathring{\mathcal{Q}} = \{(\tilde{x}_1, [a_0, a_1]), \dots, (\tilde{x}_m, [a_{m-1}, a_m])\}$$

are δ-fine, and, by the very definition of $\mathcal{D}$, we have

$$S(\mathcal{D}, \mathring{\mathcal{P}}) - S(\mathcal{D}, \mathring{\mathcal{Q}}) = \sum_{j=1}^m [\mathcal{D}(x_j) - \mathcal{D}(\tilde{x}_j)](a_j - a_{j-1}) = \sum_{j=1}^m (a_j - a_{j-1}) = b - a\,.$$

Since $\delta > 0$ was taken arbitrarily, the Cauchy criterion for R-integrability does not hold, so that f cannot be R-integrable on $[a, b]$.

7.12 Two Theorems Involving Limits

Let $I = [a, b]$ and $f : I \to \mathbb{R}$ be a continuous function; recall that the integral function $\int_a^{\cdot} f$ is a primitive of f, so it is surely continuous on I. It is then possible to define the function $\Psi : \mathcal{C}(I, \mathbb{R}) \to \mathcal{C}(I, \mathbb{R})$ as

$$[\Psi(f)](x) = \int_a^x f.$$

Taking $f, g \in \mathcal{C}(I, \mathbb{R})$, for every $x \in [a, b]$ one has that

$$\left| [\Psi(f)](x) - [\Psi(g)](x) \right| = \left| \int_a^x (f - g) \right| \le \int_a^x |f - g|$$

$$\le \int_a^b |f - g| \le (b - a)\|f - g\|_\infty,$$

whence

$$\|\Psi(f) - \Psi(g)\|_\infty \le (b - a)\|f - g\|_\infty.$$

This implies that Ψ is a continuous function. We will use this fact in the following two theorems involving limits.

We consider the situation where a sequence of continuous functions $(f_n)_n$ converges pointwise to a function f, i.e., for every $x \in I$,

$$\lim_n f_n(x) = f(x).$$

The question is whether f is integrable on I, with

$$\int_I f = \lim_n \int_I f_n,$$

i.e., whether

$$\int_I \lim_n f_n = \lim_n \int_I f_n.$$

In other words, we wonder if it is possible to commute the operations of integral and limit.

Example Let us first show that in some cases the answer could be no. Consider the functions $f_n : [0, \pi] \to \mathbb{R}$, with $n = 1, 2, \ldots$, defined by

$$f_n(x) = \begin{cases} n \sin(nx) & \text{if } x \in [0, \frac{\pi}{n}], \\ 0 & \text{otherwise.} \end{cases}$$

For every $x \in [0, \pi]$ we have $\lim_n f_n(x) = 0$, whereas

$$\int_0^\pi f_n(x)\, dx = \int_0^{\pi/n} n \sin(nx)\, dx = \int_0^\pi \sin(t)\, dt = 2 \,.$$

Hence, in this case,

$$\int_0^\pi \lim_n f_n = 0 \neq 2 = \lim_n \int_0^\pi f_n \,.$$

In the following theorem, which will be generalized later (Theorem 9.13), the answer to the foregoing question is positive, provided we assume the convergence to be uniform.

Theorem 7.24 *Let $(f_n)_n$ be a uniformly convergent sequence in $\mathcal{C}([a, b], \mathbb{R})$. Then*

$$\int_a^b \lim_n f_n = \lim_n \int_a^b f_n \,.$$

Proof Let $\lim_n f_n = f : I \to \mathbb{R}$. Since the convergence is uniform, we know that $f \in \mathcal{C}(I, \mathbb{R})$. Moreover, since Ψ is continuous, we have that $\lim_n \Psi(f_n) = \Psi(f)$, i.e.,

$$\lim_n [\Psi(f_n)](x) = [\Psi(f)](x) \,, \quad \text{uniformly in } x \in [a, b] \,.$$

In particular, taking $x = b$,

$$\lim_n \int_a^b f_n = \int_a^b f \,,$$

which is what we wanted to prove. ∎

In the second theorem, an analogous question concerning the possibility of commuting the operations of derivative and limit is analyzed.

Theorem 7.25 *Let $x_0 \in I$, $y_0 \in \mathbb{R}$, $(f_n)_n$ be a sequence in $C^1(I, \mathbb{R})$ and $g \in C(I, \mathbb{R})$ be such that*

$$\lim_n f_n(x_0) = y_0 \quad and \quad \lim_n f_n' = g \ uniformly \ on \ I.$$

Then $(f_n)_n$ converges uniformly to some function f. Moreover, $f \in C^1(I, \mathbb{R})$ and $f' = g$. Consequently, we can write

$$\frac{d}{dx}\left(\lim_n f_n(x) \right) = \lim_n \left(\frac{d}{dx} f_n(x) \right).$$

Proof Let us define the function $f : I \to \mathbb{R}$ as

$$f(x) = y_0 + \int_{x_0}^{x} g(t)\, dt.$$

Since g is continuous, the function f is differentiable and $f'(x) = g(x)$ for every $x \in I$. In particular, $f \in C^1(I, \mathbb{R})$. The proof will be completed showing that $(f_n)_n$ converges uniformly to f.

By the Fundamental Theorem, for every $n \in \mathbb{N}$ and $x \in I$ we can write

$$f_n(x) = f_n(x_0) + \int_{x_0}^{x} f_n'(t)\, dt,$$

i.e.,

$$f_n(x) = f_n(x_0) + [\Psi(f_n')](x) - [\Psi(f_n')](x_0).$$

Since $\lim_n f_n' = g$ in $C(I, \mathbb{R})$, we have that $\lim_n \Psi(f_n') = \Psi(g)$ in $C(I, \mathbb{R})$, i.e.,

$$\lim_n [\Psi(f_n')](x) = [\Psi(g)](x), \quad \text{uniformly in } x \in I.$$

Hence, since also $\lim_n f_n(x_0) = y_0$, we have that

$$\lim_n f_n(x) = y_0 + [\Psi(g)](x) - [\Psi(g)](x_0) = y_0 + \int_{x_0}^{x} g(t)\, dt, \quad \text{uniformly in } x \in I.$$

We have thus proved that $(f_n)_n$ converges uniformly to f. ∎

7.13   Integration on Noncompact Intervals

We begin by considering a function $f : [a, b[\to \mathbb{R}$, where $b \le +\infty$. Assume that f is integrable on every compact interval of the type $[a, c]$, with $c \in]a, b[$. This happens, for instance, when f is continuous on $[a, b[$.

Definition 7.26 We say that a function $f : [a, b[\to \mathbb{R}$ is "integrable" if f is integrable on $[a, c]$ for every $c \in]a, b[$, and the limit

$$\lim_{c \to b^-} \int_a^c f$$

exists and is finite. In that case, the preceding limit is called the "integral" of f on $[a, b[$ and it is denoted by $\int_a^b f$, or by $\int_a^b f(x)\, dx$.

In particular, if $b = +\infty$, then we will write $\int_a^{+\infty} f$, or $\int_a^{+\infty} f(x)\, dx$.

Examples Let $a > 0$; it is readily seen that the function $f : [a, +\infty[\to \mathbb{R}$, defined by $f(x) = x^{-\alpha}$, is integrable if and only if $\alpha > 1$, in which case we have

$$\int_a^{+\infty} \frac{dx}{x^\alpha} = \frac{a^{1-\alpha}}{\alpha - 1}.$$

Consider now the case $a < b < +\infty$. It can be verified that the function $f : [a, b[\to \mathbb{R}$, defined by $f(x) = (b - x)^{-\beta}$, is integrable if and only if $\beta < 1$, in which case we have

$$\int_a^b \frac{dx}{(b - x)^\beta} = \frac{(b - a)^{1-\beta}}{1 - \beta}.$$

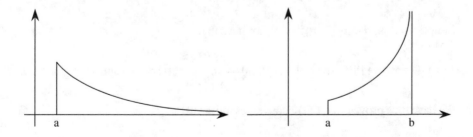

We also say that *the integral converges* if the function f is integrable on $[a, b[$, i.e., when the limit $\lim_{c \to b^-} \int_a^c f$ exists and is finite. If the limit does not exist, we say *the integral is undetermined*. If it exists and equals $+\infty$ or $-\infty$, we say that *the integral diverges* to $+\infty$ or to $-\infty$, respectively.

It is clear that the convergence of the integral depends solely on the behavior of the function "near" the point b. In other words, if the function is modified outside a neighborhood of b, the convergence of the integral is by no means compromised.

Let us now state the **Cauchy criterion**.

Theorem 7.27 *Let $f : [a, b[\to \mathbb{R}$ be a function that is integrable on $[a, c]$, for every $c \in]a, b[$. Then f is integrable on $[a, b[$ if and only if for every $\varepsilon > 0$ there is a $\bar{c} \in]a, b[$ such that, for any c' and c'' in $[\bar{c}, b[$, we have that*

$$\left| \int_{c'}^{c''} f \right| \le \varepsilon \,.$$

Proof It is a direct consequence of Theorem 4.15, when applied to the function $F : [a, b[\to \mathbb{R}$ defined as $F(c) = \int_a^c f$. ∎

From the Cauchy criterion we deduce the following **comparison criterion**.

Theorem 7.28 *Let $f : [a, b[\to \mathbb{R}$ be a function that is integrable on $[a, c]$, for every $c \in]a, b[$. If there is an integrable function $g : [a, b[\to \mathbb{R}$ such that, for every $x \in [a, b[$,*

$$|f(x)| \le g(x) \,,$$

then f is integrable on $[a, b[$, too.

Proof Once $\varepsilon > 0$ is fixed, there is a $\bar{c} \in]a, b[$ such that, taking arbitrarily c', c'' in $[\bar{c}, b[$, one has that $|\int_{c'}^{c''} g| \le \varepsilon$. If, for example, $c' \le c''$, since $-g \le f \le g$, we have

$$-\int_{c'}^{c''} g \le \int_{c'}^{c''} f \le \int_{c'}^{c''} g \,,$$

and therefore

$$\left| \int_{c'}^{c''} f \right| \le \int_{c'}^{c''} g \le \varepsilon \,.$$

The Cauchy criterion then applies, whence the conclusion. ∎

Note that it would have been sufficient to assume the inequality $|f(x)| \le g(x)$ on $[\bar{c}, b[$. As an immediate consequence, we have the following corollary.

Corollary 7.29 *Let* $f : [a, b[\rightarrow \mathbb{R}$ *be a function that is integrable on* $[a, c]$ *for every* $c \in]a, b[$ *. If* $|f|$ *is integrable on* $[a, b[$ *, then* f *is, too, and*

$$\left| \int_a^b f \right| \leq \int_a^b |f| \,.$$

A function satisfying the assumption of the preceding corollary will be said to be L-integrable. Let us now state a corollary of the comparison criterion that is often used in practice.

Corollary 7.30 *Let* $f, g : [a, b[\rightarrow \mathbb{R}$ *be two functions with positive values that are integrable on* $[a, c]$ *for every* $c \in]a, b[$ *. Assume that the limit*

$$L = \lim_{x \rightarrow b^-} \frac{f(x)}{g(x)}$$

exists. Then the following conclusions hold:

(a) *If* $L \in]0, +\infty[$ *, then* f *is integrable on* $[a, b[$ *if and only if* g *is.*
(b) *If* $L = 0$ *and* g *is integrable on* $[a, b[$ *, then* f *is as well.*
(c) *If* $L = +\infty$ *and* g *is not integrable on* $[a, b[$ *, then neither* f *is.*

Proof Case (a). If $L \in]0, +\infty[$, then there exists a $\bar{c} \in]a, b[$ such that

$$x \in [\bar{c}, b[\quad \Rightarrow \quad \frac{L}{2} \leq \frac{f(x)}{g(x)} \leq \frac{3L}{2} \,,$$

i.e.,

$$x \in [\bar{c}, b[\quad \Rightarrow \quad f(x) \leq \frac{3L}{2} g(x) \quad \text{and} \quad g(x) \leq \frac{2}{L} f(x) \,.$$

The conclusion then follows from the comparison criterion.

Case (b). If $L = 0$, then there exists a $\bar{c} \in]a, b[$ such that, if $x \in [\bar{c}, b[$, then $f(x) \leq g(x)$, and the comparison criterion applies.

Case (c). If $L = +\infty$, then we reduce this to case (b) by exchanging the roles of f and g. ∎

Example Consider the function $f : [0, +\infty[\rightarrow \mathbb{R}$ defined by

$$f(x) = e^{1/(x^2+1)} - 1 \,.$$

As a comparison function, we take

$$g(x) = \frac{1}{x^2 + 1}.$$

It is integrable on $[0, +\infty[$, with

$$\int_0^{+\infty} \frac{1}{x^2 + 1}\, dx = \lim_{c \to +\infty} [\arctan x]_0^c = \frac{\pi}{2}.$$

Since

$$\lim_{x \to +\infty} \frac{f(x)}{g(x)} = \lim_{t \to 0^+} \frac{e^t - 1}{t} = 1,$$

we conclude that f is integrable on $[0, +\infty[$ as well.

We now consider the case of a function $f :]a, b] \to \mathbb{R}$, with $a \geq -\infty$. There is an analogous definition of its integral.

Definition 7.31 We say that a function $f :]a, b] \to \mathbb{R}$ is "integrable" if f is integrable on $[c, b]$ for every $c \in]a, b[$, and the limit

$$\lim_{c \to a^+} \int_c^b f$$

exists and is finite. In that case, the preceding limit is called the "integral" of f on $]a, b]$, and it is denoted by $\int_a^b f$ or $\int_a^b f(x)\, dx$.

Given the function $f :]a, b] \to \mathbb{R}$, it is possible to consider the function $g : [a', b'[\to \mathbb{R}$, with $a' = -b$ and $b' = -a$, defined by $g(x) = f(-x)$. It is easy to see that f is integrable on $]a, b]$ if and only if g is integrable on $[a', b'[$. In this way, we are led back to the previous context.

We will also define the integral of a function $f :]a, b[\to \mathbb{R}$, with $-\infty \leq a < b \leq +\infty$, in the following way.

Definition 7.32 We say that $f :]a, b[\to \mathbb{R}$ is "integrable" if, once we fix a point $p \in]a, b[$, the function f is integrable on $[p, b[$ and on $]a, p]$. In that case, the "integral" of f on $]a, b[$ is defined by

$$\int_a^b f = \int_a^p f + \int_p^b f.$$

It is easy to verify that the given definition does not depend on the choice of $p \in]a, b[$.

Examples If $a, b \in \mathbb{R}$, one can verify that the function

$$f(x) = \frac{1}{[(x-a)(b-x)]^\beta}$$

is integrable on $]a, b[$ if and only if $\beta < 1$. In this case, it is possible to choose, for instance, $p = (a+b)/2$.

Another case arises when $a = -\infty$ and $b = +\infty$. For example, we can easily verify that the function $f(x) = (x^2 + 1)^{-1}$ is integrable on $] - \infty, +\infty[$. Taking, for instance, $p = 0$, we have

$$\int_{-\infty}^{+\infty} \frac{1}{x^2 + 1} \, dx = \int_{-\infty}^{0} \frac{1}{x^2 + 1} \, dx + \int_{0}^{+\infty} \frac{1}{x^2 + 1} \, dx = \pi \, .$$

Another case that might be encountered in the applications is when a function happens not to be defined in an interior point of an interval.

Definition 7.33 Given $a < q < b$, we say that a function $f : [a, b] \setminus \{q\} \to \mathbb{R}$ is integrable if f is integrable on both $[a, q[$ and $]q, b]$. In that case, we set

$$\int_a^b f = \int_a^q f + \int_q^b f \, .$$

For example, if $a < 0 < b$, then the function $f(x) = \sqrt{|x|}/x$ is integrable on $[a, b] \setminus \{0\}$, and

$$\int_a^b \frac{\sqrt{|x|}}{x} \, dx = \int_a^0 \frac{-1}{\sqrt{-x}} \, dx + \int_0^b \frac{1}{\sqrt{x}} \, dx = 2\sqrt{b} - 2\sqrt{-a} \, .$$

On the other hand, the function $f(x) = 1/x$ is not integrable on $[-1, 1] \setminus \{0\}$, even if the fact that f is odd caused us to say that its integral was equal to zero. However, in that case, some important properties of the integral would be lost, for example, the additivity on subintervals.

7.14 Functions with Vector Values

We now consider a function $f : [a, b] \to \mathbb{R}^N$ with vector values. As usual, we can write

$$f(x) = (f_1(x), \ldots, f_N(x)) \, ,$$

where the functions $f_k : [a, b] \to \mathbb{R}$ are the components of f. We say that f is integrable whenever all the components are integrable functions, and in that case we can define the integral of f as

$$\int_a^b f(x)\, dx = \left(\int_a^b f_1(t)\, dt, \ldots, \int_a^b f_N(t)\, dt \right).$$

The integral is thus a vector in $\mathbb{R}^N$.

A particular case is encountered when $f : [a, b] \to \mathbb{C}$. Writing

$$f(x) = f_1(x) + i f_2(x),$$

we will have

$$\int_a^b f(x)\, dx = \int_a^b f_1(x)\, dx + i \int_a^b f_2(x)\, dx.$$

Theorem 7.34 *Assume that both* $f : [a, b] \to \mathbb{R}^N$ *and* $\|f\| : [a, b] \to \mathbb{R}$ *are integrable and* $a < b$. *Then*

$$\left\| \int_a^b f(x)\, dx \right\| \le \int_a^b \|f(x)\|\, dx.$$

Proof Set $v = \int_a^b f(x)\, dx$, i.e., $v = (v_1, \ldots, v_N)$, with $v_k = \int_a^b f_k(x)\, dx$ for $k = 1, \ldots, N$. If $v = 0$, then the statement surely holds. Assume now that $v \ne 0$. Then, using the Schwarz inequality,

$$\|v\|^2 = \sum_{k=1}^N v_k^2 = \sum_{k=1}^N v_k \int_a^b f_k(x)\, dx = \sum_{k=1}^N \int_a^b v_k f_k(x)\, dx = \int_a^b \sum_{k=1}^N v_k f_k(x)\, dx$$

$$= \int_a^b v \cdot f(x)\, dx \le \int_a^b \|v\|\, \|f(x)\|\, dx = \|v\| \int_a^b \|f(x)\|\, dx,$$

whence

$$\|v\| \le \int_a^b \|f(x)\|\, dx,$$

which is what we wanted to prove.   ∎

Now let $F : [a, b] \to \mathbb{R}^N$ be a function whose components $F_k : [a, b] \to \mathbb{R}$ are differentiable. In this case we say that F is differentiable, and, writing

$$F(x) = (F_1(x), \ldots, F_N(x)),$$

we can define its derivative at some $x_0 \in [a, b]$ as

$$F'(x_0) = \lim_{x \to x_0} \frac{F(x) - F(x_0)}{x - x_0}$$

$$= \left(\lim_{x \to x_0} \frac{F_1(x) - F_1(x_0)}{x - x_0}, \ldots, \lim_{x \to x_0} \frac{F_N(x) - F_N(x_0)}{x - x_0} \right)$$

$$= (F_1'(x_0), \ldots, F_N'(x_0)).$$

Here is a version of the Fundamental Theorem in this context.

Theorem 7.35 (Fundamental Theorem—III)   *If $F : [a, b] \to \mathbb{R}^N$ is differentiable, then $F' : [a, b] \to \mathbb{R}^N$ is integrable, and*

$$\int_a^b F'(x)\, dx = F(b) - F(a).$$

Proof Since each component $F_k : [a, b] \to \mathbb{R}$ is differentiable, by the Fundamental Theorem we know that the derivatives $F_k' : [a, b] \to \mathbb{R}$ are integrable and

$$\int_a^b F_k'(x)\, dx = F_k(b) - F_k(a), \quad \text{for every } k = 1, \ldots, N.$$

Then F' is integrable, and

$$\int_a^b F'(x)\, dx = \left(\int_a^b F_1'(x)\, dx, \ldots, \int_a^b F_N'(x)\, dx \right)$$

$$= (F_1(b) - F_1(a), \ldots, F_N(b) - F_N(a))$$

$$= \left(F_1(b), \ldots, F_N(b) \right) - \left(F_1(a), \ldots, F_N(a) \right)$$

$$= F(b) - F(a),$$

thereby completing the proof.   ∎

Part III

Further Developments

Numerical Series and Series of Functions

8

8.1   Introduction and First Properties

Let V be a normed vector space. Given a sequence $(a_k)_k$ in V, the associated "series" is the sequence $(s_n)_n$ defined by

$$s_0 = a_0 \,,$$

$$s_1 = a_0 + a_1 \,,$$

$$s_2 = a_0 + a_1 + a_2 \,,$$

$$\cdots$$

$$s_n = a_0 + a_1 + a_2 + \cdots + a_n \,,$$

$$\cdots$$

The element a_k is called the "kth term" of the series, whereas $s_n = \sum_{k=0}^{n} a_k$ is said to be the "nth partial sum" of the series. Whenever $(s_n)_n$ has a limit in V, we say that the series "converges." In that case, the limit $S = \lim_n s_n$ is said to be the "sum of the series," and we will write

$$S = \lim_n \left(\sum_{k=0}^{n} a_k \right) = \sum_{k=0}^{\infty} a_k \,,$$

and sometimes we will also use the notation

$$S = a_0 + a_1 + a_2 + \cdots + a_n + \ldots$$

A. Fonda, *A Modern Introduction to Mathematical Analysis*,
https://doi.org/10.1007/978-3-031-23713-3_8

However, by an abuse of notation, the series $(s_n)_n$ itself is often denoted by the same symbols,

$$\text{either} \quad \sum_{k=0}^{\infty} a_k, \quad \text{or} \quad a_0 + a_1 + a_2 + \cdots + a_n + \ldots.$$

Sometimes, for brevity's sake, we will simply write $\sum_k a_k$.

Let us analyze three examples, in all of which $V = \mathbb{R}$.

Example 1 For $\alpha \in \mathbb{R}$, the "geometric series"

$$1 + \alpha + \alpha^2 + \alpha^3 + \cdots + \alpha^n + \ldots$$

has as its kth term $a_k = \alpha^k$. If $\alpha \neq 1$, the nth partial sum is

$$s_n = \sum_{k=0}^{n} \alpha^k = \frac{\alpha^{n+1} - 1}{\alpha - 1},$$

whereas if $\alpha = 1$, then we have $s_n = n + 1$. Hence, the series converges if and only if $|\alpha| < 1$, in which case its sum is

$$\sum_{k=0}^{\infty} \alpha^k = \frac{1}{1 - \alpha}.$$

Notice that if $\alpha \geq 1$, then $\lim_n s_n = +\infty$, whereas if $\alpha \leq -1$, then the sequence $(s_n)_n$ has no limit since $\liminf_n s_n = -\infty$ and $\limsup_n s_n = +\infty$.

In general, if the sequence $(s_n)_n$ has no limit, then we say that the series is "undetermined." On the other hand, for real valued series we say that

- The series "diverges to $+\infty$" if $\lim_n s_n = +\infty$.
- The series "diverges to $-\infty$" if $\lim_n s_n = -\infty$.

Example 2 The series

$$\frac{1}{1 \cdot 2} + \frac{1}{2 \cdot 3} + \frac{1}{3 \cdot 4} + \cdots + \frac{1}{(n+1) \cdot (n+2)} + \cdots$$

has as its kth term $a_k = \frac{1}{(k+1)(k+2)}$. It is a "telescopic series":

$$\left(\frac{1}{1} - \frac{1}{2}\right) + \left(\frac{1}{2} - \frac{1}{3}\right) + \left(\frac{1}{3} - \frac{1}{4}\right) + \cdots + \left(\frac{1}{n+1} - \frac{1}{n+2}\right) + \cdots$$

Hence, simplifying,

$$s_n = 1 - \frac{1}{n+2},$$

leading to $\lim_n s_n = 1$. We have thus proved that the series converges and that its sum is equal to 1. We can then write

$$\sum_{k=0}^{\infty} \frac{1}{(k+1)(k+2)} = 1.$$

In the preceding example, one could use different notations in the sum like, e.g.,

$$\sum_{k=1}^{\infty} \frac{1}{k(k+1)} = 1,$$

or variants of it; for example, the letter "k" could be replaced by any other, so that, e.g., $\sum_{j=1}^{\infty} \frac{1}{j(j+1)} = 1$. These remarks apply indeed to all series.

Example 3 The "harmonic series"

$$1 + \frac{1}{2} + \frac{1}{3} + \frac{1}{4} + \cdots + \frac{1}{n+1} + \cdots$$

has as its kth term $a_k = \frac{1}{k+1}$. It diverges to $+\infty$; we can see this by writing it as

$$1 + \frac{1}{2} + \left(\frac{1}{3} + \frac{1}{4}\right) + \left(\frac{1}{5} + \frac{1}{6} + \frac{1}{7} + \frac{1}{8}\right)$$
$$+ \left(\frac{1}{9} + \frac{1}{10} + \frac{1}{11} + \frac{1}{12} + \frac{1}{13} + \frac{1}{14} + \frac{1}{15} + \frac{1}{16}\right) + \cdots,$$

gathering together the first 2 of its terms, then 4, then 8, then 16, and so on, doubling their number each time. It is easy to see that the sums in the parentheses are all greater than $\frac{1}{2}$. Hence, the sequence of partial sums must have the limit $+\infty$. We can then write

$$\sum_{k=0}^{\infty} \frac{1}{k+1} = +\infty.$$

It must be said that the explicit computation of the sum of a series is a rare event. Very often we will already be satisfied when proving that a series converges or not.

It is important to notice that the convergence of a series is not compromised if only a finite number of its terms are modified. Indeed, if the series converges, we can change, add, or delete a finite number of initial terms, and the new series thus obtained will still converge. In contrast, if the series does not converge, because either it is undetermined or diverges to $\pm\infty$, the same will be true of the modified series.

Theorem 8.1 *If a series $\sum_k a_k$ converges, then*

$$\lim_n a_n = 0.$$

Proof Let $\lim_n s_n = S \in V$. Then also $\lim_n s_{n-1} = S$, and hence

$$\lim_n a_n = \lim_n (s_n - s_{n-1}) = \lim_n s_n - \lim_n s_{n-1} = S - S = 0,$$

which is what we wanted to prove.   ∎

Let us study the behavior of series with respect to the sum and to the product by a scalar.

Theorem 8.2 *Assume that the two series $\sum_k a_k$ and $\sum_k b_k$ converge with sums A and B, respectively. Then the series $\sum_k (a_k + b_k)$ also converges, and its sum is $A + B$. Moreover, for any fixed $\alpha \in \mathbb{R}$, the series $\sum_k (\alpha a_k)$ also converges, and its sum is αA. We will briefly write*

$$\sum_{k=0}^{\infty} (a_k + b_k) = \sum_{k=0}^{\infty} a_k + \sum_{k=0}^{\infty} b_k, \qquad \sum_{k=0}^{\infty} (\alpha a_k) = \alpha \sum_{k=0}^{\infty} a_k.$$

Proof Let $s_n = \sum_{k=0}^{n} a_k$ and $s'_n = \sum_{k=0}^{n} b_k$. Then

$$s_n + s'_n = \sum_{k=0}^{n} (a_k + b_k), \qquad \alpha s_n = \sum_{k=0}^{n} (\alpha a_k),$$

and the result follows, passing to the limits.   ∎

Let us see how the Cauchy criterion adapts to series in Banach spaces.

Theorem 8.3 *If V is a Banach space, the series $\sum_k a_k$ converges if and only if*

$$\forall \varepsilon > 0 \quad \exists \bar{n} : \quad n > m \geq \bar{n} \quad \Rightarrow \quad \left\| \sum_{k=m+1}^{n} a_k \right\| < \varepsilon.$$

Proof Since V is complete, the sequence $(s_n)_n$ has a limit in V if and only if it is a Cauchy sequence, i.e.,

$$\forall \varepsilon > 0 \quad \exists \bar{n} : \quad [\, m \geq \bar{n} \text{ and } n \geq \bar{n} \,] \quad \Rightarrow \quad \| s_n - s_m \| < \varepsilon.$$

Now it is not restrictive to take $n > m$, and if we substitute $s_n = \sum_{k=0}^{n} a_k$ and $s_m = \sum_{k=0}^{m} a_k$, the conclusion follows. ∎

We now state a useful convergence criterion.

Theorem 8.4 *If V is a Banach space and the series $\sum_k \| a_k \|$ converges, then the series $\sum_k a_k$ also converges.*

In that case we say that the series $\sum_k a_k$ "converges in norm," unless V coincides with either $\mathbb{R}$ or $\mathbb{C}$, in which cases we say that the series "converges absolutely."

Proof We assume that the series $\sum_k \| a_k \|$ converges. Let $\varepsilon > 0$ be fixed. By the Cauchy criterion, there exists a $\bar{n} \in \mathbb{N}$ such that

$$n > m \geq \bar{n} \quad \Rightarrow \quad \sum_{k=m+1}^{n} \| a_k \| < \varepsilon.$$

Since

$$\left\| \sum_{k=m+1}^{n} a_k \right\| \leq \sum_{k=m+1}^{n} \| a_k \|,$$

we have that

$$n > m \geq \bar{n} \quad \Rightarrow \quad \left\| \sum_{k=m+1}^{n} a_k \right\| < \varepsilon,$$

and the conclusion follows, using the Cauchy criterion again. ∎

The convergence in norm of a series thus reinforces the interest in examining the series of positive real numbers.

8.2   Series of Real Numbers

In this section we only consider series with terms a_k in $\mathbb{R}$.

If for every k one has that $a_k \geq 0$, then the sequence $(s_n)_n$ is increasing, hence it has a limit, and we have only two possibilities: The series either converges or it diverges at $+\infty$.

The following **comparison criterion** will be very useful.

Theorem 8.5   *Let $\sum_k a_k$ and $\sum_k b_k$ be two series for which*

$$\exists \bar{k} \in \mathbb{N}: \quad k \geq \bar{k} \implies 0 \leq a_k \leq b_k.$$

Then

(a) *If $\sum_k b_k$ converges, then $\sum_k a_k$ also converges.*
(b) *If $\sum_k a_k$ diverges, then $\sum_k b_k$ also diverges.*

Proof We define

$$s_n = a_0 + a_1 + a_2 + \cdots + a_n, \quad s'_n = b_0 + b_1 + b_2 + \cdots + b_n.$$

By previous considerations, we can modify a finite number of terms in the two series and assume without loss of generality that $0 \leq a_k \leq b_k$ *for every* k. Then the two sequences $(s_n)_n$ and $(s'_n)_n$ are increasing, and $s_n \leq s'_n$ for every n. Consequently, the limits $S = \lim_n s_n$ and $S' = \lim_n s'_n$ exist, and $S \leq S' \leq +\infty$. If $\sum_k b_k$ converges, then $S' \in \mathbb{R}$, so also $S \in \mathbb{R}$, meaning $\sum_k a_k$ converges. If $\sum_k a_k$ diverges, then $S = +\infty$, so also $S' = +\infty$, meaning $\sum_k b_k$ diverges.   ∎

Example The series

$$1 + \frac{1}{2^2} + \frac{1}{3^2} + \frac{1}{4^2} + \cdots + \frac{1}{(n+1)^2} + \cdots$$

converges. This can be proved by comparing it with the series

$$1 + \frac{1}{1 \cdot 2} + \frac{1}{2 \cdot 3} + \frac{1}{3 \cdot 4} + \cdots + \frac{1}{n \cdot (n+1)} + \cdots,$$

which is a slight modification of the one already treated earlier in Example 2. All the terms of the first series are smaller than or equal to the corresponding ones of the second series, which converges.

As a first corollary, we have the **asymptotic comparison criterion**.

Corollary 8.6 *Let $\sum_k a_k$ and $\sum_k b_k$ be two series with positive terms, for which the limit*

$$\ell = \lim_k \frac{a_k}{b_k}$$

exists. We have three cases:

(a) $\ell \in \,]0, +\infty[$; *the two series either both converge or both diverge.*
(b) $\ell = 0$; *if $\sum_k b_k$ converges, then $\sum_k a_k$ also converges.*
(c) $\ell = +\infty$; *if $\sum_k b_k$ diverges, then $\sum_k a_k$ also diverges.*

Proof   Case (a). If $\ell \in \,]0, +\infty[$, then there exists a $\bar{k}$ such that

$$k \geq \bar{k} \quad \Rightarrow \quad \frac{\ell}{2} \leq \frac{a_k}{b_k} \leq \frac{3\ell}{2} \, ,$$

i.e.,

$$k \geq \bar{k} \quad \Rightarrow \quad a_k \leq \frac{3\ell}{2} b_k \quad \text{and} \quad b_k \leq \frac{2}{\ell} a_k \, .$$

The conclusion then follows from the comparison criterion.
Case (b). If $\ell = 0$, then there exists a $\bar{k}$ such that if $k \geq \bar{k}$, then $a_k \leq b_k$, and the comparison criterion applies.
Case (c). If $\ell = +\infty$ we have the analogue of Case 2 with the roles of a_k and b_k interchanged.   ∎

The following corollary provides us with the "root test."

Corollary 8.7 *Let $\sum_k a_k$ be a series with nonnegative terms. If*

$$\limsup_k \sqrt[k]{a_k} < 1 \, ,$$

then the series converges.

Proof Set $\ell = \limsup_k \sqrt[k]{a_k}$, and let $\alpha \in \,]\ell, 1[$ be an arbitrarily fixed number. Then there exists a $\bar{k}$ such that

$$k \geq \bar{k} \quad \Rightarrow \quad \sqrt[k]{a_k} \leq \alpha \, ,$$

i.e.,

$$k \geq \bar{k} \quad \Rightarrow \quad a_k \leq \alpha^k \, .$$

The conclusion follows by comparison with the geometric series $\sum_k \alpha^k$, which converges, since $0 < \alpha < 1$.   ∎

Recalling Proposition 3.32, as an immediate consequence we have the "ratio test."

Corollary 8.8 *Let $\sum_k a_k$ be a series with positive terms. If*

$$\limsup_k \frac{a_{k+1}}{a_k} < 1,$$

then the series converges.

We now present the **condensation criterion**, which we already implicitly used earlier when dealing with the harmonic series of Example 3.

Theorem 8.9 *Let $(a_k)_k$ be a decreasing sequence of nonnegative numbers. Then the two series*

$$\sum_{k=0}^{\infty} a_k, \qquad \sum_{k=0}^{\infty} 2^k a_{2^k}$$

either both converge or both diverge.

Proof For simplicity, we delete the first term a_0 from the first series. Let the series $\sum_k 2^k a_{2^k}$ converge. Then

$$a_1 + (a_2 + a_3) \le a_1 + 2a_2,$$
$$a_1 + (a_2 + a_3) + (a_4 + a_5 + a_6 + a_7) \le a_1 + 2a_2 + 4a_4,$$
$$a_1 + (a_2 + a_3) + (a_4 + a_5 + a_6 + a_7) +$$
$$+(a_8 + a_9 + a_{10} + a_{11} + a_{12} + a_{13} + a_{14} + a_{15}) \le$$
$$\le a_1 + 2a_2 + 4a_4 + 8a_8,$$

$$\dots$$

leading to the inequality

$$\sum_{k=1}^{2^{n+1}-1} a_k \le \sum_{k=0}^{n} 2^k a_{2^k}, \qquad \text{for every } n \in \mathbb{N}.$$

By comparison, since $\sum_k 2^k a_{2^k}$ converges, then $\sum_k a_k$ also converges.

Assume now, conversely, that the series $\sum_k a_k$ converges. Then

$$a_1 + 2a_2 \le 2(a_1 + a_2),$$
$$a_1 + 2a_2 + 4a_4 \le 2(a_1 + a_2 + (a_3 + a_4)),$$
$$a_1 + 2a_2 + 4a_4 + 8a_8 \le 2(a_1 + a_2 + (a_3 + a_4) + (a_5 + a_6 + a_7 + a_8)),$$

$\ldots$

leading to

$$\sum_{k=0}^{n} 2^k a_{2^k} \le \sum_{k=1}^{2^n} 2a_k, \quad \text{for every } n \in \mathbb{N}.$$

By comparison, since $\sum_k 2a_k$ converges, then $\sum_k 2^k a_{2^k}$ also converges. ∎

Example 1 Let us consider the series

$$\sum_{k=1}^{\infty} \frac{1}{k^\beta},$$

where $\beta > 0$ is a fixed real number. The sequence $(a_k)_k$, with $a_k = 1/k^\beta$, is decreasing. The "condensed series"

$$\sum_{k=1}^{\infty} 2^k a_{2^k} = \sum_{k=1}^{\infty} 2^k \frac{1}{(2^k)^\beta} = \sum_{k=1}^{\infty} (2^{1-\beta})^k$$

is a geometric series of the type $\sum_k \alpha^k$, with $\alpha = 2^{1-\beta}$. It converges if and only if $|\alpha| < 1$; hence,

$$\sum_{k=1}^{\infty} \frac{1}{k^\beta} \quad \text{converges} \quad \Leftrightarrow \quad \beta > 1.$$

Example 2 Let us now examine the series

$$\sum_{k=2}^{\infty} \frac{1}{k(\ln k)^\beta}$$

for some $\beta > 0$. By some use of differential calculus, it is rather easy to see that the sequence $(a_k)_k$, with $a_k = 1/k(\ln k)^\beta$, is decreasing. The "condensed series" is

$$\sum_{k=2}^{\infty} 2^k a_{2^k} = \sum_{k=2}^{\infty} 2^k \frac{1}{2^k (\ln 2^k)^\beta} = \frac{1}{(\ln 2)^\beta} \sum_{k=2}^{\infty} \frac{1}{k^\beta}.$$

Looking back at the previous example, we conclude that

$$\sum_{k=2}^{\infty} \frac{1}{k (\ln k)^\beta} \quad \text{converges} \quad \Leftrightarrow \quad \beta > 1.$$

Till now in this section we have only considered series with nonnegative terms. Now we shift to series having alternating signs. Consider a series of the type

$$a_0 - a_1 + a_2 - a_3 + \cdots + (-1)^n a_n + \ldots,$$

where all a_k are positive. What follows is the **Leibniz criterion**.

Theorem 8.10 *If $(a_k)_k$ is a decreasing sequence of positive numbers and*

$$\lim_k a_k = 0,$$

then the series $\sum_k (-1)^k a_k$ converges.

Proof Let

$$s_n = a_0 - a_1 + a_2 - a_3 + \cdots + (-1)^n a_n,$$

and consider the sequence $(s_n)_n$ of partial sums. We divide it into two subsequences, one with even indices and the other one with odd indices. Since $(a_k)_k$ is positive and decreasing, we see that

$$s_1 \leq s_3 \leq s_5 \leq s_7 \leq \quad \cdots \quad \leq s_6 \leq s_4 \leq s_2 \leq s_0.$$

Hence, the subsequence $(s_{2m+1})_m$, the one with odd indices, is increasing and bounded from above, whereas the subsequence $(s_{2m})_m$, the one having even indices, is decreasing and bounded from below. Then both subsequences have a finite limit, and we can write

$$\lim_m s_{2m+1} = \ell_1, \qquad \lim_m s_{2m} = \ell_2.$$

On the other hand,

$$\ell_2 - \ell_1 = \lim_m(s_{2m} - s_{2m+1}) = \lim_m a_{2m+1} = 0,$$

hence $\ell_1 = \ell_2$. Because the two subsequences $(s_{2m+1})_m$ and $(s_{2m})_m$ have the same limit, we can be sure that the sequence $(s_n)_n$ will have the same limit. ∎

8.3 Series of Complex Numbers

When we consider a series $\sum_k a_k$ whose terms are complex numbers $a_k = x_k + iy_k$, where $x_k = \Re(a_k)$ and $y_k = \Im(a_k)$, we can write its partial sums as

$$s_n = \sum_{k=0}^n a_k = \sum_{k=0}^n (x_k + iy_k) = \sum_{k=0}^n x_k + i\sum_{k=0}^n y_k = \sigma_n + i\tau_n,$$

where $\sigma_n = \Re(s_n)$ and $\tau_n = \Im(s_n)$. We thus have a sequence $(\sigma_n, \tau_n)_n$ in $\mathbb{R}^2$. Recalling that such a sequence has a limit in $\mathbb{R}^2$ if and only if both its components have a limit in $\mathbb{R}$, we obtain the following statement.

Theorem 8.11 *If $a_k = x_k + iy_k$, with x_k and y_k being real numbers, the series $\sum_k a_k$ converges if and only if both series $\sum_k x_k$ and $\sum_k y_k$ converge. In that case,*

$$\sum_{k=0}^\infty a_k = \sum_{k=0}^\infty x_k + i\sum_{k=0}^\infty y_k.$$

Example Let us consider the series

$$\sum_{k=0}^\infty \frac{i^k}{k+1} = 1 + \frac{i}{2} - \frac{1}{3} - \frac{i}{4} + \frac{1}{5} + \frac{i}{6} - \frac{1}{7} - \frac{i}{8} + \cdots + \frac{i^n}{n+1} + \cdots.$$

The real part is

$$1 - \frac{1}{3} + \frac{1}{5} - \frac{1}{7} + \cdots,$$

whereas the imaginary part is

$$\frac{1}{2} - \frac{1}{4} + \frac{1}{6} - \frac{1}{8} + \cdots.$$

Both of them converge by the Leibniz criterion on series with alternating signs, so the given series also converges.

Note that, in the previous example, the series does not converge absolutely since

$$\sum_{k=0}^{\infty} \left| \frac{i^k}{k+1} \right| = \sum_{k=0}^{\infty} \frac{1}{k+1}$$

is the harmonic series, which we know to be divergent.

We now define the "Cauchy product" of two series $\sum_{k=0}^{\infty} a_k$ and $\sum_{k=0}^{\infty} b_k$. It is the series

$$\sum_{k=0}^{\infty} \left(\sum_{j=0}^{k} a_{k-j} b_j \right).$$

However, some care is needed concerning its convergence. Indeed, it is not true in general that if the two series converge, then their Cauchy product series also converges. The following theorem states that this will be true if at least one of them converges absolutely.

Theorem 8.12 (Mertens' Theorem) *Assume that the series $\sum_k a_k$ and $\sum_k b_k$ converge, with sums A and B, respectively. If at least one of them converges absolutely, then their Cauchy product series converges with sum AB.*

Proof To fix ideas, let $\sum_{k=0}^{\infty} a_k$ converge absolutely, and set $\bar{A} = \sum_{k=0}^{\infty} |a_k|$. We denote by

$$c_k = \sum_{j=0}^{k} a_{k-j} b_j$$

the kth term of the Cauchy product series. Let

$$s_n = \sum_{k=0}^{n} a_k, \quad s_n' = \sum_{k=0}^{n} b_k, \quad s_n'' = \sum_{k=0}^{n} c_k.$$

Moreover, let $r_n' = B - s_n'$. Then

$$
\begin{aligned}
s_n'' &= a_0 b_0 + (a_1 b_0 + a_0 b_1) + \cdots + (a_n b_0 + a_{n-1} b_1 + \cdots + a_1 b_{n-1} + a_0 b_n) \\
&= a_0 s_n' + a_1 s_{n-1}' + \cdots + a_{n-1} s_1' + a_n s_0' \\
&= a_0 (B - r_n') + a_1 (B - r_{n-1}') + \cdots + a_{n-1}(B - r_1') + a_n (B - r_0') \\
&= s_n B - (a_0 r_n' + a_1 r_{n-1}' + \cdots + a_{n-1} r_1' + a_n r_0').
\end{aligned}
$$

Since $\lim\limits_n s_n B = AB$, the proof will be completed if

$$\lim_n (a_0 r'_n + a_1 r'_{n-1} + \cdots + a_{n-1} r'_1 + a_n r'_0) = 0 \,.$$

Let $\varepsilon > 0$ be fixed. Since $\lim\limits_n r'_n = 0$, there exists a $\bar{n}_1$ such that

$$n \geq \bar{n}_1 \quad \Rightarrow \quad |r'_n| < \varepsilon \,.$$

Let us set $\overline{R} = \max\{|r'_n| : n \in \mathbb{N}\}$. By the Cauchy criterion, there exists a $\bar{n}_2 \geq \bar{n}_1$ such that

$$n \geq \bar{n}_2 \quad \Rightarrow \quad |a_{n-\bar{n}_1+1}| + |a_{n-\bar{n}_1+2}| + \cdots + |a_n| < \varepsilon \,.$$

Then, if $n \geq \bar{n}_2$,

$$|a_0 r'_n + a_1 r'_{n-1} + \cdots + a_{n-1} r'_1 + a_n r'_0| \leq$$
$$\leq |a_0|\,|r'_n| + \cdots + |a_{n-\bar{n}_1}|\,|r'_{\bar{n}_1}| + |a_{n-\bar{n}_1+1}|\,|r'_{\bar{n}_1-1}| + \cdots + |a_n|\,|r'_0|$$
$$\leq \varepsilon(|a_0| + \cdots + |a_{n-\bar{n}_1}|) + \overline{R}(|a_{n-\bar{n}_1+1}| + \cdots + |a_n|)$$
$$\leq \varepsilon \bar{A} + \overline{R}\,\varepsilon = (\bar{A} + \overline{R})\,\varepsilon \,,$$

thereby completing the proof. ■

8.4 Series of Functions

Let E be a metric space and F a normed vector space. If we have a sequence of functions $f_k : E \to F$, for any $x \in E$ we can ask ourselves whether or not the series $\sum_k f_k(x)$ converges. If there is a subset $U \subseteq E$ and a function $f : U \to F$ such that

$$\sum_{k=0}^{\infty} f_k(x) = f(x) \,, \qquad \text{for every } x \in U \,,$$

we say that the series "converges pointwise" to f on U; this happens when, setting $s_n(x) = \sum_{k=0}^{n} f_k(x)$, the sequence $(s_n)_n$ converges pointwise to f on U. We say that the series "converges uniformly" to f on U if the convergence of $(s_n)_n$ to f is uniform on U, meaning

$$\forall \varepsilon > 0 \quad \exists \bar{n} \in \mathbb{N} : \quad \forall x \in U \quad n \geq \bar{n} \ \Rightarrow \ \left\| \sum_{k=0}^{n} f_k(x) - f(x) \right\| < \varepsilon \,.$$

Theorem 8.13 *If every function* $f_k : E \to F$ *is continuous on* $U \subseteq E$ *and the series* $\sum_{k=0}^{\infty} f_k$ *converges uniformly to* f *on* U, *then* f *is also continuous on* U.

Proof It is a direct consequence of Theorem 4.17.   ■

Theorem 8.14 *Let* E *be compact,* F *a Banach space, and all functions* $f_k : E \to F$ *continuous. If the series* $\sum_{k=0}^{\infty} \| f_k \|_\infty$ *converges, then the series* $\sum_{k=0}^{\infty} f_k$ *converges uniformly to a continuous function on* E.

Proof We know that $V = \mathcal{C}(E, F)$ is a Banach space, and $\sum_{k=0}^{\infty} f_k$ is a series in V that converges in norm. Then it converges in V, meaning that it converges uniformly.   ■

Example Let $E = [a, b] \subseteq \mathbb{R}$ and $F = \mathbb{R}$, and let us consider the series

$$\sum_{k=1}^{\infty} \frac{1}{k^2} \sin(e^{3kx-1} + \arctan(x^2 + \sqrt{k})) \,.$$

We will examine the series of the norms in $\mathcal{C}([a, b], \mathbb{R})$. We have that

$$\sup\left\{ \left| \frac{1}{k^2} \sin(e^{3kx-1} + \arctan(x^2 + \sqrt{k})) \right| : x \in [a, b] \right\} \leq \frac{1}{k^2} \,,$$

and the series $\sum_k \frac{1}{k^2}$ converges. Then the series converges in norm and, hence, uniformly.

We now adapt the two theorems in Sect. 7.12 to the context of series.

Theorem 8.15 *Let* $(f_k)_k$ *be a sequence in* $\mathcal{C}([a, b], \mathbb{R})$ *such that the series* $\sum_k f_k$ *is uniformly convergent. Then*

$$\int_a^b \sum_{k=0}^{\infty} f_k(t)\, dt = \sum_{k=0}^{\infty} \int_a^b f_k(t)\, dt \,.$$

Proof It is a direct consequence of Theorem 7.24 when applied to the sequence of functions $(s_n)_n$.   ■

Theorem 8.16 *Let* $(f_k)_k$ *be a sequence in* $\mathcal{C}^1([a, b], \mathbb{R})$. *Assume that the series* $\sum_k f_k$ *and the series of the derivatives* $\sum_k f_k'$ *converge uniformly to some functions* $f : [a, b] \to \mathbb{R}$ *and* $g : [a, b] \to \mathbb{R}$, *respectively. Then* f *is of the class* $\mathcal{C}^1$, *and*

$f' = g$. *Consequently, we can write*

$$\frac{d}{dx} \sum_{k=0}^{\infty} f_k(x) = \sum_{k=0}^{\infty} \frac{d}{dx} f_k(x).$$

Proof Consider the sequence of partial sums $s_n = \sum_{k=0}^{n} f_k$. Then $(s_n)_n$ is in $C^1(I, \mathbb{R})$, and $s_n' = \sum_{k=0}^{n} f_k'$. By assumption, $\lim_n s_n = f$ and $\lim_n s_n' = g$, uniformly on I. Hence, by Theorem 7.25, it must be that $f \in C^1(I, \mathbb{R})$ and $f' = g$. ∎

Iterating the same argument, we can easily generalize the preceding theorem.

Theorem 8.17 *Let $(f_k)_k$ be a sequence in $C^m([a, b], \mathbb{R})$. Assume that the series*

$$\sum_k f_k, \quad \sum_k f_k', \quad \sum_k f_k'', \quad \ldots, \quad \sum_k f_k^{(m)}$$

converge uniformly on $[a, b]$ to some functions $f, g_1, g_2, \ldots, g_m$, respectively. Then f is of the class C^m, and

$$f' = g_1, \quad f'' = g_2, \quad \ldots, \quad f^{(m)} = g_m.$$

8.4.1 Power Series

An important example of a series of functions is provided by the "power series"

$$(PS)_{\mathbb{C}} \qquad \qquad \sum_{k=0}^{\infty} a_k z^k,$$

whose terms are the functions $f_k : \mathbb{C} \to \mathbb{C}$ defined as $f_k(z) = a_k z^k$, for some given coefficients $a_k \in \mathbb{C}$. Let us first analyze the pointwise convergence.

Theorem 8.18 *Setting*

$$L = \limsup_k \sqrt[k]{|a_k|},$$

we have the following possibilities:

(a) *If $L = +\infty$, then the series $(PS)_\mathbb{C}$ converges only for $z = 0$.*
(b) *If $L = 0$, then the series $(PS)_\mathbb{C}$ converges for every $z \in \mathbb{C}$.*
(c) *If $L \in {]0, +\infty[}$, then the series $(PS)_\mathbb{C}$* $\begin{cases} \text{converges if } |z| < \frac{1}{L}, \\ \text{does not converge if } |z| > \frac{1}{L}. \end{cases}$

Proof If $L = +\infty$ and $z \neq 0$, then $\sqrt[k]{|a_k|} > \frac{1}{|z|}$ for infinitely many k, hence

$$|a_k z^k| = (\sqrt[k]{|a_k|}\, |z|)^k > 1, \qquad \text{for infinitely many } k\,.$$

If the series were to converge, then it should be $\lim_k a_k z^k = 0$, but this is not so. Hence, if $L = +\infty$, then the series converges only for $z = 0$.

If $L = 0$, then for any $z \in \mathbb{C}$ we have that

$$\limsup_k \sqrt[k]{|a_k z^k|} = |z| \limsup_k \sqrt[k]{|a_k|} = 0\,,$$

so, by the root test, the series converges absolutely.

Assume now $L \in {]0, +\infty[}$. If $|z| < \frac{1}{L}$, then

$$\limsup_k \sqrt[k]{|a_k z^k|} = |z| \limsup_k \sqrt[k]{|a_k|} = |z|\, L < 1\,,$$

and, by the root test, the series converges absolutely. In contrast, if $|z| > \frac{1}{L}$, i.e., $L > \frac{1}{|z|}$, then $\sqrt[k]{|a_k|} > \frac{1}{|z|}$ for infinitely many k, and hence

$$|a_k z^k| = (\sqrt[k]{|a_k|}|z|)^k > 1, \qquad \text{for infinitely many } k\,.$$

If the series were to converge, then we would have $\lim_k a_k z^k = 0$, but this is not the case. ∎

We define the "convergence radius" r of the series $(PS)_\mathbb{C}$ as follows:

$$r = \begin{cases} 0 & \text{if } L = +\infty\,, \\ +\infty & \text{if } L = 0\,, \\ \dfrac{1}{L} & \text{if } L \in {]0, +\infty[}\,. \end{cases}$$

If $r > 0$, we will say that $B(0, r)$ is the "convergence disk" of the series. We emphasize that it is an open ball. If $r = +\infty$, then we set $B(0, r) = \mathbb{C}$.

If $r > 0$, then the series $(PS)_\mathbb{C}$ converges pointwise in $B(0, r)$. However, this convergence could not be uniform. We now see that the convergence will be uniform on any smaller disk.

Theorem 8.19 *Assume $r > 0$. Then for every $\rho \in]0, r[$ the series $(PS)_\mathbb{C}$ converges in norm, hence uniformly, on $\overline{B}(0, \rho)$.*

Proof Let $\rho \in]0, r[$ be fixed. Then

$$\sup\{|a_k z^k| : |z| \le \rho\} = |a_k|\rho^k,$$

and since

$$\limsup_k \sqrt[k]{|a_k|\rho^k} = \rho \limsup_k \sqrt[k]{|a_k|} = \rho L < rL = 1,$$

then, by the root test, the series $\sum_k |a_k|\rho^k$ converges. We have thus seen that the series $(PS)_\mathbb{C}$ converges in norm on $\overline{B}(0, \rho)$. ∎

Corollary 8.20 *If $r > 0$, and $f : B(0, r) \to \mathbb{C}$ is defined as*

$$f(z) = \sum_{k=0}^{\infty} a_k z^k,$$

then f is continuous on $B(0, r)$.

Proof From the previous theorem, for every $\rho \in]0, r[$ the convergence is uniform on $\overline{B}(0, \rho)$; hence, f is continuous on $\overline{B}(0, \rho)$. Since $\rho \in]0, r[$ is arbitrary, f is thus continuous at every point of $B(0, r)$. ∎

Remark 8.21 The above theory can be easily generalized to series of the type

$$\sum_{k=0}^{\infty} a_k (z - z_0)^k$$

for some fixed point $z_0 \in \mathbb{C}$. (Indeed, the change of variables $u = z - z_0$ leads back to the previously considered case.) The convergence disk in this case is $B(z_0, r) = \{z \in \mathbb{C} : |z - z_0| < r\}$.

8.4.2 The Complex Exponential Function

Let us now examine, for every $z \in \mathbb{C}$, the series

$$\sum_{k=0}^{\infty} \frac{z^k}{k!} \, .$$

It is a power series, which converges absolutely, since

$$\sum_{k=0}^{\infty} \left| \frac{z^k}{k!} \right| = \sum_{k=0}^{\infty} \frac{|z|^k}{k!} = e^{|z|} \, .$$

It is therefore possible to define a function $\mathcal{F} : \mathbb{C} \to \mathbb{C}$ by

$$\mathcal{F}(z) = \sum_{k=0}^{\infty} \frac{z^k}{k!} \, .$$

Recall that if $z = x \in \mathbb{R}$, then we have proved that $\mathcal{F}(x)$ is equal to $\exp(x)$, i.e.,

$$e^x = \sum_{k=0}^{\infty} \frac{x^k}{k!} \, .$$

We can then interpret this function $\mathcal{F}$ as an extension of the exponential function to the complex plane $\mathbb{C}$. For this reason, we will call $\mathcal{F}$ the "complex exponential function" and write either $\exp(z)$ or e^z instead of $\mathcal{F}(z)$.

Theorem 8.22 *For every z_1 and z_2 in $\mathbb{C}$ we have that*

$$\exp(z_1 + z_2) = \exp(z_1) + \exp(z_2) \, .$$

Proof The series $\sum_{k=0}^{\infty} \frac{z_1^k}{k!}$ and $\sum_{k=0}^{\infty} \frac{z_2^k}{k!}$ converge absolutely, and their sums are $\exp(z_1)$ and $\exp(z_2)$, respectively. Then, by Mertens' Theorem 8.12, the Cauchy product series converges, and its sum is $\exp(z_1) \exp(z_2)$. On the other hand, the Cauchy product series is

$$\sum_{k=0}^{\infty} \left(\sum_{j=0}^{k} \frac{z_1^{k-j}}{(k-j)!} \frac{z_2^j}{j!} \right) = \sum_{k=0}^{\infty} \frac{1}{k!} \left(\sum_{j=0}^{k} \binom{k}{j} z_1^{k-j} z_2^j \right) = \sum_{k=0}^{\infty} \frac{(z_1 + z_2)^k}{k!} \, ,$$

and its sum is $\exp(z_1 + z_2)$, whence the conclusion. ∎

Writing $z = x + iy$, we obtain

$$\exp(x + iy) = \exp(x) \exp(iy).$$

Moreover,

$$\exp(iy) = \sum_{k=0}^{\infty} \frac{(iy)^k}{k!} = \lim_n q_n(y),$$

where

$$q_n(y) = \sum_{k=0}^{n} \frac{(iy)^k}{k!}$$

$$= 1 + iy - \frac{y^2}{2!} - i\frac{y^3}{3!} + \frac{y^4}{4!} + i\frac{y^5}{5!} - \frac{y^6}{6!} - i\frac{y^7}{7!} + \cdots + i^n\frac{y^n}{n!}.$$

We thus have that

$$q_n(y) = q_n^{(1)}(y) + i\, q_n^{(2)}(y),$$

where

$$q_n^{(1)}(y) = 1 - \frac{y^2}{2!} + \frac{y^4}{4!} - \frac{y^6}{6!} + \cdots + (-1)^m \frac{y^{2m}}{(2m)!}$$

if either $n = 2m$ or $n = 2m + 1$, whereas

$$q_n^{(2)}(y) = y - \frac{y^3}{3!} + \frac{y^5}{5!} - \frac{y^7}{7!} + \cdots + (-1)^m \frac{y^{2m+1}}{(2m+1)!}$$

if either $n = 2m + 1$ or $n = 2m + 2$. Since

$$\lim_{n \to +\infty} q_n^{(1)}(y) = \cos y, \qquad \lim_{n \to +\infty} q_n^{(2)}(y) = \sin y,$$

we conclude that

$$\lim_n q_n(y) = \left(\lim_n q_n^{(1)}(y), \lim_n q_n^{(2)}(y)\right) = (\cos y, \sin y),$$

hence

$$e^{x+iy} = e^x(\cos y + i \sin y).$$

This is the **Euler formula**.

It can be easily verified that

$$\cos t = \frac{e^{it} + e^{-it}}{2}, \qquad \sin t = \frac{e^{it} - e^{-it}}{2i}.$$

These formulas can be used to extend the functions cos and sin to the complex field by simply taking $t \in \mathbb{C}$. The hyperbolic functions can also be extended to $\mathbb{C}$ by the formulas

$$\cosh z = \frac{e^z + e^{-z}}{2}, \qquad \sinh z = \frac{e^z - e^{-z}}{2}.$$

Note that

$$\cos t = \cosh(it), \qquad \sin t = -i \sinh(it).$$

The analogies between the trigonometric and the hyperbolic functions are now surely better understood.

8.4.3   Taylor Series

Let us now consider the power series

$$(PS)_{\mathbb{R}} \qquad\qquad \sum_{k=0}^{\infty} a_k x^k,$$

where $x \in \mathbb{R}$ and all coefficients a_k are also real numbers. We are thus considering the series $\sum_k f_k$, where the functions $f_k : \mathbb{R} \to \mathbb{R}$ are defined by $f_k(x) = a_k x^k$. Hence, if $r > 0$, then the convergence disk $B(0, r)$ is now reduced the interval $] - r, r[$, and if $r = +\infty$, then it is the whole real line $\mathbb{R}$. In these cases, we might wonder whether the sum of the series $(PS)_{\mathbb{R}}$ was differentiable on $] - r, r[$.

Theorem 8.23 *Let $r > 0$ be the convergence radius of the series $(PS)_{\mathbb{R}}$, and let*

$$f(x) = \sum_{k=0}^{\infty} a_k x^k, \quad \textit{for every } x \in]-r, r[\,.$$

Then the series

$$\sum_{k=1}^{\infty} k a_k x^{k-1}$$

has the same convergence radius r. Moreover, the function $f :]-r, r[\to \mathbb{R}$ is differentiable, and

$$f'(x) = \sum_{k=1}^{\infty} k a_k x^{k-1}, \quad \text{for every } x \in]-r, r[.$$

Proof Since

$$\limsup_{k} \sqrt[k]{|k a_k|} = \lim_{k} \sqrt[k]{k} \ \limsup_{k} \sqrt[k]{|a_k|} = \limsup_{k} \sqrt[k]{|a_k|},$$

we see that the convergence radius for the series $\sum_k k a_k x^{k-1}$ is equal to r. We can then define the new function $g :]-r, r[\to \mathbb{R}$ as $g(x) = \sum_{k=1}^{\infty} k a_k x^{k-1}$. For any fixed $\rho \in]0, r[$ we know that the convergence of the series is uniform on $[-\rho, \rho]$.

Setting $f_k(x) = a_k x^k$, we have that $f = \sum_k f_k$ and $g = \sum_k f_k'$, and the convergence is uniform on $[-\rho, \rho]$. By Theorem 8.16, f is differentiable on $[-\rho, \rho]$ and $f'(x) = g(x)$ for every $x \in [-\rho, \rho]$. The conclusion follows since $\rho \in]0, r[$ is arbitrary. ∎

Iterating the same argument and making use of Theorem 8.17 we easily obtain the following generalization.

Theorem 8.24 *Let $r > 0$ be the convergence radius of the series $(PS)_{\mathbb{R}}$, and let*

$$f(x) = \sum_{k=0}^{\infty} a_k x^k, \quad \text{for every } x \in]-r, r[.$$

Then the series

$$\sum_{k=1}^{\infty} k a_k x^{k-1}, \quad \sum_{k=2}^{\infty} k(k-1) a_k x^{k-2}, \quad \sum_{k=3}^{\infty} k(k-1)(k-2) a_k x^{k-3}, \quad \cdots,$$

$$\sum_{k=m}^{\infty} k(k-1)(k-2) \cdots (k-m+1) a_k x^{k-m}, \quad \cdots$$

all have the same convergence radius r. Moreover, the function $f :]-r, r[\to \mathbb{R}$ is infinitely differentiable and, for every positive integer j,

$$f^{(j)}(x) = \sum_{k=j}^{\infty} k(k-1)(k-2) \cdots (k-j+1) a_k x^{k-j}, \quad \text{for every } x \in]-r, r[.$$

Note now that, taking $x = 0$ in the previous formula, we obtain

$$f^{(j)}(0) = j! \, a_j$$

for every $j \in \mathbb{N}$ (recalling that $f^{(0)} = f$). Then

$$f(x) = \sum_{k=0}^{\infty} a_k x^k = \sum_{k=0}^{\infty} \frac{1}{k!} f^{(k)}(0) \, x^k \,.$$

This is the "Taylor series" associated with the function f at $x_0 = 0$. We have thus proved that any power series with positive convergence radius r defines a function f that is analytic on $] - r, r[$.

Remark 8.25 Referring to Remark 8.21, we can also extend the foregoing considerations, made for the series $(PS)_{\mathbb{R}}$, to power series of the type

$$\sum_{k=0}^{\infty} a_k (x - x_0)^k$$

for some fixed point x_0. If the convergence radius r is positive, then the convergence disk is $]x_0 - r, x_0 + r[$, and the function $f :]x_0 - r, x_0 + r[\to \mathbb{R}$ defined by the sum of the series can be expressed by

$$f(x) = \sum_{k=0}^{\infty} \frac{1}{k!} f^{(k)}(x_0)(x - x_0)^k \,,$$

i.e., $f(x) = \lim_n p_n(x)$, where $p_n(x)$ is the Taylor polynomial defined by (6.3).

8.4.4 Fourier Series

Let us now consider the "trigonometric polynomials" having some fixed period $T > 0$. They are defined as

$$f_n(t) = c_0 + \sum_{k=1}^{n} \left(a_k \cos\left(\frac{2\pi k}{T} t\right) + b_k \sin\left(\frac{2\pi k}{T} t\right) \right) ,$$

where c_0, a_k, and b_k are some real constants. We are interested in examining the convergence of the sequence of functions $(f_n)_n$.

Theorem 8.26 *If there exists a function* $f : [0, T] \to \mathbb{R}$ *such that*

$$\lim_n f_n(t) = f(t), \quad \text{uniformly on } [0, T],$$

then necessarily

$$c_0 = \frac{1}{T} \int_0^T f(t)\, dt,$$

$$a_k = \frac{2}{T} \int_0^T f(t) \cos\left(\frac{2\pi k}{T} t\right) dt,$$

$$b_k = \frac{2}{T} \int_0^T f(t) \sin\left(\frac{2\pi k}{T} t\right) dt.$$

Proof By Theorem 8.15,

$$\int_0^T f(t)\, dt = \int_0^T \left(c_0 + \sum_{k=1}^{\infty} \left(a_k \cos\left(\frac{2\pi k}{T} t\right) + b_k \sin\left(\frac{2\pi k}{T} t\right) \right) \right) dt$$

$$= c_0 T + \sum_{k=1}^{\infty} \int_0^T \left(a_k \cos\left(\frac{2\pi k}{T} t\right) + b_k \sin\left(\frac{2\pi k}{T} t\right) \right) dt = c_0 T,$$

whence the formula for c_0. Analogously, for any integer $j \geq 1$,

$$\int_0^T f(t) \cos\left(\frac{2\pi j}{T} t\right) dt =$$

$$= \int_0^T \left(c_0 + \sum_{k=1}^{\infty} \left(a_k \cos\left(\frac{2\pi k}{T} t\right) + b_k \sin\left(\frac{2\pi k}{T} t\right) \right) \right) \cos\left(\frac{2\pi j}{T} t\right) dt$$

$$= \sum_{k=1}^{\infty} \int_0^T \left(a_k \cos\left(\frac{2\pi k}{T} t\right) + b_k \sin\left(\frac{2\pi k}{T} t\right) \right) \cos\left(\frac{2\pi j}{T} t\right) dt.$$

On the other hand, integrating by parts twice, we see that, for any positive integer $k \neq j$,

$$\int_0^T \sin\left(\frac{2\pi k}{T} t\right) \cos\left(\frac{2\pi j}{T} t\right) dt = 0, \quad \int_0^T \cos\left(\frac{2\pi k}{T} t\right) \cos\left(\frac{2\pi j}{T} t\right) dt = 0,$$

whereas, if $k = j$, then

$$\int_0^T \sin\left(\frac{2\pi j}{T}t\right) \cos\left(\frac{2\pi j}{T}t\right) dt = 0, \quad \int_0^T \cos^2\left(\frac{2\pi j}{T}t\right) dt = \frac{T}{2}.$$

Hence,

$$\int_0^T f(t) \cos\left(\frac{2\pi j}{T}t\right) dt = \frac{T}{2}a_j,$$

yielding the formula for a_j. Similarly one can see that

$$\int_0^T f(t) \sin\left(\frac{2\pi j}{T}t\right) dt = \frac{T}{2}b_j,$$

providing the formula for b_j.   ∎

For any given continuous function $f : [0, T] \to \mathbb{R}$, we define its "Fourier coefficients"

$$a_k = \frac{2}{T}\int_0^T f(t) \cos\left(\frac{2\pi k}{T}t\right) dt, \quad b_k = \frac{2}{T}\int_0^T f(t) \sin\left(\frac{2\pi k}{T}t\right) dt$$

and its "Fourier series"

$$\frac{a_0}{2} + \sum_{k=1}^{\infty}\left(a_k \cos\left(\frac{2\pi k}{T}t\right) + b_k \sin\left(\frac{2\pi k}{T}t\right)\right).$$

The problem is: Does this series converge for every $t \in [0, T]$?

The answer is, in general, no: There exist continuous functions $f : [0, T] \to \mathbb{R}$ for which the Fourier series fails to converge at some points $t \in [0, T]$. However, there are many ways to overcome this difficulty. We will very briefly review some of them.

Let us define the partial sums of the Fourier series as

$$f_n(t) = \frac{a_0}{2} + \sum_{k=1}^{n}\left(a_k \cos\left(\frac{2\pi k}{T}t\right) + b_k \sin\left(\frac{2\pi k}{T}t\right)\right).$$

We can now also define the "Cesaro means"

$$\sigma_n(t) = \frac{1}{n+1}[f_0(t) + f_1(t) + \cdots + f_n(t)],$$

so as to be able to state, without proof, the following theorem.

Theorem 8.27 (Fejer Theorem) *If $f : \mathbb{R} \to \mathbb{R}$ is continuous and T-periodic, then*

$$\lim_n \sigma_n(t) = f(t)$$

uniformly for every $t \in \mathbb{R}$.

Here is a direct consequence.

Corollary 8.28 *Let $f, \tilde{f} : \mathbb{R} \to \mathbb{R}$ be continuous and T-periodic functions. If the respective Fourier coefficients are such that $a_k = \tilde{a}_k$ and $b_k = \tilde{b}_k$ for every k, then f and $\tilde{f}$ coincide.*

Proof With the notations adapted to this situation, we will have that $\sigma_n(t) = \tilde{\sigma}_n(t)$, for every n, and hence

$$f(t) - \tilde{f}(t) = \lim_n (\sigma_n(t) - \tilde{\sigma}_n(t)) = 0$$

for every $t \in \mathbb{R}$. ∎

We could also define the "complex Fourier coefficients"

$$c_k = \frac{1}{T} \int_0^T f(t) e^{-i\frac{2\pi k}{T}t} \, dt$$

for $k \in \mathbb{Z}$. Setting $b_0 = 0$, we see that

$$c_k = \begin{cases} \frac{1}{2}(a_{-k} + i b_{-k}) & \text{if } k < 0, \\ \frac{1}{2}(a_k - i b_k) & \text{if } k \geq 0, \end{cases}$$

so that

$$f_n(t) = \sum_{k=-n}^{n} c_k e^{i\frac{2\pi k}{T}t} .$$

In what follows, we use the notation

$$\sum_{k=-\infty}^{\infty} |c_k| = \lim_{n \to \infty} \left(\sum_{k=-n}^{n} |c_k| \right) .$$

Corollary 8.29 *Let* $f : \mathbb{R} \to \mathbb{R}$ *be a continuous and* T-*periodic function. If the series* $\sum_{k=-\infty}^{\infty} |c_k|$ *converges, then*

$$\lim_n f_n(t) = f(t)$$

uniformly for every $t \in \mathbb{R}$.

Proof Observe that

$$\left| c_k e^{i\frac{2\pi k}{T}t} \right| = |c_k|.$$

Hence, if the series $\sum_{k=-\infty}^{\infty} |c_k|$ converges, by Theorem 8.14 the sequence $(f_n)_n$ uniformly converges to some continuous function $\tilde{f} : \mathbb{R} \to \mathbb{R}$, which is T-periodic. On the other hand, for this function,

$$
\begin{aligned}
\tilde{c}_k &= \frac{1}{T} \int_0^T \tilde{f}(t) e^{-i\frac{2\pi k}{T}t} \, dt = \frac{1}{T} \int_0^T \left(\lim_{n\to\infty} f_n(t) \right) e^{-i\frac{2\pi k}{T}t} \, dt \\
&= \lim_{n\to\infty} \frac{1}{T} \int_0^T f_n(t) e^{-i\frac{2\pi k}{T}t} \, dt = \lim_{n\to\infty} \frac{1}{T} \int_0^T \sum_{j=-n}^n c_j e^{i\frac{2\pi j}{T}t} e^{-i\frac{2\pi k}{T}t} \, dt \\
&= \lim_{n\to\infty} \sum_{j=-n}^n \frac{1}{T} \int_0^T c_j e^{i\frac{2\pi(j-k)}{T}t} \, dt = \lim_{n\to\infty} \frac{1}{T} \int_0^T c_k \, dt = c_k
\end{aligned}
$$

for every $k \in \mathbb{Z}$. By Corollary 8.28, the two functions f and $\tilde{f}$ coincide, thereby completing the proof. ∎

In the following theorem, the function f could be discontinuous at some points.

Theorem 8.30 (Dirichlet Theorem) *Let* $f : \mathbb{R} \to \mathbb{R}$ *be a* T-*periodic function. Assume that there is a finite number of points* $t_0, t_1, t_2, \ldots, t_N$, *with*

$$0 = t_0 < t_1 < t_2 < \cdots < t_N = T$$

having the property that f *is continuously differentiable on every interval* $]t_{j-1}, t_j[$, *with* $j = 1, 2, \ldots, N$. *At the points* t_j *(where the function could either fail to be continuous, or, if continuous, it could fail to be differentiable), the following finite limits must exist:*

$$\lim_{s \to t_j^-} f(s), \quad \lim_{s \to t_j^+} f(s), \quad \lim_{s \to t_j^-} f'(s), \quad \lim_{s \to t_j^+} f'(s).$$

Then, for every $t \in [0, T]$,

$$\frac{a_0}{2} + \sum_{k=1}^{\infty} \left(a_k \cos\left(\frac{2\pi k}{T} t\right) + b_k \sin\left(\frac{2\pi k}{T} t\right) \right) = \frac{1}{2} \left(\lim_{s \to t^-} f(s) + \lim_{s \to t^+} f(s) \right).$$

Moreover, the convergence is uniform on every compact interval where f is continuous.

Note that if f is continuous at t, then

$$f(t) = \frac{1}{2} \left(\lim_{s \to t^-} f(s) + \lim_{s \to t^+} f(s) \right).$$

For the proofs of Theorems 8.27 and 8.30, we refer the reader to the book by Körner [4].

We now provide two examples of applications of the preceding theorem.

Example 1 Let $f : \mathbb{R} \to \mathbb{R}$ be the 2π-periodic function defined as

$$f(t) = t, \quad \text{if } t \in [-\pi, \pi[.$$

It is readily seen that the assumptions of the Dirichlet theorem are satisfied. We compute

$$c_0 = \frac{1}{2\pi} \int_0^{2\pi} f(t) \, dt = \frac{1}{2\pi} \int_{-\pi}^{\pi} t \, dt = 0,$$

$$a_k = \frac{2}{2\pi} \int_0^{2\pi} f(t) \cos(kt) \, dt = \frac{1}{\pi} \int_{-\pi}^{\pi} t \cos(kt) \, dt = 0,$$

since $t \mapsto t \cos(kt)$ is an odd function; and, integrating by parts,

$$b_k = \frac{2}{2\pi} \int_0^{2\pi} f(t) \sin(kt) \, dt = \frac{1}{\pi} \int_{-\pi}^{\pi} t \sin(kt) \, dt$$

$$= \frac{1}{\pi} \left(\left[-t \, \frac{\cos(kt)}{k} \right]_{-\pi}^{\pi} + \int_{-\pi}^{\pi} \frac{\cos(kt)}{k} \, dt \right) = \frac{2 \, (-1)^{k+1}}{k}.$$

We can thus state that

$$f(t) = \sum_{k=1}^{\infty} \frac{2 \, (-1)^{k+1}}{k} \sin(kt) \quad \text{for every } t \in] -\pi, \pi[.$$

As a particular case, taking $t = \frac{\pi}{2}$, we obtain the nice formula

$$\frac{\pi}{4} = 1 - \frac{1}{3} + \frac{1}{5} - \frac{1}{7} + \dots$$

Example 2 Let $f : \mathbb{R} \to \mathbb{R}$ be the 2π-periodic function defined as

$$f(t) = t^2, \quad \text{if } t \in [-\pi, \pi[\, .$$

It is readily seen that the assumptions of the Dirichlet theorem are satisfied. We compute

$$c_0 = \frac{1}{2\pi} \int_0^{2\pi} f(t)\, dt = \frac{1}{2\pi} \int_{-\pi}^{\pi} t^2\, dt = \frac{\pi^2}{3} ,$$

and, integrating by parts,

$$a_k = \frac{2}{2\pi} \int_0^{2\pi} f(t) \cos(kt)\, dt = \frac{1}{\pi} \int_{-\pi}^{\pi} t^2 \cos(kt)\, dt =$$

$$= \frac{1}{\pi} \left(\left[t^2 \frac{\sin(kt)}{k} \right]_{-\pi}^{\pi} - \int_{-\pi}^{\pi} 2t \frac{\sin(kt)}{k}\, dt \right) = -\frac{2}{\pi k} \int_{-\pi}^{\pi} t \sin(kt)\, dt =$$

$$= -\frac{2}{\pi k} \left(\left[-t \frac{\cos(kt)}{k} \right]_{-\pi}^{\pi} + \int_{-\pi}^{\pi} \frac{\cos(kt)}{k}\, dt \right) = \frac{4\,(-1)^k}{k^2} .$$

On the other hand,

$$b_k = \frac{2}{2\pi} \int_0^{2\pi} f(t) \sin(kt)\, dt = \frac{1}{\pi} \int_{-\pi}^{\pi} t^2 \sin(kt)\, dt = 0$$

since $t \mapsto t^2 \sin(kt)$ is an odd function. Since $f : \mathbb{R} \to \mathbb{R}$ is continuous, we can then state that

$$f(t) = \frac{\pi^2}{3} + \sum_{k=1}^{\infty} \frac{4\,(-1)^k}{k^2} \cos(kt), \quad \text{for every } t \in \mathbb{R} .$$

Let us focus on two interesting cases. If $t = \pi$, we obtain the formula

$$\pi^2 = \frac{\pi^2}{3} + \sum_{k=1}^{\infty} \frac{4}{k^2} ,$$

yielding

$$\sum_{k=1}^{\infty} \frac{1}{k^2} = \frac{\pi^2}{6} \ ;$$

if $t = 0$, we have

$$0 = \frac{\pi^2}{3} + \sum_{k=1}^{\infty} \frac{4 \, (-1)^k}{k^2} \ ,$$

giving us the formula

$$\sum_{k=1}^{\infty} \frac{(-1)^{k+1}}{k^2} = \frac{\pi^2}{12} \ .$$

8.5   Series and Integrals

We now prove a theorem that shows the close connection between the theory of numerical series and that of the integral.

Theorem 8.31 *Let $f : [1, +\infty[\to \mathbb{R}$ be a function that is positive, decreasing, and integrable on $[1, c]$ for every $c > 1$. Then the series $\sum_{k=1}^{\infty} f(k)$ converges if and only if f is integrable on $[1, +\infty[$. Moreover, we have*

$$\int_{1}^{+\infty} f \le \sum_{k=1}^{\infty} f(k) \le f(1) + \int_{1}^{+\infty} f \ .$$

Proof For $x \in [k, k+1]$, it must be that $f(k+1) \le f(x) \le f(k)$, hence

$$f(k+1) \le \int_{k}^{k+1} f \le f(k) \ .$$

Summing up, we obtain

$$\sum_{k=1}^{n} f(k+1) \le \int_{1}^{n+1} f \le \sum_{k=1}^{n} f(k) \ .$$

Since f is positive, the sequence $(\sum_{k=1}^{n} f(k))_n$ and the function $c \mapsto \int_{1}^{c} f$ are both increasing and, therefore, have a limit. The conclusion now follows from the comparison theorem for limits. ■

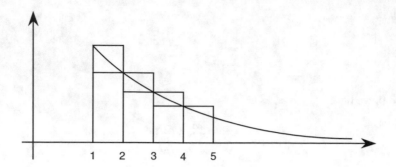

It should be clear that the choice of the starting point $a = 1$ for both the integral and the series is by no way mandatory.

Example Consider the series $\sum_{k=1}^{\infty} k^{-3}$; in this case,

$$\int_1^{+\infty} \frac{1}{x^3}\,dx \le \sum_{k=1}^{\infty} \frac{1}{k^3} \le 1 + \int_1^{+\infty} \frac{1}{x^3}\,dx,$$

and then

$$\frac{1}{2} \le \sum_{k=1}^{\infty} \frac{1}{k^3} \le \frac{3}{2}.$$

A greater accuracy is easily attained by computing the sum of the first few terms and then using the estimate given by the integral. For example, separating the first two terms, we have that

$$\sum_{k=1}^{\infty} \frac{1}{k^3} = 1 + \frac{1}{8} + \sum_{k=3}^{\infty} \frac{1}{k^3},$$

with

$$\int_3^{+\infty} \frac{1}{x^3}\,dx \le \sum_{k=3}^{\infty} \frac{1}{k^3} \le \frac{1}{27} + \int_3^{+\infty} \frac{1}{x^3}\,dx.$$

We thus have proved that

$$\frac{255}{216} \le \sum_{k=1}^{\infty} \frac{1}{k^3} \le \frac{263}{216}.$$

More on the Integral

9

9.1 Saks–Henstock Theorem

Let us further analyze the definition of integral for a function $f : I \to \mathbb{R}$ when $I = [a, b]$ is a compact interval.

The function f is integrable on I with the integral $\int_I f$ if, for every $\varepsilon > 0$, there is a gauge δ on I such that, for every δ-fine tagged partition

$$\mathring{\mathcal{P}} = \{(x_1, [a_0, a_1]), \ldots, (x_m, [a_{m-1}, a_m])\}$$

of I, we have that $\left| S(f, \mathring{\mathcal{P}}) - \int_I f \right| \le \varepsilon$. Then, since

$$S(f, \mathring{\mathcal{P}}) = \sum_{j=1}^{m} f(x_j)(a_j - a_{j-1}), \qquad \int_I f = \sum_{j=1}^{m} \int_{a_{j-1}}^{a_j} f,$$

we have that

$$\left| \sum_{j=1}^{m} \left(f(x_j)(a_j - a_{j-1}) - \int_{a_{j-1}}^{a_j} f \right) \right| \le \varepsilon.$$

This fact tells us that the sum of all "errors" $(f(x_j)(a_j - a_{j-1}) - \int_{a_{j-1}}^{a_j} f)$ is arbitrarily small, provided that the tagged partition is sufficiently fine. Note that those "errors" may be either positive or negative, so that in the sum they could compensate for one another. The following theorem tells us that even the sum of all *absolute values* of those "errors" can be made arbitrarily small.

Theorem 9.1 (Saks–Henstock Theorem—I) *Let $f : I \to \mathbb{R}$ be an integrable function, and let δ be a gauge on I such that, for every δ-fine tagged partition $\mathring{\mathcal{P}}$*

of I, it happens that $|S(f, \mathring{\mathcal{P}}) - \int_I f| \le \varepsilon$. Then for all such tagged partitions
$\mathring{\mathcal{P}} = \{(x_1, [a_0, a_1]), \dots, (x_m, [a_{m-1}, a_m])\}$ we also have that

$$\sum_{j=1}^{m} \left| f(x_j)(a_j - a_{j-1}) - \int_{a_{j-1}}^{a_j} f \right| \le 4\varepsilon.$$

Proof We consider separately in the sum the positive and negative terms. Let us prove that the sum of the positive terms is less than or equal to 2ε. In an analogous way one can proceed for the negative terms. Rearranging the terms in the sum, we can assume that the positive ones are the first q terms $(f(x_j)(a_j - a_{j-1}) - \int_{a_{j-1}}^{a_j} f)$, with $j = 1, \dots, q$, i.e.,

$$f(x_1)(a_1 - a_0) - \int_{a_0}^{a_1} f , \dots, \quad f(x_q)(a_q - a_{q-1}) - \int_{a_{q-1}}^{a_q} f .$$

Consider the remaining $m - q$ intervals $[a_{k-1}, a_k]$, with $k = q + 1, \dots, m$, i.e.,

$$[a_q, a_{q+1}], \dots, [a_{m-1}, a_m].$$

Since f is integrable on these intervals, there exist some gauges δ_k on $[a_{k-1}, a_k]$, respectively, which we can choose such that $\delta_k(x) \le \delta(x)$ for every $x \in [a_{k-1}, a_k]$, for which

$$\left| S(f, \mathring{\mathcal{P}}_k) - \int_{a_{k-1}}^{a_k} f \right| \le \frac{\varepsilon}{m - q}$$

for every δ_k-fine tagged partition $\mathring{\mathcal{P}}_k$ of $[a_{k-1}, a_k]$. Consequently, the family $\mathring{\mathcal{Q}}$ made by the couples $(x_1, [a_0, a_1]), \dots, (x_q, [a_{q-1}, a_q])$ and by the elements of the families $\mathring{\mathcal{P}}_k$, with k varying from $q + 1$ to m, is a δ-fine tagged partition of I such that

$$S(f, \mathring{\mathcal{Q}}) = \sum_{j=1}^{q} f(x_j)(a_j - a_{j-1}) + \sum_{k=q+1}^{m} S(f, \mathring{\mathcal{P}}_k).$$

Then we have

$$\sum_{j=1}^{q} \left(f(x_j)(a_j - a_{j-1}) - \int_{a_{j-1}}^{a_j} f \right) = \sum_{j=1}^{q} f(x_j)(a_j - a_{j-1}) - \sum_{j=1}^{q} \int_{a_{j-1}}^{a_j} f$$

$$= \left(S(f, \mathring{\mathcal{Q}}) - \sum_{k=q+1}^{m} S(f, \mathring{\mathcal{P}}_k) \right) - \left(\int_I f - \sum_{k=q+1}^{m} \int_{a_{k-1}}^{a_k} f \right)$$

$$\leq \left| S(f, \mathring{Q}) - \int_I f \right| + \sum_{k=q+1}^{m} \left| S(f, \mathring{P}_k) - \int_{a_{k-1}}^{a_k} f \right|$$

$$\leq \varepsilon + (m - q) \frac{\varepsilon}{m - q} = 2\varepsilon.$$

Proceeding similarly for the negative terms, the conclusion follows. ∎

The following corollary will be useful in the next section to study the integrability of the absolute value of an integrable function.

Corollary 9.2 *Let* $f : I \to \mathbb{R}$ *be an integrable function, and let* δ *be a gauge on* I *such that, for every* δ*-fine tagged partition* $\mathring{P}$ *of* I*, it happens that* $|S(f, \mathring{P}) - \int_I f| \leq \varepsilon$*. Then for such tagged partitions* $\mathring{P} = \{(x_1, [a_0, a_1]), \ldots, (x_m, [a_{m-1}, a_m])\}$ *we also have*

$$\left| S(|f|, \mathring{P}) - \sum_{j=1}^{m} \left| \int_{a_{j-1}}^{a_j} f \right| \right| \leq 4\varepsilon.$$

Proof Using the well-known inequalities for the absolute value, by Theorem 9.1,

$$\left| S(|f|, \mathring{P}) - \sum_{j=1}^{m} \left| \int_{a_{j-1}}^{a_j} f \right| \right| = \left| \sum_{j=1}^{m} \left[|f(x_j)|(a_j - a_{j-1}) - \left| \int_{a_{j-1}}^{a_j} f \right| \right] \right|$$

$$\leq \sum_{j=1}^{m} \left| |f(x_j)(a_j - a_{j-1})| - \left| \int_{a_{j-1}}^{a_j} f \right| \right|$$

$$\leq \sum_{j=1}^{m} \left| f(x_j)(a_j - a_{j-1}) - \int_{a_{j-1}}^{a_j} f \right| \leq 4\varepsilon.$$

This completes the proof. ∎

The Saks–Henstock Theorem 9.1 can be generalized to tagged subpartitions. Here is the statement.

Theorem 9.3 (Saks–Henstock Theorem—II) *Let* $f : I \to \mathbb{R}$ *be an integrable function, and let* δ *be a gauge on* I *such that, for every* δ*-fine tagged partition* $\mathring{P}$ *of* I*, it happens that* $|S(f, \mathring{P}) - \int_I f| \leq \varepsilon$*. Then for every* δ*-fine tagged subpartition* $\Xi = \{(\xi_j, [\alpha_j, \beta_j]) : j = 1, \ldots, m\}$ *of* I *we have*

$$\sum_{j=1}^{m} \left| f(\xi_j)(\beta_j - \alpha_j) - \int_{\alpha_j}^{\beta_j} f \right| \leq 4\varepsilon.$$

As a consequence, for any such δ-fine tagged subpartition,

$$\left| S(f, \Xi) - \sum_{j=1}^{m} \int_{\alpha_j}^{\beta_j} f \right| \leq 4\varepsilon .$$

Proof By the Cousin theorem, it is possible to extend any δ-fine tagged subpartition Ξ of I to a whole δ-fine tagged partition $\overset{\circ}{\mathcal{P}}$ of I. Hence, the Saks–Henstock Theorem 9.1 applies, proving the first part of the statement.

Concerning the second part, we have

$$\left| S(f, \Xi) - \sum_{j=1}^{m} \int_{\alpha_j}^{\beta_j} f \right| = \left| \sum_{j=1}^{m} f(\xi_j)(\beta_j - \alpha_j) - \sum_{j=1}^{m} \int_{\alpha_j}^{\beta_j} f \right|$$

$$= \left| \sum_{j=1}^{m} \left(f(\xi_j)(\beta_j - \alpha_j) - \int_{\alpha_j}^{\beta_j} f \right) \right|$$

$$\leq \sum_{j=1}^{m} \left| f(\xi_j)(\beta_j - \alpha_j) - \int_{\alpha_j}^{\beta_j} f \right| \leq 4\varepsilon ,$$

thereby completing the proof.   ∎

9.2   *L*-Integrable Functions

In this section, we introduce another important class of integrable functions on the interval $I = [a, b]$.

Definition 9.4 We say that an integrable function $f : I \to \mathbb{R}$ is "*L*-integrable" (or "integrable according to Lebesgue") if even $|f|$ happens to be integrable on I.

It is clear that every positive integrable function is *L*-integrable. Moreover, every continuous function on $[a, b]$ is *L*-integrable there since $|f|$ is still continuous. We have the following characterization of *L*-integrability.

Proposition 9.5 *Let $f : I \to \mathbb{R}$ be an integrable function, and consider the set $\mathcal{S}$ of all real numbers*

$$\sum_{i=1}^{q} \left| \int_{c_{i-1}}^{c_i} f \right| ,$$

obtained choosing $c_0, c_1, \ldots, c_q$ in I in such a way that $a = c_0 < c_1 < \cdots < c_q = b$. The function f is L-integrable on I if and only if $\mathcal{S}$ is bounded from above. In *that case, we have*

$$\int_I |f| = \sup \mathcal{S}.$$

Proof Assume first that f is L-integrable on I. If $a = c_0 < c_1 < \cdots < c_q = b$, then f and $|f|$ are integrable on every subinterval $[c_{i-1}, c_i]$, and we have

$$\sum_{i=1}^{q} \left| \int_{c_{i-1}}^{c_i} f \right| \le \sum_{i=1}^{q} \int_{c_{i-1}}^{c_i} |f| = \int_I |f|.$$

Consequently, the set $\mathcal{S}$ is bounded from above: $\sup \mathcal{S} \le \int_I |f| < +\infty$.

Conversely, assume now that $\mathcal{S}$ is bounded from above, and let us prove that in that case $|f|$ is integrable on I and $\int_I |f| = \sup \mathcal{S}$. Fix $\varepsilon > 0$. Let δ_1 be a gauge such that, for every δ_1-fine tagged partition $\mathring{\mathcal{P}}$ of I, we have

$$\left| S(f, \mathring{\mathcal{P}}) - \int_I f \right| \le \frac{\varepsilon}{8}.$$

On the other hand, letting $\mathcal{J} = \sup \mathcal{S}$, by the properties of the supremum there surely are $a = c_0 < c_1 < \cdots < c_q = b$ such that

$$\mathcal{J} - \frac{\varepsilon}{2} \le \sum_{i=1}^{q} \left| \int_{c_{i-1}}^{c_i} f \right| \le \mathcal{J}.$$

We construct the gauge δ_2 in such a way that, for every $x \in I$, it must be that $[x - \delta_2(x), x + \delta_2(x)]$ meets only those intervals $[c_{i-1}, c_i]$ to which x belongs. In this way,

- If x belongs to the interior of one of the intervals $[c_{i-1}, c_i]$, we have that $[x - \delta_2(x), x + \delta_2(x)]$ is contained in $]c_{i-1}, c_i[$.
- If x coincides with one of the c_i in the interior of $[a, b]$, then $[x - \delta_2(x), x + \delta_2(x)]$ is contained in $]c_{i-1}, c_{i+1}[$.
- If $x = a$, then $[x, x + \delta_2(x)]$ is contained in $[a, c_1[$.
- If $x = b$, then $[x - \delta_2(x), x]$ is contained in $]c_{q-1}, b]$.

Define $\delta(x) = \min\{\delta_1(x), \delta_2(x)\}$ for every $x \in I$. Once we take a δ-fine tagged partition $\mathring{\mathcal{P}} = \{(x_1, [a_0, a_1]), \ldots, (x_m, [a_{m-1}, a_m])\}$ of I, consider the intervals (possibly empty or reduced to a point)

$$I_{j,i} = [a_{j-1}, a_j] \cap [c_{i-1}, c_i].$$

The choice of the gauge δ_2 yields that, if $I_{j,i}$ has a positive length, then $x_j \in I_{j,i}$. Indeed, if $x_j \notin [c_{i-1}, c_i]$, then

$$[a_{j-1}, a_j] \cap [c_{i-1}, c_i] \subseteq [x_j - \delta_2(x_j), x_j + \delta_2(x_j)] \cap [c_{i-1}, c_i] = \emptyset.$$

Therefore, if we take those $I_{j,i}$, then the set

$$\mathring{\mathcal{Q}} = \{(x_j, I_{j,i}) : j = 1, \ldots, m , \ i = 1, \ldots, q , \ \mu(I_{j,i}) > 0\}$$

is a δ-fine tagged partition of I, and we have

$$S(|f|, \mathring{\mathcal{P}}) = \sum_{j=1}^{m} |f(x_j)|(a_j - a_{j-1}) = \sum_{j=1}^{m} \sum_{i=1}^{q} |f(x_j)| \mu(I_{j,i}) = S(|f|, \mathring{\mathcal{Q}}).$$

Moreover,

$$\mathcal{J} - \frac{\varepsilon}{2} \le \sum_{i=1}^{q} \left| \int_{c_{i-1}}^{c_i} f \right| = \sum_{i=1}^{q} \left| \sum_{j=1}^{m} \int_{I_{j,i}} f \right| \le \sum_{i=1}^{q} \sum_{j=1}^{m} \left| \int_{I_{j,i}} f \right| \le \mathcal{J},$$

and by Corollary 9.2,

$$\left| S(|f|, \mathring{\mathcal{Q}}) - \sum_{i=1}^{q} \sum_{j=1}^{m} \left| \int_{I_{j,i}} f \right| \right| \le 4 \frac{\varepsilon}{8} = \frac{\varepsilon}{2}.$$

Consequently, we have

$$|S(|f|, \mathring{\mathcal{P}}) - \mathcal{J}| = |S(|f|, \mathring{\mathcal{Q}}) - \mathcal{J}|$$

$$\le \left| S(|f|, \mathring{\mathcal{Q}}) - \sum_{i=1}^{q} \sum_{j=1}^{m} \left| \int_{I_{j,i}} f \right| \right| + \left| \sum_{i=1}^{q} \sum_{j=1}^{m} \left| \int_{I_{j,i}} f \right| - \mathcal{J} \right|$$

$$\le \frac{\varepsilon}{2} + \frac{\varepsilon}{2} = \varepsilon,$$

which is what we wanted to prove.   ∎

We have a series of corollaries.

Corollary 9.6 *Let $f, g : I \to \mathbb{R}$ be two integrable functions such that, for every $x \in I$,*

$$|f(x)| \le g(x) ;$$

then f is L-integrable on I.

Proof Take $c_0, c_1, \ldots, c_q$ in I so that $a = c_0 < c_1 < \cdots < c_q = b$. Since $-g(x) \leq f(x) \leq g(x)$ for every $x \in I$, we have that

$$-\int_{c_{i-1}}^{c_i} g \leq \int_{c_{i-1}}^{c_i} f \leq \int_{c_{i-1}}^{c_i} g \,,$$

i.e.,

$$\left| \int_{c_{i-1}}^{c_i} f \right| \leq \int_{c_{i-1}}^{c_i} g \,,$$

for every $1 \leq i \leq q$. Hence,

$$\sum_{i=1}^{q} \left| \int_{c_{i-1}}^{c_i} f \right| \leq \sum_{i=1}^{q} \int_{c_{i-1}}^{c_i} g = \int_I g \,.$$

Then the set $\mathcal{S}$ is bounded above by $\int_I g$, so that f is L-integrable on I. ∎

Corollary 9.7 *Let $f, g : I \to \mathbb{R}$ be two L-integrable functions, and let $\alpha \in \mathbb{R}$ be a constant. Then $f + g$ and αf are L-integrable on I.*

Proof By assumption, $f, |f|$ and $g, |g|$ are integrable on I. Then $f + g$, $|f| + |g|$, αf, and $|\alpha| |f|$ are, too. On the other hand, for every $x \in I$,

$$|(f + g)(x)| \leq |f(x)| + |g(x)| \,, \qquad |\alpha f(x)| \leq |\alpha| \, |f(x)| \,.$$

Corollary 9.6 then guarantees that $f + g$ and αf are L-integrable on I. ∎

We have thus proved that the L-integrable functions make up a vector subspace of the space of integrable functions.

Corollary 9.8 *Let $f_1, f_2 : I \to \mathbb{R}$ be two L-integrable functions. Then $\min\{f_1, f_2\}$ and $\max\{f_1, f_2\}$ are L-integrable on I.*

Proof It follows immediately from the formulas

$$\min\{f_1, f_2\} = \frac{1}{2}(f_1 + f_2 - |f_1 - f_2|) \,, \quad \max\{f_1, f_2\} = \frac{1}{2}(f_1 + f_2 + |f_1 - f_2|) \,,$$

and from Corollary 9.7. ∎

Corollary 9.9 *A function $f : I \to \mathbb{R}$ is L-integrable if and only if both its positive part $f^+ = \max\{f, 0\}$ and its negative part $f^- = \max\{-f, 0\}$ are integrable on I. In that case, $\int_I f = \int_I f^+ - \int_I f^-$.*

Proof It follows immediately from Corollary 9.8 and the formulas $f = f^+ - f^-$, $|f| = f^+ + f^-$.   ∎

We want to see now an example of an integrable function that is not L-integrable. Let $f : [0, 1] \to \mathbb{R}$ be defined by

$$f(x) = \begin{cases} \dfrac{1}{x} \sin\left(\dfrac{1}{x^2}\right) & \text{if } x \neq 0, \\ 0 & \text{if } x = 0. \end{cases}$$

Let us define the two auxiliary functions $g : [0, 1] \to \mathbb{R}$ and $h : [0, 1] \to \mathbb{R}$ as

$$g(x) = \begin{cases} \dfrac{1}{x} \sin\left(\dfrac{1}{x^2}\right) + x \cos\left(\dfrac{1}{x^2}\right) & \text{if } x \neq 0, \\ 0 & \text{if } x = 0, \end{cases}$$

$$h(x) = \begin{cases} -x \cos\left(\dfrac{1}{x^2}\right) & \text{if } x \neq 0, \\ 0 & \text{if } x = 0. \end{cases}$$

It is easily seen that g is primitivable on $[0, 1]$ and that one of its primitives $G : [0, 1] \to \mathbb{R}$ is given by

$$G(x) = \begin{cases} \dfrac{x^2}{2} \cos\left(\dfrac{1}{x^2}\right) & \text{if } x \neq 0, \\ 0 & \text{if } x = 0. \end{cases}$$

Moreover, h is continuous on $[0, 1]$, so it is primitivable there, too. Hence, even the function $f = g+h$ is primitivable on $[0, 1]$. By the Fundamental Theorem, f is then integrable on $[0, 1]$. We will show now that $|f|$ is not integrable on $[0, 1]$. Consider the intervals $[((k+1)\pi)^{-1/2}, (k\pi)^{-1/2}]$, with $k \geq 1$. The function $|f|$ is continuous on these intervals, so it is primitivable there. By the substitution $y = 1/x^2$, we obtain

$$\int_{((k+1)\pi)^{-1/2}}^{(k\pi)^{-1/2}} \frac{1}{x} \left| \sin \frac{1}{x^2} \right| dx = \int_{k\pi}^{(k+1)\pi} \frac{1}{2y} |\sin y| \, dy.$$

On the other hand,

$$\int_{k\pi}^{(k+1)\pi} \frac{1}{2y} |\sin y| \, dy \geq \frac{1}{2(k+1)\pi} \int_{k\pi}^{(k+1)\pi} |\sin y| \, dy = \frac{1}{(k+1)\pi}.$$

If $|f|$ were integrable on $[0, 1]$, we would have that, for every $n \geq 1$,

$$\int_0^1 |f| = \int_0^{((n+1)\pi)^{-1/2}} |f| + \sum_{k=1}^n \int_{((k+1)\pi)^{-1/2}}^{(k\pi)^{-1/2}} |f| + \int_{\pi^{-1/2}}^1 |f|$$

$$\geq \sum_{k=1}^n \int_{((k+1)\pi)^{-1/2}}^{(k\pi)^{-1/2}} |f| \geq \sum_{k=1}^n \frac{1}{(k+1)\pi},$$

which is impossible since the series $\sum_{k=1}^\infty \frac{1}{k+1}$ diverges. Hence, f is not L-integrable on $[0, 1]$.

9.3 Monotone Convergence Theorem

In this section and the next, we will consider the situation where a sequence of integrable functions $(f_n)_n$ converges pointwise to a function f, i.e., for every $x \in I$,

$$\lim_n f_n(x) = f(x).$$

The question is whether f is integrable on I, with

$$\int_I f = \lim_n \int_I f_n,$$

i.e., whether the following formula holds:

$$\int_I \lim_n f_n = \lim_n \int_I f_n.$$

This problem has already been faced in Theorem 7.24, involving continuous functions and uniform convergence. We will see now that the formula holds true if the sequence of functions is monotone. Let us state the following result, due to Beppo Levi.

Theorem 9.10 (Monotone Convergence Theorem—I) *We are given a function* $f : I \to \mathbb{R}$ *and a sequence of functions* $f_n : I \to \mathbb{R}$, *with* $n \in \mathbb{N}$, *verifying the following conditions:*

(a) *The sequence* $(f_n)_n$ *converges pointwise to* f.
(b) *The sequence* $(f_n)_n$ *is monotone.*
(c) *Each function* f_n *is integrable on* I.
(d) *The real sequence* $(\int_I f_n)_n$ *has a finite limit.*

Then f is integrable on I, and

$$\int_I f = \lim_n \int_I f_n \, .$$

Proof We assume for definiteness that the sequence $(f_n)_n$ is increasing, i.e.,

$$f_n(x) \le f_{n+1}(x) \le f(x) \, ,$$

for every $n \in \mathbb{N}$ and every $x \in I$. Let us set

$$\mathcal{J} = \lim_n \int_I f_n \, .$$

We will prove that f is integrable on I and that $\mathcal{J}$ is its integral. Fix $\varepsilon > 0$. Since every f_n is integrable on I, there are some gauges δ_n^* on I such that if $\overset{\circ}{\mathcal{P}}_n$ is a δ_n^*-fine tagged partition of I; then

$$\left| S(f_n, \overset{\circ}{\mathcal{P}}_n) - \int_I f_n \right| \le \frac{\varepsilon}{3 \cdot 2^{n+3}} \, .$$

Moreover, there is a $\bar{n} \in \mathbb{N}$ such that, for every $n \ge \bar{n}$, it is

$$0 \le \mathcal{J} - \int_I f_n \le \frac{\varepsilon}{3} \, ,$$

and since the sequence $(f_n)_n$ converges pointwise on I to f, for every $x \in I$ there is a natural number $n(x) \ge \bar{n}$ such that, for every $n \ge n(x)$, one has

$$|f_n(x) - f(x)| \le \frac{\varepsilon}{3(b-a)} \, .$$

Let us define the gauge δ in the following way. For every $x \in I$,

$$\delta(x) = \delta_{n(x)}^*(x) \, .$$

Now let $\overset{\circ}{\mathcal{P}} = \{(x_1, [a_0, a_1]), \dots, (x_m, [a_{m-1}, a_m])\}$ be a δ-fine tagged partition of I. We have

$$|S(f, \overset{\circ}{\mathcal{P}}) - \mathcal{J}| = \left| \sum_{j=1}^m f(x_j)(a_j - a_{j-1}) - \mathcal{J} \right|$$

$$\le \left| \sum_{j=1}^m [f(x_j) - f_{n(x_j)}(x_j)](a_j - a_{j-1}) \right|$$

$$+ \left| \sum_{j=1}^{m} \left[f_{n(x_j)}(x_j)(a_j - a_{j-1}) - \int_{a_{j-1}}^{a_j} f_{n(x_j)} \right] \right|$$

$$+ \left| \sum_{j=1}^{m} \int_{a_{j-1}}^{a_j} f_{n(x_j)} - \mathcal{J} \right|.$$

Estimation of the first term gives

$$\left| \sum_{j=1}^{m} [f(x_j) - f_{n(x_j)}(x_j)](a_j - a_{j-1}) \right| \le \sum_{j=1}^{m} |f(x_j) - f_{n(x_j)}(x_j)|(a_j - a_{j-1})$$

$$\le \sum_{j=1}^{m} \frac{\varepsilon}{3(b-a)}(a_j - a_{j-1}) = \frac{\varepsilon}{3}.$$

To estimate the second term, set

$$r = \min_{1 \le j \le m} n(x_j), \quad s = \max_{1 \le j \le m} n(x_j),$$

and note that, putting together the terms whose indices $n(x_j)$ coincide with the same value k, by the second statement of Saks–Henstock Theorem 9.3, we obtain

$$\left| \sum_{j=1}^{m} \left[f_{n(x_j)}(x_j)(a_j - a_{j-1}) - \int_{a_{j-1}}^{a_j} f_{n(x_j)} \right] \right|$$

$$= \left| \sum_{k=r}^{s} \left\{ \sum_{\{1 \le j \le m \,:\, n(x_j)=k\}} \left[f_k(x_j)(a_j - a_{j-1}) - \int_{a_{j-1}}^{a_j} f_k \right] \right\} \right|$$

$$\le \sum_{k=r}^{s} \sum_{\{1 \le j \le m \,:\, n(x_j)=k\}} \left| f_k(x_j)(a_j - a_{j-1}) - \int_{a_{j-1}}^{a_j} f_k \right|$$

$$\le \sum_{k=r}^{s} 4 \frac{\varepsilon}{3 \cdot 2^{k+3}} \le \frac{\varepsilon}{3}.$$

Concerning the third term, since $r \geq \bar{n}$, using the monotonicity of the sequence $(f_n)_n$ we have

$$0 \leq \mathcal{J} - \int_I f_s = \mathcal{J} - \sum_{j=1}^{m} \int_{a_{j-1}}^{a_j} f_s \leq$$

$$\leq \mathcal{J} - \sum_{j=1}^{m} \int_{a_{j-1}}^{a_j} f_{n(x_j)} \leq$$

$$\leq \mathcal{J} - \sum_{j=1}^{m} \int_{a_{j-1}}^{a_j} f_r = \mathcal{J} - \int_I f_r \leq \frac{\varepsilon}{3},$$

from which

$$\left| \sum_{j=1}^{m} \int_{a_{j-1}}^{a_j} f_{n(x_j)} - \mathcal{J} \right| \leq \frac{\varepsilon}{3}.$$

Hence,

$$|S(f, \mathring{\mathcal{P}}) - \mathcal{J}| \leq \frac{\varepsilon}{3} + \frac{\varepsilon}{3} + \frac{\varepsilon}{3} = \varepsilon,$$

and the proof is thus completed.   ∎

As an immediate consequence of the Monotone Convergence Theorem 9.10, we have an analogous statement for a series of functions.

Corollary 9.11 *We are given a function* $f : I \to \mathbb{R}$ *and a sequence of functions* $f_k : I \to \mathbb{R}$, *with* $k \in \mathbb{N}$, *verifying the following conditions:*

(a) *The series* $\sum_k f_k$ *converges pointwise to* f.
(b) *For every* $k \in \mathbb{N}$ *and every* $x \in I$, *we have* $f_k(x) \geq 0$.
(c) *Each function* f_k *is integrable on* I.
(d) *The series* $\sum_k (\int_I f_k)$ *converges.*

Then f *is integrable on* I, *and*

$$\int_I f = \sum_k \int_I f_k.$$

We can then write

$$\int_I \sum_k f_k = \sum_k \int_I f_k \,.$$

Example Consider the Taylor series associated with the function $f(x) = e^{x^2}$,

$$e^{x^2} = \sum_{k=0}^{\infty} \frac{x^{2k}}{k!} \,.$$

The functions $f_k(x) = \frac{x^{2k}}{k!}$ satisfy the first three assumptions of Corollary 9.11, with $I = [a, b]$ and

$$\int_a^b f_k(x)\, dx = \left[\frac{x^{2k+1}}{(2k+1)k!} \right]_a^b = \frac{b^{2k+1} - a^{2k+1}}{(2k+1)k!} \,,$$

so it can be seen that the series $\sum_k (\int_I f_k)$ converges. It is then possible to apply the corollary, thereby obtaining

$$\int_a^b e^{x^2}\, dx = \sum_{k=0}^{\infty} \frac{b^{2k+1} - a^{2k+1}}{(2k+1)k!} \,.$$

In particular, considering the integral function $\int_0^{\cdot} f$, we find an expression for the primitives of e^{x^2}, i.e.,

$$\int e^{x^2}\, dx = \sum_{k=0}^{\infty} \frac{x^{2k+1}}{(2k+1)k!} + c \,.$$

9.4   Dominated Convergence Theorem

We start by proving the following preliminary result.

Lemma 9.12 *Let $f_1, f_2, \ldots, f_n : I \to \mathbb{R}$ be integrable functions. If there exists an integrable function $g : I \to \mathbb{R}$ such that*

$$g(x) \le f_k(x), \quad \text{for every } x \in I \text{ and } k \in \{1, \ldots, n\},$$

then $\min\{f_1, f_2, \ldots, f_n\}$ and $\max\{f_1, f_2, \ldots, f_n\}$ are integrable on I.

Proof Consider the case $n = 2$. The functions $f_1 - g$ and $f_2 - g$, being integrable and nonnegative, are L-integrable. Hence, $\min\{f_1 - g, f_2 - g\}$ and $\max\{f_1 - g, f_2 - g\}$ are L-integrable, by Corollary 9.8. The conclusion then follows from the fact that

$$\min\{f_1, f_2\} = \min\{f_1 - g, f_2 - g\} + g,$$
$$\max\{f_1, f_2\} = \max\{f_1 - g, f_2 - g\} + g.$$

The general case can be easily obtained by induction. ∎

We are now ready to state and prove the following important extension of Theorem 7.24 due to Henri Lebesgue.

Theorem 9.13 (Dominated Convergence Theorem–I) *We are given a function $f : I \to \mathbb{R}$ and a sequence of functions $f_n : I \to \mathbb{R}$, with $n \in \mathbb{N}$, verifying the following conditions:*

(a) The sequence $(f_n)_n$ converges pointwise to f.
(b) Each function f_n is integrable on I.
(c) There are two integrable functions $g, h : I \to \mathbb{R}$ for which

$$g(x) \le f_n(x) \le h(x)$$

for every $n \in \mathbb{N}$ and $x \in I$.

Then the sequence $\left(\int_I f_n \right)_n$ has a finite limit, f is integrable on I, and

$$\int_I f = \lim_n \int_I f_n.$$

Proof For any couple of natural numbers n, ℓ, define the functions

$$\phi_{n,\ell} = \min\{f_n, f_{n+1}, \ldots, f_{n+\ell}\}, \quad \Phi_{n,\ell} = \max\{f_n, f_{n+1}, \ldots, f_{n+\ell}\}.$$

By Lemma 9.12, all $\phi_{n,\ell}$ and $\Phi_{n,\ell}$ are integrable on I. Moreover, for any fixed n, the sequence $(\phi_{n,\ell})_\ell$ is decreasing and bounded from below by g, and the sequence $(\Phi_{n,\ell})_\ell$ is increasing and bounded from above by h. Hence, these sequences converge to the two functions ϕ_n and Φ_n, respectively:

$$\lim_\ell \phi_{n,\ell} = \phi_n = \inf\{f_n, f_{n+1}, \ldots\}, \quad \lim_\ell \Phi_{n,\ell} = \Phi_n = \sup\{f_n, f_{n+1}, \ldots\}.$$

Furthermore, the sequence $(\int_I \phi_{n,\ell})_\ell$ is decreasing and bounded from below by $\int_I g$, whereas the sequence $(\int_I \Phi_{n,\ell})_\ell$ is increasing and bounded from above by $\int_I h$. The Monotone Convergence Theorem 9.10 then guarantees that the functions ϕ_n and Φ_n are integrable on I.

Now the sequence $(\phi_n)_n$ is increasing, and the sequence $(\Phi_n)_n$ is decreasing; as $\lim_n f_n = f$, we must have

$$\lim_n \phi_n = \liminf_n f_n = f , \quad \lim_n \Phi_n = \limsup_n f_n = f .$$

Moreover, the sequence $(\int_I \phi_n)_n$ is increasing and bounded from above by $\int_I h$, whereas the sequence $(\int_I \Phi_n)_n$ is decreasing and bounded from below by $\int_I g$. We can then apply again the Monotone Convergence Theorem 9.10, from which we deduce that f is integrable on I and

$$\int_I f = \lim_n \int_I \phi_n = \lim_n \int_I \Phi_n .$$

Since $\phi_n \le f_n \le \Phi_n$, we have $\int_I \phi_n \le \int_I f_n \le \int_I \Phi_n$, and the conclusion follows by the Squeeze Theorem 3.10. ∎

Example Consider, for $n \ge 1$, the functions $f_n : [0, 3] \to \mathbb{R}$ defined by $f_n(x) = \arctan\left(nx - \frac{n^2}{n+1}\right)$. We have the following situation:

$$\lim_n f_n(x) = \begin{cases} -\dfrac{\pi}{2} & \text{if } x \in [0, 1[, \\[2mm] \dfrac{\pi}{4} & \text{if } x = 1 , \\[2mm] \dfrac{\pi}{2} & \text{if } x \in]1, 3] . \end{cases}$$

Moreover,

$$|f_n(x)| \le \frac{\pi}{2} , \quad \text{for every } n \in \mathbb{N} \text{ and } x \in [0, 3] .$$

The assumptions of the Dominated Convergence Theorem 9.13 are then satisfied, taking the two constant functions $g(x) = -\frac{\pi}{2}$, $h(x) = \frac{\pi}{2}$. We can then conclude that

$$\lim_n \int_0^3 \arctan\left(nx - \frac{n^2}{n+1}\right) dx = -\frac{\pi}{2} + 2\frac{\pi}{2} = \frac{\pi}{2} .$$

9.5   Hake's Theorem

Recall that a function $f : [a, b[\to \mathbb{R}$ is said to be integrable if it is integrable on $[a, c]$ for every $c \in]a, b[$, and the limit

$$\lim_{c \to b^-} \int_a^c f$$

exists and is finite. We want to prove the following result by Heinrich Hake.

Theorem 9.14 (Hake's Theorem) *Let $b < +\infty$, and assume that $f : [a, b[\to \mathbb{R}$ is a function that is integrable on $[a, c]$, for every $c \in]a, b[$. Then the function f is integrable on $[a, b[$ if and only if it is the restriction of an integrable function $\bar{f} : [a, b] \to \mathbb{R}$. In that case,*

$$\int_a^b \bar{f} = \int_a^b f.$$

Proof Assume first that f is the restriction to $[a, b[$ of an integrable function $\bar{f} : [a, b] \to \mathbb{R}$. Fix $\varepsilon > 0$; we want to find a $\gamma > 0$ such that, if $c \in]a, b[$ and $b - c \le \gamma$, then

$$\left| \int_a^c f - \int_a^b \bar{f} \right| \le \varepsilon.$$

Let δ be a gauge such that, for every δ-fine tagged partition of $[a, b]$, we have $|S(\bar{f}, \mathcal{P}) - \int_a^b \bar{f}| \le \frac{\varepsilon}{8}$. We choose a positive constant $\gamma \le \delta(b)$ such that $\gamma |\bar{f}(b)| \le \frac{\varepsilon}{2}$. If $c \in]a, b[$ and $b - c \le \gamma$, then, by the Saks–Henstock Theorem 9.1, taking the δ-fine tagged subpartition $\mathring{\mathcal{P}} = \{(b, [c, b])\}$, we have

$$\left| \bar{f}(b)(b - c) - \int_c^b \bar{f} \right| \le 4\frac{\varepsilon}{8} = \frac{\varepsilon}{2},$$

and hence

$$\left| \int_a^c f - \int_a^b \bar{f} \right| = \left| \int_c^b \bar{f} \right| \le \left| \int_c^b \bar{f} - \bar{f}(b)(b - c) \right| + |\bar{f}(b)(b - c)|$$

$$\le \frac{\varepsilon}{2} + |\bar{f}(b)|\gamma \le \frac{\varepsilon}{2} + \frac{\varepsilon}{2} = \varepsilon.$$

Let us prove now the other implication. Assume that f is integrable on $[a, b[$, and let $\mathcal{J}$ be its integral, i.e.,

$$\mathcal{J} = \lim_{c \to b^-} \int_a^c f.$$

We extend f to a function $\bar{f}$ defined on the whole interval $[a, b]$ by setting, for instance, $\bar{f}(b) = 0$. To prove that $\bar{f}$ is integrable on $[a, b]$ with integral $\mathcal{J}$, fix $\varepsilon > 0$. By the preceding limit, there is a $\gamma > 0$ such that, if $c \in]a, b[$ and $b - c \le \gamma$, then

$$\left| \int_a^c f - \mathcal{J} \right| \le \frac{\varepsilon}{2}.$$

Consider the sequence $(c_i)_i$ of points in $[a, b[$ given by

$$c_i = b - \frac{b - a}{i + 1}.$$

Note that it is strictly increasing, it converges to b, and it is $c_0 = a$. Since f is integrable on each interval $[c_{i-1}, c_i]$, we can consider, for each $i \geq 1$, a gauge δ_i on $[c_{i-1}, c_i]$ such that, for every δ_i-fine tagged partition $\mathring{\mathcal{P}}_i$ of $[c_{i-1}, c_i]$, we have

$$\left| S(f, \mathring{\mathcal{P}}_i) - \int_{c_{i-1}}^{c_i} f \right| \leq \frac{\varepsilon}{2^{i+4}}.$$

We define a gauge δ on $[a, b]$ by setting

$$\delta(x) = \begin{cases} \min\left\{\delta_i(x), \dfrac{x - c_{i-1}}{2}, \dfrac{c_i - x}{2}\right\} & \text{if } x \in]c_{i-1}, c_i[, \\[2mm] \min\left\{\delta_1(a), \dfrac{c_1 - a}{2}\right\} & \text{if } x = a, \\[2mm] \min\left\{\delta_i(c_i), \delta_{i+1}(c_i), \dfrac{c_i - c_{i-1}}{2}, \dfrac{c_{i+1} - c_i}{2}\right\} & \text{if } x = c_i \text{ and } i \geq 1; \\[2mm] \gamma & \text{if } x = b. \end{cases}$$

Let $\mathring{\mathcal{P}} = \{(x_j, [a_{j-1}, a_j]) : j = 1, \ldots, m\}$ be a δ-fine tagged partition of $[a, b]$. Denote by q the smallest integer for which $c_{q+1} \geq a_{m-1}$. The choice of the gauge allows us to split the Riemann sum, much like in the proof of Theorem 7.18 on the additivity of the integral on subintervals, so that the sum $S(\bar{f}, \mathring{\mathcal{P}})$ will contain

- the Riemann sums on $[c_{i-1}, c_i]$, with $i = 1, \ldots, q$;
- a Riemann sum on $[c_q, a_{m-1}]$;
- a last term $\bar{f}(x_m)(b - a_{m-1})$.

(The first line disappears if $q = 0$.) Let $\mathring{\mathcal{P}}_i$ be the tagged partition of $[c_{i-1}, c_i]$ and $\mathring{\mathcal{Q}}$ be the tagged partition of $[c_q, a_{m-1}]$ whose intervals are those of $\mathring{\mathcal{P}}$. Then

$$S(\bar{f}, \mathring{\mathcal{P}}) = \sum_{i=1}^{q} S(f, \mathring{\mathcal{P}}_i) + S(f, \mathring{\mathcal{Q}}) + \bar{f}(x_m)(b - a_{m-1}).$$

To better clarify what was just said, assume, for example, that $q = 2$; then there must be a $\bar{j}_1$ for which $x_{\bar{j}_1} = c_1$ and a $\bar{j}_2$ for which $x_{\bar{j}_2} = c_2$. Then

$$S(\bar{f}, \mathring{\mathcal{P}}) = [f(x_1)(a_1 - a) + \ldots + f(x_{\bar{j}_1 - 1})(a_{\bar{j}_1 - 1} - a_{\bar{j}_1 - 2}) + f(c_1)(c_1 - a_{\bar{j}_1 - 1})]$$

$$+ [f(c_1)(a_{\bar{j}_1} - c_1) + \ldots + f(x_{\bar{j}_2 - 1})(a_{\bar{j}_2 - 1} - a_{\bar{j}_2 - 2}) + f(c_2)(c_2 - a_{\bar{j}_2 - 1})]$$

$$+ [f(c_2)(a_{\bar{\jmath}_2} - c_2) + \cdots + f(x_{m-1})(a_{m-1} - a_{m-2})]$$
$$+ \bar{f}(x_m)(b - a_{m-1}).$$

Note that $\mathring{\mathcal{P}}_i$ is a δ_i-fine tagged partition of $[c_{i-1}, c_i]$ and that $\mathring{\mathcal{Q}}$ is a δ_{q+1}-fine tagged subpartition of $[c_q, c_{q+1}]$. Moreover, by the choice of the gauge δ, it must be that $x_m = b$ and, hence, $\bar{f}(x_m) = 0$ and $b - a_{m-1} \le \delta(b) = \gamma$. Using the fact that

$$\int_a^{a_{m-1}} f = \sum_{i=1}^q \int_{c_{i-1}}^{c_i} f + \int_{c_q}^{a_{m-1}} f,$$

by the Saks–Henstock Theorem 9.3 we have

$$|S(\bar{f}, \mathring{\mathcal{P}}) - \mathcal{J}| \le \left| S(\bar{f}, \mathring{\mathcal{P}}) - \int_a^{a_{m-1}} f \right| + \left| \int_a^{a_{m-1}} f - \mathcal{J} \right|$$

$$\le \sum_{i=1}^q \left| S(f, \mathring{\mathcal{P}}_i) - \int_{c_{i-1}}^{c_i} f \right| + \left| S(f, \mathring{\mathcal{Q}}) - \int_{c_q}^{a_{m-1}} f \right| + \left| \int_a^{a_{m-1}} f - \mathcal{J} \right|$$

$$\le \sum_{i=1}^q \frac{\varepsilon}{2^{i+3}} + 4 \frac{\varepsilon}{2^{q+4}} + \frac{\varepsilon}{2}$$

$$\le \frac{\varepsilon}{8} + \frac{\varepsilon}{4} + \frac{\varepsilon}{2} < \varepsilon,$$

and the proof is thus completed. ∎

The above theorem suggests that even for a function $f : [a, +\infty[\to \mathbb{R}$ the definition of the integral could be reduced to that of a usual integral. Indeed, fixing arbitrarily $b > a$, we could define a continuously differentiable strictly increasing auxiliary function $\varphi : [a, b[\to \mathbb{R}$ such that $\varphi(a) = a$ and $\lim_{u \to b^-} \varphi(u) = +\infty$; for example, take $\varphi(u) = a + \ln \frac{b-a}{b-u}$. A formal change of variables then gives

$$\int_a^{+\infty} f(x)\,dx = \int_a^b f(\varphi(u))\varphi'(u)\,du,$$

and Hake's theorem applies to this last integral.

With this idea in mind, it is possible to prove that $f : [a, +\infty[\to \mathbb{R}$ is integrable and its integral is a real number $\mathcal{J}$ if and only if for every $\varepsilon > 0$ there is a gauge δ, defined on $[a, +\infty[$, and a positive constant α such that, if

$$a = a_0 < a_1 < \cdots < a_{m-1}, \quad \text{with} \quad a_{m-1} \ge \alpha,$$

and, for every $j = 1, \ldots, m - 1$, the points $x_j \in [a_{j-1}, a_j]$ satisfy

$$x_j - a_{j-1} \leq \delta(x_j) \quad \text{and} \quad a_j - x_j \leq \delta(x_j),$$

then

$$\left| \sum_{j=1}^{m-1} f(x_j)(a_j - a_{j-1}) - \mathcal{J} \right| \leq \varepsilon.$$

We refer to Bartle's book [1] for a complete treatment of this case.

Needless to say, similar considerations can be made in the case where the function f is defined on an interval of the type $]a, b]$, with $a \geq -\infty$.

Part IV

Differential and Integral Calculus in $\mathbb{R}^N$

The Differential

10

Let $\mathcal{O} \subseteq \mathbb{R}^N$ be an open set, $\boldsymbol{x}_0$ a point of $\mathcal{O}$, and $f : \mathcal{O} \to \mathbb{R}^M$ a given function. We want to extend the notion of derivative of f at $\boldsymbol{x}_0$ already known in the case $M = N = 1$. The definition, inspired by Theorem 6.2, follows.

Definition 10.1 We say that f is "differentiable" at $\boldsymbol{x}_0$ if there exists a linear function $\ell : \mathbb{R}^N \to \mathbb{R}^M$ for which we can write

$$f(\boldsymbol{x}) = f(\boldsymbol{x}_0) + \ell(\boldsymbol{x} - \boldsymbol{x}_0) + r(\boldsymbol{x}),$$

where r is a function satisfying

$$\lim_{\boldsymbol{x} \to \boldsymbol{x}_0} \frac{r(\boldsymbol{x})}{\|\boldsymbol{x} - \boldsymbol{x}_0\|} = \boldsymbol{0}.$$

If f is differentiable at $\boldsymbol{x}_0$, then the linear function ℓ is called the "differential" of f at $\boldsymbol{x}_0$ and is denoted by

$$df(\boldsymbol{x}_0).$$

Following the tradition for linear functions, taking $\boldsymbol{h} \in \mathbb{R}^N$, we will often write $df(\boldsymbol{x}_0)\boldsymbol{h}$ instead of $df(\boldsymbol{x}_0)(\boldsymbol{h})$.

Assuming that $\mathcal{O}$ is an open set is not really necessary, but guarantees the uniqueness of the differential and it simplifies many issues. In what follows, however, we will sometimes encounter situations where the domain is not open. More care will be needed in these cases.

We will now first concentrate for a while on the simpler case $M = 1$.

© The Author(s), under exclusive license to Springer Nature Switzerland AG 2023
A. Fonda, *A Modern Introduction to Mathematical Analysis*,
https://doi.org/10.1007/978-3-031-23713-3_10

10.1   The Differential of a Scalar-Valued Function

Assume, for simplicity, that $M = 1$. We start by fixing a "direction," i.e., a vector $\boldsymbol{v} \in \mathbb{R}^N$, with $\|\boldsymbol{v}\| = 1$, also called a "unit vector." Whenever it exists, we call "directional derivative" of f at $\boldsymbol{x}_0$ in the direction $\boldsymbol{v}$ the limit

$$\lim_{t \to 0} \frac{f(\boldsymbol{x}_0 + t\boldsymbol{v}) - f(\boldsymbol{x}_0)}{t},$$

which will be denoted by

$$\frac{\partial f}{\partial \boldsymbol{v}}(\boldsymbol{x}_0).$$

If $\boldsymbol{v}$ coincides with an element $\boldsymbol{e}_j$ of the canonical basis $(\boldsymbol{e}_1, \boldsymbol{e}_2, \ldots, \boldsymbol{e}_N)$ of $\mathbb{R}^N$, the directional derivative is called the jth "partial derivative" of f at $\boldsymbol{x}_0$ and is denoted by

$$\frac{\partial f}{\partial x_j}(\boldsymbol{x}_0).$$

If $\boldsymbol{x}_0 = (x_1^0, x_2^0, \ldots, x_N^0)$, then

$$\frac{\partial f}{\partial x_j}(\boldsymbol{x}_0) = \lim_{t \to 0} \frac{f(\boldsymbol{x}_0 + t\boldsymbol{e}_j) - f(\boldsymbol{x}_0)}{t}$$

$$= \lim_{t \to 0} \frac{f(x_1^0, x_2^0, \ldots, x_j^0 + t, \ldots, x_N^0) - f(x_1^0, x_2^0, \ldots, x_j^0, \ldots, x_N^0)}{t},$$

so that it is commonly called "partial derivative with respect to the jth variable."

The following theorem shows, among other things, that the differential is unique.

Theorem 10.2 *If f is differentiable at $\boldsymbol{x}_0$, then f is continuous at $\boldsymbol{x}_0$. Moreover, all the directional derivatives of f at $\boldsymbol{x}_0$ exist: For every direction $\boldsymbol{v} \in \mathbb{R}^N$ we have*

$$\frac{\partial f}{\partial \boldsymbol{v}}(\boldsymbol{x}_0) = df(\boldsymbol{x}_0)\boldsymbol{v}.$$

Proof We know that the function $\ell = df(\boldsymbol{x}_0)$, being linear, is continuous, and $\ell(\boldsymbol{0}) = 0$. Then

$$\lim_{\boldsymbol{x} \to \boldsymbol{x}_0} f(\boldsymbol{x}) = \lim_{\boldsymbol{x} \to \boldsymbol{x}_0} [f(\boldsymbol{x}_0) + \ell(\boldsymbol{x} - \boldsymbol{x}_0) + r(\boldsymbol{x})]$$

$$= f(\boldsymbol{x}_0) + \ell(\boldsymbol{0}) + \lim_{\boldsymbol{x} \to \boldsymbol{x}_0} r(\boldsymbol{x})$$

$$= f(\boldsymbol{x}_0) + \lim_{\boldsymbol{x} \to \boldsymbol{x}_0} \frac{r(\boldsymbol{x})}{\|\boldsymbol{x} - \boldsymbol{x}_0\|} \lim_{\boldsymbol{x} \to \boldsymbol{x}_0} \|\boldsymbol{x} - \boldsymbol{x}_0\|$$

$$= f(\boldsymbol{x}_0),$$

showing that f is continuous at $\boldsymbol{x}_0$. Concerning the directional derivatives, we have

$$\lim_{t \to 0} \frac{f(\boldsymbol{x}_0 + t\boldsymbol{v}) - f(\boldsymbol{x}_0)}{t} = \lim_{t \to 0} \frac{df(\boldsymbol{x}_0)(t\boldsymbol{v}) + r(\boldsymbol{x}_0 + t\boldsymbol{v})}{t}$$

$$= \lim_{t \to 0} \frac{t\, df(\boldsymbol{x}_0)\boldsymbol{v} + r(\boldsymbol{x}_0 + t\boldsymbol{v})}{t}$$

$$= df(\boldsymbol{x}_0)\boldsymbol{v} + \lim_{t \to 0} \frac{r(\boldsymbol{x}_0 + t\boldsymbol{v})}{t}.$$

On the other hand, since $\|\boldsymbol{v}\| = 1$, the change of variables formula (3.1) gives us

$$\lim_{t \to 0} \left| \frac{r(\boldsymbol{x}_0 + t\boldsymbol{v})}{t} \right| = \lim_{\boldsymbol{x} \to \boldsymbol{x}_0} \frac{|r(\boldsymbol{x})|}{\|\boldsymbol{x} - \boldsymbol{x}_0\|} = 0,$$

whence the conclusion.   ∎

In particular, if $\boldsymbol{v}$ coincides with an element $\boldsymbol{e}_j$ of the canonical basis $(\boldsymbol{e}_1, \boldsymbol{e}_2, \ldots, \boldsymbol{e}_N)$, then

$$\frac{\partial f}{\partial x_j}(\boldsymbol{x}_0) = df(\boldsymbol{x}_0)\boldsymbol{e}_j.$$

Writing the vector $\boldsymbol{h} \in \mathbb{R}^N$ as $\boldsymbol{h} = h_1\boldsymbol{e}_1 + h_2\boldsymbol{e}_2 + \cdots + h_N\boldsymbol{e}_N$, by linearity we have

$$df(\boldsymbol{x}_0)\boldsymbol{h} = h_1 df(\boldsymbol{x}_0)\boldsymbol{e}_1 + h_2 df(\boldsymbol{x}_0)\boldsymbol{e}_2 + \cdots + h_N df(\boldsymbol{x}_0)\boldsymbol{e}_N$$

$$= h_1 \frac{\partial f}{\partial x_1}(\boldsymbol{x}_0) + h_2 \frac{\partial f}{\partial x_2}(\boldsymbol{x}_0) + \cdots + h_N \frac{\partial f}{\partial x_N}(\boldsymbol{x}_0),$$

i.e.,

$$df(\boldsymbol{x}_0)\boldsymbol{h} = \sum_{j=1}^{N} \frac{\partial f}{\partial x_j}(\boldsymbol{x}_0)h_j.$$

If we define the "gradient" of f at $\boldsymbol{x}_0$ as the vector

$$\nabla f(\boldsymbol{x}_0) = \left(\frac{\partial f}{\partial x_1}(\boldsymbol{x}_0), \ldots, \frac{\partial f}{\partial x_N}(\boldsymbol{x}_0) \right),$$

we can then write

$$df(\boldsymbol{x}_0)\boldsymbol{h} = \nabla f(\boldsymbol{x}_0) \cdot \boldsymbol{h}.$$

Remark 10.3 The mere existence of the directional derivatives at some point $\boldsymbol{x}_0$ does not guarantee differentiability there. For example, the function $f : \mathbb{R}^2 \to \mathbb{R}$, defined as

$$f(x, y) = \begin{cases} \dfrac{x^4 y^2}{(x^4 + y^2)^2} & \text{if } (x, y) \neq (0, 0), \\ 0 & \text{if } (x, y) = (0, 0), \end{cases}$$

has all its directional derivatives at $\boldsymbol{x}_0 = (0, 0)$ equal to 0. However, it is not even continuous there since its restriction to the parabola $\{(x, y) : y = x^2\}$ is constantly equal to $\frac{1}{4}$.

Here is a result showing that the existence of the partial derivatives is sufficient for the differentiability, provided that they are continuous.

Theorem 10.4 *If f has partial derivatives defined in a neighborhood of $\boldsymbol{x}_0$, and they are continuous at $\boldsymbol{x}_0$, then f is differentiable at $\boldsymbol{x}_0$.*

Proof To simplify the notations, we will assume that $N = 2$. We define the function $\ell : \mathbb{R}^2 \to \mathbb{R}$ associating to every vector $\boldsymbol{h} = (h_1, h_2)$ the real number

$$\ell(\boldsymbol{h}) = \frac{\partial f}{\partial x_1}(\boldsymbol{x}_0)h_1 + \frac{\partial f}{\partial x_2}(\boldsymbol{x}_0)h_2.$$

We will prove that ℓ is indeed the differential of f at $\boldsymbol{x}_0$. First of all, it is readily verified that it is linear. Moreover, writing $\boldsymbol{x}_0 = (x_1^0, x_2^0)$ and $\boldsymbol{x} = (x_1, x_2)$, by the Lagrange Mean Value Theorem 6.11 we have

$$\begin{aligned} f(\boldsymbol{x}) - f(\boldsymbol{x}_0) &= (f(x_1, x_2) - f(x_1^0, x_2)) + (f(x_1^0, x_2) - f(x_1^0, x_2^0)) \\ &= \frac{\partial f}{\partial x_1}(\xi_1, x_2)(x_1 - x_1^0) + \frac{\partial f}{\partial x_2}(x_1^0, \xi_2)(x_2 - x_2^0) \end{aligned}$$

for some $\xi_1 \in {]}x_1^0, x_1{[}$ and $\xi_2 \in {]}x_2^0, x_2{[}$. Hence,

$$\begin{aligned} r(\boldsymbol{x}) &= f(\boldsymbol{x}) - f(\boldsymbol{x}_0) - \ell(\boldsymbol{x} - \boldsymbol{x}_0) \\ &= \left[\frac{\partial f}{\partial x_1}(\xi_1, x_2) - \frac{\partial f}{\partial x_1}(x_1^0, x_2^0) \right](x_1 - x_1^0) \\ &\quad + \left[\frac{\partial f}{\partial x_2}(x_1^0, \xi_2) - \frac{\partial f}{\partial x_2}(x_1^0, x_2^0) \right](x_2 - x_2^0). \end{aligned}$$

Then, since $|x_1 - x_1^0| \le \|\boldsymbol{x} - \boldsymbol{x}_0\|$ and $|x_2 - x_2^0| \le \|\boldsymbol{x} - \boldsymbol{x}_0\|$,

$$\frac{|r(\boldsymbol{x})|}{\|\boldsymbol{x} - \boldsymbol{x}_0\|} \le \left| \frac{\partial f}{\partial x_1}(\xi_1, x_2) - \frac{\partial f}{\partial x_1}(x_1^0, x_2^0) \right| + \left| \frac{\partial f}{\partial x_2}(x_1^0, \xi_2) - \frac{\partial f}{\partial x_2}(x_1^0, x_2^0) \right|.$$

Letting $\boldsymbol{x}$ tend to $\boldsymbol{x}_0$, we have that $(\xi_1, x_2) \to (x_1^0, x_2^0)$ and $(x_1^0, \xi_2) \to (x_1^0, x_2^0)$ so that, since $\frac{\partial f}{\partial x_1}$ and $\frac{\partial f}{\partial x_2}$ are continuous at $\boldsymbol{x}_0 = (x_1^0, x_2^0)$, it must be that

$$\lim_{\boldsymbol{x} \to \boldsymbol{x}_0} \frac{|r(\boldsymbol{x})|}{\|\boldsymbol{x} - \boldsymbol{x}_0\|} = 0,$$

whence the conclusion. ■

We say that $f : \mathcal{O} \to \mathbb{R}$ is "differentiable" if it is so at every point of $\mathcal{O}$; it is "of class $\mathcal{C}^1$" or "a $\mathcal{C}^1$-function" if it has partial derivatives that are continuous on the whole domain $\mathcal{O}$. From the previous theorem we have that a function of class $\mathcal{C}^1$ is surely differentiable.

Assume now that f is defined on some domain $\mathcal{D}$ that is not an open set. In this case, we say that f is "differentiable" if it is the restriction of some differentiable function defined on an open set $\mathcal{O}$ containing $\mathcal{D}$, and similarly when f is "of class $\mathcal{C}^1$."

10.2 Some Computational Rules

Let us start with some simple propositions.

Proposition 10.5 *If $f : \mathcal{O} \to \mathbb{R}$ is constant, then $df(\boldsymbol{x}_0) = 0$ for every $\boldsymbol{x}_0 \in \mathcal{O}$.*

Proof Let $f(\boldsymbol{x}) = c$ for every $\boldsymbol{x} \in \mathcal{O}$. Then, setting $\ell(\boldsymbol{h}) = 0$ for every $\boldsymbol{h} \in \mathbb{R}^N$,

$$f(\boldsymbol{x}) - f(\boldsymbol{x}_0) - \ell(\boldsymbol{x} - \boldsymbol{x}_0) = c - c - 0 = 0$$

for every $\boldsymbol{x} \in \mathcal{O}$. ■

Proposition 10.6 *If $\mathcal{A} : \mathbb{R}^N \to \mathbb{R}$ is linear, then $d\mathcal{A}(\boldsymbol{x}_0) = \mathcal{A}$ for every $\boldsymbol{x}_0 \in \mathcal{O}$.*

Proof Let $f(\boldsymbol{x}) = \mathcal{A}\boldsymbol{x}$ for every $\boldsymbol{x} \in \mathbb{R}^N$. Then, setting $\ell(\boldsymbol{h}) = \mathcal{A}\boldsymbol{h}$, by linearity,

$$f(\boldsymbol{x}) - f(\boldsymbol{x}_0) - \ell(\boldsymbol{x} - \boldsymbol{x}_0) = \mathcal{A}\boldsymbol{x} - \mathcal{A}\boldsymbol{x}_0 - \mathcal{A}(\boldsymbol{x} - \boldsymbol{x}_0) = 0$$

for every $\boldsymbol{x} \in \mathcal{O}$. ■

Proposition 10.7 *If $\mathcal{B} : \mathbb{R}^{N_1} \times \mathbb{R}^{N_2} \to \mathbb{R}$ is bilinear, writing $\boldsymbol{x}_0 = (x_0, y_0)$ with $x_0 \in \mathbb{R}^{N_1}$, $y_0 \in \mathbb{R}^{N_2}$, and and $\boldsymbol{h} = (h, k)$, with $h \in \mathbb{R}^{N_1}$, $k \in \mathbb{R}^{N_2}$, we have*

$$d\mathcal{B}(\boldsymbol{x}_0)(\boldsymbol{h}) = \mathcal{B}(h, y_0) + \mathcal{B}(x_0, k).$$

Proof Writing $\boldsymbol{x} = (x, y)$ with $x \in \mathbb{R}^{N_1}$, $y \in \mathbb{R}^{N_2}$, let $f(\boldsymbol{x}) = \mathcal{B}(x, y)$. Then, setting $\ell(\boldsymbol{h}) = \mathcal{B}(h, y_0) + \mathcal{B}(x_0, k)$, we compute

$$
\begin{aligned}
r(\boldsymbol{x}) &= f(\boldsymbol{x}) - f(\boldsymbol{x}_0) - \ell(\boldsymbol{x} - \boldsymbol{x}_0) \\
&= \mathcal{B}(x, y) - \mathcal{B}(x_0, y_0) - \mathcal{B}(x - x_0, y_0) - \mathcal{B}(x_0, y - y_0) \\
&= \mathcal{B}(x - x_0, y - y_0) = \mathcal{B}(\boldsymbol{x} - \boldsymbol{x}_0).
\end{aligned}
$$

Denoting by $e_1, \ldots, e_{N_1}$ the vectors of the canonical basis of $\mathbb{R}^{N_1}$ and by $\hat{e}_1, \ldots, \hat{e}_{N_2}$ those of the canonical basis of $\mathbb{R}^{N_2}$, for every $x \in \mathbb{R}^{N_1}$ and $y \in \mathbb{R}^{N_2}$ we have that

$$\mathcal{B}(x, y) = \mathcal{B}\left(\sum_{i=1}^{N_1} x_i e_i, \sum_{j=1}^{N_2} y_j \hat{e}_j \right) = \sum_{i=1}^{N_1} \sum_{j=1}^{N_2} x_i y_j \mathcal{B}(e_i, \hat{e}_j),$$

hence there is a constant C such that

$$|\mathcal{B}(x, y)| \leq C \|x\| \|y\| \quad \text{for every } x \in \mathbb{R}^{N_1}, y \in \mathbb{R}^{N_2}.$$

Then

$$|\mathcal{B}(\boldsymbol{x} - \boldsymbol{x}_0)| = |\mathcal{B}(x - x_0, y - y_0)| \leq C \|x - x_0\| \|y - y_0\| \leq C \|\boldsymbol{x} - \boldsymbol{x}_0\|^2,$$

whence, if $\boldsymbol{x} \neq \boldsymbol{x}_0$, then

$$\frac{|r(\boldsymbol{x})|}{\|\boldsymbol{x} - \boldsymbol{x}_0\|} = \frac{|\mathcal{B}(\boldsymbol{x} - \boldsymbol{x}_0)|}{\|\boldsymbol{x} - \boldsymbol{x}_0\|} \leq C \|\boldsymbol{x} - \boldsymbol{x}_0\|,$$

and finally

$$\lim_{\boldsymbol{x} \to \boldsymbol{x}_0} \frac{|r(\boldsymbol{x})|}{\|\boldsymbol{x} - \boldsymbol{x}_0\|} = 0.$$

The statement is thus proved.   ∎

We now compute the differential of the sum of two functions and the product with some constants.

Proposition 10.8 *If $f, g : \mathcal{O} \to \mathbb{R}$ are differentiable at x_0 and α, β are two real numbers, then*

$$d(\alpha f + \beta g)(x_0) = \alpha df(x_0) + \beta dg(x_0).$$

Proof Writing

$$f(x) = f(x_0) + df(x_0)(x - x_0) + r_1(x),$$
$$g(x) = g(x_0) + dg(x_0)(x - x_0) + r_2(x),$$

we have that

$$(\alpha f + \beta g)(x) = (\alpha f + \beta g)(x_0) + (\alpha df(x_0) + \beta dg(x_0))(x - x_0) + r(x),$$

with $r(x) = \alpha r_1(x) + \beta r_2(x)$, and

$$\lim_{x \to x_0} \frac{r(x)}{\|x - x_0\|} = \alpha \lim_{x \to x_0} \frac{r_1(x)}{\|x - x_0\|} + \beta \lim_{x \to x_0} \frac{r_2(x)}{\|x - x_0\|} = 0.$$

Hence, $\alpha f + \beta g$ is differentiable at x_0 with differential $\alpha df(x_0) + \beta dg(x_0)$. ∎

10.3 Twice Differentiable Functions

Let $\mathcal{O}$ be an open subset of $\mathbb{R}^N$, and $f : \mathcal{O} \to \mathbb{R}$ be a differentiable function. We want to extend the notion of "second derivative," which is well known in the case $N = 1$. For simplicity's sake, let us deal with the case $N = 2$. If the partial derivatives $\frac{\partial f}{\partial x_1}, \frac{\partial f}{\partial x_2} : \mathcal{O} \to \mathbb{R}$ have themselves partial derivatives at a point x_0, these are said to be "second-order partial derivatives" of f at x_0 and are denoted by

$$\frac{\partial^2 f}{\partial x_1^2}(x_0) = \frac{\partial}{\partial x_1} \frac{\partial f}{\partial x_1}(x_0), \qquad \frac{\partial^2 f}{\partial x_2 \partial x_1}(x_0) = \frac{\partial}{\partial x_2} \frac{\partial f}{\partial x_1}(x_0),$$

$$\frac{\partial^2 f}{\partial x_1 \partial x_2}(x_0) = \frac{\partial}{\partial x_1} \frac{\partial f}{\partial x_2}(x_0), \qquad \frac{\partial^2 f}{\partial x_2^2}(x_0) = \frac{\partial}{\partial x_2} \frac{\partial f}{\partial x_2}(x_0).$$

Here is a relation involving the "mixed derivatives."

Theorem 10.9 (Schwarz Theorem) *If the second-order mixed partial derivatives $\frac{\partial^2 f}{\partial x_2 \partial x_1}, \frac{\partial^2 f}{\partial x_1 \partial x_2}$ exist in a neighborhood of x_0 and they are continuous at x_0, then*

$$\frac{\partial^2 f}{\partial x_2 \partial x_1}(x_0) = \frac{\partial^2 f}{\partial x_1 \partial x_2}(x_0).$$

Proof Let $\rho > 0$ be such that $B(\boldsymbol{x}_0, \rho) \subseteq \mathcal{O}$. We write $\boldsymbol{x}_0 = (x_1^0, x_2^0)$, and we take an $\boldsymbol{x} = (x_1, x_2) \in B(\boldsymbol{x}_0, \rho)$ such that $x_1 \neq x_1^0$ and $x_2 \neq x_2^0$. It is then possible to define

$$g(x_1, x_2) = \frac{f(x_1, x_2) - f(x_1, x_2^0)}{x_2 - x_2^0}, \quad h(x_1, x_2) = \frac{f(x_1, x_2) - f(x_1^0, x_2)}{x_1 - x_1^0}.$$

We can verify that

$$\frac{g(x_1, x_2) - g(x_1^0, x_2)}{x_1 - x_1^0} = \frac{h(x_1, x_2) - h(x_1, x_2^0)}{x_2 - x_2^0}.$$

By the Lagrange Mean Value Theorem 6.11, there is a $\xi_1 \in]x_1^0, x_1[$ such that

$$\frac{g(x_1, x_2) - g(x_1^0, x_2)}{x_1 - x_1^0} = \frac{\partial g}{\partial x_1}(\xi_1, x_2) = \frac{\frac{\partial f}{\partial x_1}(\xi_1, x_2) - \frac{\partial f}{\partial x_1}(\xi_1, x_2^0)}{x_2 - x_2^0},$$

and there is a $\xi_2 \in]x_2^0, x_2[$ such that

$$\frac{h(x_1, x_2) - h(x_1, x_2^0)}{x_2 - x_2^0} = \frac{\partial h}{\partial x_2}(x_1, \xi_2) = \frac{\frac{\partial f}{\partial x_2}(x_1, \xi_2) - \frac{\partial f}{\partial x_2}(x_1^0, \xi_2)}{x_1 - x_1^0}.$$

Again by the Lagrange Mean Value Theorem 6.11, there is a $\eta_2 \in]x_2^0, x_2[$ such that

$$\frac{\frac{\partial f}{\partial x_1}(\xi_1, x_2) - \frac{\partial f}{\partial x_1}(\xi_1, x_2^0)}{x_2 - x_2^0} = \frac{\partial^2 f}{\partial x_2 \partial x_1}(\xi_1, \eta_2),$$

and there is a $\eta_1 \in]x_1^0, x_1[$ such that

$$\frac{\frac{\partial f}{\partial x_2}(x_1, \xi_2) - \frac{\partial f}{\partial x_2}(x_1^0, \xi_2)}{x_1 - x_1^0} = \frac{\partial^2 f}{\partial x_1 \partial x_2}(\eta_1, \xi_2).$$

Hence,

$$\frac{\partial^2 f}{\partial x_2 \partial x_1}(\xi_1, \eta_2) = \frac{\partial^2 f}{\partial x_1 \partial x_2}(\eta_1, \xi_2).$$

Taking the limit, as $\boldsymbol{x} = (x_1, x_2)$ tends to $\boldsymbol{x}_0 = (x_1^0, x_2^0)$, we have that both (ξ_1, η_2) and (η_1, ξ_2) converge to $\boldsymbol{x}_0$, and the continuity of the second-order partial derivatives leads to the conclusion. ∎

We say that $f : \mathcal{O} \to \mathbb{R}$ is " of class $\mathcal{C}^2$ " or "a $\mathcal{C}^2$-function" if all its second-order partial derivatives exist and are continuous on $\mathcal{O}$.

It is useful to consider the "Hessian matrix" of f at $\boldsymbol{x}_0$:

$$Hf(\boldsymbol{x}_0) = \begin{pmatrix} \frac{\partial^2 f}{\partial x_1^2}(\boldsymbol{x}_0) & \frac{\partial^2 f}{\partial x_2 \partial x_1}(\boldsymbol{x}_0) \\ \\ \frac{\partial^2 f}{\partial x_1 \partial x_2}(\boldsymbol{x}_0) & \frac{\partial^2 f}{\partial x_2^2}(\boldsymbol{x}_0) \end{pmatrix};$$

if f is of class $\mathcal{C}^2$, then this is a symmetric matrix.

What was just said extends without difficulty for any $N \geq 2$; if f is of class $\mathcal{C}^2$, then the Hessian matrix is an $N \times N$ symmetric matrix.

One can further define by induction the nth-order partial derivatives. It is said that $f : \mathcal{O} \to \mathbb{R}$ is " of class $\mathcal{C}^n$ " or "a $\mathcal{C}^n$-function" if all its nth-order partial derivatives exist and are continuous on $\mathcal{O}$.

10.4 Taylor Formula

Let $\mathcal{O}$ be an open subset of $\mathbb{R}^N$, and assume that $f : \mathcal{O} \to \mathbb{R}$ is a function of class $\mathcal{C}^{n+1}$ for some $n \geq 1$.

As previously, for simplicity we will deal with the case $N = 2$. Let us introduce the following notations:

$$D_{x_1} = \frac{\partial}{\partial x_1}, \quad D_{x_2} = \frac{\partial}{\partial x_2},$$

$$D_{x_1}^2 = \frac{\partial^2}{\partial x_1^2}, \quad D_{x_1} D_{x_2} = \frac{\partial^2}{\partial x_1 \partial x_2}, \quad D_{x_2}^2 = \frac{\partial^2}{\partial x_2^2},$$

and so on for the higher-order derivatives. Note that for any vector $\boldsymbol{h} = (h_1, h_2) \in \mathbb{R}^2$,

$$df(\boldsymbol{x}_0)\boldsymbol{h} = h_1 D_{x_1} f(\boldsymbol{x}_0) + h_2 D_{x_2} f(\boldsymbol{x}_0),$$

which can also be written

$$df(\boldsymbol{x}_0)\boldsymbol{h} = [h_1 D_{x_1} + h_2 D_{x_2}] f(\boldsymbol{x}_0).$$

In this way, we can think that the function f is transformed by the "operator" $[h_1 D_{x_1} + h_2 D_{x_2}]$ into the new function

$$[h_1 D_{x_1} + h_2 D_{x_2}] f = h_1 D_{x_1} f + h_2 D_{x_2} f.$$

Given two points $x_0 \neq x$ in $\mathbb{R}^N$, the "segment" joining them is defined by

$$[x_0, x] = \{x_0 + t(x - x_0) : t \in [0, 1]\};$$

similarly, we will write

$$]x_0, x[= \{x_0 + t(x - x_0) : t \in]0, 1[\}.$$

Assume now that the segment $[x_0, x]$ is contained in $\mathcal{O}$, and consider the function $\phi : [0, 1] \to \mathbb{R}$ defined as

$$\phi(t) = f(x_0 + t(x - x_0)).$$

We will prove that ϕ is $n + 1$ times differentiable on $[0, 1]$. For any $t \in [0, 1]$, since f is differentiable at $u_0 = x_0 + t(x - x_0)$, we have that

$$f(u) = f(u_0) + df(u_0)(u - u_0) + r(u),$$

with

$$\lim_{u \to u_0} \frac{r(u)}{\|u - u_0\|} = 0.$$

Hence,

$$
\begin{aligned}
\lim_{s \to t} \frac{\phi(s) - \phi(t)}{s - t} &= \lim_{s \to t} \frac{f(x_0 + s(x - x_0)) - f(x_0 + t(x - x_0))}{s - t} \\
&= \lim_{s \to t} \frac{df(x_0 + t(x - x_0))((s - t)(x - x_0)) + r(x_0 + s(x - x_0))}{s - t} \\
&= df(x_0 + t(x - x_0))(x - x_0) + \lim_{s \to t} \frac{r(x_0 + s(x - x_0))}{s - t},
\end{aligned}
$$

and since

$$\lim_{s \to t} \left| \frac{r(x_0 + s(x - x_0))}{s - t} \right| = \lim_{u \to u_0} \frac{|r(u)|}{\|u - u_0\|} \|x - x_0\| = 0,$$

we have that

$$\phi'(t) = \lim_{s \to t} \frac{\phi(s) - \phi(t)}{s - t} = df(x_0 + t(x - x_0))(x - x_0).$$

With the new notations, setting $x - x_0 = h = (h_1, h_2)$, we can write

$$\phi'(t) = [h_1 D_{x_1} + h_2 D_{x_2}]f(x_0 + t(x - x_0)) = g(x_0 + t(x - x_0)),$$

where g is the function $[h_1 D_{x_1} + h_2 D_{x_2}]f$. We can then iterate the procedure and compute the second derivative

$$\phi''(t) = [h_1 D_{x_1} + h_2 D_{x_2}]g(\boldsymbol{x}_0 + t(\boldsymbol{x} - \boldsymbol{x}_0))$$
$$= [h_1 D_{x_1} + h_2 D_{x_2}][h_1 D_{x_1} + h_2 D_{x_2}]f(\boldsymbol{x}_0 + t(\boldsymbol{x} - \boldsymbol{x}_0)).$$

For briefness, we will write

$$\phi''(t) = [h_1 D_{x_1} + h_2 D_{x_2}]^2 f(\boldsymbol{x}_0 + t(\boldsymbol{x} - \boldsymbol{x}_0)).$$

Notice that, by the linearity of the partial derivatives and the equality of the second-order mixed derivatives (Schwarz Theorem 10.9),

$$[h_1 D_{x_1} + h_2 D_{x_2}]^2 f = h_1^2 D_{x_1}^2 f + 2h_1 h_2 D_{x_1} D_{x_2} f + h_2^2 D_{x_2}^2 f$$
$$= [h_1^2 D_{x_1}^2 + 2h_1 h_2 D_{x_1} D_{x_2} + h_2^2 D_{x_2}^2]f.$$

We now observe that the equality

$$[h_1 D_{x_1} + h_2 D_{x_2}]^2 = [h_1^2 D_{x_1}^2 + 2h_1 h_2 D_{x_1} D_{x_2} + h_2^2 D_{x_2}^2]$$

is formally obtained as the square of a binomial. Proceeding in this way, we can prove by induction that, for $k = 1, 2, \ldots, n + 1$, the formula for the kth derivative of ϕ is

$$\phi^{(k)}(t) = [h_1 D_{x_1} + h_2 D_{x_2}]^k f(\boldsymbol{x}_0 + t(\boldsymbol{x} - \boldsymbol{x}_0)),$$

and, using the binomial formula

$$(a_1 + a_2)^k = \sum_{j=0}^{k} \binom{k}{j} a_1^{k-j} a_2^j,$$

we formally have that

$$[h_1 D_{x_1} + h_2 D_{x_2}]^k = \left[\sum_{j=0}^{k} \binom{k}{j} h_1^{k-j} h_2^j D_{x_1}^{k-j} D_{x_2}^j \right]$$

(in this formula, the symbols $D_{x_1}^0$ and $D_{x_2}^0$ simply denote the identity operator).
To write the Taylor formula, let us introduce the notation

$$d^k f(\boldsymbol{x}_0)\boldsymbol{h}^k = [h_1 D_{x_1} + h_2 D_{x_2}]^k f(\boldsymbol{x}_0).$$

Theorem 10.10 (Taylor Theorem—III) *Let $f : \mathcal{O} \to \mathbb{R}$ be a function of class C^{n+1} and $[\boldsymbol{x}_0, \boldsymbol{x}]$ be a segment contained in $\mathcal{O}$. Then there exists a $\boldsymbol{\xi} \in]\boldsymbol{x}_0, \boldsymbol{x}[$ such that*

$$f(\boldsymbol{x}) = p_n(\boldsymbol{x}) + r_n(\boldsymbol{x}),$$

where

$$p_n(\boldsymbol{x}) = f(\boldsymbol{x}_0) + df(\boldsymbol{x}_0)(\boldsymbol{x} - \boldsymbol{x}_0) + \frac{1}{2!}d^2 f(\boldsymbol{x}_0)(\boldsymbol{x} - \boldsymbol{x}_0)^2$$

$$+ \cdots + \frac{1}{n!}d^n f(\boldsymbol{x}_0)(\boldsymbol{x} - \boldsymbol{x}_0)^n$$

is the "nth-order Taylor polynomial associated with the function f at the point x_0", and

$$r_n(\boldsymbol{x}) = \frac{1}{(n+1)!}d^{n+1} f(\boldsymbol{\xi})(\boldsymbol{x} - \boldsymbol{x}_0)^{n+1}$$

is the "Lagrange form of the remainder."

Proof Applying the Taylor formula to the function ϕ, we have that

$$\phi(t) = \phi(0) + \phi'(0)t + \frac{1}{2!}\phi''(0)t^2 + \cdots + \frac{1}{n!}\phi^{(n)}(0)t^n + \frac{1}{(n+1)!}\phi^{(n+1)}(\xi)t^{n+1}$$

for some $\xi \in]0, t[$. We thus directly conclude the proof taking $t = 1$ and substituting the values of the derivatives of ϕ computed earlier. ∎

The Taylor polynomial can be expressed as

$$p_n(\boldsymbol{x}) = \sum_{k=0}^{n} \frac{1}{k!}d^k f(\boldsymbol{x}_0)(\boldsymbol{x} - \boldsymbol{x}_0)^k,$$

with the convention that $d^0 f(\boldsymbol{x}_0)(\boldsymbol{x} - \boldsymbol{x}_0)^0$, the first addend in the sum, is simply $f(\boldsymbol{x}_0)$. Hence,

$$p_n(\boldsymbol{x}) = \sum_{k=0}^{n} \frac{1}{k!}\left[\left(x_1 - x_1^0\right)D_{x_1} + \left(x_2 - x_2^0\right)D_{x_2}\right]^k f(\boldsymbol{x}_0)$$

$$= \sum_{k=0}^{n} \frac{1}{k!}\left(\sum_{j=0}^{k} \binom{k}{j} \frac{\partial^k f}{\partial x_1^{k-j}\partial x_2^j}(\boldsymbol{x}_0)\,(x_1 - x_1^0)^{k-j}(x_2 - x_2^0)^j\right).$$

Here is a useful expression for the second-order polynomial:

$$p_2(\boldsymbol{x}) = f(\boldsymbol{x}_0) + \nabla f(\boldsymbol{x}_0) \cdot (\boldsymbol{x} - \boldsymbol{x}_0) + \tfrac{1}{2}\Big(Hf(\boldsymbol{x}_0)(\boldsymbol{x} - \boldsymbol{x}_0)\Big) \cdot (\boldsymbol{x} - \boldsymbol{x}_0).$$

The foregoing proved theorem remains valid for any dimension N when the notations are properly interpreted. For example, for any vector $\boldsymbol{h} = (h_1, h_2, \ldots, h_N)$,

$$d^k f(\boldsymbol{x}_0)\boldsymbol{h}^k = [h_1 D_{x_1} + h_2 D_{x_2} + \cdots + h_N D_{x_N}]^k f(\boldsymbol{x}_0).$$

In this case, when writing explicitly the Taylor polynomial, the following generalization of the binomial formula will be useful:

$$(a_1 + a_2 + \cdots + a_N)^k = \sum_{m_1 + m_2 + \cdots + m_N = k} \frac{k!}{m_1! m_2! \cdots m_N!} a_1^{m_1} a_2^{m_2} \cdots a_N^{m_N}.$$

10.5 The Search for Maxima and Minima

As earlier, let $\mathcal{O} \subseteq \mathbb{R}^N$, the domain of our function $f : \mathcal{O} \to \mathbb{R}$, be an open set. Recall that $\boldsymbol{x}_0 \in \mathcal{O}$ is a "local maximum point" for f if there exists a neighborhood $U \subseteq \mathcal{O}$ of $\boldsymbol{x}_0$ such that $f(U)$ has a maximum and $f(\boldsymbol{x}_0) = \max f(U)$. A similar definition holds for "local minimum point."

Theorem 10.11 (Fermat's Theorem—II) *Assume that $\mathcal{O}$ is an open set and $f : \mathcal{O} \to \mathbb{R}$ is differentiable at $\boldsymbol{x}_0 \in \mathcal{O}$. If, moreover, $\boldsymbol{x}_0$ is a local maximum or minimum point for f, then $\nabla f(\boldsymbol{x}_0) = 0$.*

Proof If $\boldsymbol{x}_0$ is a local maximum point, then for every direction $\boldsymbol{v} \in \mathbb{R}^N$ there is a $\delta > 0$ for which

$$\frac{f(\boldsymbol{x}_0 + t\boldsymbol{v}) - f(\boldsymbol{x}_0)}{t} \quad \begin{cases} \geq 0 & \text{if } -\delta < t < 0, \\ \leq 0 & \text{if } 0 < t < \delta. \end{cases}$$

Since f is differentiable at $\boldsymbol{x}_0$, we necessarily have that

$$\frac{\partial f}{\partial \boldsymbol{v}}(\boldsymbol{x}_0) = \lim_{t \to 0} \frac{f(\boldsymbol{x}_0 + t\boldsymbol{v}) - f(\boldsymbol{x}_0)}{t} = 0.$$

In particular, all partial derivatives are equal to zero, hence $\nabla f(\boldsymbol{x}_0) = 0$. When $\boldsymbol{x}_0$ is a local minimum point, the proof is similar. ∎

A point where the gradient vanishes is called a "stationary point." We know already from the case $N = 1$ that such a point could be neither a local maximum nor a local minimum point.

We will now show how the Taylor formula provides a criterion establishing when a stationary point is either a local maximum or a local minimum point. Let us start with a definition.

We say that a symmetric $N \times N$ matrix $\mathbb{A}$ is "positive definite" if

$$[\mathbb{A}h] \cdot h > 0, \quad \text{for every } h \in \mathbb{R}^N \setminus \{0\}.$$

In contrast, we say that $\mathbb{A}$ is "negative definite" if the opposite inequality holds, i.e., when $-\mathbb{A}$ is positive definite.

Theorem 10.12 *If x_0 is a stationary point and f is of the class C^2, with a positive definite Hessian matrix $Hf(x_0)$, then x_0 is a local minimum point. In contrast, if $Hf(x_0)$ is negative definite, then x_0 is a local maximum point.*

Proof By the Taylor formula, for any $x \neq x_0$ in a neighborhood of x_0 there exists a $\xi \in]x_0, x[$ for which

$$f(x) = f(x_0) + \nabla f(x_0) \cdot (x - x_0) + \tfrac{1}{2}\Big(Hf(\xi)(x - x_0)\Big) \cdot (x - x_0).$$

If $\mathbb{A} = Hf(x_0)$ is positive definite, there is a constant $c > 0$ such that, for every $v \in \mathbb{R}^N$ with $\|v\| = 1$,

$$[\mathbb{A}v] \cdot v \geq c.$$

(We have used Weierstrass' Theorem 4.10 and the fact that the sphere $\{v \in \mathbb{R}^N : \|v\| = 1\}$ is a compact set.) Hence,

$$\left(Hf(x_0)\frac{x - x_0}{\|x - x_0\|}\right) \cdot \frac{x - x_0}{\|x - x_0\|} \geq c.$$

Recalling the continuity of the second derivatives, if $x \neq x_0$ is sufficiently near x_0, then

$$\left(Hf(\xi)\frac{x - x_0}{\|x - x_0\|}\right) \cdot \frac{x - x_0}{\|x - x_0\|} \geq \tfrac{1}{2}c > 0.$$

(This can be proved by contradiction using the compactness of the sphere again.) Since $\nabla f(x_0) = 0$, for such x we have that

$$\begin{aligned} f(x) &= f(x_0) + \tfrac{1}{2}\Big(Hf(\xi)(x - x_0)\Big) \cdot (x - x_0) \\ &\geq f(x_0) + \tfrac{1}{2}c\|x - x_0\|^2 > f(x_0), \end{aligned}$$

hence x_0 is a local minimum point.

The proof of the second statement is analogous.   ∎

We now state (without proof) two useful criteria for determining when a symmetric $N \times N$ matrix $\mathbb{A}$ is positive definite or negative definite. We recall that all the eigenvalues of a symmetric matrix are real.

First Criterion The symmetric matrix $\mathbb{A}$ is positive definite if and only if all its eigenvalues are positive. It is negative definite if and only if all its eigenvalues are negative.

Second Criterion The symmetric matrix $\mathbb{A} = (a_{ij})_{ij}$ is positive definite if and only if

$$a_{11} > 0,$$

$$\det \begin{pmatrix} a_{11} & a_{12} \\ a_{21} & a_{22} \end{pmatrix} > 0,$$

$$\det \begin{pmatrix} a_{11} & a_{12} & a_{13} \\ a_{21} & a_{22} & a_{23} \\ a_{31} & a_{32} & a_{33} \end{pmatrix} > 0, \ldots$$

$$\det \begin{pmatrix} a_{11} & a_{12} & \cdots & a_{1N} \\ a_{21} & a_{22} & \cdots & a_{2N} \\ \vdots & \vdots & \cdots & \vdots \\ a_{N1} & a_{N2} & \cdots & a_{NN} \end{pmatrix} > 0.$$

It is negative definite if and only if the foregoing written determinants have an alternating sign: those of the $M \times M$ submatrices with M odd are negative, while those with M even are positive.

10.6 Implicit Function Theorem: First Statement

We are now concerned with a problem involving a general equation of the type

$$g(x, y) = 0.$$

The question is whether or not for the solutions (x, y) of this equation it is possible to derive y as a function of x, say, $y = \eta(x)$. As a typical example, let $g(x, y) = x^2 + y^2 - 1$, so that the equation becomes

$$x^2 + y^2 = 1,$$

whose solutions lie on the unitary circle S^1. The answer to the preceding question, in this case, could be positive provided that we restrict our analysis to a small neighborhood of some particular solution (x_0, y_0), with $y_0 \neq 0$. Indeed, if $y_0 > 0$, we will obtain $\eta(x) = \sqrt{1 - x^2}$, whereas if $y_0 < 0$, we will take $\eta(x) = -\sqrt{1 - x^2}$.

In general, we will show that the same conclusion holds if we take any point (x_0, y_0) for which $g(x_0, y_0) = 0$, provided that $\frac{\partial g}{\partial y}(x_0, y_0) \neq 0$. In such a case, there exists a small neighborhood of (x_0, y_0) where

$$g(x, y) = 0 \quad \Leftrightarrow \quad y = \eta(x)$$

for some function η, which thus happens to be "implicitly defined."

This important result, due to Ulisse Dini, will be later generalized to any finite-dimensional setting.

Theorem 10.13 (Implicit Function Theorem—I) *Let $\mathcal{O} \subseteq \mathbb{R} \times \mathbb{R}$ be an open set $g : \mathcal{O} \to \mathbb{R}$ a C^1-function, and (x_0, y_0) a point in $\mathcal{O}$ for which*

$$g(x_0, y_0) = 0 \quad and \quad \frac{\partial g}{\partial y}(x_0, y_0) \neq 0 .$$

Then there exist an open neighborhood U of x_0, an open neighborhood V of y_0, and a C^1-function $\eta : U \to V$ such that $U \times V \subseteq \mathcal{O}$, and, taking $x \in U$ and $y \in V$, we have that

$$g(x, y) = 0 \quad \Leftrightarrow \quad y = \eta(x) .$$

Moreover, the function η is of class C^1, and the following formula holds:

$$\eta'(x) = - \left(\frac{\partial g}{\partial y}(x, \eta(x)) \right)^{-1} \frac{\partial g}{\partial x}(x, \eta(x)) .$$

Proof Assume, for instance, that $\frac{\partial g}{\partial y}(x_0, y_0) > 0$. By the continuity of $\frac{\partial g}{\partial y}$, there is a $\delta > 0$ such that $[x_0 - \delta, x_0 + \delta] \times [y_0 - \delta, y_0 + \delta] \subseteq \mathcal{O}$ and, if $|x - x_0| \leq \delta$ and $|y - y_0| \leq \delta$, then $\frac{\partial g}{\partial y}(x, y) > 0$. Hence, for every $x \in [x_0 - \delta, x_0 + \delta]$, the function $g(x, \cdot)$ is strictly increasing on $[y_0 - \delta, y_0 + \delta]$. Since $g(x_0, y_0) = 0$, we have that

$$g(x_0, y_0 - \delta) < 0 < g(x_0, y_0 + \delta) .$$

By continuity again, there is a $\delta' > 0$ such that, if $x \in [x_0 - \delta', x_0 + \delta']$, then

$$g(x, y_0 - \delta) < 0 < g(x, y_0 + \delta) .$$

We define $U =]x_0 - \delta', x_0 + \delta'[$, and $V =]y_0 - \delta, y_0 + \delta[$. Hence, for every $x \in U$, since $g(x, \cdot)$ is strictly increasing, there is exactly one $y \in]y_0 - \delta, y_0 + \delta[$ for which $g(x, y) = 0$; we call $\eta(x)$ such a y. We have thus defined a function $\eta : U \to V$

such that, taking $x \in U$ and $y \in V$,

$$g(x, y) = 0 \quad \Leftrightarrow \quad y = \eta(x).$$

To verify the continuity of η, let us fix a $\bar{x} \in U$ and prove that η is continuous at $\bar{x}$. With $x \in U$ and considering the function $\gamma : [0, 1] \to U \times V$ defined as

$$\gamma(t) = (\bar{x} + t(x - \bar{x}), \eta(\bar{x}) + t(\eta(x) - \eta(\bar{x}))),$$

the Lagrange Mean Value Theorem 6.11 applied to $g \circ \gamma$ tells us that there is a $\xi \in]0, 1[$ for which

$$g(x, \eta(x)) - g(\bar{x}, \eta(\bar{x})) = \frac{\partial g}{\partial x}(\gamma(\xi))(x - \bar{x}) + \frac{\partial g}{\partial y}(\gamma(\xi))(\eta(x) - \eta(\bar{x})).$$

Since $g(x, \eta(x)) = g(\bar{x}, \eta(\bar{x})) = 0$, we have that

$$|\eta(x) - \eta(\bar{x})| = \frac{1}{|\frac{\partial g}{\partial y}(\gamma(\xi))|} \left| \frac{\partial g}{\partial x}(\gamma(\xi))(x - \bar{x}) \right|.$$

Since the partial derivatives of g are continuous and $\frac{\partial g}{\partial y}$ is not zero on the compact set $\overline{U} \times \overline{V}$, we have that there is a constant $c > 0$ for which

$$\frac{1}{|\frac{\partial g}{\partial y}(\gamma(\xi))|} \left| \frac{\partial g}{\partial x}(\gamma(\xi))(x - \bar{x}) \right| \leq c|x - \bar{x}|.$$

As a consequence, η is continuous at $\bar{x}$.

We now prove the differentiability. Taking $\bar{x} \in U$ and proceeding as previously, for h small enough we have

$$\frac{\eta(\bar{x} + h) - \eta(\bar{x})}{h} = -\frac{\frac{\partial g}{\partial x}(\gamma(\xi))}{\frac{\partial g}{\partial y}(\gamma(\xi))},$$

with $\gamma(\xi)$ belonging to the segment joining $(\bar{x}, \eta(\bar{x}))$ to $(\bar{x}+h, \eta(\bar{x}+h))$. If h tends to 0, we have that $\gamma(\xi)$ tends to $(\bar{x}, \eta(\bar{x}))$, and hence

$$\eta'(\bar{x}) = \lim_{h \to 0} \frac{\eta(\bar{x} + h) - \eta(\bar{x})}{h} = -\frac{\frac{\partial g}{\partial x}(\bar{x}, \eta(\bar{x}))}{\frac{\partial g}{\partial y}(\bar{x}, \eta(\bar{x}))}.$$

This implies that η is of class $\mathcal{C}^1$, and

$$\eta'(x) = -\,\frac{\frac{\partial g}{\partial x}(x, \eta(x))}{\frac{\partial g}{\partial y}(x, \eta(x))}\,, \qquad \text{for every } x \in U\,.$$

We have thus completed the proof.   ■

10.7   The Differential of a Vector-Valued Function

Let us recall the definition given at the beginning of the chapter. The differential of a function $f : \mathcal{O} \to \mathbb{R}^M$ at a point $\boldsymbol{x}_0 \in \mathcal{O}$ is a linear function $\ell : \mathbb{R}^N \to \mathbb{R}^M$ for which one can write

$$f(\boldsymbol{x}) = f(\boldsymbol{x}_0) + \ell(\boldsymbol{x} - \boldsymbol{x}_0) + r(\boldsymbol{x})\,,$$

with

$$\lim_{\boldsymbol{x} \to \boldsymbol{x}_0} \frac{r(\boldsymbol{x})}{\|\boldsymbol{x} - \boldsymbol{x}_0\|} = \boldsymbol{0}\,.$$

This linear function ℓ, when it exists, is denoted by $df(\boldsymbol{x}_0)$.

When $M \geq 2$, let $f_k : \mathcal{O} \to \mathbb{R}$ be the components of the function $f : \mathcal{O} \to \mathbb{R}^M$, with $k = 1, 2, \ldots, M$, so that

$$f(\boldsymbol{x}) = (f_1(\boldsymbol{x}), f_2(\boldsymbol{x}), \ldots, f_M(\boldsymbol{x}))\,.$$

Theorem 10.14 *The function f is differentiable at $\boldsymbol{x}_0$ if and only if all its components are. In this case, for any vector $\boldsymbol{h} \in \mathbb{R}^N$,*

$$df(\boldsymbol{x}_0)\boldsymbol{h} = (df_1(\boldsymbol{x}_0)\boldsymbol{h}, df_2(\boldsymbol{x}_0)\boldsymbol{h}, \ldots, df_M(\boldsymbol{x}_0)\boldsymbol{h})\,.$$

Proof Considering the components in the equation

$$f(\boldsymbol{x}) = f(\boldsymbol{x}_0) + \ell(\boldsymbol{x} - \boldsymbol{x}_0) + r(\boldsymbol{x})\,,$$

we can write

$$f_k(\boldsymbol{x}) = f_k(\boldsymbol{x}_0) + \ell_k(\boldsymbol{x} - \boldsymbol{x}_0) + r_k(\boldsymbol{x})\,,$$

with $k = 1, 2, \ldots, M$, and we know that

$$\lim_{\boldsymbol{x} \to \boldsymbol{x}_0} \frac{r(\boldsymbol{x})}{\|\boldsymbol{x} - \boldsymbol{x}_0\|} = \boldsymbol{0} \quad \Leftrightarrow \quad \lim_{\boldsymbol{x} \to \boldsymbol{x}_0} \frac{r_k(\boldsymbol{x})}{\|\boldsymbol{x} - \boldsymbol{x}_0\|} = 0 \quad \text{for every } k = 1, 2, \ldots, M\,,$$

whence the conclusion.   ■

The preceding theorem permits us to recover all the computational rules obtained in the case $M = 1$. Moreover, the function $f : \mathcal{O} \to \mathbb{R}^M$ is said to be "differentiable" or "of class $\mathcal{C}^1$" if all its components are. This definition naturally extends to functions of class $\mathcal{C}^n$.

Note that when $N = 1$, the differential $df(x_0) : \mathbb{R} \to \mathbb{R}^M$ is the linear function that associates to any $h \in \mathbb{R}$ the vector

$$df(x_0)(h) = h\, df(x_0)(1).$$

This last vector $df(x_0)(1) \in \mathbb{R}^M$ is called the "derivative" of f at x_0 and is usually denoted simply by $f'(x_0)$. Using the preceding definition, one readily sees that

$$f'(x_0) = \lim_{x \to x_0} \frac{f(x) - f(x_0)}{x - x_0},$$

thereby recovering the definition given in Sect. 7.14 and the usual definition given when $N = M = 1$.

It is useful to consider the matrix associated with the linear function $\ell = df(\boldsymbol{x}_0)$ given by

$$\begin{pmatrix} \ell_1(e_1) & \ell_1(e_2) & \dots & \ell_1(e_N) \\ \ell_2(e_1) & \ell_2(e_2) & \dots & \ell_2(e_N) \\ \vdots & \vdots & & \vdots \\ \ell_M(e_1) & \ell_M(e_2) & \dots & \ell_M(e_N) \end{pmatrix},$$

where $e_1, e_2, \dots e_N$ are the vectors of the canonical basis of $\mathbb{R}^N$. This matrix is called the "Jacobian matrix" associated with the function f at $\boldsymbol{x}_0$ and is denoted by one of the symbols

$$Jf(\boldsymbol{x}_0), \qquad f'(\boldsymbol{x}_0).$$

Recalling that

$$\frac{\partial f_k}{\partial x_j}(\boldsymbol{x}_0) = df_k(\boldsymbol{x}_0)e_j,$$

with $k = 1, 2, \dots, M$ and $j = 1, 2, \dots, N$, we see that

$$Jf(\boldsymbol{x}_0) = \begin{pmatrix} \frac{\partial f_1}{\partial x_1}(\boldsymbol{x}_0) & \frac{\partial f_1}{\partial x_2}(\boldsymbol{x}_0) & \cdots & \frac{\partial f_1}{\partial x_N}(\boldsymbol{x}_0) \\ \frac{\partial f_2}{\partial x_1}(\boldsymbol{x}_0) & \frac{\partial f_2}{\partial x_2}(\boldsymbol{x}_0) & \cdots & \frac{\partial f_2}{\partial x_N}(\boldsymbol{x}_0) \\ \vdots & \vdots & & \vdots \\ \frac{\partial f_M}{\partial x_1}(\boldsymbol{x}_0) & \frac{\partial f_M}{\partial x_2}(\boldsymbol{x}_0) & \cdots & \frac{\partial f_M}{\partial x_N}(\boldsymbol{x}_0) \end{pmatrix}.$$

Remark 10.15 Note that when $M = 1$, i.e., when $f : \mathcal{O} \to \mathbb{R}$, then its gradient is

$$\nabla f(\boldsymbol{x}_0) = Jf(\boldsymbol{x}_0)^T,$$

the transpose of the row matrix $Jf(\boldsymbol{x}_0)$. (Recall that a vector is always a column matrix.)

10.8   The Chain Rule

We now examine the differentiability of the composition of functions. As usual, $\mathcal{O}$ denotes an open subset of $\mathbb{R}^N$, and $\boldsymbol{x}_0$ is a point in $\mathcal{O}$.

Theorem 10.16 *If $f : \mathcal{O} \to \mathbb{R}^M$ is differentiable at $\boldsymbol{x}_0$, while $\mathcal{O}' \subseteq \mathbb{R}^M$ is an open set containing $f(\mathcal{O})$ and $g : \mathcal{O}' \to \mathbb{R}^L$ is differentiable at $f(\boldsymbol{x}_0)$, then $g \circ f$ is differentiable at $\boldsymbol{x}_0$, and*

$$d(g \circ f)(\boldsymbol{x}_0) = dg(f(\boldsymbol{x}_0)) \circ df(\boldsymbol{x}_0).$$

Proof Setting $\boldsymbol{y}_0 = f(\boldsymbol{x}_0)$, we have

$$f(\boldsymbol{x}) = f(\boldsymbol{x}_0) + df(\boldsymbol{x}_0)(\boldsymbol{x} - \boldsymbol{x}_0) + r_1(\boldsymbol{x}),$$
$$g(\boldsymbol{y}) = g(\boldsymbol{y}_0) + dg(\boldsymbol{y}_0)(\boldsymbol{y} - \boldsymbol{y}_0) + r_2(\boldsymbol{y}),$$

with

$$\lim_{\boldsymbol{x} \to \boldsymbol{x}_0} \frac{r_1(\boldsymbol{x})}{\|\boldsymbol{x} - \boldsymbol{x}_0\|} = \boldsymbol{0}, \qquad \lim_{\boldsymbol{y} \to \boldsymbol{y}_0} \frac{r_2(\boldsymbol{y})}{\|\boldsymbol{y} - \boldsymbol{y}_0\|} = \boldsymbol{0}.$$

Let us introduce the auxiliary function $R_2 : \mathcal{O}' \to \mathbb{R}^L$, defined as

$$R_2(\boldsymbol{y}) = \begin{cases} \dfrac{r_2(\boldsymbol{y})}{\|\boldsymbol{y} - \boldsymbol{y}_0\|} & \text{if } \boldsymbol{y} \neq \boldsymbol{y}_0, \\ \boldsymbol{0} & \text{if } \boldsymbol{y} = \boldsymbol{y}_0. \end{cases}$$

Note that R_2 is continuous at $\boldsymbol{y}_0$ and

$$r_2(\boldsymbol{y}) = \|\boldsymbol{y} - \boldsymbol{y}_0\| R_2(\boldsymbol{y}), \quad \text{for every } \boldsymbol{y} \in \mathcal{O}'.$$

Then

$$\begin{aligned} g(f(\boldsymbol{x})) &= g(f(\boldsymbol{x}_0)) + dg(f(\boldsymbol{x}_0))[f(\boldsymbol{x}) - f(\boldsymbol{x}_0)] + r_2(f(\boldsymbol{x})) \\ &= g(f(\boldsymbol{x}_0)) + dg(f(\boldsymbol{x}_0))[df(\boldsymbol{x}_0)(\boldsymbol{x} - \boldsymbol{x}_0) + r_1(\boldsymbol{x})] + r_2(f(\boldsymbol{x})) \\ &= g(f(\boldsymbol{x}_0)) + [dg(f(\boldsymbol{x}_0)) \circ df(\boldsymbol{x}_0)](\boldsymbol{x} - \boldsymbol{x}_0) + r_3(\boldsymbol{x}), \end{aligned}$$

where

$$r_3(\boldsymbol{x}) = dg(f(\boldsymbol{x}_0))(r_1(\boldsymbol{x})) + r_2(f(\boldsymbol{x}))$$
$$= dg(f(\boldsymbol{x}_0))(r_1(\boldsymbol{x})) + \|f(\boldsymbol{x}) - f(\boldsymbol{x}_0)\| R_2(f(\boldsymbol{x}))$$
$$= dg(f(\boldsymbol{x}_0))(r_1(\boldsymbol{x})) + \|df(\boldsymbol{x}_0)(\boldsymbol{x} - \boldsymbol{x}_0) + r_1(\boldsymbol{x})\| R_2(f(\boldsymbol{x})) \,.$$

Hence,

$$\frac{\|r_3(\boldsymbol{x})\|}{\|\boldsymbol{x} - \boldsymbol{x}_0\|} \leq \left\| dg(f(\boldsymbol{x}_0)) \left(\frac{r_1(\boldsymbol{x})}{\|\boldsymbol{x} - \boldsymbol{x}_0\|} \right) \right\| +$$
$$+ \left(\left\| df(\boldsymbol{x}_0) \left(\frac{\boldsymbol{x} - \boldsymbol{x}_0}{\|\boldsymbol{x} - \boldsymbol{x}_0\|} \right) \right\| + \frac{\|r_1(\boldsymbol{x})\|}{\|\boldsymbol{x} - \boldsymbol{x}_0\|} \right) \| R_2(f(\boldsymbol{x})) \| \,.$$

We can see that all this tends to 0 as $\boldsymbol{x} \to \boldsymbol{x}_0$. Indeed, if $\boldsymbol{x}$ tends to $\boldsymbol{x}_0$, the first summand tends to 0, since $dg(f(\boldsymbol{x}_0)) : \mathbb{R}^N \to \mathbb{R}^L$ is linear, hence continuous, and

$$\lim_{\boldsymbol{x} \to \boldsymbol{x}_0} \frac{r_1(\boldsymbol{x})}{\|\boldsymbol{x} - \boldsymbol{x}_0\|} = \boldsymbol{0} \,. \tag{10.1}$$

On the other hand, since f is continuous at $\boldsymbol{x}_0$ and R_2 is continuous at $\boldsymbol{y}_0 = f(\boldsymbol{x}_0)$, with $R_2(\boldsymbol{y}_0) = \boldsymbol{0}$, we have that

$$\lim_{\boldsymbol{x} \to \boldsymbol{x}_0} \| R_2(f(\boldsymbol{x})) \| = 0 \,.$$

Finally, since $df(\boldsymbol{x}_0) : \mathbb{R}^N \to \mathbb{R}^M$ is linear, hence continuous, it is bounded on the compact set $\overline{B}(\boldsymbol{0}, 1)$, by Weierstrass' Theorem 4.10. Therefore, using also (10.1),

$$\left(\left\| df(\boldsymbol{x}_0) \left(\frac{\boldsymbol{x} - \boldsymbol{x}_0}{\|\boldsymbol{x} - \boldsymbol{x}_0\|} \right) \right\| + \frac{\|r_1(\boldsymbol{x})\|}{\|\boldsymbol{x} - \boldsymbol{x}_0\|} \right) \quad \text{is bounded} \,.$$

Summing up,

$$\lim_{\boldsymbol{x} \to \boldsymbol{x}_0} \frac{\|r_3(\boldsymbol{x})\|}{\|\boldsymbol{x} - \boldsymbol{x}_0\|} = 0 \,,$$

and we can conclude that $g \circ f$ is differentiable at $\boldsymbol{x}_0$, with differential $dg(f(\boldsymbol{x}_0)) \circ df(\boldsymbol{x}_0)$. ∎

It is well known that the matrix associated with the composition of two linear functions is the product of the two respective matrices. From the preceding theorem we then have the following formula for the Jacobian matrices:

$$J(g \circ f)(\boldsymbol{x}_0) = Jg(f(\boldsymbol{x}_0)) \cdot Jf(\boldsymbol{x}_0) \,;$$

this means that the matrix

$$
\begin{pmatrix}
\frac{\partial (g \circ f)_1}{\partial x_1}(\boldsymbol{x}_0) & \cdots & \frac{\partial (g \circ f)_1}{\partial x_N}(\boldsymbol{x}_0) \\
\vdots & \cdots & \vdots \\
\frac{\partial (g \circ f)_L}{\partial x_1}(\boldsymbol{x}_0) & \cdots & \frac{\partial (g \circ f)_L}{\partial x_N}(\boldsymbol{x}_0)
\end{pmatrix}
$$

is equal to the product

$$
\begin{pmatrix}
\frac{\partial g_1}{\partial y_1}(f(\boldsymbol{x}_0)) & \cdots & \frac{\partial g_1}{\partial y_M}(f(\boldsymbol{x}_0)) \\
\vdots & \cdots & \vdots \\
\frac{\partial g_L}{\partial y_1}(f(\boldsymbol{x}_0)) & \cdots & \frac{\partial g_L}{\partial y_M}(f(\boldsymbol{x}_0))
\end{pmatrix}
\begin{pmatrix}
\frac{\partial f_1}{\partial x_1}(\boldsymbol{x}_0) & \cdots & \frac{\partial f_1}{\partial x_N}(\boldsymbol{x}_0) \\
\vdots & \cdots & \vdots \\
\frac{\partial f_M}{\partial x_1}(\boldsymbol{x}_0) & \cdots & \frac{\partial f_M}{\partial x_N}(\boldsymbol{x}_0)
\end{pmatrix}.
$$

We thus obtain the formula for the partial derivatives of the composition of functions, usually called the **chain rule**:

$$
\frac{\partial (g \circ f)_i}{\partial x_j}(\boldsymbol{x}_0) =
$$

$$
= \frac{\partial g_i}{\partial y_1}(f(\boldsymbol{x}_0))\frac{\partial f_1}{\partial x_j}(\boldsymbol{x}_0) + \frac{\partial g_i}{\partial y_2}(f(\boldsymbol{x}_0))\frac{\partial f_2}{\partial x_j}(\boldsymbol{x}_0) + \cdots + \frac{\partial g_i}{\partial y_M}(f(\boldsymbol{x}_0))\frac{\partial f_M}{\partial x_j}(\boldsymbol{x}_0)
$$

$$
= \sum_{k=1}^{M} \frac{\partial g_i}{\partial y_k}(f(\boldsymbol{x}_0))\frac{\partial f_k}{\partial x_j}(\boldsymbol{x}_0) ,
$$

where $i = 1, 2, \ldots, L$ and $j = 1, 2, \ldots, N$.

Remark 10.17 When $L = 1$, i.e., when $g : \mathcal{O}' \to \mathbb{R}$, in view of Remark 10.15 we obtain the formula

$$
\nabla (g \circ f)(\boldsymbol{x}_0) = J f(\boldsymbol{x}_0)^T \nabla g(f(\boldsymbol{x}_0)) .
$$

Let us now prove the following generalization of the formula for the derivative of a product of two functions.

Theorem 10.18 *Let $f : \mathcal{O} \to \mathbb{R}^{N_1}$ and $g : \mathcal{O} \to \mathbb{R}^{N_2}$ be two functions, differentiable at some $\boldsymbol{x}_0$. Let $F : \mathcal{O} \to \mathbb{R}^L$ be defined as*

$$
F(\boldsymbol{x}) = \mathcal{B}(f(\boldsymbol{x}), g(\boldsymbol{x})) ,
$$

where $\mathcal{B} : \mathbb{R}^{N_1} \times \mathbb{R}^{N_2} \to \mathbb{R}^L$ is a bilinear function. Then, for every $\boldsymbol{h} \in \mathbb{R}^N$,

$$
d F(\boldsymbol{x}_0)\boldsymbol{h} = \mathcal{B}(d f(\boldsymbol{x}_0)\boldsymbol{h}, g(\boldsymbol{x}_0)) + \mathcal{B}(f(\boldsymbol{x}_0), d g(\boldsymbol{x}_0)\boldsymbol{h}) .
$$

Proof First define the function $\varphi : \mathcal{O} \to \mathbb{R}^{N_1} \times \mathbb{R}^{N_2}$ as $\varphi(\boldsymbol{x}) = (f(\boldsymbol{x}), g(\boldsymbol{x}))$, and note that $d\varphi(\boldsymbol{x}_0)\boldsymbol{h} = (df(\boldsymbol{x}_0)\boldsymbol{h}, dg(\boldsymbol{x}_0)\boldsymbol{h})$ for every $\boldsymbol{h} \in \mathbb{R}^N$. Then it is sufficient to apply Theorem 10.16, in view of Proposition 10.7.   ∎

The following two examples with $\mathcal{O} \subseteq \mathbb{R}$, involving the scalar product and the cross product of two functions, are direct consequences of the preceding formula. Assume that $f, g : \mathcal{O} \to \mathbb{R}^M$ are differentiable at some $x_0 \in \mathcal{O}$. Then

$$(f \cdot g)'(x_0) = f'(x_0) \cdot g(x_0) + f(x_0) \cdot g'(x_0).$$

Moreover, if $M = 3$, then

$$(f \times g)'(x_0) = f'(x_0) \times g(x_0) + f(x_0) \times g'(x_0).$$

10.9   Mean Value Theorem

Lagrange's Theorem 6.11 does not extend directly to functions having vector values. For example, taking $a = 0$ and $b = 2\pi$, the function $f : [a, b] \to \mathbb{R}^2$ defined as $f(x) = (\cos x, \sin x)$ is such that $f(b) - f(a) = (0, 0)$, but there is no $\xi \in]a, b[$ for which $f(b) - f(a) = f'(\xi)(b - a)$ since $f'(\xi) = (-\sin \xi, \cos \xi) \neq (0, 0)$. We will nevertheless try to find a substitute for this theorem, which will be useful in what follows.

We first need the following lemma.

Lemma 10.19 *Let $\varphi : [a, b] \to \mathbb{R}^M$ be a differentiable function for which there is a constant $C \geq 0$ such that*

$$\|\varphi'(t)\| \leq C, \quad \text{for every } t \in [a, b].$$

Then

$$\|\varphi(b) - \varphi(a)\| \leq C(b - a).$$

Proof We set $I_0 = [a, b]$. Assume by contradiction that

$$\|\varphi(b) - \varphi(a)\| - C(b - a) = \mu > 0.$$

We divide the interval $[a, b]$ into two equal parts, taking the midpoint $m = \frac{a+b}{2}$. Then it can be seen that one of the two following inequalities holds:

$$\|\varphi(m) - \varphi(a)\| - C(m - a) \geq \frac{\mu}{2}, \qquad \|\varphi(b) - \varphi(m)\| - C(b - m) \geq \frac{\mu}{2}.$$

If the first one holds, we set $I_1 = [a, m]$; otherwise, we set $I_1 = [m, b]$. In the same way, we proceed now to the definition of I_2, then I_3, and so on. We thus obtain a sequence of compact intervals $I_n = [a_n, b_n]$, with

$$I_0 \supseteq I_1 \supseteq I_2 \supseteq I_3 \supseteq \dots$$

such that

$$\|\varphi(b_n) - \varphi(a_n)\| - C(b_n - a_n) \geq \frac{\mu}{2^n}$$

for every $n \in \mathbb{N}$. By Cantor's Theorem 1.9, there is a $c \in \mathbb{R}$ such that $a_n \leq c \leq b_n$ for every $n \in \mathbb{N}$, and since

$$b_n - a_n = \frac{b - a}{2^n},$$

we have that $\lim_n a_n = \lim_n b_n = c$. Since φ is differentiable at c, we can write

$$\varphi(t) = \varphi(c) + \varphi'(c)(t - c) + r(t),$$

with

$$\lim_{t \to c} \frac{r(t)}{t - c} = 0.$$

Let $\varepsilon \in \,]0, \frac{\mu}{b-a}[$. If n is sufficiently large, we have

$$
\begin{aligned}
\mu &\leq 2^n \big(\|\varphi(b_n) - \varphi(a_n)\| - C(b_n - a_n) \big) \\
&\leq 2^n \big(\|\varphi(b_n) - \varphi(c)\| + \|\varphi(c) - \varphi(a_n)\| - C(b_n - a_n) \big) \\
&= 2^n \big(\|\varphi'(c)(b_n - c) + r(b_n)\| + \|\varphi'(c)(a_n - c) + r(a_n)\| - C(b_n - a_n) \big) \\
&\leq 2^n \big(\|\varphi'(c)\| \, |b_n - c| + \|r(b_n)\| + \|\varphi'(c)\| \, |a_n - c| + \|r(a_n)\| - C(b_n - a_n) \big) \\
&\leq 2^n \big(C(b_n - c) + \|r(b_n)\| + C(c - a_n) + \|r(a_n)\| - C(b_n - a_n) \big) \\
&= 2^n \big(\|r(b_n)\| + \|r(a_n)\| \big) \\
&\leq 2^n \big(\varepsilon|b_n - c| + \varepsilon|a_n - c| \big) = 2^n \varepsilon(b_n - a_n) = \varepsilon(b - a),
\end{aligned}
$$

a contradiction, which completes the proof.   ∎

It will now be useful to introduce the norm of a linear function $\mathcal{A} : \mathbb{R}^N \to \mathbb{R}^M$ as

$$\|\mathcal{A}\| = \max\{\|\mathcal{A}\boldsymbol{x}\| : \|\boldsymbol{x}\| = 1\}.$$

Such a maximum exists by Weierstrass' Theorem 4.10 since the function $\mathcal{A}$, being linear, is continuous. The reader might like to check that we have indeed defined a norm, verifying the following properties:

(a) $\|\mathcal{A}\| \geq 0$.
(b) $\|\mathcal{A}\| = 0 \Leftrightarrow x = 0$.
(c) $\|\alpha \mathcal{A}\| = |\alpha| \|\mathcal{A}\|$.
(d) $\|\mathcal{A} + \mathcal{A}'\| \leq \|\mathcal{A}\| + \|\mathcal{A}'\|$.

Moreover, we have that

$$\|\mathcal{A}\boldsymbol{x}\| \leq \|\mathcal{A}\| \|\boldsymbol{x}\|, \quad \text{for every } \boldsymbol{x} \in \mathbb{R}^N.$$

We are now ready to state our extension of Lagrange's Mean Value Theorem 6.11. Let $\mathcal{O}$ be an open set in $\mathbb{R}^N$, and let $f : \mathcal{O} \to \mathbb{R}^M$ be a differentiable function.

Theorem 10.20 (Mean Value Theorem)   *If $[\boldsymbol{x}_0, \boldsymbol{x}]$ is a segment contained in $\mathcal{O}$, then*

$$\|f(\boldsymbol{x}) - f(\boldsymbol{x}_0)\| \leq \sup \left\{ \|df(\boldsymbol{v})\| : \boldsymbol{v} \in [\boldsymbol{x}_0, \boldsymbol{x}] \right\} \|\boldsymbol{x} - \boldsymbol{x}_0\|.$$

Proof   If the supremum is equal to $+\infty$, there is nothing to be proved. Suppose, then, that

$$\sup \left\{ \|df(\boldsymbol{v})\| : \boldsymbol{v} \in [\boldsymbol{x}_0, \boldsymbol{x}] \right\} = C \in \mathbb{R}.$$

We consider the function $\varphi : [0, 1] \to \mathbb{R}^M$, defined as $\varphi(t) = f(\boldsymbol{x}_0 + t(\boldsymbol{x} - \boldsymbol{x}_0))$. Then

$$\begin{aligned} \|\varphi'(t)\| &= \|df(\boldsymbol{x}_0 + t(\boldsymbol{x} - \boldsymbol{x}_0))(\boldsymbol{x} - \boldsymbol{x}_0)\| \\ &\leq \|df(\boldsymbol{x}_0 + t(\boldsymbol{x} - \boldsymbol{x}_0))\| \|\boldsymbol{x} - \boldsymbol{x}_0\| \\ &\leq C \|\boldsymbol{x} - \boldsymbol{x}_0\| \end{aligned}$$

for every $t \in [0, 1]$. By Lemma 10.19,

$$\|f(\boldsymbol{x}) - f(\boldsymbol{x}_0)\| = \|\varphi(1) - \varphi(0)\| \leq C \|\boldsymbol{x} - \boldsymbol{x}_0\|(1 - 0) = C \|\boldsymbol{x} - \boldsymbol{x}_0\|,$$

which is exactly what we wanted to prove.   ∎

10.10 Implicit Function Theorem: General Statement

We will now generalize the Implicit Function Theorem 10.13 in its general finite-dimensional context. Let $\mathcal{O}$ be an open subset of $\mathbb{R}^M \times \mathbb{R}^N$ and $g : \mathcal{O} \to \mathbb{R}^N$ a C^1-function. Hence, g has N components

$$g(\boldsymbol{x}, \boldsymbol{y}) = (g_1(\boldsymbol{x}, \boldsymbol{y}), \ldots, g_N(\boldsymbol{x}, \boldsymbol{y})).$$

Here $\boldsymbol{x} = (x_1, \ldots, x_M) \in \mathbb{R}^M$, and $\boldsymbol{y} = (y_1, \ldots, y_N) \in \mathbb{R}^N$. We will use the following notation for the Jacobian matrices:

$$\frac{\partial g}{\partial \boldsymbol{x}}(\boldsymbol{x}, \boldsymbol{y}) = \begin{pmatrix} \frac{\partial g_1}{\partial x_1}(\boldsymbol{x}, \boldsymbol{y}) & \cdots & \frac{\partial g_1}{\partial x_M}(\boldsymbol{x}, \boldsymbol{y}) \\ \vdots & \cdots & \vdots \\ \frac{\partial g_N}{\partial x_1}(\boldsymbol{x}, \boldsymbol{y}) & \cdots & \frac{\partial g_N}{\partial x_M}(\boldsymbol{x}, \boldsymbol{y}) \end{pmatrix},$$

$$\frac{\partial g}{\partial \boldsymbol{y}}(\boldsymbol{x}, \boldsymbol{y}) = \begin{pmatrix} \frac{\partial g_1}{\partial y_1}(\boldsymbol{x}, \boldsymbol{y}) & \cdots & \frac{\partial g_1}{\partial y_N}(\boldsymbol{x}, \boldsymbol{y}) \\ \vdots & \cdots & \vdots \\ \frac{\partial g_N}{\partial y_1}(\boldsymbol{x}, \boldsymbol{y}) & \cdots & \frac{\partial g_N}{\partial y_N}(\boldsymbol{x}, \boldsymbol{y}) \end{pmatrix}.$$

Theorem 10.21 (Implicit Function Theorem—II) *Let $\mathcal{O} \subseteq \mathbb{R}^M \times \mathbb{R}^N$ be an open set, $g : \mathcal{O} \to \mathbb{R}^N$ a C^1-function, and $(\boldsymbol{x}_0, \boldsymbol{y}_0)$ a point in $\mathcal{O}$ for which*

$$g(\boldsymbol{x}_0, \boldsymbol{y}_0) = 0 \quad and \quad \det \frac{\partial g}{\partial \boldsymbol{y}}(\boldsymbol{x}_0, \boldsymbol{y}_0) \neq 0.$$

Then there exist an open neighborhood U of $\boldsymbol{x}_0$, an open neighborhood V of $\boldsymbol{y}_0$, and a C^1-function $\eta : U \to V$ such that $U \times V \subseteq \mathcal{O}$, and, taking $\boldsymbol{x} \in U$ and $\boldsymbol{y} \in V$, we have that

$$g(\boldsymbol{x}, \boldsymbol{y}) = 0 \quad \Leftrightarrow \quad \boldsymbol{y} = \eta(\boldsymbol{x}).$$

Moreover, the function η is of class C^1, and the following formula holds true:

$$J\eta(\boldsymbol{x}) = -\left(\frac{\partial g}{\partial \boldsymbol{y}}(\boldsymbol{x}, \eta(\boldsymbol{x}))\right)^{-1} \frac{\partial g}{\partial \boldsymbol{x}}(\boldsymbol{x}, \eta(\boldsymbol{x})).$$

Proof In the case where $N = 1$, the definition of η is almost the same as the one given in the proof of Theorem 10.13. It will be sufficient to replace the interval $[x_0 - \delta, x_0 + \delta]$ with the ball $\overline{B}(\boldsymbol{x}_0, \delta)$ and to replace $]x_0 - \delta', x_0 + \delta'[$ with $B(\boldsymbol{x}_0, \delta')$. Once the function $\eta : U \to V$ has been defined, let us see how to prove its continuity and its differentiablility.

To verify the continuity of η, let us fix a $\bar{\boldsymbol{x}} \in U$ and prove that η is continuous at $\bar{\boldsymbol{x}}$. If we take $\boldsymbol{x} \in U$ and consider the function $\gamma : [0, 1] \to U \times V$, defined as

$$\gamma(t) = (\bar{\boldsymbol{x}} + t(\boldsymbol{x} - \bar{\boldsymbol{x}}), \eta(\bar{\boldsymbol{x}}) + t(\eta(\boldsymbol{x}) - \eta(\bar{\boldsymbol{x}}))),$$

Lagrange's Mean Value Theorem 6.11 applied to $g \circ \gamma$ tells us that there is a $\xi \in \,]0, 1[$ for which

$$g(\boldsymbol{x}, \eta(\boldsymbol{x})) - g(\bar{\boldsymbol{x}}, \eta(\bar{\boldsymbol{x}})) = \frac{\partial g}{\partial \boldsymbol{x}}(\gamma(\xi))(\boldsymbol{x} - \bar{\boldsymbol{x}}) + \frac{\partial g}{\partial y}(\gamma(\xi))(\eta(\boldsymbol{x}) - \eta(\bar{\boldsymbol{x}})).$$

Since $g(\boldsymbol{x}, \eta(\boldsymbol{x})) = g(\bar{\boldsymbol{x}}, \eta(\bar{\boldsymbol{x}})) = 0$, we have that

$$|\eta(\boldsymbol{x}) - \eta(\bar{\boldsymbol{x}})| = \frac{1}{|\frac{\partial g}{\partial y}(\gamma(\xi))|} \left| \frac{\partial g}{\partial \boldsymbol{x}}(\gamma(\xi))(\boldsymbol{x} - \bar{\boldsymbol{x}}) \right|.$$

Since the partial derivatives of g are continuous and $\frac{\partial g}{\partial y}$ is not zero on the compact set $\overline{U} \times \overline{V}$, we have that there is a constant $c > 0$ for which

$$\frac{1}{|\frac{\partial g}{\partial y}(\gamma(\xi))|} \left| \frac{\partial g}{\partial \boldsymbol{x}}(\gamma(\xi))(\boldsymbol{x} - \bar{\boldsymbol{x}}) \right| \le c \|\boldsymbol{x} - \bar{\boldsymbol{x}}\|.$$

As a consequence, η is continuous at $\bar{\boldsymbol{x}}$.

We now prove the differentiability. Taking $\bar{\boldsymbol{x}} = (\bar{x}_1, \bar{x}_2, \ldots, \bar{x}_M)$, let $\boldsymbol{x} = (\bar{x}_1 + h, \bar{x}_2, \ldots, \bar{x}_M)$; proceeding as previously, for h small enough we have

$$\frac{\eta(\bar{x}_1 + h, \bar{x}_2, \ldots, \bar{x}_M) - \eta(\bar{x}_1, \bar{x}_2, \ldots, \bar{x}_M)}{h} = -\frac{\frac{\partial g}{\partial x_1}(\gamma(\xi))}{\frac{\partial g}{\partial y}(\gamma(\xi))},$$

with $\gamma(\xi)$ belonging to the segment joining $(\bar{\boldsymbol{x}}, \eta(\bar{\boldsymbol{x}}))$ to $(\boldsymbol{x}, \eta(\boldsymbol{x}))$. If h tends to 0, we have that $\gamma(\xi)$ tends to $(\bar{\boldsymbol{x}}, \eta(\bar{\boldsymbol{x}}))$, and hence

$$\frac{\partial \eta}{\partial x_1}(\bar{\boldsymbol{x}}) = \lim_{h \to 0} \frac{\eta(\bar{x}_1 + h, \bar{x}_2, \ldots, \bar{x}_M) - \eta(\bar{x}_1, \bar{x}_2, \ldots, \bar{x}_M)}{h} = -\frac{\frac{\partial g}{\partial x_1}(\bar{\boldsymbol{x}}, \eta(\bar{\boldsymbol{x}}))}{\frac{\partial g}{\partial y}(\bar{\boldsymbol{x}}, \eta(\bar{\boldsymbol{x}}))}.$$

The partial derivatives with respect to $x_2, \ldots, x_M$ are computed similarly, thereby yielding that η is of class C^1 and

$$J\eta(\boldsymbol{x}) = -\frac{1}{\frac{\partial g}{\partial y}(\boldsymbol{x}, \eta(\boldsymbol{x}))} \frac{\partial g}{\partial \boldsymbol{x}}(\boldsymbol{x}, \eta(\boldsymbol{x})) \quad \text{for every } \boldsymbol{x} \in U.$$

We have thus proved the theorem in the case $N = 1$.

We now assume that the statement holds till $N - 1$ for some $N \geq 2$ (and any $M \geq 1$) and prove that it then also holds for N. We will use the notation

$$\boldsymbol{y}_1 = (y_1, \ldots, y_{N-1}),$$

and we will write $\boldsymbol{y} = (\boldsymbol{y}_1, y_N)$. Since

$$\det \begin{pmatrix} \frac{\partial g_1}{\partial y_1}(\boldsymbol{x}_0, \boldsymbol{y}_0) & \cdots & \frac{\partial g_1}{\partial y_N}(\boldsymbol{x}_0, \boldsymbol{y}_0) \\ \vdots & \cdots & \vdots \\ \frac{\partial g_N}{\partial y_1}(\boldsymbol{x}_0, \boldsymbol{y}_0) & \cdots & \frac{\partial g_N}{\partial y_N}(\boldsymbol{x}_0, \boldsymbol{y}_0) \end{pmatrix} \neq 0,$$

at least one of the elements in the last column is different from zero. We can assume without loss of generality, possibly changing the rows, that $\frac{\partial g_N}{\partial y_N}(\boldsymbol{x}_0, \boldsymbol{y}_0) \neq 0$. Writing $\boldsymbol{y}_0 = (\boldsymbol{y}_1^0, y_N^0)$, with $\boldsymbol{y}_1^0 = (y_1^0, \ldots, y_{N-1}^0)$, we then have

$$g_N(\boldsymbol{x}_0, \boldsymbol{y}_1^0, y_N^0) = 0 \quad \text{and} \quad \frac{\partial g_N}{\partial y_N}(\boldsymbol{x}_0, \boldsymbol{y}_1^0, y_N^0) \neq 0.$$

Then, by the already proved one-dimensional case, there are an open neighborhood U_1 of $(\boldsymbol{x}_0, \boldsymbol{y}_1^0)$, an open neighborhood V_N of y_N^0, and a C^1-function $\eta_1 : U_1 \to V_N$ such that $U_1 \times V_N \subseteq \mathcal{O}$, with the following properties. If $(\boldsymbol{x}, \boldsymbol{y}_1) \in U_1$ and $y_N \in V_N$, then

$$g_N(\boldsymbol{x}, \boldsymbol{y}_1, y_N) = 0 \quad \Leftrightarrow \quad y_N = \eta_1(\boldsymbol{x}, \boldsymbol{y}_1)$$

and

$$J\eta_1(\boldsymbol{x}, \boldsymbol{y}_1) = -\frac{1}{\frac{\partial g_N}{\partial y_N}(\boldsymbol{x}, \boldsymbol{y}_1, \eta_1(\boldsymbol{x}, \boldsymbol{y}_1)))} \frac{\partial g_N}{\partial(\boldsymbol{x}, \boldsymbol{y}_1)}(\boldsymbol{x}, \boldsymbol{y}_1, \eta_1(\boldsymbol{x}, \boldsymbol{y}_1)).$$

We may assume that U_1 is of the type $\widetilde{U} \times \widetilde{V}_1$, with $\widetilde{U}$ being an open neighborhood of $\boldsymbol{x}_0$ and $\widetilde{V}_1$ an open neighborhood of $\boldsymbol{y}_1^0$.

Let us define the function $\phi : \widetilde{U} \times \widetilde{V}_1 \to \mathbb{R}^{N-1}$ by setting

$$\phi(\boldsymbol{x}, \boldsymbol{y}_1) = (g_1(\boldsymbol{x}, \boldsymbol{y}_1, \eta_1(\boldsymbol{x}, \boldsymbol{y}_1)), \ldots, g_{N-1}(\boldsymbol{x}, \boldsymbol{y}_1, \eta_1(\boldsymbol{x}, \boldsymbol{y}_1))).$$

For brevity's sake we will write

$$g_{(1,\ldots,N-1)}(\boldsymbol{x}, \boldsymbol{y}) = (g_1(\boldsymbol{x}, \boldsymbol{y}), \ldots, g_{N-1}(\boldsymbol{x}, \boldsymbol{y})),$$

so that

$$\phi(\boldsymbol{x}, \boldsymbol{y}_1) = g_{(1,\ldots,N-1)}(\boldsymbol{x}, \boldsymbol{y}_1, \eta_1(\boldsymbol{x}, \boldsymbol{y}_1)).$$

Note that, since $\eta_1(\boldsymbol{x}_0, \boldsymbol{y}_1^0) = y_N^0$, we have that

$$\phi(\boldsymbol{x}_0, \boldsymbol{y}_1^0) = g_{(1,\dots,N-1)}(\boldsymbol{x}_0, \boldsymbol{y}_0) = \boldsymbol{0}$$

and

$$\frac{\partial \phi}{\partial \boldsymbol{y}_1}(\boldsymbol{x}_0, \boldsymbol{y}_1^0) = \frac{\partial g_{(1,\dots,N-1)}}{\partial \boldsymbol{y}_1}(\boldsymbol{x}_0, \boldsymbol{y}_0) + \frac{\partial g_{(1,\dots,N-1)}}{\partial y_N}(\boldsymbol{x}_0, \boldsymbol{y}_0)\frac{\partial \eta_1}{\partial \boldsymbol{y}_1}(\boldsymbol{x}_0, \boldsymbol{y}_1^0).$$
$$(10.2)$$

Moreover, since $g_N(\boldsymbol{x}, \boldsymbol{y}_1, \eta_1(\boldsymbol{x}, \boldsymbol{y}_1)) = 0$ for every $(\boldsymbol{x}, \boldsymbol{y}_1) \in U_1$, differentiating with respect to $\boldsymbol{y}_1$ we see that

$$0 = \frac{\partial g_N}{\partial \boldsymbol{y}_1}(\boldsymbol{x}_0, \boldsymbol{y}_0) + \frac{\partial g_N}{\partial y_N}(\boldsymbol{x}_0, \boldsymbol{y}_0)\frac{\partial \eta_1}{\partial \boldsymbol{y}_1}(\boldsymbol{x}_0, \boldsymbol{y}_1^0).$$
$$(10.3)$$

Let us write the identity

$$\det \frac{\partial \phi}{\partial \boldsymbol{y}_1}(\boldsymbol{x}_0, \boldsymbol{y}_1^0) = \frac{1}{\frac{\partial g_N}{\partial y_N}(\boldsymbol{x}_0, \boldsymbol{y}_0)} \det \left(\begin{array}{c|c} \frac{\partial \phi}{\partial \boldsymbol{y}_1}(\boldsymbol{x}_0, \boldsymbol{y}_1^0) & \frac{\partial g_{(1,\dots,N-1)}}{\partial y_N}(\boldsymbol{x}_0, \boldsymbol{y}_0) \\ \hline 0 & \frac{\partial g_N}{\partial y_N}(\boldsymbol{x}_0, \boldsymbol{y}_0) \end{array} \right).$$

Substituting the two equalities (10.2) and (10.3), we have that

$$\det \left(\begin{array}{c|c} \frac{\partial \phi}{\partial \boldsymbol{y}_1}(\boldsymbol{x}_0, \boldsymbol{y}_1^0) & \frac{\partial g_{(1,\dots,N-1)}}{\partial y_N}(\boldsymbol{x}_0, \boldsymbol{y}_0) \\ \hline 0 & \frac{\partial g_N}{\partial y_N}(\boldsymbol{x}_0, \boldsymbol{y}_0) \end{array} \right) =$$

$$= \det \left(\begin{array}{c|c} \frac{\partial g_{(1,\dots,N-1)}}{\partial \boldsymbol{y}_1}(\boldsymbol{x}_0, \boldsymbol{y}_0) + \frac{\partial g_{(1,\dots,N-1)}}{\partial y_N}(\boldsymbol{x}_0, \boldsymbol{y}_0)\frac{\partial \eta_1}{\partial \boldsymbol{y}_1}(\boldsymbol{x}_0, \boldsymbol{y}_1^0) & \frac{\partial g_{(1,\dots,N-1)}}{\partial y_N}(\boldsymbol{x}_0, \boldsymbol{y}_0) \\ \hline \frac{\partial g_N}{\partial \boldsymbol{y}_1}(\boldsymbol{x}_0, \boldsymbol{y}_0) + \frac{\partial g_N}{\partial y_N}(\boldsymbol{x}_0, \boldsymbol{y}_0)\frac{\partial \eta_1}{\partial \boldsymbol{y}_1}(\boldsymbol{x}_0, \boldsymbol{y}_1^0) & \frac{\partial g_N}{\partial y_N}(\boldsymbol{x}_0, \boldsymbol{y}_0) \end{array} \right)$$

$$= \det \left(\frac{\partial g}{\partial \boldsymbol{y}_1}(\boldsymbol{x}_0, \boldsymbol{y}_0) + \frac{\partial g}{\partial y_N}(\boldsymbol{x}_0, \boldsymbol{y}_0)\frac{\partial \eta_1}{\partial \boldsymbol{y}_1}(\boldsymbol{x}_0, \boldsymbol{y}_1^0) \;\middle|\; \frac{\partial g}{\partial y_N}(\boldsymbol{x}_0, \boldsymbol{y}_0) \right)$$

$$= \det \left[\frac{\partial g}{\partial \boldsymbol{y}}(\boldsymbol{x}_0, \boldsymbol{y}_0) + \left(\frac{\partial g}{\partial y_N}(\boldsymbol{x}_0, \boldsymbol{y}_0)\frac{\partial \eta_1}{\partial \boldsymbol{y}_1}(\boldsymbol{x}_0, \boldsymbol{y}_1^0) \;\middle|\; \frac{\partial g}{\partial y_N}(\boldsymbol{x}_0, \boldsymbol{y}_0) \right) \right].$$

We now recall that adding a scalar multiple of one column to another column of a matrix does not change the value of its determinant. Hence, since

$$
\left(\frac{\partial g}{\partial y_N}(x_0, y_0) \frac{\partial \eta_1}{\partial y_1}(x_0, y_1^0) \;\middle|\; \frac{\partial g}{\partial y_N}(x_0, y_0) \right) =
$$

$$
= \begin{pmatrix}
\dfrac{\partial g_1}{\partial y_N}(x_0, y_0)\dfrac{\partial \eta_1}{\partial y_1}(x_0, y_1^0) & \cdots & \dfrac{\partial g_1}{\partial y_N}(x_0, y_0)\dfrac{\partial \eta_1}{\partial y_{N-1}}(x_0, y_1^0) & \dfrac{\partial g_1}{\partial y_N}(x_0, y_0) \\[2ex]
\vdots & \cdots & \vdots & \vdots \\[2ex]
\dfrac{\partial g_N}{\partial y_N}(x_0, y_0)\dfrac{\partial \eta_1}{\partial y_1}(x_0, y_1^0) & \cdots & \dfrac{\partial g_N}{\partial y_N}(x_0, y_0)\dfrac{\partial \eta_1}{\partial y_{N-1}}(x_0, y_1^0) & \dfrac{\partial g_N}{\partial y_N}(x_0, y_0)
\end{pmatrix},
$$

it must be that

$$
\det\left[\frac{\partial g}{\partial y}(x_0, y_0) + \left(\frac{\partial g}{\partial y_N}(x_0, y_0)\frac{\partial \eta_1}{\partial y_1}(x_0, y_1^0) \;\middle|\; \frac{\partial g}{\partial y_N}(x_0, y_0) \right) \right]
$$

$$
= \det \frac{\partial g}{\partial y}(x_0, y_0) \,.
$$

Thus,

$$
\det \frac{\partial \phi}{\partial y_1}(x_0, y_1^0) = \frac{1}{\frac{\partial g_N}{\partial y_N}(x_0, y_0)} \det \frac{\partial g}{\partial y}(x_0, y_0) \,,
$$

and finally we have

$$
\phi(x_0, y_1^0) = 0 \quad \text{and} \quad \det \frac{\partial \phi}{\partial y_1}(x_0, y_1^0) \neq 0 \,.
$$

By the inductive assumption, there are an open neighborhood U of x_0, an open neighborhood V_1 of y_1^0, and a C^1-function $\eta_2 : U \to V_1$ such that $U \times V_1 \subseteq \tilde{U} \times \tilde{V}_1$, and the following holds: For every $x \in U$ and $y_1 \in V_1$,

$$
\phi(x, y_1) = 0 \quad \Leftrightarrow \quad y_1 = \eta_2(x) \,.
$$

In conclusion, for $x \in U$ and $y = (y_1, y_N) \in V_1 \times V_N$, we have that

$$
g(x, y) = 0 \quad \Leftrightarrow \quad \begin{cases} g_{(1,\dots,N-1)}(x, y_1, y_N) = 0 \\ g_N(x, y_1, y_N) = 0 \end{cases}
$$

$$
\Leftrightarrow \quad \begin{cases} g_{(1,\dots,N-1)}(x, y_1, y_N) = 0 \\ y_N = \eta_1(x, y_1) \end{cases}
$$

$$\Leftrightarrow \quad \begin{cases} \phi(\boldsymbol{x}, \boldsymbol{y}_1) = \boldsymbol{0} \\ y_N = \eta_1(\boldsymbol{x}, \boldsymbol{y}_1) \end{cases}$$

$$\Leftrightarrow \quad \begin{cases} \boldsymbol{y}_1 = \eta_2(\boldsymbol{x}) \\ y_N = \eta_1(\boldsymbol{x}, \boldsymbol{y}_1) \end{cases}$$

$$\Leftrightarrow \quad \boldsymbol{y} = (\eta_2(\boldsymbol{x}), \eta_1(\boldsymbol{x}, \eta_2(\boldsymbol{x}))).$$

Setting $V = V_1 \times V_N$, we may then define the function $\eta : U \to V$ as

$$\eta(\boldsymbol{x}) = (\eta_2(\boldsymbol{x}), \eta_1(\boldsymbol{x}, \eta_2(\boldsymbol{x}))).$$

This function is of class C^1 since both η_1 and η_2 are as well. Since $g(\boldsymbol{x}, \eta(\boldsymbol{x})) = 0$ for every $\boldsymbol{x} \in U$, we easily deduce that

$$\frac{\partial g}{\partial \boldsymbol{x}}(\boldsymbol{x}, \eta(\boldsymbol{x})) + \frac{\partial g}{\partial \boldsymbol{y}}(\boldsymbol{x}, \eta(\boldsymbol{x})) J\eta(\boldsymbol{x}) = 0,$$

whence the formula for $J\eta(\boldsymbol{x})$. ∎

Clearly, the following analogous statement holds true, where the roles of $\boldsymbol{x}$ and $\boldsymbol{y}$ are interchanged.

Theorem 10.22 (Implicit Function Theorem—III) *Let $\mathcal{O} \subseteq \mathbb{R}^M \times \mathbb{R}^N$ be an open set, $g : \mathcal{O} \to \mathbb{R}^M$ a C^1-function, and $(\boldsymbol{x}_0, \boldsymbol{y}_0)$ a point in $\mathcal{O}$ for which*

$$g(\boldsymbol{x}_0, \boldsymbol{y}_0) = \boldsymbol{0} \quad and \quad \det \frac{\partial g}{\partial \boldsymbol{x}}(\boldsymbol{x}_0, \boldsymbol{y}_0) \neq 0.$$

Then there exist an open neighborhood U of $\boldsymbol{x}_0$, an open neighborhood V of $\boldsymbol{y}_0$, and a C^1-function $\eta : V \to U$ such that $U \times V \subseteq \mathcal{O}$, and, taking $\boldsymbol{x} \in U$ and $\boldsymbol{y} \in V$, we have that

$$g(\boldsymbol{x}, \boldsymbol{y}) = \boldsymbol{0} \quad \Leftrightarrow \quad \boldsymbol{x} = \eta(\boldsymbol{y}).$$

Moreover, the function η is of class C^1, and the following formula holds:

$$J\eta(\boldsymbol{y}) = -\left(\frac{\partial g}{\partial \boldsymbol{x}}(\eta(\boldsymbol{y}), \boldsymbol{y})\right)^{-1} \frac{\partial g}{\partial \boldsymbol{y}}(\eta(\boldsymbol{y}), \boldsymbol{y}).$$

10.11 Local Diffeomorphisms

Let us introduce the notion of "diffeomorphism."

Definition 10.23 Given A and B, two open subsets of $\mathbb{R}^N$, a function $\varphi : A \to B$ is said to be a "diffeomorphism" if it is of class C^1, it is a bijection, and its inverse $\varphi^{-1} : B \to A$ is also of class C^1.

Let us state the following important consequence of the Implicit Function Theorem.

Theorem 10.24 (Local Diffeomorphism Theorem) *Let A be an open subset of $\mathbb{R}^N$, and let $\varphi : A \to \mathbb{R}^N$ be a C^1-function. If, for some $\boldsymbol{x}_0 \in A$, we have that $\det J\varphi(\boldsymbol{x}_0) \neq 0$, then there exist an open neighborhood U of $\boldsymbol{x}_0$ contained in A and an open neighborhood V of $\varphi(\boldsymbol{x}_0)$ such that $\varphi(U) = V$, and the restricted function $\varphi|_U : U \to V$ is a diffeomorphism.*

Proof We consider the function $g : A \times \mathbb{R}^N \to \mathbb{R}^N$ defined as

$$g(\boldsymbol{x}, \boldsymbol{y}) = \varphi(\boldsymbol{x}) - \boldsymbol{y} \,.$$

Setting $\boldsymbol{y}_0 = \varphi(\boldsymbol{x}_0)$, we have that

$$g(\boldsymbol{x}_0, \boldsymbol{y}_0) = \boldsymbol{0} \quad \text{and} \quad \det \frac{\partial g}{\partial \boldsymbol{x}}(\boldsymbol{x}_0, \boldsymbol{y}_0) = \det J\varphi(\boldsymbol{x}_0) \neq 0 \,.$$

By the Implicit Function Theorem 10.22, there exist an open neighborhood V of $\boldsymbol{y}_0$, an open neighborhood U of $\boldsymbol{x}_0$, and a C^1-function $\eta : V \to U$ such that $U \subseteq A$ and, taking $\boldsymbol{y} \in V$ and $\boldsymbol{x} \in U$,

$$\varphi(\boldsymbol{x}) = \boldsymbol{y} \quad \Leftrightarrow \quad g(\boldsymbol{x}, \boldsymbol{y}) = \boldsymbol{0} \quad \Leftrightarrow \quad \boldsymbol{x} = \eta(\boldsymbol{y}) \,.$$

Hence, $\eta = \varphi|_U^{-1}$, and the proof is thus completed.   ∎

The following corollary will be useful.

Corollary 10.25 *Let $A \subseteq \mathbb{R}^N$ be an open set and $\sigma : A \to \mathbb{R}^N$ an injective C^1-function such that $\det J\sigma(\boldsymbol{x}) \neq 0$ for every $\boldsymbol{x} \in A$. Then the set $B = \sigma(A)$ is open, and the function $\varphi : A \to B$ defined as $\varphi(\boldsymbol{x}) = \sigma(\boldsymbol{x})$ is a diffeomorphism.*

Proof For every $\boldsymbol{y}_0 \in \sigma(A)$ there is a unique $\boldsymbol{x}_0 \in A$ such that $\sigma(\boldsymbol{x}_0) = \boldsymbol{y}_0$, and we know that $\det J\sigma(\boldsymbol{x}_0) \neq 0$. Hence, by the Local Diffeomorphism Theorem 10.24, there exist an open neighborhood U of $\boldsymbol{x}_0$ contained in A and an open neighborhood V of $\boldsymbol{y}_0$ such that $\sigma(U) = V$, and the restricted function $\sigma|_U : U \to V$ is a diffeomorphism. Then $V = \sigma(U) \subseteq \sigma(A)$, thereby proving that $\sigma(A)$ is an open set.

In conclusion, the function $\varphi : A \to \sigma(A)$ defined as $\varphi(\boldsymbol{x}) = \sigma(\boldsymbol{x})$ is bijective and of class C^1, and, being a local diffeomorphism, its inverse $\varphi^{-1} : \sigma(A) \to A$ is of class C^1 as well. ∎

We now derive the formula for the differential of the inverse function.

Theorem 10.26 *Let* $\varphi : A \to B$ *be a diffeomorphism, take* $\boldsymbol{x}_0 \in A$, *and let* $\boldsymbol{y}_0 = \varphi(\boldsymbol{x}_0)$. *Then* $d\varphi(\boldsymbol{x}_0)$ *is invertible, and*

$$d\varphi^{-1}(\boldsymbol{y}_0) = d\varphi(\boldsymbol{x}_0)^{-1},$$

whence

$$J\varphi^{-1}(\boldsymbol{y}_0) = [J\varphi(\boldsymbol{x}_0)]^{-1}.$$

Proof Observe that $\varphi^{-1} \circ \varphi : A \to A$ is the identity function I on A, and $\varphi \circ \varphi^{-1} : B \to B$ is the identity function I on B. Then their differentials at $\boldsymbol{x}_0$ and at $\boldsymbol{y}_0$, respectively, are also identity functions, and hence

$$I = d(\varphi^{-1} \circ \varphi)(\boldsymbol{x}_0) = d\varphi^{-1}(\varphi(\boldsymbol{x}_0)) \circ d\varphi(\boldsymbol{x}_0) = d\varphi^{-1}(\boldsymbol{y}_0) \circ d\varphi(\boldsymbol{x}_0),$$

$$I = d(\varphi \circ \varphi^{-1})(\boldsymbol{y}_0) = d\varphi(\varphi^{-1}(\boldsymbol{y}_0)) \circ d\varphi^{-1}(\boldsymbol{y}_0) = d\varphi(\boldsymbol{x}_0) \circ d\varphi^{-1}(\boldsymbol{y}_0).$$

This proves that $d\varphi(\boldsymbol{x}_0)$ is invertible, and $d\varphi^{-1}(\boldsymbol{y}_0)$ is its inverse. The equality for the Jacobian matrices is a consequence of the fact that the matrix associated with the inverse of a linear function is the inverse of the matrix of that linear function. ∎

10.12 *M*-Surfaces

We often hear talk of "curves" and "surfaces" without a precise definition of what, in fact, they are. We now begin examining these objects from a dynamical point of view. The motivation comes from a typical situation in physics when one wants to describe the trajectory of a moving object. Assuming that the object is a point, surely enough we would not be satisfied if we were only told, for example, that its trajectory was a circle. We would also like to know how the object moves on this circle: Is its speed constant or varying? Is it moving clockwise or counterclockwise? Or is it oscillating back and forth?

To satisfy the need to know precisely how the object is moving, we introduce a function, defined on some interval $[a, b]$, which to each instant of time $t \in [a, b]$ associates its position in space, say, $\sigma(t)$. Such a function $\sigma : [a, b] \to \mathbb{R}^N$, if it is sufficiently regular, will be called a "curve" in $\mathbb{R}^N$.

Similar observations can be made for a "surface," which will be a function defined on some rectangle $[a_1, b_1] \times [a_2, b_2]$. (The choice of a rectangular domain is made for simplicity.) These two situations will now be generalized to an arbitrary dimension M, leading to the concept of "M-surface."

We denote by I a "rectangle" in $\mathbb{R}^M$, i.e., a set of the type

$$I = [a_1, b_1] \times \cdots \times [a_M, b_M].$$

This word is surely familiar in the case of $M = 2$. If $M = 1$, a rectangle happens to be a compact interval, whereas if $M = 3$, we usually prefer to call it a "rectangular parallelepiped" or "cuboid."

Definition 10.27 Let $1 \leq M \leq N$. We call "M-surface" in $\mathbb{R}^N$ a function $\sigma : I \to \mathbb{R}^N$ of class C^1. If $M = 1$, then σ is also said to be a "curve"; if $M = 2$, then we will simply call it a "surface." The set $\sigma(I)$ is the "image" of the M-surface σ. We will say that the M-surface σ is "regular" if, for every $\boldsymbol{u} \in \overset{\circ}{I}$, the Jacobian matrix $J\sigma(\boldsymbol{u})$ has rank M.

A curve in $\mathbb{R}^N$ is a function $\sigma : [a, b] \to \mathbb{R}^N$, with

$$\sigma(t) = (\sigma_1(t), \ldots, \sigma_N(t)).$$

The curve is regular if, for every $t \in]a, b[$, the vector $\sigma'(t) = (\sigma_1'(t), \ldots, \sigma_N'(t))$ is different from zero, i.e., $\sigma'(t) \neq (0, \ldots, 0)$. In that case, it is possible to define the "tangent unit vector" at the point $\sigma(t)$,

$$\tau_\sigma(t) = \frac{\sigma'(t)}{\|\sigma'(t)\|}.$$

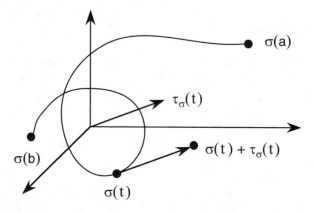

Example The curve $\sigma : [0, 2\pi] \to \mathbb{R}^3$, defined by

$$\sigma(t) = (R\cos(2t), R\sin(2t), 0),$$

has as its image the circle

$$\{(x, y, z) : x^2 + y^2 = R^2, z = 0\}$$

(which is covered twice). Since $\sigma'(t) = (-2R\sin(2t), 2R\cos(2t), 0)$, it is a regular curve, and

$$\tau_\sigma(t) = (-\sin(2t), \cos(2t), 0).$$

A surface in $\mathbb{R}^3$ is a function $\sigma : [a_1, b_1] \times [a_2, b_2] \to \mathbb{R}^3$. The surface is regular if, for every $(u, v) \in]a_1, b_1[\times]a_2, b_2[$, the vectors $\frac{\partial\sigma}{\partial u}(u, v)$, $\frac{\partial\sigma}{\partial v}(u, v)$ are linearly independent. In that case, they determine a plane, called the "tangent plane" to the surface at the point $\sigma(u, v)$, and it is possible to define the "normal unit vector"

$$v_\sigma(u, v) = \frac{\frac{\partial\sigma}{\partial u}(u, v) \times \frac{\partial\sigma}{\partial v}(u, v)}{\left\|\frac{\partial\sigma}{\partial u}(u, v) \times \frac{\partial\sigma}{\partial v}(u, v)\right\|},$$

which is visualized in the following figure.

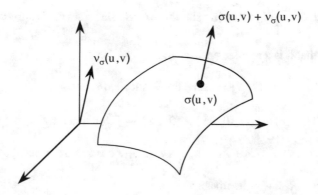

Example 1 The surface $\sigma : [0, \pi] \times [0, \pi] \to \mathbb{R}^3$, defined by

$$\sigma(\phi, \theta) = (R\sin\phi\cos\theta, R\sin\phi\sin\theta, R\cos\phi),$$

has as its image the hemisphere

$$\{(x, y, z) : x^2 + y^2 + z^2 = R^2, y \geq 0\}.$$

Since

$$\frac{\partial \sigma}{\partial \phi}(\phi, \theta) = (R \cos \phi \cos \theta,\, R \cos \phi \sin \theta,\, -R \sin \phi)\,,$$

$$\frac{\partial \sigma}{\partial \theta}(\phi, \theta) = (-R \sin \phi \sin \theta,\, R \sin \phi \cos \theta,\, 0)\,,$$

we compute

$$\frac{\partial \sigma}{\partial \phi}(\phi, \theta) \times \frac{\partial \sigma}{\partial \theta}(\phi, \theta) = (R^2 \sin^2 \phi \cos \theta,\, R^2 \sin^2 \phi \sin \theta,\, R^2 \sin \phi \cos \phi)\,.$$

We thus see that it is a regular surface, and

$$\nu_\sigma(\phi, \theta) = (\sin \phi \cos \theta,\, \sin \phi \sin \theta,\, \cos \phi)\,.$$

Example 2 The surface $\sigma : [0, 2\pi] \times [0, 2\pi] \to \mathbb{R}^3$, defined by

$$\sigma(u, v) = ((R + r \cos u) \cos v,\, (R + r \cos u) \sin v,\, r \sin u)\,,$$

where $0 < r < R$, has as its image a torus

$$\left\{ (x, y, z) : \left(\sqrt{x^2 + y^2} - R \right)^2 + z^2 = r^2 \right\}\,.$$

Even in this case, one can verify that it is a regular surface.

A 3-surface in $\mathbb{R}^3$ is also called a "volume."

Example The function $\sigma : [0, R] \times [0, \pi] \times [0, 2\pi] \to \mathbb{R}^3$, defined by

$$\sigma(\rho, \phi, \theta) = (\rho \sin \phi \cos \theta, \rho \sin \phi \sin \theta, \rho \cos \phi),$$

has as image the closed ball

$$\{(x, y, z) : x^2 + y^2 + z^2 \leq R^2\}.$$

In this case, $\det J\sigma(\rho, \phi, \theta) = \rho^2 \sin \phi$, so that it is a regular volume.

The best way to describe a set $\mathcal{M}$ in $\mathbb{R}^N$ is to find a parametrization. Let us explain precisely what this means.

Definition 10.28 An M-surface $\sigma : I \to \mathbb{R}^N$ is an "M-parametrization" of a set $\mathcal{M}$ if it is regular and injective on $\overset{\circ}{I}$ and $\sigma(I) = \mathcal{M}$. We say that a subset of $\mathbb{R}^N$ is "M-parametrizable" if there is an M-parametrization of it.

Examples The circle $\mathcal{M} = \{(x, y) \in \mathbb{R}^2 : x^2 + y^2 = 1\}$ is 1-parametrizable, and $\sigma : [0, 2\pi] \to \mathbb{R}^2$, given by $\sigma(t) = (\cos t, \sin t)$, is a 1-parametrization of it.

A 2-parametrization of the sphere $\mathcal{M} = \{(x, y, z) \in \mathbb{R}^3 : x^2 + y^2 + z^2 = 1\}$ is, for example, $\sigma : [0, \pi] \times [0, 2\pi] \to \mathbb{R}^3$, defined by

$$\sigma(\phi, \theta) = (\sin \phi \cos \theta, \sin \phi \sin \theta, \cos \phi).$$

10.13 Local Analysis of M-Surfaces

Sometimes geometrical objects are given by an equation like $y = x^2$ (a parabola) or $x^2 + y^2 = 1$ (a circle) or $x^2 + y^2 + z^2 = 1$ (a sphere). We will now show that, under reasonable assumptions, these kinds of objects can be locally described by a curve, a surface, or, in general, an M-surface, which, we recall, is a C^1-function defined on a rectangle, with values in $\mathbb{R}^N$. We now assume $1 \leq M < N$.

We thus have in mind a geometrical object described by an equation like

$$g(\boldsymbol{x}) = 0\,.$$

We will focus our attention at a point $\boldsymbol{x}_0$ and describe locally our object by some C^1-function defined on some rectangle of the type

$$\overline{B}[r] = [-r, r] \times \cdots \times [-r, r]\,.$$

Theorem 10.29 *Let $\mathcal{O} \subseteq \mathbb{R}^N$ be an open set, $\boldsymbol{x}_0$ a point of $\mathcal{O}$, and $g : \mathcal{O} \to \mathbb{R}^{N-M}$ a function of class C^1 such that*

$$g(\boldsymbol{x}_0) = 0\,, \quad \text{and} \quad Jg(\boldsymbol{x}_0) \text{ has rank } N - M\,.$$

Then there exist a neighborhood U of $\boldsymbol{x}_0$ and a regular and injective M-surface $\sigma : \overline{B}[r] \to \mathbb{R}^N$ for some $r > 0$ such that $\sigma(\boldsymbol{0}) = \boldsymbol{x}_0$ and

$$\{\boldsymbol{x} \in U : g(\boldsymbol{x}) = 0\} = \sigma(\overline{B}[r])\,.$$

Proof Assume, for instance, that the matrix

$$\frac{\partial g}{\partial \tilde{\boldsymbol{x}}}(\boldsymbol{x}_0) = \begin{pmatrix} \frac{\partial g_1}{\partial x_{M+1}}(\boldsymbol{x}_0) & \cdots & \frac{\partial g_1}{\partial x_N}(\boldsymbol{x}_0) \\ \vdots & \cdots & \vdots \\ \frac{\partial g_{N-M}}{\partial x_{M+1}}(\boldsymbol{x}_0) & \cdots & \frac{\partial g_{N-M}}{\partial x_N}(\boldsymbol{x}_0) \end{pmatrix}$$

is invertible. (If not, it will be sufficient to shift the columns of the matrix $Jg(\boldsymbol{x}_0)$ to reduce to this case.) Let us write each $\boldsymbol{x} \in \mathcal{O}$ as $(\hat{\boldsymbol{x}}, \tilde{\boldsymbol{x}})$, where $\hat{\boldsymbol{x}} \in \mathbb{R}^M$ and $\tilde{\boldsymbol{x}} \in \mathbb{R}^{N-M}$. Since

$$g(\hat{\boldsymbol{x}}_0, \tilde{\boldsymbol{x}}_0) = 0 \quad \text{and} \quad \det \frac{\partial g}{\partial \tilde{\boldsymbol{x}}}(\hat{\boldsymbol{x}}_0, \tilde{\boldsymbol{x}}_0) \neq 0\,,$$

by the Implicit Function Theorem 10.21, there exist an open neighborhood $\hat{U}$ of $\hat{\boldsymbol{x}}_0$, an open neighborhood $\tilde{U}$ of $\tilde{\boldsymbol{x}}_0$, and a C^1-function $\eta : \hat{U} \to \tilde{U}$ such that $\hat{U} \times \tilde{U} \subseteq \mathcal{O}$, and, taking $\hat{\boldsymbol{x}} \in \hat{U}$ and $\tilde{\boldsymbol{x}} \in \tilde{U}$, we have that

$$g(\hat{\boldsymbol{x}}, \tilde{\boldsymbol{x}}) = 0 \quad \Leftrightarrow \quad \tilde{\boldsymbol{x}} = \eta(\hat{\boldsymbol{x}})\,.$$

Let $r > 0$ be chosen such that $\overline{B}[\hat{\boldsymbol{x}}_0, r] \subseteq \hat{U}$, let $U = \overline{B}[\hat{\boldsymbol{x}}_0, r] \times \tilde{U}$, and let $\sigma : \overline{B}[r] \to \mathbb{R}^N$ be defined as $\sigma(\boldsymbol{u}) = (\boldsymbol{u} + \hat{\boldsymbol{x}}_0, \eta(\boldsymbol{u} + \hat{\boldsymbol{x}}_0))$. Since $J\sigma(\boldsymbol{u})$ has as a submatrix the identity $M \times M$ matrix, surely σ is regular. Moreover, σ is injective since its first component $u \mapsto u + \hat{x}_0$ is. Finally, if $x = (\hat{x}, \tilde{x}) \in U$, then

$$g(\hat{x}, \tilde{x}) = 0 \quad \Leftrightarrow \quad \tilde{x} = \eta(\hat{x}) \quad \Leftrightarrow \quad (\hat{x}, \tilde{x}) = \sigma(\hat{x} - \hat{x}_0),$$

yielding the conclusion.   ∎

The M-surface σ appearing in the statement of the previous theorem is called a "local M-parametrization."

Let us analyze in greater detail three interesting cases. We start by considering a planar curve, i.e., the case $M = 1, N = 2$.

Corollary 10.30 *Let $\mathcal{O} \subseteq \mathbb{R}^2$ be an open set, (x_0, y_0) a point of $\mathcal{O}$, and $g : \mathcal{O} \to \mathbb{R}$ a function of class C^1 such that*

$$g(x_0, y_0) = 0 \quad and \quad \nabla g(x_0, y_0) \neq \mathbf{0}.$$

Then there exist a neighborhood U of (x_0, y_0) and a regular and injective curve $\sigma : [-r, r] \to \mathbb{R}^2$ for some $r > 0$ such that $\sigma(0) = (x_0, y_0)$ and

$$\{(x, y) \in U : g(x, y) = 0\} = \sigma([-r, r]).$$

Let us now examine the case of a surface in $\mathbb{R}^3$, i.e., the case $M = 2, N = 3$.

Corollary 10.31 *Let $\mathcal{O} \subseteq \mathbb{R}^3$ be an open set, (x_0, y_0, z_0) a point of $\mathcal{O}$, and $g : \mathcal{O} \to \mathbb{R}$ a function of class C^1 such that*

$$g(x_0, y_0, z_0) = 0 \quad and \quad \nabla g(x_0, y_0, z_0) \neq \mathbf{0}.$$

Then there exist a neighborhood U of (x_0, y_0, z_0) and a regular and injective surface $\sigma : [-r, r] \times [-r, r] \to \mathbb{R}^3$ for some $r > 0$ such that $\sigma(0, 0) = (x_0, y_0, z_0)$ and

$$\{(x, y, z) \in U : g(x, y, z) = 0\} = \sigma([-r, r] \times [-r, r]).$$

We conclude with the case of a curve in $\mathbb{R}^3$, i.e., the case $M = 1, N = 3$.

Corollary 10.32 *Let $\mathcal{O} \subseteq \mathbb{R}^3$ be an open set, (x_0, y_0, z_0) a point of $\mathcal{O}$, and $g_1, g_2 : \mathcal{O} \to \mathbb{R}$ two functions of class C^1, such that*

$$g_1(x_0, y_0, z_0) = g_2(x_0, y_0, z_0) = 0 \quad and \quad \nabla g_1(x_0, y_0, z_0) \times \nabla g_2(x_0, y_0, z_0) \neq \mathbf{0}.$$

Then there exist a neighborhood U of (x_0, y_0, z_0) and a regular and injective curve $\sigma : [-r, r] \to \mathbb{R}^3$ for some $r > 0$ such that $\sigma(0) = (x_0, y_0, z_0)$ and

$$\{(x, y, z) \in U : g_1(x, y, z) = g_2(x, y, z) = 0\} = \sigma([-r, r]).$$

10.14 Lagrange Multipliers

We are now interested in finding local minimum or local maximum points for f when its domain is constrained to a set defined by some vector valued function g.

Theorem 10.33 (Lagrange Multiplier Theorem) *Let $\mathcal{O} \subseteq \mathbb{R}^N$ be an open set and x_0 a point of $\mathcal{O}$. Let $g : \mathcal{O} \to \mathbb{R}^{N-M}$ be a function of class C^1 such that*

$$g(x_0) = 0 , \quad and \quad Jg(x_0) \ has \ rank \ N - M ,$$

and let $f : \mathcal{O} \to \mathbb{R}$ be differentiable at x_0. Setting

$$S = \{x \in \mathcal{O} : g(x) = 0\} ,$$

if x_0 is either a local minimum or a local maximum point for $f|_S$ (the restriction of f to S), then there exist $(N - M)$ real numbers $\lambda_1, \ldots, \lambda_{N-M}$ such that

$$\nabla f(x_0) = \sum_{j=1}^{N-M} \lambda_j \nabla g_j(x_0) .$$

The numbers $\lambda_1, \ldots, \lambda_{N-M}$ are called "Lagrange multipliers."

Proof By Theorem 10.29, there exist a neighborhood U of x_0 and a regular and injective M-surface $\sigma : \overline{B}[r] \to \mathbb{R}^N$ for some $r > 0$ such that $\sigma(0) = x_0$ and

$$S \cap U = \sigma(\overline{B}[r]) .$$

Consider the function $F : \overline{B}[r] \to \mathbb{R}$ defined as $F(u) = f(\sigma(u))$. Then 0 is either a local minimum or a local maximum point for F, hence $\nabla F(0) = 0$, i.e.,

$$0 = JF(0) = Jf(x_0) J\sigma(0) = \nabla f(x_0)^T J\sigma(0) .$$

As a consequence,

$$\nabla f(x_0) \cdot \frac{\partial \sigma}{\partial u_1}(0) = 0 , \quad \ldots , \nabla f(x_0) \cdot \frac{\partial \sigma}{\partial u_M}(0) = 0 ,$$

i.e.,

$$\nabla f(x_0) \ is \ orthogonal \ to \ \frac{\partial \sigma}{\partial u_1}(0) , \ \ldots , \ \frac{\partial \sigma}{\partial u_M}(0) .$$

Moreover, since $g(\sigma(\boldsymbol{u})) = 0$ for every $\boldsymbol{u} \in \overline{B}[r]$, we have that

$$Jg(\boldsymbol{x}_0)J\sigma(\boldsymbol{0}) = 0\,,$$

hence, the vectors

$$\nabla g_1(\boldsymbol{x}_0)\,, \ \dots\,, \ \nabla g_{N-M}(\boldsymbol{x}_0) \ \textit{are all orthogonal to} \ \frac{\partial\sigma}{\partial u_1}(\boldsymbol{0})\,, \ \dots\,, \ \frac{\partial\sigma}{\partial u_M}(\boldsymbol{0}).$$

By assumption, $J\sigma(\boldsymbol{0})$ has rank M, i.e.,

$$\textit{the real vector space } \mathcal{T} \textit{ generated by}$$

$$\frac{\partial\sigma}{\partial u_1}(\boldsymbol{0})\,, \ \dots\,, \ \frac{\partial\sigma}{\partial u_M}(\boldsymbol{0}) \textit{ has dimension } M.$$

Therefore, the orthogonal space $\mathcal{T}^{\perp}$ has dimension $N - M$. The vectors $\nabla g_1(\boldsymbol{x}_0)$, $\dots$, $\nabla g_{N-M}(\boldsymbol{x}_0)$ are linearly independent and, as we saw earlier, they belong to $\mathcal{T}^{\perp}$, so these vectors form a basis for $\mathcal{T}^{\perp}$. Since $\nabla f(\boldsymbol{x}_0)$ also belongs to $\mathcal{T}^{\perp}$, it must be a linear combination of the vectors of the basis.   ∎

As in the previous section, we analyze in detail three interesting cases. We start by considering the case $M = 1$, $N = 2$.

Corollary 10.34 *Let $\mathcal{O} \subseteq \mathbb{R}^2$ be an open set and (x_0, y_0) a point of $\mathcal{O}$. Let $g : \mathcal{O} \to \mathbb{R}$ be a function of class C^1 such that*

$$g(x_0, y_0) = 0\,, \quad \textit{and} \quad \nabla g(x_0, y_0) \neq \boldsymbol{0}\,,$$

and let $f : \mathcal{O} \to \mathbb{R}$ be differentiable at (x_0, y_0). Setting

$$S = \{(x, y) \in \mathcal{O} : g(x, y) = 0\}\,,$$

if (x_0, y_0) is either a local minimum or a local maximum point for $f|_S$, then there exists a real number λ such that

$$\nabla f(x_0, y_0) = \lambda \nabla g(x_0, y_0)\,.$$

Example Among all rectangles in the plane with a given perimeter p, we want to find those that maximize the area. Let us denote by x and y the lengths of the sides of a rectangle and define the area function

$$f(x, y) = xy\,.$$

We are looking for the maximum points of the function f over the set

$$K = \{(x, y) \in \mathbb{R}^2 : x \geq 0, \ y \geq 0, \ 2x + 2y = p\}.$$

This set is compact, so that f, being continuous, surely has a maximum point in K. Taking $(x, y) \in K$, note that $f(x, y) = 0$ only when $x = 0$ or $y = 0$; otherwise, $f(x, y) > 0$. Define now the function

$$g(x, y) = 2x + 2y - p.$$

Then

$$\nabla f(x, y) = \lambda \nabla g(x, y) \quad \Leftrightarrow \quad (y, x) = \lambda(2, 2).$$

By the preceding considerations and Corollary 10.34, the maximum point (x_0, y_0) must be such that $x_0 = y_0$; hence, the rectangle must be a square.

Now we move to the case $M = 2$, $N = 3$.

Corollary 10.35 *Let $\mathcal{O} \subseteq \mathbb{R}^3$ be an open set and (x_0, y_0, z_0) a point of $\mathcal{O}$. Let $g : \mathcal{O} \to \mathbb{R}$ be a function of class C^1 such that*

$$g(x_0, y_0, z_0) = 0, \quad \text{and} \quad \nabla g(x_0, y_0, z_0) \neq \mathbf{0},$$

and let $f : \mathcal{O} \to \mathbb{R}$ be differentiable at (x_0, y_0, z_0). Setting

$$S = \{(x, y, z) \in \mathcal{O} : g(x, y, z) = 0\},$$

if (x_0, y_0, z_0) is either a local minimum or a local maximum point for $f|_S$, then there exists a real number λ such that

$$\nabla f(x_0, y_0, z_0) = \lambda \nabla g(x_0, y_0, z_0).$$

Example Among all cuboids with a given area a, we want to find those that maximize the volume. Let us denote by x, y, and z the lengths of the sides of a cuboid, and define the volume function

$$f(x, y, z) = xyz.$$

We are looking for the maximum points of the function f over the set

$$K = \{(x, y, z) \in \mathbb{R}^2 : x \geq 0, \ y \geq 0, z \geq 0, \ 2xy + 2xz + 2yz = a\}.$$

Taking $(x, y, z) \in K$, note that $f(x, y, z) = 0$ only when $x = 0$ or $y = 0$ or $z = 0$; otherwise, $f(x, y, z) > 0$. Everything then seems as in the previous example, but there is a difficulty now. The set K is unbounded, hence not compact, and the argument of the previous example needs to be modified. First of all we note that

$$\left(\sqrt{\frac{a}{6}}, \sqrt{\frac{a}{6}}, \sqrt{\frac{a}{6}}\right) \in K, \quad \text{and} \quad f\left(\sqrt{\frac{a}{6}}, \sqrt{\frac{a}{6}}, \sqrt{\frac{a}{6}}\right) = \left(\frac{a}{6}\right)^{3/2}.$$

Hence, if (x_0, y_0, z_0) is a maximum point of f on K, it must be that

$$f(x_0, y_0, z_0) \geq \left(\frac{a}{6}\right)^{3/2}.$$

We now prove that it must be that

$$0 \leq x_0 \leq 3\sqrt{\frac{3a}{2}}, \quad 0 \leq y_0 \leq 3\sqrt{\frac{3a}{2}}, \quad 0 \leq z_0 \leq 3\sqrt{\frac{3a}{2}}.$$

Let us prove the first one, the others being analogous. By contradiction, if $x_0 > 3\sqrt{\frac{3a}{2}}$, then, since $2x_0 y_0 \leq a$ and $2x_0 z_0 \leq a$, it must be that

$$x_0 y_0 z_0 \leq \frac{a}{2} z_0 \leq \frac{a^2}{4x_0} < \frac{a^2}{4} \frac{1}{3} \sqrt{\frac{2}{3a}} = \left(\frac{a}{6}\right)^{3/2},$$

in contrast to $f(x_0, y_0, z_0) \geq \left(\frac{a}{6}\right)^{3/2}$.

We can then restrict the search of a maximum point of f on the set

$$\widetilde{K} = \left\{(x, y, z) \in \mathbb{R}^2 : 0 \leq x \leq 3\sqrt{\frac{3a}{2}}, 0 \leq y \leq 3\sqrt{\frac{3a}{2}}, 0 \leq z \leq 3\sqrt{\frac{3a}{2}}, \right.$$
$$\left. 2xy + 2xz + 2yz = a\right\},$$

which is compact; hence, the point of maximum exists. Now define the function

$$g(x, y, z) = 2xy + 2xz + 2yz - a.$$

Then

$$\nabla f(x, y, z) = \lambda \nabla g(x, y, z) \quad \Leftrightarrow \quad (yz, xz, xy) = \lambda(2y + 2z, 2x + 2z, 2x + 2y).$$

If (x, y, z) solves the preceding equation with $x > 0$, $y > 0$, and $z > 0$, then

$$x(y + z) = y(x + z) = z(x + y),$$

and hence $x = y = z$. By the foregoing considerations and Corollary 10.35, the maximum point (x_0, y_0, z_0) must be such that $x_0 = y_0 = z_0$, so the cuboid must be a cube.

Let us conclude with the case $M = 1$, $N = 3$.

Corollary 10.36 *Let $\mathcal{O} \subseteq \mathbb{R}^3$ be an open set and (x_0, y_0, z_0) a point of $\mathcal{O}$. Let $g_1, g_2 : \mathcal{O} \to \mathbb{R}$ be two functions of class $\mathcal{C}^1$ such that*

$$g_1(x_0, y_0, z_0) = g_2(x_0, y_0, z_0) = 0 \ \text{ and } \ \nabla g_1(x_0, y_0, z_0) \times \nabla g_2(x_0, y_0, z_0) \neq \mathbf{0},$$

and let $f : \mathcal{O} \to \mathbb{R}$ be differentiable at (x_0, y_0, z_0). Setting

$$S = \{(x, y, z) \in U : g_1(x, y, z) = 0, \ g_2(x, y, z) = 0\},$$

if (x_0, y_0, z_0) is either a local minimum or a local maximum point for $f|_S$, then there exist two real numbers λ_1, λ_2 such that

$$\nabla f(x_0, y_0, z_0) = \lambda_1 \nabla g_1(x_0, y_0, z_0) + \lambda_2 \nabla g_2(x_0, y_0, z_0).$$

Example We want to find the minimum and maximum points of the function $f(x, y, z) = z$ on the set

$$S = \{(x, y, z) \in \mathbb{R}^3 : x^2 + y^2 + z^2 = 1, \ x + y + z = 1\}.$$

Note that S is compact, so the minimum and maximum of f on S exist. Define $g_1(x, y, z) = x^2 + y^2 + z^2 - 1$ and $g_2(x, y, z) = x + y + z - 1$. Then

$$\nabla g_1(x, y, z) = (2x, 2y, 2z), \quad \nabla g_2(x, y, z) = (1, 1, 1),$$

hence

$$\nabla g_1(x, y, z) \times \nabla g_2(x, y, z) = (2(y - z), 2(z - x), 2(x - y)).$$

Note that

$$\nabla g_1(x, y, z) \times \nabla g_2(x, y, z) = \mathbf{0} \quad \Leftrightarrow \quad x = y = z,$$

which implies that $(x, y, z) \notin S$. Now, a simple computation shows that if $(x, y, z) \in S$, then we have that

$$\nabla f(x, y, z) = \lambda_1 \nabla g_1(x, y, z) + \lambda_2 \nabla g_2(x, y, z)$$

if and only if either

$$(x, y, z) = (0, 0, 1), \quad \lambda_1 = \frac{1}{2}, \quad \lambda = 0$$

or

$$(x, y, z) = \left(\frac{2}{3}, \frac{2}{3}, -\frac{1}{3}\right), \quad \lambda_1 = -\frac{1}{2}, \quad \lambda_2 = \frac{2}{3}.$$

Since $f(0, 0, 1) = 1$ and $f\left(\frac{2}{3}, \frac{2}{3}, -\frac{1}{3}\right) = -\frac{1}{3}$, by the preceding considerations and Corollary 10.36, we conclude that $(0, 0, 1)$ is a maximum point and $\left(\frac{2}{3}, \frac{2}{3}, -\frac{1}{3}\right)$ is a minimum point of f on S.

10.15 Differentiable Manifolds

There is an alternative way of looking at some geometrical objects such as "curves" and "surfaces." The intuitive idea is that they locally "look the same" as a straight line or a plane. In other words, when observing these objects from a very small distance, they look "almost flat."

We will now make this idea precise, in a general finite-dimensional context. Thus, let $\mathcal{M}$ be a subset of $\mathbb{R}^N$.

Definition 10.37 The set $\mathcal{M}$ is an "M-dimensional differentiable manifold," with $1 \le M \le N$ (or a "M-manifold" for short) if, taking a point $\boldsymbol{x}$ in $\mathcal{M}$, there are an open neighborhood A of $\boldsymbol{x}$, an open neighborhood B of $\mathbf{0}$ in $\mathbb{R}^N$, and a diffeomorphism $\varphi : A \to B$ such that $\varphi(\boldsymbol{x}) = \mathbf{0}$ and either

(a) $\varphi(A \cap \mathcal{M}) = \{y = (y_1, \ldots, y_N) \in B : y_{M+1} = \cdots = y_N = 0\}$
   or
(b) $\varphi(A \cap \mathcal{M}) = \{y = (y_1, \ldots, y_N) \in B : y_{M+1} = \cdots = y_N = 0 \text{ and } y_M \ge 0\}$.

It can be seen that (a) and (b) cannot hold at the same time. The points $\boldsymbol{x}$ for which (b) is verified make up the "boundary" of $\mathcal{M}$, which we denote by $\partial \mathcal{M}$. If $\partial \mathcal{M}$ is empty, we are speaking of an M-manifold without boundary; otherwise, $\mathcal{M}$ is sometimes said to be an M-manifold with boundary.

First, note that the boundary of a differentiable manifold is itself a differentiable manifold, with a lower dimension.

Theorem 10.38 *The set $\partial \mathcal{M}$ is a $(M - 1)$-manifold without boundary, i.e.,*

$$\partial(\partial \mathcal{M}) = \emptyset.$$

Proof Taking a point $\boldsymbol{x}$ in $\partial \mathcal{M}$, there are an open neighborhood A of $\boldsymbol{x}$, an open neighborhood B of $\mathbf{0}$ in $\mathbb{R}^N$, and a diffeomorphism $\varphi : A \to B$ such that $\varphi(\boldsymbol{x}) = \mathbf{0}$ and

$$\varphi(A \cap \mathcal{M}) = \{y = (y_1, \ldots, y_N) \in B : y_{M+1} = \cdots = y_N = 0 \text{ and } y_M \ge 0\}.$$

Based on the fact that the conditions (a) and (b) of the definition cannot hold simultaneously for any point of $\mathcal{M}$, it is possible to prove that

$$\varphi(A \cap \partial\mathcal{M}) = \{y = (y_1, \ldots, y_N) \in B : y_M = y_{M+1} = \cdots = y_N = 0\}.$$

This completes the proof.   ∎

There are many examples of manifolds: Circles, spheres, and toruses are manifolds without boundary. A hemisphere is a 2-manifold whose boundary is a circle. However, a cone is not a manifold because of a single point, its vertex. Notice that any open set in $\mathbb{R}^N$ is an N-manifold (without boundary).

Let us now see that, given an M-manifold $\mathcal{M}$, corresponding to each of its points $\boldsymbol{x}$ it is possible to find a local M-parametrization.

Theorem 10.39 *For every* $\boldsymbol{x} \in \mathcal{M}$ *there is a neighborhood* A' *of* $\boldsymbol{x}$ *such that* $A' \cap \mathcal{M}$ *can be M-parametrized with an injective function* $\sigma : I \to \mathbb{R}^N$, *where I is a rectangle of $\mathbb{R}^M$ of the type*

$$I = \begin{cases} [-\alpha, \alpha]^M & \text{if } \boldsymbol{x} \notin \partial\mathcal{M}, \\ [-\alpha, \alpha]^{M-1} \times [0, \alpha] & \text{if } \boldsymbol{x} \in \partial\mathcal{M}, \end{cases}$$

and $\sigma(0) = \boldsymbol{x}$. *Moreover, if* $\boldsymbol{x}$ *is a point of the boundary* $\partial\mathcal{M}$, *the M-parametrization* σ *is such that the interior points of a single face of rectangle I are sent on* $\partial\mathcal{M}$.

Proof Consider the diffeomorphism $\varphi : A \to B$ given by the preceding definition, and take an $\alpha > 0$ such that the rectangle $B' = [-\alpha, \alpha]^N$ is contained in B. Setting $A' = \varphi^{-1}(B')$, we have that A' is a neighborhood of $\boldsymbol{x}$ (indeed, the set $B'' =]-\alpha, \alpha[^N$ is open and, hence, also $A'' = \varphi^{-1}(B'')$ is open, and $\boldsymbol{x} \in A'' \subseteq A'$). We can then take rectangle I as in the statement and define $\sigma(\boldsymbol{u}) = \varphi^{-1}(\boldsymbol{u}, 0)$. It is readily seen that σ is injective and $\sigma(I) = A' \cap \mathcal{M}$. Moreover, $\varphi_{(1,\ldots,M)}(\sigma(\boldsymbol{u})) = \boldsymbol{u}$ for every $\boldsymbol{u} \in I$; hence, $J\varphi_{(1,\ldots,M)}(\sigma(\boldsymbol{u})) \cdot J\sigma(\boldsymbol{u})$ is the identity matrix, so that $J\sigma(\boldsymbol{u})$ has rank M for every $\boldsymbol{u} \in I$.

Finally, if $\boldsymbol{x} \in \partial\mathcal{M}$, then

$$[-\alpha, \alpha]^{M-1} \times \{0\} = \sigma^{-1}(A' \cap \partial\mathcal{M}),$$

thereby completing the proof.   ∎

Notice that the function σ is indeed defined on an open set containing I, and it is injective there.

The Integral

11

In this chapter we extend the theory of the integral to functions of several variables defined on subsets of $\mathbb{R}^N$, with values in $\mathbb{R}$. For simplicity, in the exposition we will first focus our attention on the case $N = 2$ and later provide all the results in the case of a generic dimension N.

11.1   Integrability on Rectangles

We begin by considering the case of functions defined on rectangles. We recall that a "rectangle" of $\mathbb{R}^N$ is a set of the type $[a_1, b_1] \times \cdots \times [a_N, b_N]$. In the following exposition, we concentrate for simplicity on the two-dimensional case. The general case is largely identical and does not involve greater difficulties, except for the notations.

We consider the rectangle $I = [a_1, b_1] \times [a_2, b_2] \subseteq \mathbb{R}^2$ and define its measure

$$\mu(I) = (b_1 - a_1)(b_2 - a_2).$$

As a particular case, given $\boldsymbol{x} = (x, y) \in I$ and $r > 0$, we have

$$\overline{B}[\boldsymbol{x}, r] = [x - r, x + r] \times [y - r, y + r];$$

it is the square centered at $\boldsymbol{x}$ having r as half of the length of its sides. We say that two rectangles are "nonoverlapping" if their interiors are disjoint.

A "tagged partition" of the rectangle I is a set

$$\mathring{\mathcal{P}} = \{(\boldsymbol{x}_1, I_1), (\boldsymbol{x}_2, I_2), \dots, (\boldsymbol{x}_m, I_m)\},$$

where the I_j are nonoverlapping rectangles whose union is I and, for every $j = 1, \dots, m$, the point $\boldsymbol{x}_j = (x_j, y_j)$ belongs to I_j.

© The Author(s), under exclusive license to Springer Nature Switzerland AG 2023   301
A. Fonda, *A Modern Introduction to Mathematical Analysis*,
https://doi.org/10.1007/978-3-031-23713-3_11

Example If $I = [0, 10] \times [0, 6]$, a possible tagged partition is the following:

$$\mathring{\mathcal{P}} = \{((1, 1), [0, 7] \times [0, 2]), ((0, 5), [0, 3] \times [2, 6]),$$
$$((5, 4), [3, 10] \times [4, 6]), ((10, 0), [7, 10] \times [0, 4]), ((5, 3), [3, 7] \times [2, 4])\}.$$

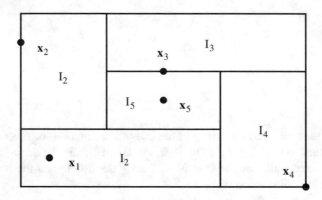

Let us now consider a function f defined on the rectangle I, with values in $\mathbb{R}$, and let $\mathring{\mathcal{P}} = \{(\boldsymbol{x}_j, I_j) : j = 1, \ldots, m\}$ be a tagged partition of I. We call "Riemann sum" associated with f and $\mathring{\mathcal{P}}$ the real number $S(f, \mathring{\mathcal{P}})$ defined by

$$S(f, \mathring{\mathcal{P}}) = \sum_{j=1}^{m} f(\boldsymbol{x}_j)\mu(I_j).$$

Whenever f happens to be positive, this number is the sum of the volumes of the parallelepipeds having as base I_j and height $[0, f(\boldsymbol{x}_j)]$.

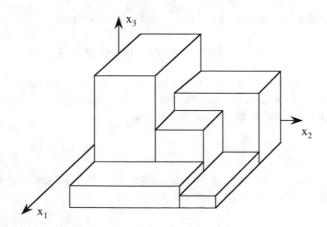

We call a "gauge" on I every *positive* function $\delta : I \to \mathbb{R}$. Given a gauge δ on I, we say that the tagged partition $\mathring{\mathcal{P}}$ introduced previously is "δ-fine" if, for every $j = 1, \dots, m$,

$$I_j \subseteq \overline{B}[\boldsymbol{x}_j, \delta(\boldsymbol{x}_j)].$$

Example Let $I = [0, 1] \times [0, 1]$ and δ be the gauge defined as follows:

$$\delta(x, y) = \begin{cases} \dfrac{x + y}{3} & \text{if } (x, y) \neq (0, 0), \\ \dfrac{1}{2} & \text{if } (x, y) = (0, 0). \end{cases}$$

We want to find a δ-fine tagged partition of I. Much like what we saw at the end of Sect. 7.2, in this case one of the points $\boldsymbol{x}_j$ necessarily must coincide with $(0, 0)$. We can then choose, for example,

$$\mathring{\mathcal{P}} = \left\{ \left((0, 0), \left[0, \tfrac{1}{2}\right] \times \left[0, \tfrac{1}{2}\right]\right), \left(\left(\tfrac{1}{2}, 1\right), \left[0, \tfrac{1}{2}\right] \times \left[\tfrac{1}{2}, 1\right]\right), \right.$$
$$\left. \left(\left(1, \tfrac{1}{2}\right), \left[\tfrac{1}{2}, 1\right] \times \left[0, \tfrac{1}{2}\right]\right), \left((1, 1), \left[\tfrac{1}{2}, 1\right] \times \left[\tfrac{1}{2}, 1\right]\right) \right\}.$$

It is interesting to observe that it is not always possible to construct δ-fine tagged partitions by only joining points on the edges of I. The reader may become convinced when attempting to do this using the following gauge:

$$\delta(x, y) = \begin{cases} \dfrac{x + y}{16} & \text{if } (x, y) \neq (0, 0), \\ \dfrac{1}{2} & \text{if } (x, y) = (0, 0). \end{cases}$$

As in the one-dimensional case, one can prove that for every gauge δ on I there exists a δ-fine tagged partition of I (see Cousin's Theorem 7.1). The following definition is identical to the one given in Chap. 7.

Definition 11.1 A function $f : I \to \mathbb{R}$ is said to be "integrable" (on the rectangle I) if there is a real number $\mathcal{J}$ with the following property. Given $\varepsilon > 0$, it is possible to find a gauge δ on I such that, for every δ-fine tagged partition $\mathring{\mathcal{P}}$ of I,

$$|S(f, \mathring{\mathcal{P}}) - \mathcal{J}| \leq \varepsilon.$$

We briefly overview all the properties that can be obtained from the given definition in the same way as was done in the case of a function of a single variable.

First of all, there is at most one $\mathcal{J} \in \mathbb{R}$ that verifies the conditions of the definition. Such a number is called the "integral" of f on I and is denoted by one of the following symbols:

$$\int_I f\,, \qquad \int_I f(\boldsymbol{x})\,d\boldsymbol{x}\,, \qquad \int_I f(x, y)\,dx\,dy\,.$$

The set of integrable functions is a real vector space, and the integral is a linear function on it:

$$\int_I (f + g) = \int_I f + \int_I g\,, \qquad \int_I (\alpha f) = \alpha \int_I f$$

(with $\alpha \in \mathbb{R}$). It preserves the order

$$f \le g \quad \Rightarrow \quad \int_I f \le \int_I g\,.$$

The Cauchy criterion of integrability holds.

Theorem 11.2 (Cauchy Criterion) *A function $f : I \to \mathbb{R}$ is integrable if and only if for every $\varepsilon > 0$ there is a gauge $\delta : I \to \mathbb{R}$ such that, taking two δ-fine tagged partitions $\mathring{\mathcal{P}}, \mathring{\mathcal{Q}}$ of I, we have*

$$|S(f, \mathring{\mathcal{P}}) - S(f, \mathring{\mathcal{Q}})| \le \varepsilon\,.$$

Moreover, we have the following property of "additivity on subrectangles."

Theorem 11.3 *Let $f : I \to \mathbb{R}$ be a function and $K_1, K_2, \ldots, K_l$ be nonoverlapping subrectangles of I whose union is I. Then f is integrable on I if and only if it is integrable on each of the K_i. In that case, we have*

$$\int_I f = \sum_{i=1}^{l} \int_{K_i} f\,.$$

In particular, if a function is integrable on a rectangle, it is still on every subrectangle. The proof of the theorem is similar to that of Theorem 7.18 and is based on the possibility of constructing a gauge that would allow us to split the Riemann sums on I into Riemann sums on the single subrectangles.

We say that an integrable function on I is "R-integrable" there (or integrable according to Riemann) if, among all possible gauges δ which verify the definition of integrability, it is always possible to choose one that is constant on I. The set of R-integrable functions is a vector subspace of the space of integrable functions and contains the subspace of continuous functions.

We say that an integrable function $f : I \to \mathbb{R}$ is "L-integrable" (or integrable according to Lebesgue) if $|f|$ is integrable on I as well. The L-integrable functions make up a vector subspace of the space of integrable functions. If f and g are two L-integrable functions on I, then the functions $\min\{f, g\}$ and $\max\{f, g\}$ are L-integrable on I, too. A function f is L-integrable if and only if both its positive part $f^+ = \max\{f, 0\}$ and its negative part $f^- = \max\{-f, 0\}$ are integrable.

The Saks–Henstock Theorem 9.1, Monotone Convergence Theorem 9.10, and Dominated Convergence Theorem 9.13 extend to the integrable functions on a rectangle, with statements and proofs perfectly analogous to those provided in Chap. 9.

11.2 Integrability on a Bounded Set

We will now provide the definition of the integral on an arbitrary bounded domain. Given a bounded set E and a function $f : E \to \mathbb{R}$, we define the function $f_E : \mathbb{R}^N \to \mathbb{R}$ as follows:

$$f_E(\boldsymbol{x}) = \begin{cases} f(\boldsymbol{x}) & \text{if } \boldsymbol{x} \in E, \\ 0 & \text{if } \boldsymbol{x} \notin E. \end{cases}$$

We are thus led to the following definition.

Definition 11.4 Given a bounded set E, we say that the function $f : E \to \mathbb{R}$ is "integrable" (on E) if there is a rectangle I containing the set E on which f_E is integrable. In that case, we set

$$\int_E f = \int_I f_E .$$

To verify the consistency of the preceding definition, we will now show that when f is integrable on E, we have that f_E is integrable on *any* rectangle containing the set E, and the integral of f_E remains the same on each such rectangle.

Proposition 11.5 *Let I and J be two rectangles containing the set E. Then f_E is integrable on I if and only if it is integrable on J. In that case, we have $\int_I f_E = \int_J f_E$.*

Proof We consider for simplicity the case $N = 2$. Assume that f_E is integrable on I. Let K be a rectangle containing both I and J. We can construct some nonoverlapping rectangles $K_1, \ldots, K_r$, also nonoverlapping with I such that $I \cup K_1 \cup \cdots \cup K_r = K$. We now prove that f_E is integrable on each of the subrectangles $K_1, \ldots, K_r$ and that the integrals $\int_{K_1} f_E, \ldots, \int_{K_r} f_E$ are all equal to zero. Notice that f_E, restricted to each of these subrectangles, is zero everywhere except perhaps on one of their edges. We are thus led to prove the following lemma, which will permit us to conclude the proof. ∎

Lemma 11.6 *Let $\mathcal{K}$ be a rectangle and $g : \mathcal{K} \to \mathbb{R}$ be a function which is zero everywhere except perhaps on one edge of $\mathcal{K}$. Then g is integrable on $\mathcal{K}$ and $\int_{\mathcal{K}} g = 0$.*

Proof We first assume that the function g is bounded on $\mathcal{K}$, i.e., that there is a constant $C > 0$ for which

$$|g(x, y)| \leq C$$

for every $(x, y) \in \mathcal{K}$. Fix $\varepsilon > 0$. Let L be the edge of rectangle $\mathcal{K}$ on which g may be nonzero, and denote by ℓ its length. Define the constant gauge $\delta = \frac{\varepsilon}{C\ell}$. Then, for every δ-fine tagged partition $\overset{\circ}{\mathcal{P}} = \{(x_1, I_1), \ldots, (x_m, I_m)\}$ of $\mathcal{K}$, we have

$$|S(g, \overset{\circ}{\mathcal{P}})| \leq \sum_{j=1}^{m} |g(x_j)| \mu(I_j) = \sum_{\{j \,:\, x_j \in L\}} |g(x_j)| \mu(I_j)$$

$$\leq \sum_{\{j \,:\, x_j \in L\}} C \mu(I_j) \leq C \delta \ell = \varepsilon.$$

This proves that g is integrable on $\mathcal{K}$ and $\int_{\mathcal{K}} g = 0$ in the case where g is bounded on $\mathcal{K}$. If g is not bounded on $\mathcal{K}$, assume that it has nonnegative values. Define the following sequence $(g_n)_n$ of functions as

$$g_n(x) = \min\{g(x), n\}.$$

Since the functions g_n are bounded, for what we saw earlier we have $\int_{\mathcal{K}} g_n = 0$, for every n. It is easily seen that the sequence thus defined satisfies the conditions of the Monotone Convergence Theorem 9.10 and converges pointwise to g. It then follows that g is integrable on $\mathcal{K}$ and

$$\int_{\mathcal{K}} g = \lim_n \int_{\mathcal{K}} g_n = 0.$$

If g does not have only nonnegative values, it is always possible to consider g^+ and g^-. From the preceding discussion, $\int_{\mathcal{K}} g^+ = \int_{\mathcal{K}} g^- = 0$, and then $\int_{\mathcal{K}} g = \int_{\mathcal{K}} g^+ - \int_{\mathcal{K}} g^- = 0$, which is what we wanted to prove. ∎

End of Proof of Proposition 11.5 Having proved that f_E is integrable on each of the $K_1, \ldots, K_r$ and that the integrals $\int_{K_1} f_E, \ldots, \int_{K_r} f_E$ are equal to zero, by the theorem of additivity on subrectangles we have that, since f_E is integrable on I, it is also integrable on K, and

$$\int_K f_E = \int_I f_E + \int_{K_1} f_E + \cdots + \int_{K_r} f_E = \int_I f_E.$$

But then f_E is integrable on every subinterval of K, and in particular on J. We can now construct, analogously to what was just done for I, some nonoverlapping rectangles $J_1, \ldots, J_s$, also nonoverlapping with J, such that $J \cup J_1 \cup \cdots \cup J_s = K$. Similarly, we will have

$$\int_K f_E = \int_J f_E + \int_{J_1} f_E + \cdots + \int_{J_s} f_E = \int_J f_E \,,$$

which proves that $\int_I f_E = \int_J f_E$. To see that the condition is necessary and sufficient, simply exchange the roles of I and J in the foregoing proof.

With the given definition, all the properties of the integral seen earlier easily extend to this setting. *There is an exception concerning the additivity* since it is not true in general that a function that is integrable on a bounded set remains integrable on any of its subsets. Indeed, take a function $f : E \to \mathbb{R}$, which is integrable but not L-integrable. We consider the subset

$$E' = \{x \in E : f(x) \geq 0\} \,,$$

and we claim that f cannot be integrable on E'. If it were, then f^+ would be integrable on E. But then $f^- = f^+ - f$ would also be integrable on E, and therefore f should be L-integrable on E, in contradiction with the assumption.

We will see that, with respect to additivity, the L-integrable functions have a somewhat better behavior.

11.3   The Measure

We now give the definition of "measure" for a bounded subset of $\mathbb{R}^N$.

Definition 11.7 A bounded set E is said to be "measurable" if the constant function 1 is integrable on E. In that case, the number $\int_E 1$ is said to be the "measure" of E and is denoted by $\mu(E)$.

The measure of a measurable set is thus a nonnegative number. The empty set is assumed to be measurable, and its measure is equal to 0. In the case of a subset of $\mathbb{R}^2$, its measure is also called the "area" of the set. If $E = [a_1, b_1] \times [a_2, b_2]$ is a rectangle, it is easily seen that

$$\mu(E) = \int_E 1 = (b_1 - a_1)(b_2 - a_2) \,,$$

so that the notation is in accordance with the one already introduced for rectangles. For a subset of $\mathbb{R}^3$, the measure is also called the "volume" of the set.

Let us analyze some properties of the measure. It is useful to introduce the characteristic function of a set E, defined by

$$\chi_E(x) = \begin{cases} 1 & \text{if } x \in E, \\ 0 & \text{if } x \notin E. \end{cases}$$

If I is a rectangle containing the set E, we thus have

$$\mu(E) = \int_I \chi_E.$$

Proposition 11.8 *Let A and B be two measurable bounded sets. The following properties hold:*

(a) If $A \subseteq B$, then $B \setminus A$ is measurable, and

$$\mu(B \setminus A) = \mu(B) - \mu(A);$$

in particular, $\mu(A) \leq \mu(B)$.
(b) $A \cup B$ and $A \cap B$ are measurable, and

$$\mu(A \cup B) + \mu(A \cap B) = \mu(A) + \mu(B);$$

in particular, if A and B are disjoint, then $\mu(A \cup B) = \mu(A) + \mu(B)$.

Proof Let I be a rectangle containing $A \cup B$. If $A \subseteq B$, then $\chi_{B \setminus A} = \chi_B - \chi_A$, and property (a) follows integrating on I.

Since $\chi_{A \cup B} = \max\{\chi_A, \chi_B\}$ and $\chi_{A \cap B} = \min\{\chi_A, \chi_B\}$, we have that $\chi_{A \cup B}$ and $\chi_{A \cap B}$ are integrable on I. Moreover,

$$\chi_{A \cup B} + \chi_{A \cap B} = \chi_A + \chi_B,$$

and integrating on I, we have (b).   ∎

The following proposition states the **Complete Additivity** property of the measure.

Proposition 11.9 *If $(A_k)_{k \geq 1}$ is a sequence of measurable bounded sets whose union $A = \cup_{k \geq 1} A_k$ is bounded, then A is measurable, and*

$$\mu(A) \leq \sum_{k=1}^{\infty} \mu(A_k).$$

If the sets A_k are pairwise disjoint, then equality holds.

Proof Assume first that the sets A_k are pairwise disjoint. Let I be a rectangle containing their union A. Then, for every $x \in I$,

$$\chi_A(x) = \sum_{k=1}^{\infty} \chi_{A_k}(x).$$

Moreover, since for every positive integer q we have

$$\sum_{k=1}^{q} \mu(A_k) = \mu\left(\bigcup_{k=1}^{q} A_k\right) \le \mu(I),$$

the series $\sum_{k=1}^{\infty} \int_I \chi_{A_k} = \sum_{k=1}^{\infty} \mu(A_k)$ converges. By Corollary 9.11, we have that A is measurable and

$$\mu(A) = \int_I \chi_A = \int_I \sum_{k=1}^{\infty} \chi_{A_k} = \sum_{k=1}^{\infty} \int_I \chi_{A_k} = \sum_{k=1}^{\infty} \mu(A_k).$$

When the sets A_k are not pairwise disjoint, consider the sets $B_1 = A_1$, $B_2 = A_2 \setminus A_1$ and, in general, $B_k = A_k \setminus (A_1 \cup \cdots \cup A_{k-1})$. The sets B_k are measurable and pairwise disjoint, and $\cup_{k \ge 1} B_k = \cup_{k \ge 1} A_k$. The conclusion then follows from what was proved earlier. ∎

We have a similar proposition concerning the intersection of a countable family of sets.

Proposition 11.10 *If $(A_k)_{k \ge 1}$ is a sequence of measurable bounded sets, their intersection $A = \cap_{k \ge 1} A_k$ is a measurable set.*

Proof Let I be a rectangle containing the set A. Then, since $A = \cap_{k \ge 1}(A_k \cap I)$, we have that

$$\bigcap_{k \ge 1} A_k = I \setminus \left(\bigcup_{k \ge 1} (I \setminus (A_k \cap I)) \right),$$

and the conclusion follows from the two previous propositions. ∎

The following two propositions will provide us with a large class of measurable sets.

Proposition 11.11 *Every open and bounded set is measurable.*

Proof Consider for simplicity the case $N = 2$. Let A be an open set contained in a rectangle I. We divide the rectangle I into four rectangles of equal areas using the axes of its edges. Then we proceed analogously with each of these four rectangles, thereby obtaining 16 smaller rectangles, and so on. Since A is open, for every $\boldsymbol{x} \in A$ there is a small rectangle among those just constructed that contains $\boldsymbol{x}$ and is contained in A. In this way, it is seen that set A is covered by a countable family of rectangles; being the union of a countable family of measurable sets, it is therefore measurable.   ∎

Proposition 11.12 *Every compact set is measurable.*

Proof Let B be a compact set, and let I be a rectangle whose interior $\overset{\circ}{I}$ contains B. Since $\overset{\circ}{I}$ and $\overset{\circ}{I} \setminus B$ are open and, hence, measurable, we have that $B = \overset{\circ}{I} \setminus (\overset{\circ}{I} \setminus B)$ is measurable.   ∎

Example The set

$$E = \{(x, y) \in \mathbb{R}^2 : 1 < x^2 + y^2 \le 4\}$$

is measurable, since it is the difference of the closed disks with radius 2 and 1 centered at the origin, i.e.,

$$E = \{(x, y) \in \mathbb{R}^2 : x^2 + y^2 \le 4\} \setminus \{(x, y) \in \mathbb{R}^2 : x^2 + y^2 \le 1\}.$$

11.4   Negligible Sets

Definition 11.13 We say that a bounded set is "negligible" if it is measurable and its measure is equal to zero.

Every set made of a single point is negligible. Consequently, all finite or countable bounded sets are negligible. The edge of a rectangle in $\mathbb{R}^2$ is a negligible set, as shown by Lemma 11.6.

By the complete additivity of the measure, the union of any sequence of negligible sets, if it is bounded, is always a negligible set.

Theorem 11.14 *If E is a bounded set and $f : E \to \mathbb{R}$ is equal to zero except for a negligible set, then f is integrable on E and $\int_E f = 0$.*

Proof Let T be the negligible set on which f is different from zero. Assume first that the function f is bounded, i.e., that there is a constant $C > 0$ such that

$$|f(\boldsymbol{x})| \le C$$

for every $\boldsymbol{x} \in E$. We consider a rectangle I containing E and prove that $\int_I f_E = 0$. Fix $\varepsilon > 0$. Since T has zero measure, there is a gauge δ such that, for every δ-fine tagged partition $\overset{\circ}{\mathcal{P}} = \{(\boldsymbol{x}_j, I_j), j = 1, \ldots, m\}$ of I,

$$S(\chi_T, \overset{\circ}{\mathcal{P}}) = \sum_{\{j\, :\, \boldsymbol{x}_j \in T\}} \mu(I_j) \le \frac{\varepsilon}{C},$$

so that

$$|S(f_E, \overset{\circ}{\mathcal{P}})| \le \sum_{\{j\, :\, \boldsymbol{x}_j \in T\}} |f(\boldsymbol{x}_j)| \mu(I_j) \le C \sum_{\{j\, :\, \boldsymbol{x}_j \in T\}} \mu(I_j) \le \varepsilon.$$

Hence, if f is bounded, it is integrable on E and $\int_E f = 0$. If f is not bounded, assume first that it has nonnegative values. Define a sequence of functions $f_n : E \to \mathbb{R}$ as

$$f_n(\boldsymbol{x}) = \min\{f(\boldsymbol{x}), n\}.$$

Since the functions f_n are bounded and zero except on T, for what we just saw they are integrable on E, with $\int_E f_n = 0$, for every n. It is easily seen that the defined sequence satisfies the conditions of the Monotone Convergence Theorem 9.10 and converges pointwise to f. Hence, f is integrable on E, and

$$\int_E f = \lim_n \int_E f_n = 0.$$

If f does not have nonnegative values, it is sufficient to consider f^+ and f^- and apply to them what was said earlier. ∎

Here is a counterpart of the foregoing result.

Theorem 11.15 *If $f : E \to \mathbb{R}$ is an integrable function on a bounded set E, having nonnegative values, with $\int_E f = 0$, then f is equal to zero except on a negligible set.*

To prove this, we need the following **Chebyshev inequality**.

Lemma 11.16 *Let E be a bounded set and $f : E \to \mathbb{R}$ an integrable function with nonnegative values. Then, for every $r > 0$, the set*

$$E_r = \{\boldsymbol{x} \in E : f(\boldsymbol{x}) > r\}$$

is measurable, and

$$\mu(E_r) \le \frac{1}{r} \int_E f \,.$$

Proof Let I be a rectangle containing E. Once we have fixed $r > 0$, we define the functions $f_n : I \to \mathbb{R}$ as

$$f_n(x) = \min\{1, n \max\{f_E(x) - r, 0\}\} \,.$$

These make up an increasing sequence of L-integrable functions that pointwise converges to χ_{E_r}. Clearly,

$$0 \le f_n(x) \le 1 \quad \text{for every } n \text{ and every } x \in I \,.$$

The Monotone Convergence Theorem 9.10 guarantees that χ_{E_r} is integrable on I, i.e., that E_r is measurable. Since, for every $x \in E$, we have $r\chi_{E_r}(x) \le f(x)$, integrating both sides of this inequality we obtain the inequality we are looking for. ∎

Proof of Theorem 11.15 Using the Chebyshev inequality, we have that, for every positive integer k,

$$\mu(E_{\frac{1}{k}}) \le k \int_E f = 0 \,.$$

Hence, every $E_{\frac{1}{k}}$ is negligible, and since their union is just the set where f is different from zero, we have the conclusion. ∎

Definition 11.17 Let E be a bounded set. We say that a proposition is true "almost everywhere" on E (or for almost every point of E) if the set of points for which it is false is negligible.

The results just proved have the following simple consequence.

Corollary 11.18 *If two functions f and g, defined on the bounded set E, are equal almost everywhere, then f is integrable on E if and only if g is. In that case, $\int_E f = \int_E g$.*

Proof In such a case the function $g - f$ is equal to zero almost everywhere, hence $\int_E (g - f) = 0$. Then

$$\int_E f = \int_E f + \int_E (g - f) = \int_E (f + (g - f)) = \int_E g \,,$$

thereby completing the proof. ∎

This last corollary permits us to consider some functions that are *defined almost everywhere* and to define their integral.

Definition 11.19 A function f, defined almost everywhere on E, with real values, is said to be "integrable" on E if it can be extended to an integrable function $g : E \to \mathbb{R}$. In this case, we set $\int_E f = \int_E g$.

The preceding definition is consistent since the integral will not depend on the particular extension.

It can be seen that all the properties and theorems seen till now remain true for such functions. The reader is invited to verify this.

11.5   A Characterization of Measurable Bounded Sets

The following **covering lemma** will be useful in what follows.

Lemma 11.20 *Let E be a set contained in a rectangle I, and let δ be a gauge on E. Then there is a finite or countable family of nonoverlapping rectangles J_k, contained in I, whose union covers the set E, with the following property: In each of the sets J_k there is a point $\boldsymbol{x}_k$ belonging to E such that $J_k \subseteq \overline{B}[\boldsymbol{x}_k, \delta(\boldsymbol{x}_k)]$.*

Proof We consider for simplicity the case $N = 2$. Let us divide the rectangle I into four rectangles, having the same areas, by the axes of its edges. We proceed analogously with each of these four rectangles, obtaining 16 smaller rectangles, and so on. We thus obtain a countable family of smaller and smaller rectangles. For every point $\boldsymbol{x}$ of E we can choose one of these rectangles that contains $\boldsymbol{x}$ and is itself contained in $\overline{B}[\boldsymbol{x}, \delta(\boldsymbol{x})]$. These rectangles would satisfy the properties of the statement if they were nonoverlapping.

In order that the sets J_k be nonoverlapping, it is necessary to choose them carefully, and here is how to do it. We first choose those from the first four-rectangle partition, if there are any, that contain a point $\boldsymbol{x}_k$ belonging to E such that $J_k \subseteq \overline{B}[\boldsymbol{x}_k, \delta(\boldsymbol{x}_k)]$; once this choice has been made, we eliminate all the smaller rectangles contained in them. We consider then the 16-rectangle partition and, among those that remained after the first elimination procedure, we choose those, if there are any, that contain a point $\boldsymbol{x}_k$ belonging to E such that $J_k \subseteq \overline{B}[\boldsymbol{x}_k, \delta(\boldsymbol{x}_k)]$; once this choice has been made, we eliminate all the smaller rectangles contained in them; and so on. ∎

Remark 11.21 Note that if, in the assumptions of the covering lemma, it happens that E is contained in an open set that itself is contained in I, then all the rectangles J_k can be chosen so that they are all contained in that open set.

We can now prove a characterization of measurable bounded sets. In the following statement, the words in square brackets may be omitted.

Proposition 11.22 *Let E be a bounded set, contained in a rectangle I. The following three propositions are equivalent:*

(*i*) *The set E is measurable.*

(*ii*) *For every $\varepsilon > 0$ there are two finite or countable families (J_k) and (J_k'), each made of [nonoverlapping] rectangles contained in I, such that*

$$E \subseteq \bigcup_k J_k, \quad I \setminus E \subseteq \bigcup_k J_k' \quad \text{and} \quad \mu\left(\left(\bigcup_k J_k\right) \cap \left(\bigcup_k J_k'\right)\right) \leq \varepsilon.$$

(*iii*) *There are two sequences $(E_n)_{n\geq 1}$ and $(E_n')_{n\geq 1}$ of measurable bounded subsets such that*

$$E_n' \subseteq E \subseteq E_n, \quad \lim_n (\mu(E_n) - \mu(E_n')) = 0.$$

In that case, we have

$$\mu(E) = \lim_n \mu(E_n) = \lim_n \mu(E_n').$$

Proof Let us first prove that (*i*) implies (*ii*). Assume that E is measurable, and fix $\varepsilon > 0$. By the Saks–Henstock Theorem 9.3, there is a gauge δ on I such that, for every δ-fine tagged subpartition $\mathring{\mathcal{P}} = \{(\boldsymbol{x}_j, K_j) : j = 1, \ldots, m\}$ of I,

$$\sum_{j=1}^{m} \left| \chi_E(\boldsymbol{x}_j)\mu(K_j) - \int_{K_j} \chi_E \right| \leq \frac{\varepsilon}{2}.$$

By Lemma 11.20, there is a family of nonoverlapping rectangles J_k, contained in I, whose union covers E, and in each J_k there is a point $\boldsymbol{x}_k$ belonging to E such that $J_k \subseteq \overline{B}[\boldsymbol{x}_k, \delta(\boldsymbol{x}_k)]$. Let us fix a positive integer N and consider only $(\boldsymbol{x}_1, J_1), \ldots, (\boldsymbol{x}_N, J_N)$. They make up a δ-fine tagged subpartition of I. From the preceding inequality we then deduce that

$$\sum_{k=1}^{N} \left| \mu(J_k) - \int_{J_k} \chi_E \right| \leq \frac{\varepsilon}{2},$$

whence

$$\sum_{k=1}^{N} \mu(J_k) \leq \sum_{k=1}^{N} \int_{J_k} \chi_E + \frac{\varepsilon}{2} \leq \int_I \chi_E + \frac{\varepsilon}{2} = \mu(E) + \frac{\varepsilon}{2}.$$

Since this holds for every positive integer N, we have thus constructed a family (J_k) of nonoverlapping rectangles such that

$$E \subseteq \bigcup_k J_k, \quad \sum_k \mu(J_k) \leq \mu(E) + \frac{\varepsilon}{2}.$$

Consider now $I \setminus E$, which is also measurable. We can repeat the same procedure that we just followed replacing E with $I \setminus E$, thereby finding a family (J_k') of nonoverlapping rectangles, contained in I, such that

$$I \setminus E \subseteq \bigcup_k J_k', \quad \sum_k \mu(J_k') \leq \mu(I \setminus E) + \frac{\varepsilon}{2}.$$

Consequently,

$$I \setminus \left(\bigcup_k J_k' \right) \subseteq E \subseteq \bigcup_k J_k,$$

and hence

$$\mu\left(\left(\bigcup_k J_k \right) \cap \left(\bigcup_k J_k' \right) \right) = \mu\left(\left(\bigcup_k J_k \right) \setminus \left(I \setminus \left(\bigcup_k J_k' \right) \right) \right)$$

$$= \mu\left(\bigcup_k J_k \right) - \mu\left(I \setminus \left(\bigcup_k J_k' \right) \right)$$

$$= \mu\left(\bigcup_k J_k \right) - \mu(I) + \mu\left(\bigcup_k J_k' \right)$$

$$\leq \left(\mu(E) + \frac{\varepsilon}{2} \right) - \mu(I) + \left(\mu(I \setminus E) + \frac{\varepsilon}{2} \right)$$

$$= \varepsilon,$$

and the implication is thus proved.

Taking $\varepsilon = \frac{1}{n}$, it is easy to see that (ii) implies (iii). Let us prove now that (iii) implies (i). Consider the measurable sets

$$\widetilde{E}' = \bigcup_{n \geq 1} E_n', \quad \widetilde{E} = \bigcap_{n \geq 1} E_n,$$

for which it must be that

$$\widetilde{E}' \subseteq E \subseteq \widetilde{E}, \quad \mu(\widetilde{E}') = \mu(\widetilde{E}).$$

Equivalently, we have

$$\chi_{\widetilde{E}'} \leq \chi_E \leq \chi_{\widetilde{E}}, \qquad \int_I (\chi_{\widetilde{E}} - \chi_{\widetilde{E}'}) = 0,$$

so that $\chi_{\widetilde{E}'} = \chi_E = \chi_{\widetilde{E}}$ almost everywhere. Then E is measurable and $\mu(E) = \mu(\widetilde{E}') = \mu(\widetilde{E})$. Moreover,

$$0 \leq \lim_n [\mu(E) - \mu(E'_n)] \leq \lim_n [\mu(E_n) - \mu(E'_n)] = 0,$$

hence $\mu(E) = \lim_n \mu(E'_n)$. Analogously, we see that $\mu(E) = \lim_n \mu(E_n)$, and the proof is thus completed.   ∎

Proposition 11.23 *Let E be a bounded set. Then E is negligible if and only if, for every $\varepsilon > 0$, there is a finite or countable family (J_k) of [nonoverlapping] rectangles such that*

$$E \subseteq \bigcup_k J_k, \qquad \sum_k \mu(J_k) \leq \varepsilon.$$

Proof The necessary condition is proved in the first part of the previous proposition.

Let us prove the sufficiency. Once fixed $\varepsilon > 0$, assume there exists a family (J_k) with the given properties. Let I be a rectangle containing the set E. On the other hand, consider a family (J'_k) whose elements all coincide with I. The conditions of the previous proposition are then satisfied, so that E is indeed measurable. Then

$$\mu(E) \leq \mu\left(\bigcup_k J_k\right) \leq \sum_k \mu(J_k) \leq \varepsilon;$$

since ε is arbitrary, it must be that $\mu(E) = 0$.   ∎

Remark 11.24 Observe that if E is contained in an open set that is itself contained in a rectangle I, then all the rectangles J_k can be chosen in such a way that they are all contained in that open set.

As a consequence of the preceding proposition, it is not difficult to prove the following corollary.

Corollary 11.25 *If I_{N-1} is a rectangle in $\mathbb{R}^{N-1}$ and T is a negligible subset of $\mathbb{R}$, then $I_{N-1} \times T$ is negligible in $\mathbb{R}^N$.*

Proof Fix $\varepsilon > 0$, and, according to Proposition 11.23, let (J_k) be a finite or countable family of intervals in $\mathbb{R}$ such that

$$T \subseteq \bigcup_k J_k, \qquad \sum_k \mu(J_k) \leq \frac{\varepsilon}{\mu(I_{N-1})}.$$

Defining the rectangles $\tilde{J}_k = I_{N-1} \times J_k$, we have that

$$I_{N-1} \times T \subseteq \bigcup_k \tilde{J}_k, \qquad \sum_k \mu(\tilde{J}_k) = \mu(I_{N-1}) \sum_k \mu(J_k) \leq \mu(I_{N-1}) \frac{\varepsilon}{\mu(I_{N-1})} = \varepsilon,$$

and Proposition 11.23 applies. ∎

11.6   Continuous Functions and L-Integrable Functions

We begin this section by showing that the continuous functions are L-integrable on compact sets.

Theorem 11.26 *Let $E \subseteq \mathbb{R}^N$ be a compact set and $f : E \to \mathbb{R}$ be a continuous function. Then f is L-integrable on E.*

Proof We consider for simplicity the case $N = 2$. Since f is continuous on a compact set, there is a constant $C > 0$ such that

$$|f(\boldsymbol{x})| \leq C, \qquad \text{for every } \boldsymbol{x} \in E.$$

Let I be a rectangle containing E. First we divide I into four rectangles by tracing the segments joining the midpoints of its edges; we denote by $U_{1,1}, U_{1,2}, U_{1,3}, U_{1,4}$ these subrectangles. We now divide again each of these rectangles in the same way, thereby obtaining 16 smaller subrectangles, which we denote by $U_{2,1}, U_{2,2}, \ldots, U_{2,16}$. Proceeding in this way, for every n we will have a subdivision of the rectangle I into 2^{2n} small rectangles $U_{n,j}$, with $j = 1, \ldots, 2^{2n}$. Whenever E has a nonempty intersection with $\mathring{U}_{n,j}$, we choose and fix a point $\boldsymbol{x}_{n,j} \in E \cap \mathring{U}_{n,j}$. Define now the function f_n in the following way:

- If $E \cap \mathring{U}_{n,j}$ is nonempty, then f_n is constant on $\mathring{U}_{n,j}$ with value $f(\boldsymbol{x}_{n,j})$.
- If $E \cap \mathring{U}_{n,j}$ is empty, then f_n is constant on $\mathring{U}_{n,j}$ with value 0.

The functions f_n are thus defined almost everywhere on I, not being defined only on the points of the grid made up by the previously constructed segments, which form a countable family of negligible sets. The functions f_n are integrable on each subrectangle $U_{n,j}$, since they are constant in its interior. By the property of additivity on subrectangles, these functions are therefore integrable on I. Moreover,

$$|f_n(\pmb{x})| \leq C\,, \qquad \text{for almost every } \pmb{x} \in I \text{ and every } k \geq 1\,.$$

Let us see now that $(f_n)_n$ converges pointwise almost everywhere to f_E. Indeed, taking a point $\pmb{x} \in I$ not belonging to the grid, for every n there is a $j = j(n)$ for which $\pmb{x} \in \mathring{U}_{n,j(n)}$. We have two possibilities:

(a) $\pmb{x} \notin E$; in this case, since E is closed, we have that, for n sufficiently large, $\mathring{U}_{n,j(n)}$ (whose dimensions tend to zero as $n \to \infty$) will have an empty intersection with E, and then $f_n(\pmb{x}) = 0 = f_E(\pmb{x})$.

(b) $\pmb{x} \in E$; in this case, if $n \to +\infty$, we have that $\pmb{x}_{n,j(n)} \to \pmb{x}$ (again using the fact that $\mathring{U}_{n,j(n)}$ has dimensions tending to zero). By the continuity of f, we have that

$$f_n(\pmb{x}) = f(\pmb{x}_{n,j(n)}) \to f(\pmb{x}) = f_E(\pmb{x})\,.$$

The Dominated Convergence Theorem 9.13 then yields the conclusion. ∎

We now see that L-integrability is conserved on measurable subsets.

Theorem 11.27 *Let $f : E \to \mathbb{R}$ be a L-integrable function on a bounded set E. Then f is L-integrable on every measurable subset of E.*

Proof Assume first that f has nonnegative values. Let S be a measurable subset of E, and define on E the functions $f_n = \min\{f, n\chi_S\}$. They form an increasing sequence of L-integrable functions since both f and $n\chi_S$ are L-integrable, and the sequence converges pointwise to f_S. Moreover, we have

$$\int_E f_n \leq \int_E f$$

for every n. The Monotone Convergence Theorem 9.10 then guarantees that f is integrable on S in this case. In the general case, since f is L-integrable, both f^+ and f^- are L-integrable on E. Hence, based on the preceding discussion, they are both L-integrable on S, and then f is, too. ∎

Let us now prove the **complete additivity** property of the integral for L-integrable functions. We will say that two measurable bounded subsets are "nonoverlapping" if their intersection is a negligible set.

Theorem 11.28 *Let* (E_k) *be a finite or countable family of measurable nonoverlapping sets whose union is a bounded set* E. *Then* f *is L-integrable on* E *if and only if the following two conditions hold:*

(a) f *is L-integrable on each* E_k.
(b) $\sum_k \int_{E_k} |f(\boldsymbol{x})| \, d\boldsymbol{x} < +\infty$.

In that case, we have

$$\int_E f = \sum_k \int_{E_k} f \, .$$

Proof Observe that

$$f(\boldsymbol{x}) = \sum_k f_{E_k}(\boldsymbol{x}) \, , \qquad |f(\boldsymbol{x})| = \sum_k |f_{E_k}(\boldsymbol{x})|$$

for almost every $\boldsymbol{x} \in E$. If f is L-integrable on E, then from the preceding theorem (a) follows. Moreover, it is obvious that (b) holds whenever the sets E_k are finite in number. If instead they are infinite, then for any fixed n we have

$$\sum_{k=1}^n \int_{E_k} |f(\boldsymbol{x})| \, d\boldsymbol{x} = \sum_{k=1}^n \int_E |f_{E_k}(\boldsymbol{x})| \, d\boldsymbol{x} = \int_E \sum_{k=1}^n |f_{E_k}(\boldsymbol{x})| \, d\boldsymbol{x} \le \int_E |f(\boldsymbol{x})| \, d\boldsymbol{x} \, ,$$

and (b) follows.

Assume now that (a) and (b) hold. If the sets E_k are finite in number, it is sufficient to integrate on E both terms in the equation $f = \sum_k f_{E_k}$. If instead they are infinite, assume first that f has nonnegative values. In this case, Corollary 9.11, when applied to the series $\sum_k f_{E_k}$, yields the conclusion. In the general case, it is sufficient to consider, as usual, the positive and negative parts of f. ∎

11.7 Limits and Derivatives under the Integration Sign

Let X be a metric space, Y be a bounded subset of $\mathbb{R}^N$, and consider a function $f : X \times Y \to \mathbb{R}$. (For simplicity, we may think of X and Y as subsets of $\mathbb{R}$.) The first question we want to address is when the formula

$$\lim_{x \to x_0} \left(\int_Y f(x, y) \, dy \right) = \int_Y \left(\lim_{x \to x_0} f(x, y) \right) dy$$

holds. What follows is a generalization of the Dominated Convergence Theorem 9.13.

Theorem 11.29 *Let x_0 be an accumulation point of X, and let the following assumptions hold true:*

(a) *For every $x \in X \setminus \{x_0\}$, the function $f(x, \cdot)$ is integrable on Y, so that we can define the function*

$$F(x) = \int_Y f(x, y)\, dy \,.$$

(b) *For almost every $y \in Y$ the limit $\lim_{x \to x_0} f(x, y)$ exists and is finite, so that we can define almost everywhere the function*

$$\eta(y) = \lim_{x \to x_0} f(x, y) \,.$$

(c) *There are two integrable functions $g, h : Y \to \mathbb{R}$ such that*

$$g(y) \le f(x, y) \le h(y)$$

for every $x \in X \setminus \{x_0\}$ and almost every $y \in Y$.

Then η is integrable on Y, and we have

$$\lim_{x \to x_0} F(x) = \int_Y \eta(y)\, dy \,.$$

Proof Let us take a sequence $(x_n)_n$ in $X \setminus \{x_0\}$ that tends to x_0. Define, for every n, the functions $f_n : Y \to \mathbb{R}$ such that $f_n(y) = f(x_n, y)$. The assumptions allow us to apply the Dominated Convergence Theorem 9.13, so that

$$\lim_n F(x_n) = \lim_n \left(\int_Y f_n(y)\, dy \right) = \int_Y \left(\lim_n f_n(y) \right) dy = \int_Y \eta(y)\, dy \,.$$

The conclusion then follows from the characterization of the limit by the use of sequences (Proposition 4.3).   ∎

We have the following consequence of the above theorem.

Corollary 11.30 *If X is a subset of $\mathbb{R}^M$, $Y \subseteq \mathbb{R}^N$ is compact, and $f : X \times Y \to \mathbb{R}$ is continuous, then the function $F : X \to \mathbb{R}$, defined by*

$$F(x) = \int_Y f(x, y)\, dy \,,$$

is continuous.

Proof The function $F(x)$ is well defined, since $f(x, \cdot)$ is continuous on the compact set Y. Let us fix $x_0 \in X$ and prove that F is continuous at x_0. By the continuity of f,

$$\eta(y) = \lim_{x \to x_0} f(x, y) = f(x_0, y).$$

Moreover, given a compact neighborhood U of x_0, there is a constant $C > 0$ such that $|f(x, y)| \leq C$ for every $(x, y) \in U \times Y$. The previous theorem can then be applied, and we have

$$\lim_{x \to x_0} F(x) = \int_Y f(x_0, y) \, dy = F(x_0),$$

thereby proving that F is continuous at x_0.   ∎

Now let X be a subset of $\mathbb{R}$. The second question we want to address is when the formula

$$\frac{d}{dx} \left(\int_Y f(x, y) \, dy \right) = \int_Y \left(\frac{\partial f}{\partial x}(x, y) \right) dy$$

holds. Here is an answer.

Theorem 11.31 (Leibniz Rule) *Let X be an interval in $\mathbb{R}$ containing x_0, and let the following assumptions hold true:*

(a) *For every $x \in X$ the function $f(x, \cdot)$ is integrable on Y, so that we can define the function*

$$F(x) = \int_Y f(x, y) \, dy.$$

(b) *For every $x \in X$ and almost every $y \in Y$ the partial derivative $\frac{\partial f}{\partial x}(x, y)$ exists.*
(c) *There are two integrable functions $g, h : Y \to \mathbb{R}$ such that*

$$g(y) \leq \frac{\partial f}{\partial x}(x, y) \leq h(y)$$

for every $x \in X$ and almost every $y \in Y$.

Then the function $\frac{\partial f}{\partial x}(x, \cdot)$, defined almost everywhere on Y, is integrable there, the derivative of F in x_0 exists, and we have

$$F'(x_0) = \int_Y \left(\frac{\partial f}{\partial x}(x_0, y) \right) dy.$$

Proof We define, for $x \in X$ different from x_0, the function

$$\psi(x, y) = \frac{f(x, y) - f(x_0, y)}{x - x_0}.$$

For every $x \in X \setminus \{x_0\}$ the function $\psi(x, \cdot)$ is integrable on Y. Moreover, for almost every $y \in Y$ we have

$$\lim_{x \to x_0} \psi(x, y) = \frac{\partial f}{\partial x}(x_0, y).$$

By the Lagrange Mean Value Theorem 6.11, for (x, y) as previously there is a $\xi \in X$ between x_0 and x such that

$$\psi(x, y) = \frac{\partial f}{\partial x}(\xi, y).$$

By assumption (iii), we then have

$$g(y) \le \psi(x, y) \le h(y)$$

for every $x \in X \setminus \{x_0\}$ and almost every $y \in Y$. By the previous theorem, we can conclude that the function $\frac{\partial f}{\partial x}(x_0, \cdot)$, defined almost everywhere on Y, is integrable there, and

$$\lim_{x \to x_0} \left(\int_Y \psi(x, y) \, dy \right) = \int_Y \left(\frac{\partial f}{\partial x}(x_0, y) \right) dy.$$

On the other hand,

$$\lim_{x \to x_0} \left(\int_Y \psi(x, y) \, dy \right) = \lim_{x \to x_0} \left(\int_Y \frac{f(x, y) - f(x_0, y)}{x - x_0} \, dy \right)$$

$$= \lim_{x \to x_0} \frac{1}{x - x_0} \left(\int_Y f(x, y) \, dy - \int_Y f(x_0, y) \, dy \right)$$

$$= \lim_{x \to x_0} \frac{F(x) - F(x_0)}{x - x_0},$$

so that F is differentiable at x_0, and the conclusion holds.   ∎

Corollary 11.32 *If X is an interval in $\mathbb{R}$, Y is a compact subset of $\mathbb{R}^N$, and the function $f : X \times Y \to \mathbb{R}$ is continuous and has a continuous partial derivative $\frac{\partial f}{\partial x} : X \times Y \to \mathbb{R}$, then the function $F : X \to \mathbb{R}$, defined by*

$$F(x) = \int_Y f(x, y) \, dy,$$

is differentiable with a continuous derivative.

Proof The function $F(x)$ is well defined, since $f(x, \cdot)$ is continuous on the compact set Y. Taking a point $x_0 \in X$ and a nontrivial compact interval $U \subseteq X$ containing it, there is a constant $C > 0$ such that $|\frac{\partial f}{\partial x}(x, y)| \leq C$ for every $(x, y) \in U \times Y$. By the preceding theorem, F is differentiable at x_0. The same argument holds replacing x_0 with any $x \in X$, and

$$F'(x) = \int_Y \left(\frac{\partial f}{\partial x}(x, y) \right) dy.$$

The continuity of $F' : X \to \mathbb{R}$ now follows from Corollary 11.30. ∎

Example Consider, for $x \geq 0$, the function

$$f(x, y) = \frac{e^{-x^2(y^2+1)}}{y^2 + 1}.$$

We want to determine whether the corresponding function $F(x) = \int_0^1 f(x, y) \, dy$ is differentiable and, in this case, to find its derivative. We have that

$$\frac{\partial f}{\partial x}(x, y) = -2xe^{-x^2(y^2+1)},$$

which, for $y \in [0, 1]$ and $x \geq 0$, is such that

$$-\sqrt{\frac{2}{e}} \leq -2xe^{-x^2} \leq -2xe^{-x^2(y^2+1)} \leq 0.$$

We can then apply the Leibniz rule, so that

$$F'(x) = -2x \int_0^1 e^{-x^2(y^2+1)} \, dy.$$

Let us make a digression, so as to present an elegant formula. By the change of variable $t = xy$, we have

$$-2x \int_0^1 e^{-x^2(y^2+1)} \, dy = -2e^{-x^2} \int_0^x e^{-t^2} \, dt = -\frac{d}{dx} \left(\int_0^x e^{-t^2} \, dt \right)^2.$$

Taking into account that $F(0) = \pi/4$, we have

$$F(x) = \frac{\pi}{4} - \left(\int_0^x e^{-t^2} \, dt \right)^2.$$

We would like now to pass to the limit for $x \to +\infty$. Since, for $x \geq 0$, we have

$$0 \leq \frac{e^{-x^2(y^2+1)}}{y^2 + 1} \leq 1 ,$$

we can pass to the limit under the sign of integration, thereby obtaining

$$\lim_{x \to +\infty} \int_0^1 \frac{e^{-x^2(y^2+1)}}{y^2 + 1} \, dy = \int_0^1 \left(\lim_{x \to +\infty} \frac{e^{-x^2(y^2+1)}}{y^2 + 1} \right) dy = 0 .$$

Hence,

$$\left(\int_0^{+\infty} e^{-t^2} \, dt \right)^2 = \frac{\pi}{4}$$

and, by symmetry,

$$\int_{-\infty}^{+\infty} e^{-t^2} \, dt = \sqrt{\pi} ,$$

which is a very useful formula in various applications.

11.8   Reduction Formula

In this section we will prove a fundamental result that permits us to compute the integral of a function of several variables by an iterative process of integration of functions of a single variable. It will be useful to recall some notation. For any fixed x we will denote by $f(x, \cdot)$ the function $y \mapsto f(x, y)$. Similarly, for any fixed y we will denote by $f(\cdot, y)$ the function $x \mapsto f(x, y)$.

Before stating the main theorem, it will be useful to first prove a preliminary result.

Proposition 11.33 *Let $f : I \to \mathbb{R}$ be an integrable function on the rectangle $I = [a_1, b_1] \times [a_2, b_2]$. Then, for almost every $x \in [a_1, b_1]$, the function $f(x, \cdot)$ is integrable on $[a_2, b_2]$.*

Proof Let $T \subseteq [a_1, b_1]$ be the set of those $x \in [a_1, b_1]$ for which $f(x, \cdot)$ is not integrable on $[a_2, b_2]$. Let us prove that T is a negligible set. For each $x \in T$, the Cauchy condition does not hold. Hence, if we define the sets

$$T_n = \left\{ x : \begin{array}{l} \text{for every gauge } \delta_2 \text{ on } [a_2, b_2] \text{ there are two} \\ \delta_2\text{-fine tagged partitions } \mathring{\mathcal{P}}_2 \text{ and } \mathring{\mathcal{Q}}_2 \text{ of } [a_2, b_2] \text{ such that} \\ S(f(x, \cdot), \mathring{\mathcal{P}}_2) - S(f(x, \cdot), \mathring{\mathcal{Q}}_2) > \frac{1}{n} \end{array} \right\} ,$$

we have that each $x \in T$ belongs to T_n if n is sufficiently large. Thus, T is the union of all T_n, and if we prove that any T_n is negligible, then by the properties of the measure we will have that T is also negligible. To do so, let us consider a certain T_n and fix $\varepsilon > 0$. Since f is integrable on I, there is a gauge δ on I such that, given two δ-fine tagged partitions $\mathring{\mathcal{P}}$ and $\mathring{\mathcal{Q}}$ of I, we have

$$|S(f, \mathring{\mathcal{P}}) - S(f, \mathring{\mathcal{Q}})| \leq \frac{\varepsilon}{n}.$$

The gauge δ on I determines, for every $x \in [a_1, b_1]$, a gauge $\delta(x, \cdot)$ on $[a_2, b_2]$. We now associate to each $x \in [a_1, b_1]$ two $\delta(x, \cdot)$-fine tagged partitions $\mathring{\mathcal{P}}_2^x$ and $\mathring{\mathcal{Q}}_2^x$ of $[a_2, b_2]$ in the following way:

– If $x \in T_n$, we can choose $\mathring{\mathcal{P}}_2^x$ and $\mathring{\mathcal{Q}}_2^x$ such that

$$S(f(x, \cdot), \mathring{\mathcal{P}}_2^x) - S(f(x, \cdot), \mathring{\mathcal{Q}}_2^x) > \frac{1}{n}.$$

– If instead $x \notin T_n$, we take $\mathring{\mathcal{P}}_2^x$ and $\mathring{\mathcal{Q}}_2^x$ equal to each other.

Let us write the two tagged partitions $\mathring{\mathcal{P}}_2^x$ and $\mathring{\mathcal{Q}}_2^x$ thus determined:

$$\mathring{\mathcal{P}}_2^x = \{(y_j^x, K_j^x) : j = 1, \ldots, m^x\}, \quad \mathring{\mathcal{Q}}_2^x = \{(\tilde{y}_j^x, \tilde{K}_j^x) : j = 1, \ldots, \tilde{m}^x\}.$$

We define a gauge δ_1 on $[a_1, b_1]$, setting

$$\delta_1(x) = \min\{\delta(x, y_1^x), \ldots, \delta(x, y_{m^x}^x), \delta(x, \tilde{y}_1^x), \ldots, \delta(x, \tilde{y}_{\tilde{m}^x}^x)\}.$$

Now let $\mathring{\mathcal{P}}_1 = \{(x_i, J_i) : i = 1, \ldots, k\}$ be a δ_1-fine tagged partition of $[a_1, b_1]$. We want to prove that $S(\chi_{T_n}, \mathring{\mathcal{P}}_1) \leq \varepsilon$, i.e.,

$$\sum_{\{i \,:\, x_i \in T_n\}} \mu(J_i) \leq \varepsilon.$$

To this end, define the following two tagged partitions of I, which make use of the elements of $\mathring{\mathcal{P}}_1$:

$$\mathring{\mathcal{P}} = \{((x_i, y_j^{x_i}), J_i \times K_j^{x_i}) : i = 1, \ldots, k, \; j = 1, \ldots, m^{x_i}\},$$
$$\mathring{\mathcal{Q}} = \{((x_i, \tilde{y}_j^{x_i}), J_i \times \tilde{K}_j^{x_i}) : i = 1, \ldots, k, \; j = 1, \ldots, \tilde{m}^{x_i}\}.$$

They are δ-fine, and hence

$$|S(f, \mathring{\mathcal{P}}) - S(f, \mathring{\mathcal{Q}})| \leq \frac{\varepsilon}{n}.$$

On the other hand, we have

$$|S(f, \mathring{\mathcal{P}}) - S(f, \mathring{\mathcal{Q}})| =$$

$$= \left| \sum_{i=1}^{k} \sum_{j=1}^{m^{x_i}} f(x_i, y_j^{x_i}) \mu(J_i \times K_j^{x_i}) - \sum_{i=1}^{k} \sum_{j=1}^{\tilde{m}^{x_i}} f(x_i, \tilde{y}_j^{x_i}) \mu(J_i \times \tilde{K}_j^{x_i}) \right|$$

$$= \left| \sum_{i=1}^{k} \mu(J_i) \left[\sum_{j=1}^{m^{x_i}} f(x_i, y_j^{x_i}) \mu(K_j^{x_i}) - \sum_{j=1}^{\tilde{m}^{x_i}} f(x_i, \tilde{y}_j^{x_i}) \mu(\tilde{K}_j^{x_i}) \right] \right|$$

$$= \left| \sum_{i=1}^{k} \mu(J_i) \left[S(f(x_i, \cdot), \mathring{\mathcal{P}}_2^{x_i}) - S(f(x_i, \cdot), \mathring{\mathcal{Q}}_2^{x_i}) \right] \right|$$

$$= \sum_{\{i \,:\, x_i \in T_n\}} \mu(J_i)[S(f(x_i, \cdot), \mathring{\mathcal{P}}_2^{x_i}) - S(f(x_i, \cdot), \mathring{\mathcal{Q}}_2^{x_i})].$$

Recalling that

$$S(f(x_i, \cdot), \mathring{\mathcal{P}}_2^{x_i}) - S(f(x_i, \cdot), \mathring{\mathcal{Q}}_2^{x_i}) > \frac{1}{n},$$

we conclude that

$$\frac{\varepsilon}{n} \geq |S(f, \mathring{\mathcal{P}}) - S(f, \mathring{\mathcal{Q}})| > \frac{1}{n} \sum_{\{i \,:\, x_i \in T_n\}} \mu(J_i),$$

and hence $S(\chi_{T_n}, \mathring{\mathcal{P}}_1) \leq \varepsilon$, which is what we wanted to prove. All this shows that the sets T_n are negligible, and therefore T is negligible, too. ∎

The following theorem, due to Guido Fubini, permits us to compute the integral of an integrable function of two variables by performing two integrations of functions of one variable.

Theorem 11.34 (Reduction Theorem—I) *Let $f : I \rightarrow \mathbb{R}$ be an integrable function on the rectangle $I = [a_1, b_1] \times [a_2, b_2]$. Then:*

(a) *For almost every $x \in [a_1, b_1]$ the function $f(x, \cdot)$ is integrable on $[a_2, b_2]$.*
(b) *The function $\int_{a_2}^{b_2} f(\cdot, y) \, dy$, defined almost everywhere on $[a_1, b_1]$, is integrable there.*

(c) We have

$$\int_I f = \int_{a_1}^{b_1} \left(\int_{a_2}^{b_2} f(x, y) \, dy \right) dx \, .$$

Proof We already proved (a) in Proposition 11.33. Let us now prove (b) and (c). Let T be the negligible subset of $[a_1, b_1]$ such that, for $x \in T$, the function $f(x, \cdot)$ is not integrable on $[a_2, b_2]$. Since $T \times [a_2, b_2]$ is negligible in I, we can modify on that set the function f without changing the integrability properties. We can set, for example, $f = 0$ on that set. In this way, we can assume without loss of generality that T is empty. Let us define

$$F(x) = \int_{a_2}^{b_2} f(x, y) \, dy \, .$$

We want to prove that F is integrable on $[a_1, b_1]$ and that

$$\int_{a_1}^{b_1} F = \int_I f \, .$$

Let $\varepsilon > 0$ be fixed. Because of the integrability of f on I, there is a gauge δ on I such that, for every δ-fine tagged partition $\overset{\circ}{\mathcal{P}}$ of I,

$$\left| S(f, \overset{\circ}{\mathcal{P}}) - \int_I f \right| \le \frac{\varepsilon}{2} \, .$$

For every $x \in [a_1, b_1]$, since $f(x, \cdot)$ is integrable on $[a_2, b_2]$ with integral $F(x)$, there exists a gauge $\bar{\delta}^x : [a_2, b_2] \to \mathbb{R}$ such that, taking any $\bar{\delta}^x$-fine tagged partition $\overset{\circ}{\mathcal{P}}_2$ of $[a_2, b_2]$, we have that

$$|S(f(x, \cdot), \overset{\circ}{\mathcal{P}}_2) - F(x)| \le \frac{\varepsilon}{2(b_1 - a_1)} \, .$$

We can assume that $\bar{\delta}^x(y) \le \delta(x, y)$ for every $(x, y) \in I$. Then let us choose for every $x \in [a_1, b_1]$ a $\bar{\delta}^x$-fine tagged partition $\overset{\circ}{\mathcal{P}}_2^x$ of $[a_2, b_2]$ and write it explicitly as

$$\overset{\circ}{\mathcal{P}}_2^x = \{(y_j^x, K_j^x) : j = 1, \dots, m^x\} \, .$$

We will thus have that, for every $x \in [a_1, b_1]$,

$$|F(x) - S(f(x, \cdot), \overset{\circ}{\mathcal{P}}_2^x)| \le \frac{\varepsilon}{2(b_1 - a_1)} \, .$$

We define a gauge δ_1 on $[a_1, b_1]$ by setting

$$\delta_1(x) = \min\{\delta(x, y_1^x), \ldots, \delta(x, y_{m^x}^x)\}.$$

We will prove that, for every δ_1-fine tagged partition $\mathring{\mathcal{P}}_1$ of $[a_1, b_1]$,

$$\left| S(F, \mathring{\mathcal{P}}_1) - \int_I f \right| \leq \varepsilon.$$

Let us take a δ_1-fine tagged partition of $[a_1, b_1]$,

$$\mathring{\mathcal{P}}_1 = \{(x_i, J_i) : i = 1, \ldots, n\},$$

and construct, starting from it, a δ-fine tagged partition of I,

$$\mathring{\mathcal{P}} = \{((x_i, y_j^{x_i}), J_i \times K_j^{x_i}) : i = 1, \ldots, n, \ j = 1, \ldots, m^{x_i}\}.$$

We have the following inequalities:

$$\left| S(F, \mathring{\mathcal{P}}_1) - \int_I f \right| \leq |S(F, \mathring{\mathcal{P}}_1) - S(f, \mathring{\mathcal{P}})| + \left| S(f, \mathring{\mathcal{P}}) - \int_I f \right|$$

$$\leq \left| \sum_{i=1}^n F(x_i)\mu(J_i) - \sum_{i=1}^n \sum_{j=1}^{m^{x_i}} f(x_i, y_j^{x_i})\mu(J_i \times K_j^{x_i}) \right| + \frac{\varepsilon}{2}$$

$$\leq \sum_{i=1}^n \left| F(x_i) - \sum_{j=1}^{m^{x_i}} f(x_i, y_j^{x_i})\mu(K_j^{x_i}) \right| \mu(J_i) + \frac{\varepsilon}{2}$$

$$\leq \sum_{i=1}^n \frac{\varepsilon}{2(b_1 - a_1)} \mu(J_i) + \frac{\varepsilon}{2} = \varepsilon.$$

This proves that F is integrable on $[a_1, b_1]$ and

$$\int_{a_1}^{b_1} F = \int_I f.$$

The proof is thus completed.   ■

Example Consider the function $f(x, y) = x^2 \sin y$ on the rectangle $I = [-1, 1] \times [0, \pi]$. Since f is continuous on a compact set, it is integrable there, so that

$$\int_I f = \int_{-1}^{1} \left(\int_0^\pi x^2 \sin y \, dy \right) dx$$

$$= \int_{-1}^{1} x^2 [- \cos y]_0^\pi \, dx = 2 \int_{-1}^{1} x^2 \, dx = 2 \left[\frac{x^3}{3} \right]_{-1}^{1} = \frac{4}{3}.$$

Clearly, the following version of the Fubini theorem holds, which is symmetric with respect to the preceding one.

Theorem 11.35 (Reduction Theorem—II) *Let $f : I \to \mathbb{R}$ be an integrable function on the rectangle $I = [a_1, b_1] \times [a_2, b_2]$. Then:*

(a) *For almost every $y \in [a_2, b_2]$ the function $f(\cdot, y)$ is integrable on $[a_1, b_1]$.*
(b) *The function $\int_{a_1}^{b_1} f(x, \cdot) \, dx$, defined almost everywhere on $[a_2, b_2]$, is integrable there.*
(c) *We have*

$$\int_I f = \int_{a_2}^{b_2} \left(\int_{a_1}^{b_1} f(x, y) \, dx \right) dy.$$

As an immediate consequence, we have that, if f is integrable on $I = [a_1, b_1] \times [a_2, b_2]$, then

$$\int_{a_1}^{b_1} \left(\int_{a_2}^{b_2} f(x, y) \, dy \right) dx = \int_{a_2}^{b_2} \left(\int_{a_1}^{b_1} f(x, y) \, dx \right) dy.$$

Therefore, if the preceding equality does not hold, then the function f is not integrable on I.

Examples Consider the function

$$f(x, y) = \begin{cases} \dfrac{x^2 - y^2}{(x^2 + y^2)^2} & \text{if } (x, y) \neq (0, 0), \\ 0 & \text{if } (x, y) = (0, 0), \end{cases}$$

on the rectangle $I = [0, 1] \times [0, 1]$. If $x \neq 0$, then we have

$$\int_0^1 \frac{x^2 - y^2}{(x^2 + y^2)^2} \, dy = \left[\frac{y}{x^2 + y^2} \right]_{y=0}^{y=1} = \frac{1}{x^2 + 1},$$

so that

$$\int_0^1 \left(\int_0^1 \frac{x^2 - y^2}{(x^2 + y^2)^2} \, dy \right) dx = \int_0^1 \frac{1}{x^2 + 1} \, dx = [\arctan x]_0^1 = \frac{\pi}{4}.$$

Analogously, we see that

$$\int_0^1 \left(\int_0^1 \frac{x^2 - y^2}{(x^2 + y^2)^2} \, dx \right) dy = -\frac{\pi}{4},$$

and we thus conclude that f is not integrable on I.

As a further example, consider the function

$$f(x, y) = \begin{cases} \dfrac{xy}{(x^2 + y^2)^2} & \text{if } (x, y) \neq (0, 0), \\ 0 & \text{if } (x, y) = (0, 0), \end{cases}$$

on the rectangle $I = [-1, 1] \times [-1, 1]$. In this case, if $x \neq 0$, we have

$$\int_{-1}^1 \frac{xy}{(x^2 + y^2)^2} \, dy = \left[\frac{-x}{2(x^2 + y^2)} \right]_{y=-1}^{y=1} = 0,$$

so that

$$\int_{-1}^1 \left(\int_{-1}^1 \frac{xy}{(x^2 + y^2)^2} \, dy \right) dx = 0.$$

Analogously, we see that

$$\int_{-1}^1 \left(\int_{-1}^1 \frac{xy}{(x^2 + y^2)^2} \, dx \right) dy = 0.$$

Nevertheless, we are not allowed to conclude that f is integrable on I. Actually, it is not at all. Indeed, if f were integrable, it should be on every subrectangle, and in particular on $[0, 1] \times [0, 1]$. But if $x \neq 0$, then we have

$$\int_0^1 \frac{xy}{(x^2 + y^2)^2} \, dy = \left[\frac{-x}{2(x^2 + y^2)} \right]_{y=0}^{y=1} = \frac{1}{2x(x^2 + 1)},$$

which is not integrable with respect to x on $[0, 1]$.

When the function f is defined on a bounded subset E of $\mathbb{R}^2$, it is possible to state the reduction theorem for the function f_E. Let $I = [a_1, b_1] \times [a_2, b_2]$ be a rectangle containing E. Let us define the "sections" of E :

$$E_x = \{y \in [a_2, b_2] : (x, y) \in E\}, \qquad E_y = \{x \in [a_1, b_1] : (x, y) \in E\},$$

and the "projections" of E :

$$P_1 E = \{x \in [a_1, b_1] : E_x \neq \emptyset\}, \qquad P_2 E = \{y \in [a_2, b_2] : E_y \neq \emptyset\}.$$

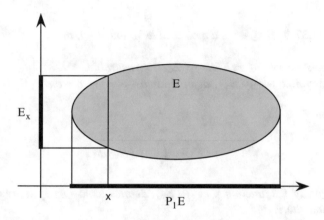

We can then reformulate the Fubini theorem in the following way.

Theorem 11.36 (Reduction Theorem—III) *Let* $f : E \to \mathbb{R}$ *be an integrable function on the bounded set* E. *Then:*

(a) *For almost every* $x \in P_1 E$ *the function* $f_E(x, \cdot)$ *is integrable on the set* E_x.
(b) *The function* $x \mapsto \int_{E_x} f(x, y)\, dy$, *defined almost everywhere on* $P_1 E$, *is integrable there.*
(c) *We have*

$$\int_E f = \int_{P_1 E} \left(\int_{E_x} f(x, y)\, dy \right) dx.$$

Analogously, the function $y \mapsto \int_{E_y} f(x, y)\, dx$, *defined almost everywhere on* $P_2 E$, *is integrable there, and*

$$\int_E f = \int_{P_2 E} \left(\int_{E_y} f(x, y)\, dx \right) dy.$$

Example Consider the function $f(x, y) = |xy|$ on the set

$$E = \{(x, y) \in \mathbb{R}^2 : 0 \leq x \leq 1, -x^2 \leq y \leq x^2\}.$$

Since f is continuous and E is compact, the theorem applies; we have $P_1 E = [0, 1]$ and, for every $x \in P_1 E$, $E_x = [-x^2, x^2]$. Hence:

$$\int_E f = \int_0^1 \left(\int_{-x^2}^{x^2} |xy| \, dy \right) dx = \int_0^1 |x| \left[\frac{y|y|}{2} \right]_{-x^2}^{x^2} dx = \int_0^1 x^5 \, dx = \left[\frac{x^6}{6} \right]_0^1 = \frac{1}{6} .$$

As a corollary, we have a method to compute the measure of a bounded measurable set.

Corollary 11.37 *If $E \subseteq \mathbb{R}^2$ is a measurable bounded set, then:*

(a) *For almost every $x \in P_1 E$ the set E_x is measurable.*
(b) *The function $x \mapsto \mu(E_x)$, defined almost everywhere on $P_1 E$, is integrable there.*
(c) *We have*

$$\mu(E) = \int_{P_1 E} \mu(E_x) \, dx .$$

Analogously, the function $y \mapsto \mu(E_y)$, defined almost everywhere on $P_2 E$, is integrable there, and

$$\mu(E) = \int_{P_2 E} \mu(E_y) \, dy .$$

Example Let us compute the area of a disk with radius $R > 0$: Let $E = \{(x, y) \in \mathbb{R}^2 : x^2 + y^2 \leq R^2\}$. Since E is a compact set, it is measurable. We have that $P_1 E = [-R, R]$ and, for every $x \in P_1 E$, $E_x = [-\sqrt{R^2 - x^2}, \sqrt{R^2 - x^2}]$. Hence:

$$\mu(E) = \int_{-R}^R 2\sqrt{R^2 - x^2} \, dx = \int_{-\pi/2}^{\pi/2} 2R^2 \cos^2 t \, dt$$

$$= R^2 [t + \cos t \sin t]_{-\pi/2}^{\pi/2} = \pi R^2 .$$

In the case of functions of more than two variables, results analogous to the preceding ones hold true, with the same proofs. One simply needs to separate the variables into two different groups, calling x the first group and y the second one, and the same formulas hold.

Example We want to compute the volume of a three-dimensional ball with radius $R > 0$. Let $E = \{(x, y, z) \in \mathbb{R}^3 : x^2 + y^2 + z^2 \leq R^2\}$. Let us group together the variables (y, z) and consider the projection on the x-axis: $P_1 E = [-R, R]$. The sections E_x then are disks of radius $\sqrt{R^2 - x^2}$, and we have

$$\mu(E) = \int_{-R}^{R} \pi(R^2 - x^2)\, dx = \pi(2R^3) - \pi\left[\frac{x^3}{3}\right]_{-R}^{R} = \frac{4}{3}\pi R^3 .$$

Another way to compute the same volume is to group the variables (x, y) and consider $P_1 E = \{(x, y) : x^2 + y^2 = R^2\}$. For every $(x, y) \in P_1 E$ we have

$$E_{(x,y)} = \left[-\sqrt{R^2 - x^2 - y^2}, \sqrt{R^2 - x^2 - y^2}\right],$$

so that

$$\mu(E) = \int_{P_1 E} 2\sqrt{R^2 - x^2 - y^2}\, dx\, dy$$

$$= \int_{-R}^{R} \left(\int_{-\sqrt{R^2-x^2}}^{\sqrt{R^2-x^2}} 2\sqrt{R^2 - x^2 - y^2}\, dy\right) dx$$

$$= \int_{-R}^{R} \left(\int_{-\pi/2}^{\pi/2} 2(R^2 - x^2) \cos^2 t\, dt\right) dx = \int_{-R}^{R} \pi(R^2 - x^2)\, dx = \frac{4}{3}\pi R^3 ,$$

by the change of variable $t = \arcsin\left(y/\sqrt{R^2 - x^2}\right)$.

Iterating the preceding reduction procedure, it is possible to prove, for a function of N variables that is integrable on a rectangle

$$I = [a_1, b_1] \times [a_2, b_2] \times \cdots \times [a_N, b_N],$$

formulas like

$$\int_I f = \int_{a_1}^{b_1} \left(\int_{a_2}^{b_2} \left(\ldots \int_{a_N}^{b_N} f(x_1, x_2, \ldots, x_N)\, dx_N \ldots\right) dx_2\right) dx_1 .$$

11.9   Change of Variables in the Integral

In this section we look for an analogue to the formula of integration by substitution, which was proved in Chap. 7 for functions of a single variable. The proof of that formula was based on the Fundamental Theorem. Since we do not have such a powerful tool for functions of several variables, actually we will not be able to completely generalize that formula.

For example, not only will the function φ be assumed to be differentiable, but we will need it to be a diffeomorphism between two open sets A and B of $\mathbb{R}^N$. In other words, $\varphi : A \to B$ will be continuously differentiable and invertible, and $\varphi^{-1} : B \to A$ will be continuously differentiable as well. It is useful to recall that, by Theorem 2.11, a diffeomorphism transforms open sets into open sets and closed sets into closed sets. Moreover, by Theorem 10.26, for every point $u \in A$ the Jacobian matrix $J\varphi(u)$ is invertible: We have

$$\det J\varphi(u) \neq 0\,.$$

From now on, we will often use a different notation for the Jacobian matrix:

$$\text{instead of } J\varphi(u)\text{, we will write } \varphi'(u)\,.$$

We will also need the following property.

Lemma 11.38 *Let $A \subseteq \mathbb{R}^N$ be an open set and $\varphi : A \to \mathbb{R}^N$ a C^1-function; if S is a subset of A of the type*

$$S = [a_1, b_1] \times \cdots \times [a_{N-1}, b_{N-1}] \times \{c\}\,,$$

then $\varphi(S)$ is negligible.

Proof For simplicity, let us concentrate on the case of a subset of $\mathbb{R}^2$ of the type

$$S = [0, 1] \times \{0\}\,.$$

For any positive integer n, consider the rectangles (actually squares)

$$J_{k,n} = \left[\frac{k-1}{n}, \frac{k}{n}\right] \times \left[-\frac{1}{2n}, \frac{1}{2n}\right],$$

with $k = 1, \ldots, n$. For n large enough, they are contained in a rectangle R, which itself is contained in A. Since R is a compact set, there is a constant $C > 0$ such that $\|\varphi'(u)\| \leq C$ for every $u \in R$. By the Mean Vale Theorem 10.20, φ is "Lipschitz continuous" on R with Lipschitz constant C, i.e.,

$$\|\varphi(u) - \varphi(v)\| \leq C\|u - v\|\,, \quad \text{for every } u, v \in R\,.$$

Since the sets $J_{k,n}$ have as diameter $\frac{1}{n}\sqrt{2}$, the sets $\varphi(J_{k,n})$ are surely contained in some squares $\tilde{J}_{k,n}$ whose sides' lengths are equal to $\frac{C}{n}\sqrt{2}$. We then have that $\varphi(S)$ is covered by the rectangles $\tilde{J}_{k,n}$ and

$$\sum_{k=1}^{n} \mu(\tilde{J}_{k,n}) \le n \left(\frac{C}{n}\sqrt{2}\right)^2 = \frac{2C^2}{n}.$$

Since this quantity can be made arbitrarily small, the conclusion follows from Corollary 11.23. ∎

As a consequence of the foregoing lemma, it is easy to see that the image of the boundary of a rectangle through a diffeomorphism φ is a negligible set. In particular, given two nonoverlapping rectangles, their images are nonoverlapping sets.

We are now ready to prove a first version of the **change of variables formula** in the integral, which will be generalized in a later section.

Theorem 11.39 (Change of Variables Theorem—I) *Let A and B be open subsets of $\mathbb{R}^N$ and $\varphi : A \to B$ a diffeomorphism. If $f : B \to \mathbb{R}$ is a continuous function, then, for every compact subset D of A,*

$$\int_{\varphi(D)} f(\boldsymbol{x}) \, d\boldsymbol{x} = \int_D f(\varphi(\boldsymbol{u})) \, |\det \varphi'(\boldsymbol{u})| \, d\boldsymbol{u} .$$

Proof Note first of all that the integrals in the formula are both meaningful, since the sets D and $\varphi(D)$ are compact and the considered functions continuous. We will proceed by induction on the dimension N. Let us first consider the case $N = 1$.

First, using the method of integration by substitution, one verifies that the formula is true when D is a compact interval $[a, b]$: It is sufficient to consider the two possible cases in which φ is increasing or decreasing and recall that every continuous function is primitivable. For instance, if φ is decreasing, then we have $\varphi([a, b]) = [\varphi(b), \varphi(a)]$, so that

$$
\begin{aligned}
\int_{\varphi([a,b])} f(x) \, dx &= \int_{\varphi(b)}^{\varphi(a)} f(x) \, dx \\
&= \int_b^a f(\varphi(u))\varphi'(u) \, du \\
&= \int_a^b f(\varphi(u))|\varphi'(u)| \, du \\
&= \int_{[a,b]} f(\varphi(u))|\varphi'(u)| \, du .
\end{aligned}
$$

Now let R be a compact subset of A whose interior $\mathring{R}$ contains D. Since both f and $(f \circ \varphi)|\varphi'|$ are continuous, they are integrable on the compact sets $\varphi(R)$ and R, respectively. The open sets $\mathring{R}$ and $\mathring{R} \setminus D$ can each be split into a countable union of nonoverlapping compact intervals whose images through φ also are nonoverlapping close intervals. By the complete additivity of the integral, the formula holds true for

$\mathring{R}$ and $\mathring{R} \setminus D$:

$$\int_{\varphi(\mathring{R})} f(x)\, dx = \int_{\mathring{R}} f(\varphi(u))|\varphi'(u)|\, du\, ,$$

$$\int_{\varphi(\mathring{R} \setminus D)} f(x)\, dx = \int_{\mathring{R} \setminus D} f(\varphi(u))|\varphi'(u)|\, du\, .$$

Hence,

$$\int_{\varphi(D)} f(x)\, dx = \int_{\varphi(\mathring{R} \setminus (\mathring{R} \setminus D))} f(x)\, dx$$

$$= \int_{\varphi(\mathring{R})} f(x)\, dx - \int_{\varphi(\mathring{R} \setminus D)} f(x)\, dx$$

$$= \int_{\mathring{R}} f(\varphi(u))|\varphi'(u)|\, du - \int_{\mathring{R} \setminus D} f(\varphi(u))|\varphi'(u)|\, du$$

$$= \int_{D} f(\varphi(u))|\varphi'(u)|\, du\, ,$$

so that the formula is proved in the case $N = 1$.

Assume now that the formula holds for the dimension N, and let us prove that it also holds for $N + 1$.[1] Once we fix a point $\bar{u} \in A$, at least one of the partial derivatives $\frac{\partial \varphi_i}{\partial u_j}(\bar{u})$ is different from zero. We can assume without loss of generality that it is $\frac{\partial \varphi_{N+1}}{\partial u_{N+1}}(\bar{u}) \neq 0$. Consider the function

$$\alpha(u_1, \ldots, u_{N+1}) = (u_1, \ldots, u_N, \varphi_{N+1}(u_1, \ldots, u_{N+1}))\, .$$

Since $\det \alpha'(\bar{u}) = \frac{\partial \varphi_{N+1}}{\partial u_{N+1}}(\bar{u}) \neq 0$, by Theorem 10.24 we have that α is a diffeomorphism between an open neighborhood U of $\bar{u}$ and an open neighborhood V of $\alpha(\bar{u})$. Assume first that D is contained in U, and set $\widetilde{D} = \alpha(D)$.

We define on V the function $\beta = \varphi \circ \alpha^{-1}$, which is of the form

$$\beta(v_1, \ldots, v_{N+1}) = (\beta_1(v_1, \ldots, v_{N+1}), \ldots, \beta_N(v_1, \ldots, v_{N+1}), v_{N+1})\, ,$$

where, for $j = 1, \ldots, N$, we have

$$\beta_j(v_1, \ldots, v_{N+1}) = \varphi_j(v_1, \ldots, v_N, [\varphi_{N+1}(v_1, \ldots, v_N, \cdot)]^{-1}(v_{N+1}))\, .$$

Such a function β is a diffeomorphism between the open sets V and $W = \varphi(U)$.

---

[1] At a first reading, it is advisable to consider the transition from $N = 1$ to $N + 1 = 2$.

Consider the sections

$$V_t = \{(v_1, \ldots, v_N) : (v_1, \ldots, v_N, t) \in V\}$$

and the projection

$$P_{N+1}V = \{t : V_t \neq \emptyset\}.$$

For $t \in P_{N+1}V$, define the function

$$\beta_t(v_1, \ldots, v_N) = (\beta_1(v_1, \ldots, v_N, t), \ldots, \beta_N(v_1, \ldots, v_N, t)),$$

which happens to be a diffeomorphism defined on the open set V_t whose image is the open set

$$W_t = \{(x_1, \ldots, x_N) : (x_1, \ldots, x_N, t) \in W\}.$$

Moreover, $\det \beta_t'(v_1, \ldots, v_N) = \det \beta'(v_1, \ldots, v_N, t)$. Consider also the sections

$$\widetilde{D}_t = \{(v_1, \ldots, v_N) : (v_1, \ldots, v_N, t) \in \widetilde{D}\}$$

and the projection

$$P_{N+1}\widetilde{D} = \{t : \widetilde{D}_t \neq \emptyset\}.$$

Analogously, we consider $\beta(\widetilde{D})_t$ and $P_{N+1}\beta(\widetilde{D})$. By the definition of β, we have

$$\beta(\widetilde{D})_t = \beta_t(\widetilde{D}_t), \qquad P_{N+1}\beta(\widetilde{D}) = P_{N+1}\widetilde{D}.$$

Using the Reduction Theorem 11.34 and the inductive assumption, we have

$$\int_{\beta(\widetilde{D})} f = \int_{P_{N+1}\beta(\widetilde{D})} \left(\int_{\beta_t(\widetilde{D}_t)} f(x_1, \ldots, x_N, t) \, dx_1 \ldots dx_N \right) dt$$

$$= \int_{P_{N+1}\widetilde{D}} \left(\int_{\widetilde{D}_t} f(\beta_t(v_1, \ldots, v_N), t) \, | \det \beta_t'(v_1, \ldots, v_N)| \, dv_1 \ldots dv_N \right) dt$$

$$= \int_{P_{N+1}\widetilde{D}} \left(\int_{\widetilde{D}_t} f(\beta(v_1, \ldots, v_N, t)) \, | \det \beta'(v_1, \ldots, v_N, t)| \, dv_1 \ldots dv_N \right) dt$$

$$= \int_{\widetilde{D}} f(\beta(v)) \, | \det \beta'(v)| \, dv.$$

Consider now the function $\tilde{f} : V \to \mathbb{R}$ defined as

$$\tilde{f}(v) = f(\beta(v)) \, | \det \beta'(v)|.$$

Define the sections

$$D_{u_1,\ldots,u_N} = \{u_{N+1} : (u_1,\ldots,u_N,u_{N+1}) \in D\}$$

and the projection

$$P_{1,\ldots,N}D = \{(u_1,\ldots,u_N) : D_{u_1,\ldots,u_N} \neq \varnothing\}\,.$$

In an analogous way we define $\alpha(D)_{u_1,\ldots,u_N}$ and $P_{1,\ldots,N}\alpha(D)$. They are all closed sets, and by the definition of α, we have

$$\alpha(D)_{u_1,\ldots,u_N} = \varphi_{N+1}(u_1,\ldots,u_N,D_{u_1,\ldots,u_N})\,, \qquad P_{1,\ldots,N}\alpha(D) = P_{1,\ldots,N}D\,.$$

Moreover, for every $(u_1,\ldots,u_N) \in P_{1,\ldots,N}D$, the function defined by

$$t \to \varphi_{N+1}(u_1,\ldots,u_N,t)$$

is a diffeomorphism of one variable between the open sets $U_{u_1,\ldots,u_N}$ and $V_{u_1,\ldots,u_N}$, sections of U and V, respectively. Using the Reduction Theorem 11.34 and the one-dimensional change of variables formula proved earlier, we have that

$$\int_{\alpha(D)} \tilde{f} = \int_{P_{1,\ldots,N}\alpha(D)} \left(\int_{\alpha(D)_{u_1,\ldots,u_N}} \tilde{f}(v_1,\ldots,v_{N+1})\,dv_{N+1} \right) dv_1 \ldots dv_N$$

$$= \int_{P_{1,\ldots,N}D} \left(\int_{\varphi_{N+1}(u_1,\ldots,u_N,D_{u_1,\ldots,u_N})} \tilde{f}(v_1,\ldots,v_{N+1})\,dv_{N+1} \right) dv_1 \ldots dv_N$$

$$= \int_{P_{1,\ldots,N}D} \left(\int_{D_{u_1,\ldots,u_N}} \tilde{f}(u_1,\ldots,u_N,\varphi_{N+1}(u_1,\ldots,u_{N+1}))\,\cdot \right.$$

$$\left. \cdot \left| \frac{\partial \varphi_{N+1}}{\partial u_{N+1}}(u_1,\ldots,u_{N+1}) \right| du_{N+1} \right) du_1 \ldots du_N$$

$$= \int_D \tilde{f}(\alpha(\boldsymbol{u}))\,|\det \alpha'(\boldsymbol{u})|\,d\boldsymbol{u}\,.$$

Hence, since $\varphi = \beta \circ \alpha$, we have

$$\int_{\varphi(D)} f(\boldsymbol{x})\,d\boldsymbol{x} = \int_{\beta(\tilde{D})} f(\boldsymbol{x})\,d\boldsymbol{x}$$

$$= \int_{\tilde{D}} f(\beta(\boldsymbol{v}))\,|\det \beta'(\boldsymbol{v})|\,d\boldsymbol{v}$$

$$= \int_{\alpha(D)} \tilde{f}(\boldsymbol{v})\,d\boldsymbol{v}$$

$$= \int_D \tilde{f}(\alpha(\boldsymbol{u})) \, |\det \alpha'(\boldsymbol{u})| \, d\boldsymbol{u}$$

$$= \int_D f(\beta(\alpha(\boldsymbol{u}))) \, |\det \beta'(\alpha(\boldsymbol{u}))| \, |\det \alpha'(\boldsymbol{u})| \, d\boldsymbol{u}$$

$$= \int_D f(\varphi(\boldsymbol{u})) \, |\det \varphi'(\boldsymbol{u})| \, d\boldsymbol{u} \, .$$

We have then proved that, for every $\boldsymbol{u} \in A$, there is a $\delta(\boldsymbol{u}) > 0$ such that the thesis holds true when D is contained in $\overline{B}[\boldsymbol{u}, \delta(\boldsymbol{u})]$. A gauge δ is thus defined on A. By Lemma 11.20, we can now cover A with a countable family $(J_k)_k$ of nonoverlapping rectangles, each contained in a rectangle of the type $\overline{B}[\boldsymbol{u}, \delta(\boldsymbol{u})]$, so that the formula holds for the closed sets contained in any of these rectangles.

At this point let us consider an arbitrary compact subset D of A. Then the formula holds for each $D \cap J_k$, and, by the complete additivity of the integral and the fact that the sets $\varphi(D \cap J_k)$ are nonoverlapping (as a consequence of Lemma 11.38), we have

$$\int_{\varphi(D)} f(\boldsymbol{x}) \, d\boldsymbol{x} = \sum_k \int_{\varphi(D \cap J_k)} f(\boldsymbol{x}) \, d\boldsymbol{x}$$

$$= \sum_k \int_{D \cap J_k} f(\varphi(\boldsymbol{u})) \, |\det \varphi'(\boldsymbol{u})| \, d\boldsymbol{u}$$

$$= \int_D f(\varphi(\boldsymbol{u})) \, |\det \varphi'(\boldsymbol{u})| \, d\boldsymbol{u} \, .$$

The theorem is thus completely proved.   ■

Remark 11.40 The change of variables formula is often written, setting $\varphi(D) = E$, in the equivalent form

$$\int_E f(\boldsymbol{x}) \, d\boldsymbol{x} = \int_{\varphi^{-1}(E)} f(\varphi(\boldsymbol{u})) \, |\det \varphi'(\boldsymbol{u})| \, d\boldsymbol{u} \, .$$

Example Consider the set

$$E = \{(x, y) \in \mathbb{R}^2 : -1 \le x \le 1, \, x^2 \le y \le x^2 + 1\},$$

and let $f(x, y) = x^2 y$ be a function on it. Defining $\varphi(u, v) = (u, v + u^2)$, we have a diffeomorphism with $\det \varphi'(u, v) = 1$. Since $\varphi^{-1}(E) = [-1, 1] \times [0, 1]$, by the change of variables formula and the use of the Fubini reduction theorem we have

$$\int_E x^2 y \, dx \, dy = \int_{-1}^1 \left(\int_0^1 u^2 (v + u^2) \, dv \right) du = \int_{-1}^1 \left(\frac{u^2}{2} + u^4 \right) du = \frac{11}{15} \, .$$

11.10 Change of Measure by Diffeomorphisms

In this section we study how a measure is changed by the action of a diffeomorphism.

Theorem 11.41 *Let A and B be open subsets of $\mathbb{R}^N$, and let $\varphi : A \to B$ be a diffeomorphism. Let $D \subseteq A$ and $\varphi(D) \subseteq B$ be bounded sets. If D is measurable, then $\varphi(D)$ is measurable, $|\det \varphi'|$ is integrable on D, and*

$$\mu(\varphi(D)) = \int_D |\det \varphi'(u)| \, du.$$

Proof By the preceding theorem, the formula holds true whenever D is compact. Since every open set can be written as the union of a countable family of nonoverlapping (closed) rectangles, by the complete additivity and the fact that A is bounded, the formula holds true even if D is an open bounded set.

Assume now that D is a measurable bounded set whose closure $\overline{D}$ is contained in A. Let R be a compact subset of A whose interior $\mathring{R}$ contains $\overline{D}$. Then there is a constant $C > 0$ such that $|\det \varphi'(u)| \le C$ for every $u \in R$. By Proposition 11.22, for every $\varepsilon > 0$ there are two finite or countable families (J_k) and (J'_k), each made of nonoverlapping rectangles contained in $\mathring{R}$ such that

$$\mathring{R} \setminus \left(\bigcup_k J'_k \right) \subseteq D \subseteq \bigcup_k J_k, \quad \mu\left(\left(\bigcup_k J_k \right) \cap \left(\bigcup_k J'_k \right) \right) \le \varepsilon.$$

Since the formula to be proved holds on both the open bounded sets and the compact sets, it certainly holds on each rectangle J_k and J'_k; then it holds on $\cup_k J_k$ and on $\cup_k J'_k$, and since it holds even on $\mathring{R}$, it must be true on $\mathring{R} \setminus (\cup_k J'_k)$ as well. Thus, we have that $\varphi(\cup_k J_k)$ and $\varphi(\mathring{R} \setminus (\cup_k J'_k))$ are measurable,

$$\varphi\left(\mathring{R} \setminus \left(\bigcup_k J'_k \right) \right) \subseteq \varphi(D) \subseteq \varphi\left(\bigcup_k J_k \right),$$

and

$$\mu\left(\varphi\left(\bigcup_k J_k \right) \right) - \mu\left(\varphi\left(\mathring{R} \setminus \left(\bigcup_k J'_k \right) \right) \right) =$$

$$= \int_{\cup_k J_k} |\det \varphi'(u)| \, du - \int_{\mathring{R} \setminus (\cup_k J'_k)} |\det \varphi'(u)| \, du$$

$$= \int_{(\cup_k J_k) \cap (\cup_k J'_k)} |\det \varphi'(u)| \, du$$

$$\leq C\mu\left(\left(\bigcup_k J_k\right) \cap \left(\bigcup_k J_k'\right)\right)$$

$$\leq C\varepsilon.$$

Taking $\varepsilon = \frac{1}{n}$, we find in this way two sequences $D_n = \cup_k J_{k,n}$ and $D_n' = \mathring{R} \setminus (\cup_k J_{k,n}')$ with the aforementioned properties. By Proposition 11.22, we have that $\varphi(D)$ is measurable and $\mu(\varphi(D)) = \lim_n \mu(\varphi(D_n)) = \lim_n \mu(\varphi(D_n'))$. Moreover, since χ_{D_n} converges almost everywhere to χ_D, by the Dominated Convergence Theorem 9.13, we have that

$$\mu(\varphi(D)) = \lim_n \mu(\varphi(D_n))$$

$$= \lim_n \int_{D_n} |\det \varphi'(\boldsymbol{u})|\, d\boldsymbol{u}$$

$$= \lim_n \int_R |\det \varphi'(\boldsymbol{u})|\chi_{D_n}(\boldsymbol{u})\, d\boldsymbol{u}$$

$$= \int_R |\det \varphi'(\boldsymbol{u})|\chi_D(\boldsymbol{u})\, d\boldsymbol{u}$$

$$= \int_D |\det \varphi'(\boldsymbol{u})|\, d\boldsymbol{u}.$$

We can now consider the case of an arbitrary measurable bounded set D in A. Since D is bounded, there is an open ball $B(\boldsymbol{0}, \rho)$ containing it. Let $A' = A \cap B(\boldsymbol{0}, \rho)$ and $B' = \varphi(A')$. Since A' is open and bounded, as in the proof of Proposition 11.11, we can consider a sequence of nonoverlapping rectangles $(K_n)_n$ whose union is equal to A'. The formula holds for each of the sets $D \cap K_n$, by the foregoing considerations. The complete additivity of the integral (Theorem 11.28) and the fact that A' is bounded then lead us to our conclusion.   ∎

Example Consider the set

$$E = \{(x, y) \in \mathbb{R}^2 : x < y < 2x, \ 3x^2 < y < 4x^2\}.$$

We see that E is measurable since it is an open set. Taking

$$\varphi(u, v) = \left(\frac{u}{v}, \frac{u^2}{v}\right),$$

we have a diffeomorphism between the set $D =]1, 2[\times]3, 4[$ and $E = \varphi(D)$. Moreover,

$$\det \varphi'(u, v) = \det \begin{pmatrix} 1/v & -u/v^2 \\ 2u/v & -u^2/v^2 \end{pmatrix} = \frac{u^2}{v^3}.$$

Applying the formula on the change of measure and the Reduction Theorem 11.34, we have that

$$\mu(E) = \int_1^2 \left(\int_3^4 \frac{u^2}{v^3} \, dv \right) du = \int_1^2 \frac{7}{288} u^2 \, du = \frac{49}{864} \, .$$

11.11   The General Theorem on Change of Variables

We are now interested in generalizing the Change of Variables Theorem 11.39 assuming f is not necessarily continuous but only L-integrable on a measurable set. To do this, it will be useful to prove the following important relationship between the integral of a function having nonnegative values and the measure of its hypograph.

Proposition 11.42 *Let E be a measurable bounded set and $f : E \to \mathbb{R}$ a bounded function with nonnegative values. Let G_f be the set thus defined:*

$$G_f = \{(\boldsymbol{x}, t) \in E \times \mathbb{R} : 0 \le t \le f(\boldsymbol{x})\} \, .$$

Then f is integrable on E if and only if G_f is measurable, in which case

$$\mu(G_f) = \int_E f \, .$$

Proof Assume first that G_f is measurable. By Fubini's Reduction Theorem 11.36, since $P_1 G_f = E$, the sections being $(G_f)_{\boldsymbol{x}} = [0, f(\boldsymbol{x})]$, we have that the function $\boldsymbol{x} \mapsto \int_0^{f(\boldsymbol{x})} 1 = f(\boldsymbol{x})$ is integrable on E and

$$\mu(G_f) = \int_{G_f} 1 = \int_E \left(\int_0^{f(\boldsymbol{x})} 1 \, dt \right) d\boldsymbol{x} = \int_E f(\boldsymbol{x}) \, d\boldsymbol{x} \, .$$

Assume now that f is integrable on E. Let $C > 0$ be a constant such that $0 \le f(\boldsymbol{x}) < C$ for every $\boldsymbol{x} \in E$. Given a positive integer n, we divide the interval $[0, C]$ into n equal parts and consider, for $j = 1, \ldots, n$, the sets

$$E_n^j = \left\{ \boldsymbol{x} \in E : \frac{j-1}{n} C \le f(\boldsymbol{x}) < \frac{j}{n} C \right\} ;$$

as a consequence of Lemma 11.16, they are measurable and nonoverlapping, and their union is E. We can then define on E the function ψ_n in the following way:

$$\psi_n = \sum_{j=1}^n \frac{j}{n} C \chi_{E_n^j} \, ,$$

and so

$$G_{\psi_n} = \bigcup_{j=1}^{n} \left(E_n^j \times \left[0, \frac{j}{n}C \right] \right).$$

By Proposition 11.22, it is easy to see that, since the sets E_n^j are measurable, the sets $E_n^j \times \left[0, \frac{j}{n}C \right]$ are, too. Consequently, the sets G_{ψ_n} are measurable. Moreover, since

$$G_f = \bigcap_{n \geq 1} G_{\psi_n} ,$$

even G_f is measurable, and the proof is thus completed.   ∎

We are now in a position to prove the second version of the theorem on the change of variables in the integral.

Theorem 11.43 (Change of Variables Theorem—II) *Let A and B be open subsets of $\mathbb{R}^N$, and let $\varphi : A \to B$ be a diffeomorphism. Let $D \subseteq A$ and $\varphi(D) \subseteq B$ be measurable bounded sets and $f : \varphi(D) \to \mathbb{R}$ a function. Then f is L-integrable on $\varphi(D)$ if and only if $(f \circ \varphi) \,|\det \varphi'|$ is L-integrable on D, in which case*

$$\int_{\varphi(D)} f(\boldsymbol{x}) \, d\boldsymbol{x} = \int_D f(\varphi(\boldsymbol{u})) \, |\det \varphi'(\boldsymbol{u})| \, d\boldsymbol{u} .$$

Proof Assume that f is L-integrable on $E = \varphi(D)$. We first consider the case where f is bounded with nonnegative values.

Let $C > 0$ be such that $0 \leq f(\boldsymbol{x}) < C$ for every $\boldsymbol{x} \in E$. We define the open sets

$$\widetilde{A} = A \times \,] - C, C[\, , \qquad \widetilde{B} = B \times \,] - C, C[$$

and the function $\widetilde{\varphi} : \widetilde{A} \to \widetilde{B}$ in the following way:

$$\widetilde{\varphi}(u_1, \ldots, u_n, t) = (\varphi_1(u_1, \ldots, u_n), \ldots, \varphi_n(u_1, \ldots, u_n), t) .$$

This function is a diffeomorphism, and $\det \widetilde{\varphi}'(\boldsymbol{u}, t) = \det \varphi'(\boldsymbol{u})$ for every $(\boldsymbol{u}, t) \in \widetilde{A}$. Let G_f be the hypograph of f :

$$G_f = \{(\boldsymbol{x}, t) \in E \times \mathbb{R} : 0 \leq t \leq f(\boldsymbol{x})\} .$$

Since f is L-integrable and E is measurable, by the preceding proposition we have that G_f is a measurable set. Moreover,

$$\widetilde{\varphi}^{-1}(G_f) = \{(\boldsymbol{u}, t) \in D \times \mathbb{R} : 0 \leq t \leq f(\varphi(\boldsymbol{u}))\} .$$

Using Theorem 11.41 and Fubini's reduction theorem, we have

$$\mu(G_f) = \int_{\widetilde{\varphi}^{-1}(G_f)} |\det \widetilde{\varphi}'(\boldsymbol{u}, t)| \, d\boldsymbol{u} \, dt$$

$$= \int_{\widetilde{\varphi}^{-1}(G_f)} |\det \varphi'(\boldsymbol{u})| \, d\boldsymbol{u} \, dt$$

$$= \int_D \left(\int_0^{f(\varphi(\boldsymbol{u}))} |\det \varphi'(\boldsymbol{u})| \, dt \right) d\boldsymbol{u}$$

$$= \int_D f(\varphi(\boldsymbol{u})) \, |\det \varphi'(\boldsymbol{u})| \, d\boldsymbol{u} \,.$$

On the other hand, by Proposition 11.42, we have that $\mu(G_f) = \int_{\varphi(D)} f$, and this proves that the formula holds in the case where f is bounded with nonnegative values.

In the case where f is not bounded but still has nonnegative values, we consider the functions

$$f_n(\boldsymbol{x}) = \min\{f(\boldsymbol{x}), n\} \,.$$

For each of them, the formula holds true, and using the Monotone Convergence Theorem 9.10 we prove that the formula holds for f even in this case.

When f does not have nonnegative values, it is sufficient to consider its positive and negative parts, apply the formula to them, and then subtract.

To obtain the opposite implication, it is sufficient to consider $(f \circ \varphi) |\det \varphi'|$ instead of f and φ^{-1} instead of φ and to apply what was just proved. ∎

We recall here the equivalent formula

$$\int_E f(\boldsymbol{x}) \, d\boldsymbol{x} = \int_{\varphi^{-1}(E)} f(\varphi(\boldsymbol{u})) \, |\det \varphi'(\boldsymbol{u})| \, d\boldsymbol{u} \,.$$

11.12 Some Useful Transformations in $\mathbb{R}^2$

Some transformations do not change the measure of any measurable set. We consider here some of those that are most frequently used in applications.

Translations We call translation by a given vector $a = (a_1, a_2) \in \mathbb{R}^2$ the transformation $\varphi : \mathbb{R}^2 \to \mathbb{R}^2$ defined by

$$\varphi(u, v) = (u + a_1, v + a_2) \,.$$

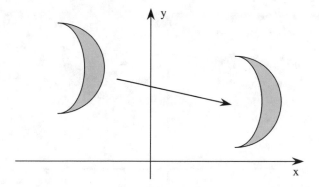

It is readily seen that φ is a diffeomorphism, with $\det \varphi' = 1$, so that, given a measurable bounded set D and an L-integrable function f on $\varphi(D)$, we have

$$\int_{\varphi(D)} f(x, y)\, dx\, dy = \int_D f(u + a_1, v + a_2)\, du\, dv.$$

Reflections A reflection with respect to one of the cartesian axes is defined by

$$\varphi(u, v) = (-u, v), \qquad \text{or} \qquad \varphi(u, v) = (u, -v).$$

Here, $\det \varphi' = -1$, so that, taking for example the first case, we have

$$\int_{\varphi(D)} f(x, y)\, dx\, dy = \int_D f(-u, v)\, du\, dv.$$

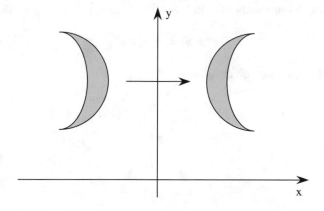

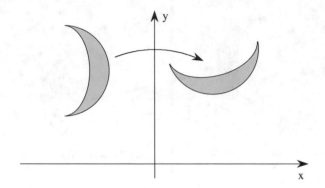

Rotations A rotation around the origin by a fixed angle α is given by

$$\varphi(u, v) = (u \cos \alpha - v \sin \alpha \, , \; u \sin \alpha + v \cos \alpha) \, .$$

It is a diffeomorphism, with

$$\det \varphi'(u, v) = \det \begin{pmatrix} \cos \alpha & -\sin \alpha \\ \sin \alpha & \cos \alpha \end{pmatrix} = (\cos \alpha)^2 + (\sin \alpha)^2 = 1 \, .$$

Hence, given a measurable bounded set D and an L-integrable function[2] f on $\varphi(D)$, we have

$$\int_{\varphi(D)} f(x, y) \, dx \, dy = \int_D f(u \cos \alpha - v \sin \alpha \, , \; u \sin \alpha + v \cos \alpha) \, du \, dv \, .$$

Homotheties A homothety of ratio $\alpha > 0$ is a function $\varphi : \mathbb{R}^2 \to \mathbb{R}^2$ defined by

$$\varphi(u, v) = (\alpha u, \alpha v) \, .$$

It is a diffeomorphism, with $\det \varphi' = \alpha^2$. Hence,

$$\int_{\varphi(D)} f(x, y) \, dx \, dy = \alpha^2 \int_D f(\alpha u, \alpha v) \, du \, dv \, .$$

---

[2] Let us mention here that Buczolich [2] found an ingenious example of an integrable function in $\mathbb{R}^2$ whose rotation by $\alpha = \pi/4$ is not integrable. This is why we have restricted our attention only to L-integrable functions.

Polar Coordinates   Another useful transformation is provided by the function ψ : $[0, +\infty[\times [0, 2\pi[\to \mathbb{R}^2$ given by

$$\psi(\rho, \theta) = (\rho \cos \theta, \rho \sin \theta),$$

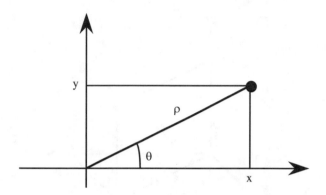

which defines the so-called "polar coordinates" in $\mathbb{R}^2$. Consider the open sets

$$A =]0, +\infty[\times]0, 2\pi[, \qquad B = \mathbb{R}^2 \setminus ([0, +\infty[\times \{0\}) .$$

The function $\varphi : A \to B$ defined by $\varphi(\rho, \theta) = \psi(\rho, \theta)$ happens to be a diffeomorphism, and it is easily seen that

$$\det \varphi'(\rho, \theta) = \rho \qquad \text{for every } (\rho, \theta) \in A .$$

Let $E \subseteq \mathbb{R}^2$ be a measurable bounded set, and consider a function $f : E \to \mathbb{R}$. We can apply the Change of Variables Theorem 11.43 to the set $\widetilde{E} = E \cap B$. Since $\widetilde{E}$ and $\varphi^{-1}(\widetilde{E})$ differ from E and $\psi^{-1}(E)$, respectively, by negligible sets, we obtain the following formula on the **change of variables in polar coordinates**:

$$\int_E f(x, y) \, dx \, dy = \int_{\psi^{-1}(E)} f(\psi(\rho, \theta)) \rho \, d\rho \, d\theta .$$

Example Let $f(x, y) = xy$ be defined on

$$E = \{(x, y) \in \mathbb{R}^2 : x \geq 0, y \geq 0, x^2 + y^2 < 9\} .$$

By the formula on the change of variables in polar coordinates, we have $\psi^{-1}(E) = [0, 3[\times [0, \frac{\pi}{2}]$; by the Reduction Theorem 11.36, we can then compute

$$\int_E f = \int_0^{\pi/2} \left(\int_0^3 \rho^3 \cos \theta \sin \theta \, d\rho \right) d\theta = \frac{81}{4} \int_0^{\pi/2} \cos \theta \sin \theta \, d\theta = \frac{81}{8} .$$

11.13 Cylindrical and Spherical Coordinates in $\mathbb{R}^3$

We consider the function $\xi : [0, +\infty[\times [0, 2\pi[\times \mathbb{R} \to \mathbb{R}^3$ defined by

$$\xi(\rho, \theta, z) = (\rho \cos \theta, \rho \sin \theta, z),$$

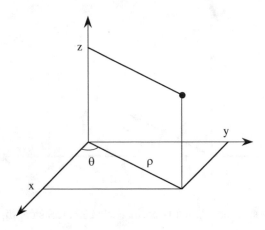

which gives us the so-called "cylindrical coordinates" in $\mathbb{R}^3$. Consider the open sets

$$A =]0, +\infty[\times]0, 2\pi[\times \mathbb{R},$$

$$B = (\mathbb{R}^2 \setminus ([0, +\infty[\times \{0\})) \times \mathbb{R}.$$

The function $\varphi : A \to B$ defined by $\varphi(\rho, \theta, z) = \xi(\rho, \theta, z)$ happens to be a diffeomorphism, and it is easily seen that

$$\det \varphi'(\rho, \theta, z) = \rho, \qquad \text{for every } (\rho, \theta, z) \in A.$$

Let $E \subseteq \mathbb{R}^3$ be a measurable bounded set, and consider a function $f : E \to \mathbb{R}$. We can then apply the Change of Variables Theorem 11.43 to the set $\widetilde{E} = E \cap B$. Since $\widetilde{E}$ and $\varphi^{-1}(\widetilde{E})$ differ from E and $\xi^{-1}(E)$, respectively, by negligible sets, we obtain the following formula on the **change of variables in cylindrical coordinates**:

$$\int_E f(x, y, z) \, dx \, dy \, dz = \int_{\xi^{-1}(E)} f(\xi(\rho, \theta, z)) \rho \, d\rho \, d\theta \, dz.$$

Example Let us compute the integral $\int_E f$, where $f(x, y, z) = x^2 + y^2$ and

$$E = \{(x, y, z) \in \mathbb{R}^3 : x^2 + y^2 \leq 1, 0 \leq z \leq x + y + \sqrt{2}\}.$$

Passing to cylindrical coordinates, we notice that

$$\rho \cos \theta + \rho \sin \theta + \sqrt{2} \geq 0$$

for every $\theta \in [0, 2\pi[$ and every $\rho \in [0, 1]$. By the Change of Variables Theorem 11.43, using also Fubini's Reduction Theorem 11.36, we compute

$$\int_E (x^2 + y^2)\, dx\, dy\, dz = \int_{\xi^{-1}(E)} \rho^3\, d\rho\, d\theta\, dz$$

$$= \int_0^1 \left(\int_0^{2\pi} \left(\int_0^{\rho \cos\theta + \rho \sin\theta + \sqrt{2}} \rho^3\, dz \right) d\theta \right) d\rho$$

$$= \int_0^1 \left(\int_0^{2\pi} \rho^3 (\rho \cos\theta + \rho \sin\theta + \sqrt{2})\, d\theta \right) d\rho$$

$$= 2\pi \int_0^1 \rho^3 \sqrt{2}\, d\rho = \frac{\pi \sqrt{2}}{2}\,.$$

Now consider the function $\sigma : [0, +\infty[\times [0, 2\pi[\times [0, \pi] \to \mathbb{R}^3$ defined by

$$\sigma(\rho, \theta, \phi) = (\rho \sin\phi \cos\theta,\, \rho \sin\phi \sin\theta,\, \rho \cos\phi)\,,$$

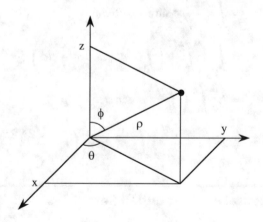

which defines the so-called "spherical coordinates" in $\mathbb{R}^3$. Consider the open sets

$$A =]0, +\infty[\times]0, 2\pi[\times]0, \pi[\,, \qquad B = \mathbb{R}^3 \setminus ([0, +\infty[\times \{0\} \times \mathbb{R})\,.$$

The function $\varphi : A \to B$ defined by $\varphi(\rho, \theta, \phi) = \sigma(\rho, \theta, \phi)$ happens to be a diffeomorphism, and it can be easily checked that

$$\det \varphi'(\rho, \theta, \phi) = -\rho^2 \sin \phi, \quad \text{for every } (\rho, \theta, z) \in A.$$

Let $E \subseteq \mathbb{R}^3$ be a measurable bounded set, and consider a function $f : E \to \mathbb{R}$. We can then apply the Change of Variables Theorem 11.43 to $\widetilde{E} = E \cap B$. Since $\widetilde{E}$ and $\varphi^{-1}(\widetilde{E})$ differ from E and $\sigma^{-1}(E)$, respectively, by negligible sets, we obtain the following formula on the **change of variables in spherical coordinates**:

$$\int_E f(x, y, z) \, dx \, dy \, dz = \int_{\sigma^{-1}(E)} f(\sigma(\rho, \theta, \phi)) \rho^2 \sin \phi \, d\rho \, d\theta \, d\phi.$$

Example Let us compute the volume of the set

$$E = \left\{ (x, y, z) \in \mathbb{R}^3 : x^2 + y^2 + z^2 \leq 1, z \geq \sqrt{x^2 + y^2} \right\}.$$

We have

$$\mu(E) = \int_E 1 \, dx \, dy \, dz$$

$$= \int_{\sigma^{-1}(E)} \rho^2 \sin \phi \, d\rho \, d\theta \, d\phi$$

$$= \int_0^1 \left(\int_0^{\pi/4} \left(\int_0^{2\pi} \rho^2 \sin \phi \, d\theta \right) d\phi \right) d\rho$$

$$= 2\pi \int_0^1 \left(\int_0^{\pi/4} \rho^2 \sin \phi \, d\phi \right) d\rho$$

$$= 2\pi \left(1 - \frac{\sqrt{2}}{2} \right) \int_0^1 \rho^2 \, d\rho = \left(1 - \frac{\sqrt{2}}{2} \right) \frac{2\pi}{3}.$$

11.14 The Integral on Unbounded Sets

When dealing with unbounded domains, there are good reasons to limit our attention only to L-integrable functions. This section extends the theory of the integral to this context.

Let E be a subset of $\mathbb{R}^N$, not necessarily bounded, and assume first that $f : E \to \mathbb{R}$ is a nonnegative function, i.e.,

$$f(x) \geq 0, \quad \text{for every } x \in E.$$

As usual, we will use the notation

$$\overline{B}[0, r] = [-r, r] \times \cdots \times [-r, r].$$

If f is integrable on $E \cap B[0, r]$ for every $r > 0$, we define

$$\int_E f = \lim_{r \to +\infty} \int_{E \cap \overline{B}[0,r]} f.$$

Notice that this limit always exists since the function $r \mapsto \int_{E \cap \overline{B}[0,r]} f$ is increasing, because of $f \geq 0$. When this limit happens to be finite (i.e., not equal to $+\infty$), we will say that f is "integrable" (on E).

It can be easily seen that the same result is obtained if, instead of $\overline{B}[0, r]$, we take the Euclidean close balls $\overline{B}(0, r)$. This is due to the fact that, for every $r > 0$,

$$\overline{B}(0, r) \subseteq \overline{B}[0, r], \quad \text{and} \quad \overline{B}[0, r] \subseteq \overline{B}(0, r\sqrt{N}).$$

The same observation can be made, of course, for many other families (S_r) of bounded sets invading $\mathbb{R}^N$, meaning that for every $r > 0$ there exists $r' > 0$ such that

$$\overline{B}[0, r] \subseteq S_{r'}.$$

In the case where the function f also has negative values, we consider both its positive part $f^+ = \max\{f, 0\}$ and its negative part $f^- = \max\{-f, 0\}$, so that $f = f^+ - f^-$. Notice that $f^+ \geq 0$ and $f^- \geq 0$. We say that f is L-integrable if both f^+ and f^- are integrable, in which case we define

$$\int_E f = \int_E f^+ - \int_E f^-.$$

Notice that, in this case, since $|f| = f^+ + f^-$, we have that

$$\int_E |f| = \int_E f^+ + \int_E f^-.$$

The fact that $|f|$ is integrable justifies the name "L-integrable" for the function f.

It is not difficult to prove that the set of L-integrable functions is a real vector space, and the integral is a linear function on it which preserves the order. Moreover, we can easily verify that a function f is L-integrable on a set E if and only if the function f_E is L-integrable on $\mathbb{R}^N$.

Definition 11.44 A set $E \subseteq \mathbb{R}^N$ is said to be "measurable" if $E \cap \overline{B}[0, r]$ is measurable for every $r > 0$. In that case, we set

$$\mu(E) = \lim_{r \to +\infty} \mu(E \cap \overline{B}[0, r]).$$

Notice that $\mu(E)$, in some cases, can be $+\infty$. It is finite if and only if the constant function 1 is L-integrable on E, i.e., the characteristic function of E is L-integrable on $\mathbb{R}^N$. The properties of measurable bounded sets extend easily to unbounded sets. In particular, all open sets and all closed sets are measurable.

The Monotone Convergence Theorem of Beppo Levi attains the following general form.

Theorem 11.45 (Monotone Convergence Theorem—II) *We are given a function* f *and a sequence of functions* f_n, *with* $n \in \mathbb{N}$, *defined almost everywhere on a subset* E *of* $\mathbb{R}^N$, *with real values, verifying the following conditions:*

(a) *The sequence* $(f_n)_n$ *converges pointwise to* f, *almost everywhere on* E.
(b) *The sequence* $(f_n)_n$ *is monotone.*
(c) *Each function* f_n *is* L-integrable on E.
(d) *The real sequence* $(\int_E f_n)_n$ *has a finite limit.*

Then f *is* L-integrable on E, *and*

$$\int_E f = \lim_n \int_E f_n .$$

Proof Assume, for definiteness, that the sequence $(f_n)_n$ is increasing. By considering the sequence $(f_n - f_0)_n$ instead of $(f_n)_n$, we can assume without loss of generality that all the functions have almost everywhere nonnegative values. Let $\mathcal{J} = \lim_n(\int_E f_n)$; for every $r > 0$ we can apply the Monotone Convergence Theorem 9.10 on the bounded set $E \cap \overline{B}[0, r]$, so that f is integrable on $E \cap \overline{B}[0, r]$ and

$$\int_{E \cap \overline{B}[0,r]} f = \lim_{n \to \infty} \int_{E \cap \overline{B}[0,r]} f_n \le \lim_{n \to \infty} \int_E f_n = \mathcal{J} .$$

Let us prove that the limit of $\int_{E \cap \overline{B}[0,r]} f$ exists, as $r \to +\infty$, and that it is equal to $\mathcal{J}$. Fix $\varepsilon > 0$; there is a $\bar{n} \in \mathbb{N}$ such that, for $n \ge \bar{n}$,

$$\mathcal{J} - \frac{\varepsilon}{2} \le \int_E f_n \le \mathcal{J} ;$$

since, moreover,

$$\int_E f_{\bar{n}} = \lim_{r \to +\infty} \int_{E \cap \overline{B}[0,r]} f_{\bar{n}} ,$$

there is a $\bar{r} > 0$ such that, for $r \ge \bar{r}$,

$$\mathcal{J} - \varepsilon \le \int_{E \cap \overline{B}[0,r]} f_{\bar{n}} \le \mathcal{J} .$$

Then, since the sequence $(f_n)_n$ is increasing, we have that, for every $n \geq \bar{n}$ and every $r \geq \bar{r}$,

$$\mathcal{J} - \varepsilon \leq \int_{E \cap \overline{B}[0,r]} f_n \leq \mathcal{J}.$$

Passing to the limit as $n \to +\infty$, we obtain, for every $r \geq \bar{r}$,

$$\mathcal{J} - \varepsilon \leq \int_{E \cap \overline{B}[0,r]} f \leq \mathcal{J}.$$

The proof is thus completed. ∎

As an immediate consequence, there is an analogous statement for the series of functions.

Corollary 11.46 *We are given a function f and a sequence of functions f_k, with $k \in \mathbb{N}$, defined almost everywhere on a subset E of $\mathbb{R}^N$, with real values, verifying the following conditions:*

(a) *The series $\sum_k f_k$ converges pointwise to f, almost everywhere on E.*
(b) *For every $k \in \mathbb{N}$ and almost every $\boldsymbol{x} \in E$, we have $f_k(\boldsymbol{x}) \geq 0$.*
(c) *Each function f_k is L-integrable on E.*
(d) *The series $\sum_k (\int_E f_k)$ converges.*

Then f is L-integrable on E and

$$\int_E f = \sum_{k=0}^{\infty} \int_E f_k.$$

From the Monotone Convergence Theorem 11.45 we deduce, in complete analogy with what we have seen for bounded sets, the Dominated Convergence Theorem of Henri Lebesgue.

Theorem 11.47 (Dominated Convergence Theorem—II) *We are given a function f and a sequence of functions f_n, with $n \in \mathbb{N}$, defined almost everywhere on a subset E of $\mathbb{R}^N$, with real values, verifying the following conditions:*

(a) *The sequence $(f_n)_n$ converges pointwise to f, almost everywhere on E.*
(b) *Each function f_n is L-integrable on E.*
(c) *There are two functions g, h, defined almost everywhere and L-integrable on E, such that*

$$g(\boldsymbol{x}) \leq f_n(\boldsymbol{x}) \leq h(\boldsymbol{x})$$

for every $n \in \mathbb{N}$ and almost every $\boldsymbol{x} \in E$.

Then the sequence $(\int_E f_n)_n$ has a finite limit, f is L-integrable on E, and

$$\int_E f = \lim_n \int_E f_n \,.$$

As a direct consequence we have the **complete additivity** property of the integral for L-integrable functions.

Theorem 11.48 *Let (E_k) be a finite or countable family of pairwise nonoverlapping measurable subsets of $\mathbb{R}^N$ whose union is a set E. Then f is L-integrable on E if and only if the following two conditions hold:*

(a) f is L-integrable on each set E_k .
(b) $\sum_k \int_{E_k} |f(x)| \, dx < +\infty.$

In that case, we have

$$\int_E f = \sum_k \int_{E_k} f \,.$$

As another consequence, we have the Leibniz rule for not necessarily bounded subsets Y of $\mathbb{R}^N$, which is stated as follows.

Theorem 11.49 (Leibniz Rule—II) *Let $f : X \times Y \to \mathbb{R}$ be a function, where X is a nontrivial interval of $\mathbb{R}$ containing x_0, and Y is a subset of $\mathbb{R}^N$, such that:*

(a) For every $x \in X$, the function $f(x, \cdot)$ is L-integrable on Y, so that we can define the function

$$F(x) = \int_Y f(x, y) \, dy \,.$$

(b) For every $x \in X$ and almost every $y \in Y$ the partial derivative $\frac{\partial f}{\partial x}(x, y)$ exists.
(c) There are two L-integrable functions $g, h : Y \to \mathbb{R}$ such that

$$g(y) \le \frac{\partial f}{\partial x}(x, y) \le h(y)$$

for every $x \in X$ and almost every $y \in Y$.

Then the function $\frac{\partial f}{\partial x}(x, \cdot)$, defined almost everywhere on Y, is L-integrable there, the derivative of F in x_0 exists, and

$$F'(x_0) = \int_Y \left(\frac{\partial f}{\partial x}(x_0, y) \right) dy \,.$$

Also the reduction theorem of Guido Fubini extends to functions defined on a not necessarily bounded subset E of $\mathbb{R}^N$. Let $N = N_1 + N_2$, and write $\mathbb{R}^N = \mathbb{R}^{N_1} \times \mathbb{R}^{N_2}$. For every $(\boldsymbol{x}, \boldsymbol{y}) \in \mathbb{R}^{N_1} \times \mathbb{R}^{N_2}$, consider the "sections" of E:

$$E_{\boldsymbol{x}} = \{\boldsymbol{y} \in \mathbb{R}^{N_2} : (\boldsymbol{x}, \boldsymbol{y}) \in E\}, \qquad E_{\boldsymbol{y}} = \{\boldsymbol{x} \in \mathbb{R}^{N_1} : (\boldsymbol{x}, \boldsymbol{y}) \in E\},$$

and the "projections" of E:

$$P_1 E = \{\boldsymbol{x} \in \mathbb{R}^{N_1} : E_{\boldsymbol{x}} \neq \emptyset\}, \qquad P_2 E = \{\boldsymbol{y} \in \mathbb{R}^{N_2} : E_{\boldsymbol{y}} \neq \emptyset\}.$$

We can then reformulate the theorem in the following form.

Theorem 11.50 (Reduction Theorem—IV) *Let $f : E \to \mathbb{R}$ be an L-integrable function. Then:*

(a) *For almost every $\boldsymbol{x} \in P_1 E$, the function $f(\boldsymbol{x}, \cdot)$ is L-integrable on the set $E_{\boldsymbol{x}}$.*
(b) *The function $\boldsymbol{x} \mapsto \int_{E_{\boldsymbol{x}}} f(\boldsymbol{x}, \boldsymbol{y})\, d\boldsymbol{y}$, defined almost everywhere on $P_1 E$, is L-integrable there.*
(c) *We have*

$$\int_E f = \int_{P_1 E} \left(\int_{E_{\boldsymbol{x}}} f(\boldsymbol{x}, \boldsymbol{y})\, d\boldsymbol{y} \right) d\boldsymbol{x}.$$

Analogously, the function $\boldsymbol{y} \mapsto \int_{E_{\boldsymbol{y}}} f(\boldsymbol{x}, \boldsymbol{y})\, d\boldsymbol{x}$, defined almost everywhere on $P_2 E$, is L-integrable there, and we have

$$\int_E f = \int_{P_2 E} \left(\int_{E_{\boldsymbol{y}}} f(\boldsymbol{x}, \boldsymbol{y})\, d\boldsymbol{x} \right) d\boldsymbol{y}.$$

Proof Consider for simplicity the case $N_1 = N_2 = 1$, the general case being perfectly analogous. Assume first that f has nonnegative values. By Fubini's Reduction Theorem 11.36 for bounded sets, once $r > 0$ is fixed, we have that, for almost every $x \in P_1 E \cap [-r, r]$, the function $f(x, \cdot)$ is L-integrable on $E_x \cap [-r, r]$; the function $g_r(x) = \int_{E_x \cap [-r,r]} f(x, y)\, dy$, defined almost everywhere on $P_1 E \cap [-r, r]$, is L-integrable there, and

$$\int_{E \cap \overline{B}[0,r]} f = \int_{P_1 E \cap [-r,r]} g_r(x)\, dx.$$

In particular,

$$\int_{P_1 E \cap [-r,r]} g_r(x)\, dx \leq \int_E f,$$

so that if $0 < s \leq r$, then we have that g_r is L-integrable on $P_1 E \cap [-s, s]$, and

$$\int_{P_1 E \cap [-s,s]} g_r(x)\, dx \leq \int_E f.$$

Keeping s fixed, we let r tend to $+\infty$. Since f has nonnegative values, $g_r(x)$ will be increasing with respect to r. Consequently, for almost every $x \in P_1 E \cap [-s, s]$, the limit $\lim_{r \to +\infty} g_r(x)$ exists (possibly infinite), and we set

$$g(x) = \lim_{r \to +\infty} g_r(x) = \lim_{r \to +\infty} \int_{E_x \cap [-r,r]} f(x, y)\, dy.$$

Let $T = \{x \in P_1 E \cap [-s, s] : g(x) = +\infty\}$; let us prove that T is negligible. We define the sets

$$E_n^r = \{x \in P_1 E \cap [-s, s] : g_r(x) > n\}.$$

By Lemma 11.16, these sets are measurable sets and

$$\mu(E_n^r) \leq \frac{1}{n} \int_{P_1 E \cap [-s,s]} g_r(x)\, dx \leq \frac{1}{n} \int_E f.$$

Hence, since the sets E_n^r increase with r, the sets $F_n = \cup_r E_n^r$ are also measurable, and we have that $\mu(F_n) \leq \frac{1}{n} \int_E f$. Since $T \subseteq \cap_n F_n$, we deduce that T is measurable, with $\mu(T) = 0$.

Hence, for almost every $x \in P_1 E \cap [-s, s]$, the function $f(x, \cdot)$ is L-integrable on the set E_x and, by definition,

$$\int_{E_x} f(x, y)\, dy = g(x).$$

Moreover, if we take r in the set of natural numbers and apply the Monotone Convergence Theorem to the functions g_r, it follows that g is L-integrable on $P_1 E \cap [-s, s]$, and

$$\int_{P_1 E \cap [-s,s]} g = \lim_{r \to \infty} \int_{P_1 E \cap [-s,s]} g_r,$$

so that

$$\int_{P_1 E \cap [-s,s]} \left(\int_{E_x} f(x, y)\, dy \right) dx \leq \int_E f.$$

Letting now s tend to $+\infty$, we see that the limit

$$\lim_{s \to +\infty} \int_{P_1 E \cap [-s,s]} \left(\int_{E_x} f(x, y)\, dy \right) dx$$

exists and is finite; therefore, the function $x \mapsto \int_{E_x} f(x, y)\, dy$, defined almost everywhere on $P_1 E$, is L-integrable there, and its integral is the preceding limit. Moreover, from the inequality proved earlier, passing to the limit, we have that

$$\int_{P_1 E} \left(\int_{E_x} f(x, y)\, dy \right) dx \le \int_E f \,.$$

On the other hand,

$$\int_{E \cap \bar{B}[0,r]} f = \int_{P_1 E \cap [-r,r]} \left(\int_{E_x \cap [-r,r]} f(x, y)\, dy \right) dx$$

$$\le \int_{P_1 E \cap [-r,r]} \left(\int_{E_x} f(x, y)\, dy \right) dx$$

$$\le \int_{P_1 E} \left(\int_{E_x} f(x, y)\, dy \right) dx \,,$$

so that, passing to the limit as $r \to +\infty$,

$$\int_E f \le \int_{P_1 E} \left(\int_{E_x} f(x, y)\, dy \right) dx \,.$$

In conclusion, equality must hold, and the proof is thus completed in the case where f has nonnegative values. In the general case, just consider f^+ and f^-, and subtract the corresponding formulas.   ∎

The analogous corollary for the computation of the measure holds.

Corollary 11.51 *Let E be a measurable set. Then E has a finite measure if and only if:*

(a) *For almost every $x \in P_1 E$ the set E_x is measurable and has a finite measure.*
(b) *The function $x \mapsto \mu(E_x)$, defined almost everywhere on $P_1 E$, is L-integrable there.*
(c) *We have*

$$\mu(E) = \int_{P_1 E} \mu(E_x)\, dx \,.$$

With a symmetric statement, if E has a finite measure, we also have

$$\mu(E) = \int_{P_2 E} \mu(E_y) \, dy.$$

The change of variables formula also extends to unbounded sets, with the same statement.

Theorem 11.52 (Change of Variables Theorem—III) *Let A and B be open subsets of $\mathbb{R}^N$ and $\varphi : A \to B$ a diffeomorphism. Let $D \subseteq A$ be a measurable set and $f : \varphi(D) \to \mathbb{R}$ a function. Then f is L-integrable on $\varphi(D)$ if and only if $(f \circ \varphi) \, | \det \varphi' |$ is L-integrable on D, in which case*

$$\int_{\varphi(D)} f(\boldsymbol{x}) \, d\boldsymbol{x} = \int_D f(\varphi(\boldsymbol{u})) \, | \det \varphi'(\boldsymbol{u}) | \, d\boldsymbol{u}.$$

Proof Assume first that f is L-integrable on $E = \varphi(D)$ with nonnegative values. Then, for every $r > 0$,

$$\int_{D \cap \overline{B}[0,r]} f(\varphi(\boldsymbol{u})) \, | \det \varphi'(\boldsymbol{u}) | \, d\boldsymbol{u} = \int_{\varphi(D \cap \overline{B}[0,r])} f(\boldsymbol{x}) \, d\boldsymbol{x} \leq \int_{\varphi(D)} f(\boldsymbol{x}) \, d\boldsymbol{x},$$

so that the limit

$$\lim_{r \to +\infty} \int_{D \cap \overline{B}[0,r]} f(\varphi(\boldsymbol{u})) \, | \det \varphi'(\boldsymbol{u}) | \, d\boldsymbol{u}$$

exists and is finite. Then $(f \circ \varphi) \, | \det \varphi' |$ is L-integrable on D, and we have

$$\int_D f(\varphi(\boldsymbol{u})) \, | \det \varphi'(\boldsymbol{u}) | \, d\boldsymbol{u} \leq \int_{\varphi(D)} f(\boldsymbol{x}) \, d\boldsymbol{x}.$$

On the other hand, for every $r > 0$,

$$\int_{E \cap \overline{B}[0,r]} f = \int_{\varphi^{-1}(E \cap \overline{B}[0,r])} (f \circ \varphi) \, | \det \varphi' | \leq \int_{\varphi^{-1}(E)} (f \circ \varphi) \, | \det \varphi' |,$$

so that, passing to the limit,

$$\int_E f(\boldsymbol{x}) \, d\boldsymbol{x} = \lim_{r \to +\infty} \int_{E \cap \overline{B}[0,r]} f(\boldsymbol{x}) \, d\boldsymbol{x} \leq \int_{\varphi^{-1}(E)} f(\varphi(\boldsymbol{u})) \, | \det \varphi'(\boldsymbol{u}) | \, d\boldsymbol{u}.$$

The formula is thus proved when f has nonnegative values. In general, just proceed as usual, considering f^+ and f^-.

To obtain the opposite implication, it is sufficient to consider $(f \circ \varphi) \, | \det \varphi' |$ instead of f and φ^{-1} instead of φ and to repeat the preceding argument.   ∎

Concerning the change of variables in polar coordinates in $\mathbb{R}^2$ or in cylindrical or spherical coordinates in $\mathbb{R}^3$, the same type of considerations we made for bounded sets extend to the general case as well.

Example Let $E = \{(x, y) \in \mathbb{R}^2 : x^2 + y^2 \geq 1\}$ and $f(x, y) = (x^2 + y^2)^{-\alpha}$, with $\alpha > 0$. We have

$$\int_E \frac{1}{(x^2 + y^2)^\alpha} \, dx \, dy = \int_0^{2\pi} \left(\int_1^{+\infty} \frac{1}{\rho^{2\alpha}} \rho \, d\rho \right) d\theta = 2\pi \int_1^{+\infty} \rho^{1-2\alpha} \, d\rho \, .$$

It is thus seen that f is integrable on E if and only if $\alpha > 1$, in which case the integral is $\frac{\pi}{\alpha-1}$.

Example Let us compute the three-dimensional measure of the set

$$E = \left\{ (x, y, z) \in \mathbb{R}^3 : x \geq 1, \sqrt{y^2 + z^2} \leq \frac{1}{x} \right\} \, .$$

Using Fubini's Reduction Theorem 11.50, grouping together the variables (y, z) we have

$$\mu(E) = \int_1^{+\infty} \pi \frac{1}{x^2} \, dx = \pi \, .$$

Example Consider the function $f(x, y) = e^{-(x^2+y^2)}$, and let us make a change of variables in polar coordinates:

$$\int_{\mathbb{R}^2} e^{-(x^2+y^2)} \, dx \, dy = \int_0^{2\pi} \left(\int_0^{+\infty} e^{-\rho^2} \rho \, d\rho \right) d\theta = 2\pi \left[-\frac{1}{2} e^{-\rho^2} \right]_0^{+\infty} = \pi \, .$$

Notice that, using Fubini's Reduction Theorem 11.50, we have

$$\int_{\mathbb{R}^2} e^{-(x^2+y^2)} \, dx \, dy = \int_{-\infty}^{+\infty} \left(\int_{-\infty}^{+\infty} e^{-x^2} e^{-y^2} \, dx \right) dy$$

$$= \left(\int_{-\infty}^{+\infty} e^{-x^2} \, dx \right) \left(\int_{-\infty}^{+\infty} e^{-y^2} \, dy \right)$$

$$= \left(\int_{-\infty}^{+\infty} e^{-x^2} \, dx \right)^2 ,$$

and we thus find again the formula

$$\int_{-\infty}^{+\infty} e^{-x^2}\, dx = \sqrt{\pi}\ .$$

11.15 The Integral on M-Surfaces

We now want to define the integral of a function $f : U \to \mathbb{R}$ on an M-surface $\sigma : I \to \mathbb{R}^N$ whose image is contained in U.

When the indices $i_1, \ldots, i_M$ vary in the set $\{1, \ldots, N\}$, then for every $\boldsymbol{u} \in I$ we can consider the $M \times M$ matrices obtained from the Jacobian matrix $J\sigma(\boldsymbol{u})$ (also denoted by $\sigma'(\boldsymbol{u})$) by selecting the corresponding lines, i.e.,

$$\sigma'_{(i_1,\ldots,i_M)}(\boldsymbol{u}) = \begin{pmatrix} \frac{\partial \sigma_{i_1}}{\partial u_1}(\boldsymbol{u}) & \cdots & \frac{\partial \sigma_{i_1}}{\partial u_M}(\boldsymbol{u}) \\ \vdots & \cdots & \vdots \\ \frac{\partial \sigma_{i_M}}{\partial u_1}(\boldsymbol{u}) & \cdots & \frac{\partial \sigma_{i_M}}{\partial u_M}(\boldsymbol{u}) \end{pmatrix},$$

and we can define the vector $\Sigma(u)$ in $\mathbb{R}^{\binom{N}{M}}$ as

$$\Sigma(\boldsymbol{u}) = \Big(\det \sigma'_{(i_1,\ldots,i_M)}(\boldsymbol{u})\Big)_{1 \le i_1 < \cdots < i_M \le N}\ .$$

Definition 11.53 The function $f : U \to \mathbb{R}$ is "integrable" on the M-surface $\sigma : I \to \mathbb{R}^N$ if $(f \circ \sigma)\|\Sigma\|$ is integrable on I. In that case, we set

$$\int_\sigma f = \int_I f(\sigma(\boldsymbol{u}))\, \|\Sigma(\boldsymbol{u})\|\, d\boldsymbol{u}\ .$$

In the case $M = 1$, we have a curve $\sigma : [a, b] \to \mathbb{R}^N$ and, given a scalar function f defined on the image of σ,

$$\int_\sigma f = \int_a^b f(\sigma(t))\, \|\sigma'(t)\|\, dt\ .$$

If $M = 2$ and $N = 3$, we have the surface $\sigma : [a_1, b_1] \times [a_2, b_2] \to \mathbb{R}^3$ and, given a scalar function f, defined on the image of σ, it can be checked that

$$\int_\sigma f = \int_{a_2}^{b_2} \int_{a_1}^{b_1} f(\sigma(u, v)) \left\| \frac{\partial \sigma}{\partial u}(u, v) \times \frac{\partial \sigma}{\partial v}(u, v) \right\| du\, dv\ .$$

It is important to see what happens to the integral when we have two equivalent M-surfaces.

Definition 11.54 Two M-surfaces $\sigma : I \to \mathbb{R}^N$ and $\tilde{\sigma} : J \to \mathbb{R}^N$ are said to be "equivalent" if they have the same image and there are two open sets $A \subseteq I$, $B \subseteq J$, and a diffeomorphism $\varphi : A \to B$ with the following properties. The sets $I \setminus A$ and $J \setminus B$ are negligible and $\sigma(u) = \tilde{\sigma}(\varphi(u))$ for every $u \in A$.

Let us prove that the integral does not differ for equivalent M-surfaces.

Theorem 11.55 *If σ and $\tilde{\sigma}$ are two equivalent M-surfaces, then*

$$\int_\sigma f = \int_{\tilde{\sigma}} f.$$

Proof With the notations introduced previously, since $\sigma = \tilde{\sigma} \circ \varphi$, with $\varphi : A \to B$, we have

$$
\begin{aligned}
\Sigma(u) &= \left(\det \sigma'_{(i_1, \dots, i_M)}(u) \right)_{1 \le i_1 < \cdots < i_M \le N} \\
&= \left(\det \left(\tilde{\sigma}'_{(i_1, \dots, i_M)}(\varphi(u)) \varphi'(u) \right) \right)_{1 \le i_1 < \cdots < i_M \le N} \\
&= \left(\det \tilde{\sigma}'_{(i_1, \dots, i_M)}(\varphi(u)) \right)_{1 \le i_1 < \cdots < i_M \le N} \det \varphi'(u) \\
&= \tilde{\Sigma}(\varphi(u)) \det \varphi'(u).
\end{aligned}
$$

Therefore, by the Change of Variables Theorem 11.43, since $I \setminus A$ and $J \setminus B$ are negligible, we have that

$$
\begin{aligned}
\int_\sigma f &= \int_A f(\sigma(u)) \, \|\Sigma(u)\| \, du \\
&= \int_A f(\tilde{\sigma}(\varphi(u))) \, \|\tilde{\Sigma}(\varphi(u))\| \, |\det \varphi'(u)| \, du \\
&= \int_B f(\tilde{\sigma}(v)) \, \|\tilde{\Sigma}(v)\| \, dv = \int_{\tilde{\sigma}} f,
\end{aligned}
$$

thereby proving the claim. ∎

The following theorem is crucial for the treatment of the measure of M-parametrizable M-surfaces. We recall that $\sigma : I \to \mathbb{R}^N$ is an "M-parametrization" of a set $\mathcal{M}$ if it is regular, injective on $\overset{\circ}{I}$, and $\sigma(I) = \mathcal{M}$.

Theorem 11.56 *Two M-parametrizations of the same set are always equivalent.*

Proof Let $\mathscr{M}$ be the subset of $\mathbb{R}^N$ taken into consideration, and let $\sigma : I \to \mathbb{R}^N$ and $\tilde{\sigma} : J \to \mathbb{R}^N$ be two of its M-parametrizations. We define the sets

$$A = \mathring{I} \cap \sigma^{-1}(\mathscr{M} \setminus (\sigma(\partial I) \cup \tilde{\sigma}(\partial J))), \qquad B = \mathring{J} \cap \tilde{\sigma}^{-1}(\mathscr{M} \setminus (\sigma(\partial I) \cup \tilde{\sigma}(\partial J))).$$

Then, for every $u \in A$, since $\sigma(u) \in \mathscr{M} \setminus (\sigma(\partial I) \cup \tilde{\sigma}(\partial J))$ and $\tilde{\sigma}(J) = \mathscr{M}$, there exists a $v \in \mathring{J}$ such that $\tilde{\sigma}(v) = \sigma(u)$. Clearly, $\tilde{\sigma}(v) \in \mathscr{M} \setminus (\sigma(\partial I) \cup \tilde{\sigma}(\partial J))$, so that $v \in B$. Moreover, since $\tilde{\sigma}$ is injective on $\mathring{J}$, there is a unique v in $\mathring{J}$ with such a property. We can thus define $\varphi : A \to B$ by setting $\varphi(u) = v$. Hence, for $u \in A$ and $v \in B$,

$$\varphi(u) = v \quad \Leftrightarrow \quad \sigma(u) = \tilde{\sigma}(v).$$

This function $\varphi : A \to B$ is invertible; a symmetrical argument may be used to define its inverse $\varphi^{-1} : B \to A$.

Let us verify that the set A is open. Since $\sigma, \tilde{\sigma}$ are continuous functions and ∂I, ∂J are compact sets, we have that $\sigma(\partial I) \cup \tilde{\sigma}(\partial J)$ is compact, hence closed. Then $\mathscr{M} \setminus (\sigma(\partial I) \cup \tilde{\sigma}(\partial J))$ is relatively open in $\mathscr{M}$, and $\sigma^{-1}(\mathscr{M} \setminus (\sigma(\partial I) \cup \tilde{\sigma}(\partial J)))$ is relatively open in I, so that its intersection with $\mathring{I}$ is an open set. In an analogous way it can be seen that B is an open set, as well.

Let us take a $v_0 \in \mathring{J}$, and set $x_0 = \tilde{\sigma}(v_0)$. The Jacobian matrix $\tilde{\sigma}'(v_0)$ has rank M, and we may assume without loss of generality that the first M lines are linearly independent. Since $\mathbb{R}^N \simeq \mathbb{R}^M \times \mathbb{R}^{N-M}$, we will write every point $x \in \mathbb{R}^N$ in the form $x = (x_1, x_2)$, with $x_1 \in \mathbb{R}^M$ and $x_2 \in \mathbb{R}^{N-M}$. However, so as not to have double indices in what follows, we will write $x_0 = (x_1^0, x_2^0)$.

Let $\Phi : J \times \mathbb{R}^{N-M} \to \mathbb{R}^N$ be defined as

$$\Phi(v, z) = \tilde{\sigma}(v) + (0, z).$$

Then $\Phi'(v_0, 0)$ is invertible, so that Φ is a local diffeomorphism: There are an open neighborhood V_0 of v_0, an open neighborhood Ω_0 of 0 in $\mathbb{R}^{N-M}$, and an open neighborhood W_0 of x_0 such that $\Phi : V_0 \times \Omega_0 \to W_0$ is a diffeomorphism. Moreover, we can assume that $V_0 \subseteq \mathring{J}$. Let $\Psi = \Phi^{-1} : W_0 \to V_0 \times \Omega_0$. We will write $\Psi(x) = (\Psi_1(x), \Psi_2(x))$, with $\Psi_1(x) \in V_0$ and $\Psi_2(x) \in \Omega_0$.

We now prove that φ is of class C^1. Take $u_0 \in A$, and set $x_0 = \sigma(u_0)$ and $v_0 = \varphi(u_0)$. Assume v_0 as previously, with $\tilde{\sigma}'(v_0)$ having the first M lines linearly independent, so that the local diffeomorphism $\Psi : W_0 \to V_0 \times \Omega_0$ can be defined. Take an open neighborhood U_0 of u_0, contained in A, such that $\sigma(U_0) \subseteq W_0$. Then, for $u \in U_0$ and $v \in B$,

$$\varphi(u) = v \quad \Leftrightarrow \quad \sigma(u) = \Phi(v, 0) \quad \Leftrightarrow \quad (v, 0) = \Psi(\sigma(u)).$$

Hence, φ coincides with $\Psi_1 \circ \sigma$ on the open set U_0, yielding that φ is continuously differentiable.

In a symmetric way, it is proved that $\varphi^{-1} : B \to A$ is of class $\mathcal{C}^1$, so that φ happens to be a diffeomorphism.

We now prove that the sets $I \setminus A$ and $J \setminus B$ are negligible. Let us consider, e.g., the second one:

$$J \setminus B = \partial J \cup (\mathring{J} \setminus B) = \partial J \cup \{v \in \mathring{J} : \tilde{\sigma}(v) \in \sigma(\partial I)\} \cup \{v \in \mathring{J} : \tilde{\sigma}(v) \in \tilde{\sigma}(\partial J)\}.$$

We know that ∂J is negligible. Let us prove that $\{v \in \mathring{J} : \tilde{\sigma}(v) \in \sigma(\partial I)\}$ is also negligible.

Let $v_0 \in \mathring{J}$ be such that $\tilde{\sigma}(v_0) \in \sigma(\partial I)$. Then there is a $u_0 \in \partial I$ such that $\sigma(u_0) = \tilde{\sigma}(v_0)$. We argue as previously and define $\Psi : W_0 \to V_0 \times \Omega_0$. Let U_0 be an open neighborhood of u_0 such that $\sigma(U_0 \cap I) \subseteq W_0$. Let us see that

$$\mathring{J} \cap \tilde{\sigma}^{-1}(\sigma(U_0 \cap \partial I)) \subseteq (\Psi_1 \circ \sigma)(U_0 \cap \partial I).$$

Indeed, taking $v \in \mathring{J} \cap \tilde{\sigma}^{-1}(\sigma(U_0 \cap \partial I))$, we have that $\tilde{\sigma}(v) \in \sigma(U_0 \cap \partial I)$. Then, since $\Phi(v, \mathbf{0}) = \tilde{\sigma}(v)$, we have that $\Psi(\tilde{\sigma}(v)) = (v, \mathbf{0}) \in V_0 \times \Omega_0$, hence $v \in \Psi_1(\sigma(U_0 \cap \partial I))$, and the inclusion is thus proved. Now, since $\Psi_1 \circ \sigma$ is of class $\mathcal{C}^1$, by Lemma 11.38 we have that $(\Psi_1 \circ \sigma)(U_0 \cap \partial I)$ is negligible. Finally, the conclusion that $\{v \in \mathring{J} : \tilde{\sigma}(v) \in \sigma(\partial I)\}$ is negligible follows from the fact that ∂I is compact, so that it can be covered by a finite number of such open sets as U_0.

It remains to be proved that $\{v \in \mathring{J} : \tilde{\sigma}(v) \in \tilde{\sigma}(\partial J)\}$ is negligible. Let $v_0 \in \mathring{J}$ be such that $\tilde{\sigma}(v_0) \in \tilde{\sigma}(\partial J)$. Then there is a $\tilde{v}_0 \in \partial J$ such that $\tilde{\sigma}(\tilde{v}_0) = \tilde{\sigma}(v_0)$. Let $\widetilde{V}_0$ be an open neighborhood of $\tilde{v}_0$ such that $\tilde{\sigma}(\widetilde{V}_0 \cap J) \subseteq W_0$. As above, one sees that $\mathring{J} \cap \tilde{\sigma}^{-1}(\tilde{\sigma}(\widetilde{V}_0 \cap \partial J)) \subseteq (\Psi_1 \circ \tilde{\sigma})(\widetilde{V}_0 \cap \partial J)$, showing that $\mathring{J} \cap \tilde{\sigma}^{-1}(\tilde{\sigma}(\widetilde{V}_0 \cap \partial J))$ is negligible. The conclusion is obtained as above, covering ∂J by a finite number of such open sets $\widetilde{V}_0$. ∎

If $\mathcal{M}$ is an M-parametrizable set, we can define the integral of f on $\mathcal{M}$ as $\int_\sigma f$, where σ is any M-parametrization of $\mathcal{M}$. We will denote it by

$$\int_{\mathcal{M}} f \, d\mu_M, \quad \text{or} \quad \int_{\mathcal{M}} f(x) \, d\mu_M(x).$$

If $M = N$, one reobtains the usual integral, i.e., $\int_{\mathcal{M}} f(x) \, dx$.

11.16 *M*-Dimensional Measure

Consider the interesting case where f is constantly equal to 1.

If $M = 1$, we have a curve $\sigma : [a, b] \to \mathbb{R}^N$. The integral

$$\iota_1(\sigma) = \int_a^b \|\sigma'(t)\| \, dt$$

is said to be the "length" (or curvilinear measure) of the curve σ.

Example Let $\sigma : [0, b] \to \mathbb{R}^3$ be defined by $\sigma(t) = (t, t^2, 0)$. Its image is an arc of parabola, and its length is given by

$$\iota_1(\sigma) = \int_0^b \sqrt{1 + (2t)^2}\, dt$$

$$= \int_{\sinh^{-1}(0)}^{\sinh^{-1}(2b)} \frac{1}{2}(\cosh u)^2\, du$$

$$= \frac{1}{2}\left[\frac{u + \sinh u \cosh u}{2}\right]_0^{\sinh^{-1}(2b)}$$

$$= \frac{1}{4}\left(\sinh^{-1}(2b) + 2b\sqrt{1 + 4b^2}\right)$$

$$= \frac{1}{4}\ln\left(2b + \sqrt{1 + 4b^2}\right) + \frac{b}{2}\sqrt{1 + 4b^2}\,.$$

If $M = 2$ and $N = 3$, then we have a surface $\sigma : [a_1, b_1] \times [a_2, b_2] \to \mathbb{R}^3$. The integral

$$\iota_2(\sigma) = \int_{a_2}^{b_2}\int_{a_1}^{b_1} \left\|\frac{\partial\sigma}{\partial u}(u, v) \times \frac{\partial\sigma}{\partial v}(u, v)\right\|\, du\, dv$$

is said to be the "area" (or surface measure) of the surface σ.

Example Let $\sigma : [0, \pi] \times [0, 2\pi] \to \mathbb{R}^3$ be defined by

$$\sigma(\phi, \theta) = (R\sin\phi\cos\theta,\ R\sin\phi\sin\theta,\ R\cos\phi)\,.$$

Its image is a sphere of radius R, and its area is given by

$$\iota_2(\sigma) = \int_0^{2\pi}\int_0^\pi \sqrt{(R^2\sin^2\phi\cos\theta)^2 + (R^2\sin^2\phi\sin\theta)^2 + (R^2\sin\phi\cos\theta)^2}\, d\phi\, d\theta$$

$$= \int_0^{2\pi}\int_0^\pi R^2\sin\phi\, d\phi\, d\theta$$

$$= 4\pi R^2\,.$$

In general, when the function f is constantly equal to 1, we have the following definition.

Definition 11.57 We call "M-superficial measure" of an M-surface $\sigma : I \to \mathbb{R}^N$ the following integral:

$$\iota_M(\sigma) = \int_I \|\Sigma(\boldsymbol{u})\| \, d\boldsymbol{u} = \int_I \left[\sum_{1 \le i_1 < \cdots < i_M \le N} \left(\det \sigma'_{(i_1,\ldots,i_M)}(\boldsymbol{u}) \right)^2 \right]^{\frac{1}{2}} d\boldsymbol{u} \, .$$

As reasonably expected, a direct consequence of Theorems 11.55 and 11.56 is the following corollary.

Corollary 11.58 *Two equivalent M-surfaces always have the same M-superficial measure. In particular, this is true of any two M-parametrizations of a given set.*

Example Consider the two curves $\sigma, \tilde{\sigma} : [0, 2\pi] \to \mathbb{R}^2$, defined by

$$\sigma(t) = (\cos(t), \sin(t)) \, , \qquad \tilde{\sigma}(t) = (\cos(2t), \sin(2t)) \, .$$

Notice that, even if they have the same image, these curves are not equivalent. Indeed, as easily seen, $\iota_1(\sigma) = 2\pi$ while $\iota_1(\tilde{\sigma}) = 4\pi$.

The foregoing considerations naturally lead to the following definition.

Definition 11.59 We call "M-dimensional measure" of an M-parametrizable set $\mathcal{M} \subseteq \mathbb{R}^N$ the M-superficial measure of any of its M-parametrizations.

We denote the M-dimensional measure of $\mathcal{M}$ by $\mu_M(\mathcal{M})$. In cases where $M = 1, 2$, the M-dimensional measure of $\mathcal{M}$ is often called the "length" or "area" of $\mathcal{M}$, respectively. We may thus consider, for example, the length of a circle or the area of a sphere. If $M = N$, it can be verified that the N-dimensional measure of the set $\mathcal{M}$ is the same as the usual measure, i.e., $\mu_N(\mathcal{M}) = \mu(\mathcal{M})$.

11.17 Length and Area

Let us first consider a curve $\sigma : [a, b] \to \mathbb{R}^N$. For any partition $\mathcal{P}$ of the interval $[a, b]$ of the type

$$a = a_0 < a_1 < \cdots < a_{m-1} < a_m = b \, ,$$

we compute the length of the polygonal curve joining the points $\sigma(a_j)$, i.e.,

$$\ell(\sigma, \mathcal{P}) = \sum_{j=1}^{n} \|\sigma(a_j) - \sigma(a_{j-1})\| \, .$$

It is rather intuitive that these lengths may be chosen as a good approximation of the length of the curve σ, provided that the points of the partition are sufficiently close to one another. What follows is the precise statement.

Theorem 11.60 *The length of the curve σ is obtained as*

$$\iota_1(\sigma) = \sup\left\{\ell(\sigma, \mathcal{P}) : \mathcal{P} \text{ partition of } [a, b]\right\}.$$

Proof For every partition $\mathcal{P}$ of $[a, b]$ one has that

$$\|\sigma(a_j) - \sigma(a_{j-1})\| = \left\| \int_{a_{j-1}}^{a_j} \sigma'(t)\, dt \right\| \leq \int_{a_{j-1}}^{a_j} \|\sigma'(t)\|\, dt\,,$$

hence

$$\ell(\sigma, \mathcal{P}) = \sum_{j=1}^{n} \|\sigma(a_j) - \sigma(a_{j-1})\| \leq \sum_{j=1}^{n} \int_{a_{j-1}}^{a_j} \|\sigma'(t)\|\, dt = \int_{a}^{b} \|\sigma'(t)\|\, dt\,.$$

Then also

$$\sup\left\{\ell(\sigma, \mathcal{P}) : \mathcal{P} \text{ partition of } [a, b]\right\} \leq \int_{a}^{b} \|\sigma'(t)\|\, dt\,.$$

Let us prove now the opposite inequality. Fix $\varepsilon > 0$. Since $\sigma' : [a, b] \to \mathbb{R}^N$ is continuous, by the Heine Theorem 4.12, it is uniformly continuous. Hence, there exists a $\delta > 0$ such that

$$|s - t| \leq \delta \quad \Rightarrow \quad \|\sigma'(s) - \sigma'(t)\| \leq \varepsilon\,.$$

Let $\mathcal{P}$ be a partition of $[a, b]$ such that $a_j - a_{j-1} \leq \delta$ for every $j = 1, \ldots, m$.
If $t \in [a_{j-1}, a_j]$, then $\|\sigma'(t) - \sigma'(a_j)\| \leq \varepsilon$, hence $\|\sigma'(t)\| \leq \|\sigma'(a_j)\| + \varepsilon$, implying that

$$\int_{a_{j-1}}^{a_j} \|\sigma'(t)\|\, dt \leq \int_{a_{j-1}}^{a_j} (\|\sigma'(a_j)\| + \varepsilon)\, dt$$

$$= \int_{a_{j-1}}^{a_j} \|\sigma'(a_j)\|\, dt + \varepsilon(a_j - a_{j-1})$$

$$= \left\| \int_{a_{j-1}}^{a_j} \sigma'(a_j)\, dt \right\| + \varepsilon(a_j - a_{j-1})$$

$$\leq \left\| \int_{a_{j-1}}^{a_j} (\sigma'(a_j) - \sigma'(t))\, dt \right\| + \left\| \int_{a_{j-1}}^{a_j} \sigma'(t)\, dt \right\| + \varepsilon(a_j - a_{j-1})$$

$$\leq \int_{a_{j-1}}^{a_j} \|\sigma'(a_j) - \sigma'(t)\| \, dt + \|\sigma(a_j) - \sigma(a_{j-1})\| + \varepsilon(a_j - a_{j-1})$$

$$\leq \varepsilon(a_j - a_{j-1}) + \|\sigma(a_j) - \sigma(a_{j-1})\| + \varepsilon(a_j - a_{j-1}) \,.$$

Therefore,

$$\int_a^b \|\sigma'(t)\| \, dt = \sum_{j=1}^m \int_{a_{j-1}}^{a_j} \|\sigma'(t)\| \, dt$$

$$\leq \sum_{j=1}^m \Big(\|\sigma(a_j) - \sigma(a_{j-1})\| + 2\varepsilon(a_j - a_{j-1}) \Big)$$

$$= \ell(\sigma, \mathcal{P}) + 2\varepsilon(b - a)$$

$$\leq \sup \{\ell(\sigma, \mathcal{P}) : \mathcal{P} \text{ partition of } [a, b]\} + 2\varepsilon(b - a) \,.$$

In view of the arbitrariness of ε, the inequality

$$\int_a^b \|\sigma'(t)\| \, dt \leq \sup \{\ell(\sigma, \mathcal{P}) : \mathcal{P} \text{ partition of } [a, b]\}$$

must hold true. The statement is thus proved. ∎

One could now try to see whether a similar construction can also be made for surfaces. Surprisingly enough, we will see that, in general, this is not possible.

Let us consider the lateral surface of a cylinder with a circular base having radius r and height h. We parametrize it in cylindrical coordinates through the function $\sigma : [0, 2\pi] \times [0, h] \to \mathbb{R}^3$ defined as

$$\sigma(\theta, z) = (r \cos \theta, r \sin \theta, z) \,.$$

Its area can be easily computed, and it is equal to

$$\iota_2(\sigma) = 2\pi r h \,.$$

Let us now construct the "Schwarz lantern." It is a polyhedron having $4mn$ triangular faces inscribed in the considered cylinder. The vertices of this polyhedron correspond to the points obtained subdividing the domain into nm subrectangles

$$\left[(j-1)\frac{2\pi}{m}, \, j\frac{2\pi}{m} \right] \times \left[(k-1)\frac{h}{n}, \, k\frac{h}{n} \right], \quad \text{with } j = 1, \ldots, m, \ k = 1, \ldots, n,$$

and then further dividing each of them, by means of their diagonals, into four equal triangles. We will denote by $A(m, n)$ the area of this polyhedron.

Using simple geometrical considerations it can be seen that each of the $4mn$ faces of the polyhedron is an isosceles triangle whose base length is equal to

$$b = 2r \sin\left(\frac{\pi}{m}\right),$$

and height length is equal to

$$h' = \sqrt{r^2 \left[1 - \cos\left(\frac{\pi}{m}\right)\right]^2 + \left(\frac{h}{2n}\right)^2}.$$

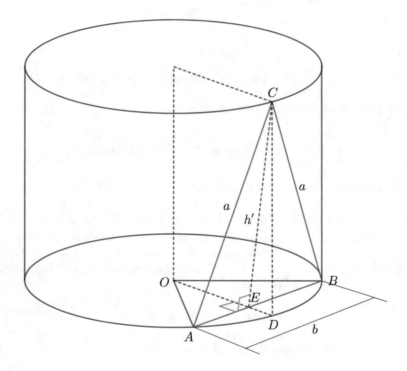

Indeed, as seen in the figure,

$$AE = r \sin\left(\frac{\pi}{m}\right), \quad OE = r \cos\left(\frac{\pi}{m}\right), \quad ED = r\left(1 - \cos\left(\frac{\pi}{m}\right)\right), \quad CD = \frac{h}{2n}.$$

Hence, the sum of the areas of the $4mn$ triangles is

$$A(m, n) = 4mn \frac{b\,h'}{2} = 2\pi r \frac{\sin\left(\frac{\pi}{m}\right)}{\frac{\pi}{m}} \sqrt{h^2 + \left[2\frac{1 - \cos\left(\frac{\pi}{m}\right)}{(\frac{\pi}{m})^2}\right]^2 \left(\frac{\pi^2 rn}{m^2}\right)^2}.$$

We see in this formula an unpleasant term: $\left(\dfrac{\pi^2 rn}{m^2}\right)^2$. If $m \to +\infty$ and $n \to +\infty$, it is not guaranteed that it tends to zero, as we would like, so as to have that $A(m, n)$ approaches $2\pi rh$. On the contrary, we must admit that

the limit of $A(m, n)$ for $(m, n) \to (+\infty, +\infty)$ does not exist!

Note, for example, that $A(m, m) \to 2\pi rh$, while $A(m, m^3) \to +\infty$ and, for every $\ell \geq 2\pi rh$, there exists a sequence $(n_m)_m$ such that $n_m \to +\infty$ and $A(m, n_m) \to \ell$.

11.18 Approximation with Smooth M-Surfaces

Let $\sigma : I \to \mathbb{R}^N$ be an M-surface. In the final chapter of the book, we will need the following approximation result.

Proposition 11.61 *There exists a sequence of C^∞-functions $\sigma_n : \mathbb{R}^M \to \mathbb{R}^N$ such that*

$$\lim_n \sigma_n(\boldsymbol{u}) = \sigma(\boldsymbol{u}), \quad \lim_n \frac{\partial \sigma_n}{\partial u_j}(\boldsymbol{u}) = \frac{\partial \sigma}{\partial u_j}(\boldsymbol{u}), \quad \textit{uniformly on } I,$$

for every $j = 1, \ldots, M$.

Proof First of all, we can extend σ to a C^1-function on a rectangle J such that $I \subseteq \mathring{J}$, and then we set $\sigma(\boldsymbol{u}) = 0$ when $\boldsymbol{u} \notin J$.

Consider the C^∞-function $\varphi : \mathbb{R}^M \to \mathbb{R}$ defined as

$$\varphi(\boldsymbol{z}) = \begin{cases} C \exp\left(\dfrac{1}{\|\boldsymbol{z}\|^2 - 1}\right) & \text{if } \|\boldsymbol{z}\| < 1, \\ 0 & \text{if } \|\boldsymbol{z}\| \geq 1, \end{cases} \tag{11.1}$$

where the constant $C > 0$ is chosen in such a way that

$$\int_{B(0,1)} \varphi(\boldsymbol{z})\, d\boldsymbol{z} = 1.$$

Define $\sigma_n : \mathbb{R}^M \to \mathbb{R}^N$ as

$$\sigma_n(\boldsymbol{u}) = n^M \int_{\mathbb{R}^M} \sigma(\boldsymbol{y})\varphi(n(\boldsymbol{u} - \boldsymbol{y}))\, d\boldsymbol{y}.$$

We see that each σ_n is a C^∞-function and, by a change of variables,

$$\sigma_n(\boldsymbol{u}) = \int_{\mathbb{R}^M} \sigma\left(\boldsymbol{u} - \frac{1}{n}\boldsymbol{z}\right)\varphi(\boldsymbol{z})\,d\boldsymbol{z} = \int_{B(0,1)} \sigma\left(\boldsymbol{u} - \frac{1}{n}\boldsymbol{z}\right)\varphi(\boldsymbol{z})\,d\boldsymbol{z}.$$

Hence,

$$\sigma_n(\boldsymbol{u}) - \sigma(\boldsymbol{u}) = \int_{B(0,1)} \left(\sigma\left(\boldsymbol{u} - \frac{1}{n}\boldsymbol{z}\right) - \sigma(\boldsymbol{u})\right)\varphi(\boldsymbol{z})\,d\boldsymbol{z}.$$

Let $\varepsilon > 0$ be fixed. Since σ is uniformly continuous on J, there exists a $\delta > 0$ such that, if $\boldsymbol{u}$ and $\boldsymbol{u}'$ belong to J and $\|\boldsymbol{u}' - \boldsymbol{u}\| \leq \delta$, then $\|\sigma(\boldsymbol{u}') - \sigma(\boldsymbol{u})\| \leq \varepsilon$. Now, it is surely true that if $\boldsymbol{u} \in I$ and n is large enough, then $\boldsymbol{u} - \frac{1}{n}\boldsymbol{z} \in J$ and $\|\frac{1}{n}\boldsymbol{z}\| \leq \delta$, whence

$$\|\sigma_n(\boldsymbol{u}) - \sigma(\boldsymbol{u})\| \leq \int_{B(0,1)} \left\|\sigma\left(\boldsymbol{u} - \frac{1}{n}\boldsymbol{z}\right) - \sigma(\boldsymbol{u})\right\|\varphi(\boldsymbol{z})\,d\boldsymbol{z} \leq \varepsilon \int_{B(0,1)} \varphi(\boldsymbol{z})\,d\boldsymbol{z} = \varepsilon.$$

This proves that $\lim_n \sigma_n = \sigma$ uniformly on I.

Now observe that, by the Leibniz Rule, for every $j = 1, \ldots, M$,

$$\frac{\partial \sigma_n}{\partial u_j}(\boldsymbol{u}) = \int_{B(0,1)} \frac{\partial \sigma}{\partial u_j}\left(\boldsymbol{u} - \frac{1}{n}\boldsymbol{z}\right)\varphi(\boldsymbol{z})\,d\boldsymbol{z}.$$

Then the same argument as above shows that $\lim_n \frac{\partial \sigma_n}{\partial u_j} = \frac{\partial \sigma}{\partial u_j}$, uniformly on I.   ∎

11.19   The Integral on a Compact Manifold

We now assume that $\mathcal{M}$ is a compact M-manifold. We will see how it is possible to define the integral of a function f on $\mathcal{M}$.

By Theorem 10.39, for every $\boldsymbol{x} \in \mathcal{M}$ there is a local parametrization by some M-surface $\sigma : I \to \mathbb{R}^N$, where I is a rectangle of $\mathbb{R}^M$ of the type

$$I = \begin{cases} [-\alpha, \alpha]^M & \text{if } \boldsymbol{x} \notin \partial\mathcal{M}, \\ [-\alpha, \alpha]^{M-1} \times [0, \alpha] & \text{if } \boldsymbol{x} \in \partial\mathcal{M}, \end{cases}$$

and $\sigma(\boldsymbol{0}) = \boldsymbol{x}$.

In the case where $f|_{\mathcal{M}}$, the restriction of f to the set $\mathcal{M}$ is equal to zero outside the image of a single local M-parametrization, we simply define

$$\int_{\mathcal{M}} f = \int_{\sigma} f.$$

Let us verify that this is a good definition.

Proposition 11.62 *If $\sigma : I \to \mathbb{R}^N$ and $\tilde{\sigma} : J \to \mathbb{R}^N$ are two local M-parametrizations and $f|_{\mathcal{M}}$ is equal to zero outside the image of σ and outside the image of $\tilde{\sigma}$, then $\int_\sigma f = \int_{\tilde{\sigma}} f$.*

Proof If this is the case, we set

$$\widehat{\mathcal{M}} = \sigma(I) \cap \tilde{\sigma}(J), \qquad \mathcal{I} = \sigma^{-1}(\widehat{\mathcal{M}}), \qquad \mathcal{J} = \tilde{\sigma}^{-1}(\widehat{\mathcal{M}}).$$

Following closely the proof of Theorem 11.56, we define the open sets

$$A = \mathring{\mathcal{I}} \cap \sigma^{-1}(\widehat{\mathcal{M}} \setminus (\sigma(\partial I) \cup \tilde{\sigma}(\partial J))), \qquad B = \mathring{\mathcal{J}} \cap \tilde{\sigma}^{-1}(\widehat{\mathcal{M}} \setminus (\sigma(\partial I) \cup \tilde{\sigma}(\partial J))).$$

For every $u \in A$ there is a unique $v \in \mathring{J}$ such that $\tilde{\sigma}(v) = \sigma(u)$, and we can then define $\varphi : A \to B$ by setting $\varphi(u) = v$. It can be seen that $\varphi : A \to B$ is a diffeomorphism such that $\sigma(u) = \tilde{\sigma}(\varphi(u))$ for every $u \in A$. Proceeding as in the proof of Theorem 11.55, since $f|_{\mathcal{M}}$ vanishes outside $\sigma(I) \cap \sigma(J)$, we have that

$$\int_\sigma f = \int_A f(\sigma(u)) \|\Sigma(u)\| \, du$$

$$= \int_A f(\tilde{\sigma}(\varphi(u))) \|\tilde{\Sigma}(\varphi(u))\| \, |\det \varphi'(u)| \, du$$

$$= \int_B f(\tilde{\sigma}(v)) \|\tilde{\Sigma}(v)\| \, dv = \int_{\tilde{\sigma}} f,$$

thereby completing the proof. ∎

In general, we saw in Theorem 10.39 that for every $x \in \mathcal{M}$ there is a neighborhood A' of x such that $A' \cap \mathcal{M}$ can be M-parametrized by some function $\sigma : I \to \mathbb{R}^N$. For every such x there is an open ball $B(x, \rho_x)$ contained in A'. We thus have an open covering of $\mathcal{M}$ made of these open balls. Since $\mathcal{M}$ is compact, by Theorem 4.9 there exists a finite open subcovering, which we denote by $A_1, \ldots, A_n$. Hence, the open set $V = A_1 \cup \cdots \cup A_n$ contains $\mathcal{M}$. We now need the following result.

Theorem 11.63 *There exist some functions $\phi_1, \ldots, \phi_n : V \to \mathbb{R}$, of class C^∞, such that, for every $x \in V$ and every $k \in \{1, \ldots, n\}$, the following properties hold:*

(a) $0 \le \phi_k(x) \le 1$,
(b) $x \notin A_k \implies \phi_k(x) = 0$,
(c) $\sum_{k=1}^n \phi_k(x) = 1$ for every $x \in \mathcal{M}$.

The functions $\phi_1, \ldots, \phi_n$ are said to be a "partition of unity" associated with the open covering $A_1, \ldots, A_n$.

Proof Let $A_k = B(x_k, \rho_{x_k})$, with $k = 1, \ldots, n$. Consider the C^∞-function $f :$ $\mathbb{R} \to \mathbb{R}$ defined by

$$f(u) = \begin{cases} \exp\left(\frac{1}{u^2-1}\right) & \text{if } |u| < 1, \\ 0 & \text{if } |u| \geq 1, \end{cases} \tag{11.2}$$

and set

$$\psi_k(x) = f\left(\frac{\|x - x_k\|}{\rho_{x_k}}\right).$$

Then, for every $x \in V$, we have $\psi_1(x) + \cdots + \psi_n(x) > 0$, and we can define the functions

$$\phi_k(x) = \frac{\psi_k(x)}{\psi_1(x) + \cdots + \psi_n(x)}.$$

The required properties are now easily verified.   ∎

Since each $\phi_k \cdot f|_{\mathcal{M}}$ vanishes outside the image of a single local M-parametrization, we can define the integral of f on $\mathcal{M}$ by

$$\int_{\mathcal{M}} f = \sum_{k=1}^{n} \int_{\mathcal{M}} \phi_k f.$$

Let us verify that this definition depends neither on the choice of the local M-parametrizations nor on the particular partition of unity.

Proposition 11.64 *If $\tilde{A}_1, \ldots, \tilde{A}_m$ is any other finite open covering of $\mathcal{M}$ and $\tilde{\phi}_1, \ldots, \tilde{\phi}_m$ is an associated partition of unity, then $\sum_{k=1}^{n} \int_{\mathcal{M}} \phi_k f = \sum_{j=1}^{m} \int_{\mathcal{M}} \tilde{\phi}_j f$.*

Proof Indeed,

$$\sum_{k=1}^{n} \int_{\mathcal{M}} \phi_k f = \sum_{k=1}^{n} \int_{\mathcal{M}} \left(\sum_{j=1}^{m} \tilde{\phi}_j\right) \phi_k f$$

$$= \sum_{k=1}^{n} \sum_{j=1}^{m} \int_{\mathcal{M}} \tilde{\phi}_j \phi_k f$$

$$= \sum_{j=1}^{m} \sum_{k=1}^{n} \int_{\mathscr{M}} \phi_k \tilde{\phi}_j f$$

$$= \sum_{j=1}^{m} \int_{\mathscr{M}} \Big(\sum_{k=1}^{n} \phi_k \Big) \tilde{\phi}_j f = \sum_{j=1}^{m} \int_{\mathscr{M}} \tilde{\phi}_j f ,$$

which is what we had to verify.   ∎

It is now possible to define the M-dimensional measure of an M-manifold $\mathscr{M}$ as the integral over $\mathscr{M}$ of the constant function 1. We can use the notation

$$\mu_M(\mathscr{M}) = \int_{\mathscr{M}} 1$$

since this quantity coincides with the already defined M-dimensional measure introduced when $\mathscr{M}$ is an M-parametrizable set. In the cases $M = 1$ or 2, the M-dimensional measure of $\mathscr{M}$ is often called the "length" or "area" of $\mathscr{M}$, respectively.

If $M = N$, then it can be verified that the N-dimensional measure of the manifold $\mathscr{M}$ coincides with the usual measure, i.e., $\mu_N(\mathscr{M}) = \mu(\mathscr{M})$.

Differential Forms

12

Let us start considering the projections in $\mathbb{R}^N$; as we have already seen, these are the functions $p_m : \mathbb{R}^N \to \mathbb{R}$ defined by

$$p_m(x_1, x_2, \ldots, x_N) = x_m .$$

However, it will now be useful to use a different notation; instead of p_m, we will write dx_m. In this way, the known formula for the differential,

$$df(\boldsymbol{x}_0)(\boldsymbol{h}) = \sum_{m=1}^{N} \frac{\partial f}{\partial x_m}(\boldsymbol{x}_0)\, h_m ,$$

where $\boldsymbol{h} = (h_1, h_2, \ldots, h_m)$ is any vector in $\mathbb{R}^N$, can be written in the form

$$df(\boldsymbol{x}_0)(\boldsymbol{h}) = \sum_{m=1}^{N} \frac{\partial f}{\partial x_m}(\boldsymbol{x}_0)\, dx_m(\boldsymbol{h})$$

or, more succinctly,

$$df(\boldsymbol{x}_0) = \sum_{m=1}^{N} \frac{\partial f}{\partial x_m}(\boldsymbol{x}_0)\, dx_m .$$

Here we have a first example of what will be called a "differential form."

12.1 An Informal Definition

Let us introduce an operation between these symbols dx_m, which looks like a "product"; we will denote it by the symbol $\wedge$. Without entering into its precise definition (which will be provided in Sect. 12.14), we will simply explain its main properties.

The crucial feature of this operation is that it is antisymmetric, i.e.,

$$dx_i \wedge dx_j = -dx_j \wedge dx_i .$$

We can also multiply these objects several times, maintaining the rule that when two of them are interchanged, there is a change in sign, e.g.,

$$dx_{i_1} \wedge \ldots dx_i \ldots dx_j \ldots \wedge dx_{i_M} = -dx_{i_1} \wedge \ldots dx_j \ldots dx_i \ldots \wedge dx_{i_M}.$$

Note that if two indices happen to be the same, then

$$dx_{i_1} \wedge \ldots dx_i \ldots dx_i \ldots \wedge dx_{i_M} = 0.$$

This is why we will usually consider the indices in a strictly increasing order, i.e.,

$$dx_{i_1} \wedge \cdots \wedge dx_{i_M} , \text{ with } 1 \leq i_1 < \cdots < i_M \leq N .$$

All the other products of M elements are either equal to 0 or can be reduced to this form after a reordering of the indices, possibly leading to a change of sign.

Let us analyze, for example, the case $N = 3$. We have here dx_1, dx_2, and dx_3. If we take two of them with the indices in strictly increasing order, we obtain

$$dx_1 \wedge dx_2, \quad dx_1 \wedge dx_3, \quad dx_2 \wedge dx_3 .$$

In the other cases, we have

$$dx_2 \wedge dx_1 = -dx_1 \wedge dx_2, \quad dx_3 \wedge dx_1 = -dx_1 \wedge dx_3, \quad dx_3 \wedge dx_2 = -dx_2 \wedge dx_3,$$

while

$$dx_1 \wedge dx_1 = dx_2 \wedge dx_2 = dx_3 \wedge dx_3 = 0 .$$

Moreover, there is a unique product of three elements with strictly increasing indices:

$$dx_1 \wedge dx_2 \wedge dx_3 .$$

Concerning the other products of three elements, we have

$$dx_2 \wedge dx_3 \wedge dx_1 = dx_3 \wedge dx_1 \wedge dx_2 = dx_1 \wedge dx_2 \wedge dx_3$$

and

$$dx_1 \wedge dx_3 \wedge dx_2 = dx_3 \wedge dx_2 \wedge dx_1 = dx_2 \wedge dx_1 \wedge dx_3 = -dx_1 \wedge dx_2 \wedge dx_3,$$

while all those having two or three coinciding indices are equal to 0.

Let $\mathcal{O}$ be an open subset of $\mathbb{R}^N$ and M a positive integer. We will call "differential form of degree M" (or "M-differential form") a function defined on $\mathcal{O}$ by an expression of the type

$$\omega(\boldsymbol{x}) = \sum_{1 \leq i_1 < \cdots < i_M \leq N} f_{i_1 \ldots i_M}(\boldsymbol{x}) \, dx_{i_1} \wedge \cdots \wedge dx_{i_M}.$$

The functions $f_{i_1 \ldots i_M} : \mathcal{O} \to \mathbb{R}$ are the "components" of ω. We will say that ω is of class C^k if all its components are of that class. The set of all M-differential forms of class C^k defined on the subset $\mathcal{O}$ of $\mathbb{R}^N$ will be denoted by

$$\mathcal{F}_M^k(\mathcal{O}, \mathbb{R}^N).$$

If $k = 0$ (i.e., when the components of the differential forms are continuous), we will simply write $\mathcal{F}_M(\mathcal{O}, \mathbb{R}^N)$. Note that $\omega(\boldsymbol{x})$ is determined by the $\binom{N}{M}$-dimensional vector

$$F(\boldsymbol{x}) = \left(f_{i_1 \ldots i_M}(\boldsymbol{x}) \right)_{1 \leq i_1 < \cdots < i_M \leq N}.$$

We will call 0-differential form any function defined on $\mathcal{O}$, with values in $\mathbb{R}$. Hence,

$$\mathcal{F}_0^k(\mathcal{O}, \mathbb{R}^N) = C^k(\mathcal{O}, \mathbb{R}^N).$$

Let us take a closer look at the case $N = 3$. Denoting by ω_M an M-differential form, with $M = 1, 2, 3$, we can write

$$\omega_1(\boldsymbol{x}) = f_1(\boldsymbol{x}) \, dx_1 + f_2(\boldsymbol{x}) \, dx_2 + f_3(\boldsymbol{x}) \, dx_3,$$

$$\omega_2(\boldsymbol{x}) = f_{12}(\boldsymbol{x}) \, dx_1 \wedge dx_2 + f_{13}(\boldsymbol{x}) \, dx_1 \wedge dx_3 + f_{23}(\boldsymbol{x}) \, dx_2 \wedge dx_3,$$

$$\omega_3(\boldsymbol{x}) = f_{123}(\boldsymbol{x}) \, dx_1 \wedge dx_2 \wedge dx_3.$$

Notice that $\omega_1(\boldsymbol{x})$ and $\omega_2(\boldsymbol{x})$ are determined by the three-dimensional vectors

$$F(\boldsymbol{x}) = (f_1(\boldsymbol{x}), f_2(\boldsymbol{x}), f_3(\boldsymbol{x})) \quad \text{and} \quad \widetilde{F}(\boldsymbol{x}) = (f_{12}(\boldsymbol{x}), f_{13}(\boldsymbol{x}), f_{23}(\boldsymbol{x})),$$

respectively, while $\omega_3(\boldsymbol{x})$ is determined by a single function $f_{123}(\boldsymbol{x})$.

Henceforth, a function like $F : \mathcal{O} \to \mathbb{R}^N$, with $\mathcal{O} \subseteq \mathbb{R}^N$, will be called a "vector field."

12.2   Algebraic Operations

To simplify the notation, we will sometimes write

$$dx_{i_1} \wedge \cdots \wedge dx_{i_M} = dx_{i_1,\ldots,i_M} .$$

Hence, the previously defined differential form will also be written as follows:

$$\omega(\boldsymbol{x}) = \sum_{1 \le i_1 < \cdots < i_M \le N} f_{i_1 \ldots i_M}(\boldsymbol{x}) \, dx_{i_1,\ldots,i_M} .$$

It is possible to define the sum of two M-differential forms: If ω and $\widetilde{\omega}$ are both defined on $\mathcal{O}$, writing

$$\widetilde{\omega}(\boldsymbol{x}) = \sum_{1 \le i_1 < \cdots < i_M \le N} g_{i_1,\ldots,i_M}(\boldsymbol{x}) \, dx_{i_1,\ldots,i_M} ,$$

we define $\omega + \widetilde{\omega}$ in a natural way as

$$(\omega + \widetilde{\omega})(\boldsymbol{x}) = \sum_{1 \le i_1 < \cdots < i_M \le N} (f_{i_1,\ldots,i_M}(\boldsymbol{x}) + g_{i_1,\ldots,i_M}(\boldsymbol{x})) \, dx_{i_1,\ldots,i_M} .$$

Moreover, if $c \in \mathbb{R}$, then we define $c\,\omega$, the product of the scalar c by the M-differential form ω, as

$$(c\,\omega)(\boldsymbol{x}) = \sum_{1 \le i_1 < \cdots < i_M \le N} c f_{i_1,\ldots,i_M}(\boldsymbol{x}) \, dx_{i_1,\ldots,i_M} .$$

With these definitions, it can be checked that the set $\mathcal{F}_M^k(\mathcal{O}, \mathbb{R}^N)$ is a real vector space.

Given two differential forms $\omega \in \mathcal{F}_M(\mathcal{O}, \mathbb{R}^N)$, $\widetilde{\omega} \in \mathcal{F}_{\widetilde{M}}(\mathcal{O}, \mathbb{R}^N)$, of degrees M and $\widetilde{M}$, respectively, we now define the differential form $\omega \wedge \widetilde{\omega}$, of degree $M + \widetilde{M}$, which is called the "exterior product" of ω and $\widetilde{\omega}$. If

$$\omega(\boldsymbol{x}) = \sum_{1 \le i_1 < \cdots < i_M \le N} f_{i_1,\ldots,i_M}(\boldsymbol{x}) \, dx_{i_1,\ldots,i_M}$$

and

$$\widetilde{\omega}(\boldsymbol{x}) = \sum_{1 \le j_1 < \cdots < j_{\widetilde{M}} \le N} g_{j_1,\ldots,j_{\widetilde{M}}}(\boldsymbol{x}) \, dx_{j_1,\ldots,j_{\widetilde{M}}} ,$$

then we set

$$(\omega \wedge \widetilde{\omega})(\boldsymbol{x}) = \sum_{\substack{1 \le i_1 < \cdots < i_M \le N \\ 1 \le j_1 < \cdots < j_{\widetilde{M}} \le N}} f_{i_1,\ldots,i_M}(\boldsymbol{x}) g_{j_1,\ldots,j_{\widetilde{M}}}(\boldsymbol{x}) \, dx_{i_1,\ldots,i_M,j_1,\ldots,j_{\widetilde{M}}} \, .$$

Usually the symbol $\wedge$ is omitted when one of the two is a 0-differential form, since the exterior product is, in this case, similar to the product with a scalar. Notice that, in the preceding sum, all elements with a repeated index will be zero. Here are some properties of the exterior product.

Proposition 12.1 *If* $\omega, \widetilde{\omega}, \widetilde{\widetilde{\omega}}$ *are three differential forms of degrees* $M,\ \widetilde{M},\ \widetilde{\widetilde{M}},$ *respectively, then*

$$\widetilde{\omega} \wedge \omega = (-1)^{M\widetilde{M}} \omega \wedge \widetilde{\omega} \, ,$$

$$(\omega \wedge \widetilde{\omega}) \wedge \widetilde{\widetilde{\omega}} = \omega \wedge (\widetilde{\omega} \wedge \widetilde{\widetilde{\omega}}) \, ;$$

if $c \in \mathbb{R}$, *then*

$$(c\,\omega) \wedge \widetilde{\omega} = \omega \wedge (c\,\widetilde{\omega}) = c(\omega \wedge \widetilde{\omega}) \, ;$$

moreover, when $M = \widetilde{M}$,

$$(\omega + \widetilde{\omega}) \wedge \widetilde{\widetilde{\omega}} = (\omega \wedge \widetilde{\widetilde{\omega}}) + (\widetilde{\omega} \wedge \widetilde{\widetilde{\omega}}) \, ,$$

$$\widetilde{\widetilde{\omega}} \wedge (\omega + \widetilde{\omega}) = (\widetilde{\widetilde{\omega}} \wedge \omega) + (\widetilde{\widetilde{\omega}} \wedge \widetilde{\omega}) \, .$$

Proof Assume that ω and $\widetilde{\omega}$ are written as above, and let

$$\widetilde{\widetilde{\omega}}(\boldsymbol{x}) = \sum_{1 \le k_1 < \cdots < k_{\widetilde{\widetilde{M}}} \le N} h_{k_1,\ldots,k_{\widetilde{\widetilde{M}}}}(\boldsymbol{x}) \, dx_{k_1,\ldots,k_{\widetilde{\widetilde{M}}}} \, .$$

The first identity is obtained observing that, in order to arrive from the sequence of indices $i_1, \ldots, i_M, j_1, \ldots, j_{\widetilde{M}}$ at $j_1, \ldots, j_{\widetilde{M}}, i_1, \ldots, i_M$, we must first move j_1 to the left making M exchanges, then do the same for j_2, and so on, until we reach $j_{\widetilde{M}}$. In the end, it is then necessary to perform $M\widetilde{M}$ exchanges of indices. Taking into account the fact that the differential form changes sign each time there is an exchange of indices, we have the formula we wanted to prove.

The proof of the second identity (associative property) shows no great difficulties, and also the identities where the constant c appears can be easily verified.

Concerning the distributive property, when $M = \tilde{M}$, we have

$$((\omega + \widetilde{\omega}) \wedge \widetilde{\widetilde{\omega}})(\boldsymbol{x}) =$$

$$= \sum_{\substack{1 \leq i_1 < \cdots < i_M \leq N \\ 1 \leq k_1 < \cdots < k_{\tilde{\widetilde{M}}} \leq N}} (f_{i_1,\ldots,i_M}(\boldsymbol{x}) + g_{i_1,\ldots,i_M}(\boldsymbol{x})) h_{k_1,\ldots,k_{\tilde{\widetilde{M}}}}(\boldsymbol{x})\, dx_{i_1,\ldots,i_M,k_1,\ldots,k_{\tilde{\widetilde{M}}}}$$

$$= \sum_{\substack{1 \leq i_1 < \cdots < i_M \leq N \\ 1 \leq k_1 < \cdots < k_{\tilde{\widetilde{M}}} \leq N}} (f_{i_1,\ldots,i_M}(\boldsymbol{x}) h_{k_1,\ldots,k_{\tilde{\widetilde{M}}}}(\boldsymbol{x}) +$$

$$+ g_{i_1,\ldots,i_M}(\boldsymbol{x}) h_{k_1,\ldots,k_{\tilde{\widetilde{M}}}}(\boldsymbol{x}))\, dx_{i_1,\ldots,i_M,k_1,\ldots,k_{\tilde{\widetilde{M}}}}$$

$$= ((\omega \wedge \widetilde{\widetilde{\omega}}) + (\widetilde{\omega} \wedge \widetilde{\widetilde{\omega}}))(\boldsymbol{x}).$$

The last identity is proved either in an analogous way or by using the first and fourth identities. ∎

12.3 The Exterior Differential

Given an M-differential form ω of class $\mathcal{C}^1$, we want to define the differential form $d_{ex}\omega$, of degree $M + 1$, which is said to be the "exterior differential" of ω.

If ω is a 0-differential form, $\omega = f : \mathcal{O} \to \mathbb{R}$, its exterior differential $d_{ex}\omega(\boldsymbol{x})$ is simply the usual differential

$$df(\boldsymbol{x}) = \sum_{m=1}^{N} \frac{\partial f}{\partial x_m}(\boldsymbol{x})\, dx_m.$$

In the general case, if

$$\omega(\boldsymbol{x}) = \sum_{1 \leq i_1 < \cdots < i_M \leq N} f_{i_1,\ldots,i_M}(\boldsymbol{x})\, dx_{i_1} \wedge \cdots \wedge dx_{i_M},$$

we set

$$d_{ex}\omega(\boldsymbol{x}) = \sum_{1 \leq i_1 < \cdots < i_M \leq N} df_{i_1,\ldots,i_M}(\boldsymbol{x}) \wedge dx_{i_1} \wedge \cdots \wedge dx_{i_M}$$

or, equivalently,

$$d_{ex}\omega(\boldsymbol{x}) = \sum_{1 \leq i_1 < \cdots < i_M \leq N} \sum_{m=1}^{N} \frac{\partial f_{i_1,\ldots,i_M}}{\partial x_m}(\boldsymbol{x})\, dx_m \wedge dx_{i_1} \wedge \cdots \wedge dx_{i_M}.$$

In what follows, to simplify the notation, we will always write $d\omega$ instead of $d_{ex}\omega$. Let us consider some properties of the exterior differential.

Proposition 12.2 *If ω and $\widetilde{\omega}$ are two differential forms of class C^1, of degrees M and $\widetilde{M}$, respectively, then*

$$d(\omega \wedge \widetilde{\omega}) = d\omega \wedge \widetilde{\omega} + (-1)^M \omega \wedge d\widetilde{\omega};$$

if $M = \widetilde{M}$ and $c \in \mathbb{R}$, then we have

$$d(\omega + \widetilde{\omega}) = d\omega + d\widetilde{\omega}, \qquad d(c\,\omega) = c\,d\omega;$$

if ω is of class C^2, then

$$d(d\omega) = 0.$$

Proof Concerning the first identity, if ω and $\widetilde{\omega}$ are as above, we have

$$d(\omega \wedge \widetilde{\omega})(\boldsymbol{x}) = \sum_{\substack{1 \le i_1 < \cdots < i_M \le N \\ 1 \le j_1 < \cdots < j_{\widetilde{M}} \le N}} \sum_{m=1}^{N} \frac{\partial}{\partial x_m}(f_{i_1,\ldots,i_M} g_{j_1,\ldots,j_{\widetilde{M}}})(\boldsymbol{x})\, dx_{m,i_1,\ldots,i_M,j_1,\ldots,j_{\widetilde{M}}}$$

$$= \sum_{\substack{1 \le i_1 < \cdots < i_M \le N \\ 1 \le j_1 < \cdots < j_{\widetilde{M}} \le N}} \sum_{m=1}^{N} \left(\frac{\partial f_{i_1,\ldots,i_M}}{\partial x_m} g_{j_1,\ldots,j_{\widetilde{M}}} + \right.$$

$$\left. + f_{i_1,\ldots,i_M} \frac{\partial g_{j_1,\ldots,j_{\widetilde{M}}}}{\partial x_m} \right)(\boldsymbol{x})\, dx_{m,i_1,\ldots,i_M,j_1,\ldots,j_{\widetilde{M}}}$$

$$= \sum_{\substack{1 \le i_1 < \cdots < i_M \le N \\ 1 \le j_1 < \cdots < j_{\widetilde{M}} \le N}} \sum_{m=1}^{N} \left(\frac{\partial f_{i_1,\ldots,i_M}}{\partial x_m} g_{j_1,\ldots,j_{\widetilde{M}}} \right)(\boldsymbol{x})\, dx_{m,i_1,\ldots,i_M,j_1,\ldots,j_{\widetilde{M}}} +$$

$$+ (-1)^M \sum_{\substack{1 \le i_1 < \cdots < i_M \le N \\ 1 \le j_1 < \cdots < j_{\widetilde{M}} \le N}} \sum_{m=1}^{N} \left(f_{i_1,\ldots,i_M} \frac{\partial g_{j_1,\ldots,j_{\widetilde{M}}}}{\partial x_m} \right)(\boldsymbol{x})\, dx_{i_1,\ldots,i_M,m,j_1,\ldots,j_{\widetilde{M}}}$$

$$= (d\omega \wedge \widetilde{\omega})(\boldsymbol{x}) + (-1)^M (\omega \wedge d\widetilde{\omega})(\boldsymbol{x}).$$

The second and third identities follow easily from the linearity of the derivative. Concerning the last identity, we can see that

$$d(d\omega)(\boldsymbol{x}) = \sum_{1 \le i_1 < \cdots < i_M \le N} \sum_{k=1}^{N} \sum_{m=1}^{N} \frac{\partial}{\partial x_k} \frac{\partial f_{i_1,\ldots,i_M}}{\partial x_m}(\boldsymbol{x})\, dx_{k,m,i_1,\ldots,i_M}.$$

Since, by Schwarz's Theorem 10.9,

$$\frac{\partial}{\partial x_k}\frac{\partial f_{i_1,\dots,i_M}}{\partial x_m} = \frac{\partial}{\partial x_m}\frac{\partial f_{i_1,\dots,i_M}}{\partial x_k},$$

taking into account the fact that $dx_k \wedge dx_m = -dx_m \wedge dx_k$, it is seen that all the terms in the sums pairwise eliminate one another, so that $d(d\omega)(\boldsymbol{x}) = 0$. ∎

12.4   Differential Forms in $\mathbb{R}^3$

In the special case $N = 3$, in view of the applications of the theory of differential forms to some important physical situations, we prefer adopting a different order in the components of a 2-differential form. Instead of the increasing ordering of the indices that has been adopted so far,

$$dx_1 \wedge dx_2, \quad dx_1 \wedge dx_3, \quad dx_2 \wedge dx_3,$$

henceforth we will prefer to take

$$dx_2 \wedge dx_3, \quad dx_3 \wedge dx_1, \quad dx_1 \wedge dx_2.$$

Let us then investigate the operation of the exterior product in this case, with the newly adopted convention.

If ω_1 and $\widetilde{\omega}_1$ are two 1-differential forms, e.g.,

$$\omega_1(\boldsymbol{x}) = f_1(\boldsymbol{x})\,dx_1 + f_2(\boldsymbol{x})\,dx_2 + f_3(\boldsymbol{x})\,dx_3,$$
$$\widetilde{\omega}_1(\boldsymbol{x}) = \tilde{f}_1(\boldsymbol{x})\,dx_1 + \tilde{f}_2(\boldsymbol{x})\,dx_2 + \tilde{f}_3(\boldsymbol{x})\,dx_3,$$

we compute

$$(\omega_1 \wedge \widetilde{\omega}_1)(\boldsymbol{x}) = (f_2(\boldsymbol{x})\tilde{f}_3(\boldsymbol{x}) - f_3(\boldsymbol{x})\tilde{f}_2(\boldsymbol{x}))\,dx_2 \wedge dx_3 +$$
$$+ (f_3(\boldsymbol{x})\tilde{f}_1(\boldsymbol{x}) - f_1(\boldsymbol{x})\tilde{f}_3(\boldsymbol{x}))\,dx_3 \wedge dx_1 +$$
$$+ (f_1(\boldsymbol{x})\tilde{f}_2(\boldsymbol{x}) - f_2(\boldsymbol{x})\tilde{f}_1(\boldsymbol{x}))\,dx_1 \wedge dx_2.$$

Hence, if $F = (f_1, f_2, f_3)$ and $\widetilde{F} = (\tilde{f}_1, \tilde{f}_2, \tilde{f}_3)$ are the vector fields corresponding to ω_1 and $\widetilde{\omega}_1$, respectively, we see that the vector field determined by $\omega_1 \wedge \widetilde{\omega}_1$ is

$$F \times \widetilde{F} = (f_2\tilde{f}_3 - f_3\tilde{f}_2, \ f_3\tilde{f}_1 - f_1\tilde{f}_3, \ f_1\tilde{f}_2 - f_2\tilde{f}_1),$$

the "cross product" of F and $\widetilde{F}$.

On the other hand, if ω_1 is a 1-differential form and $\widetilde{\omega}_2$ is a 2-differential form, e.g.,

$$\omega_1(\boldsymbol{x}) = f_1(\boldsymbol{x})dx_1 + f_2(\boldsymbol{x})dx_2 + f_3(\boldsymbol{x})dx_3\,,$$

$$\widetilde{\omega}_2(\boldsymbol{x}) = \tilde{f}_1(\boldsymbol{x})\,dx_2 \wedge dx_3 + \tilde{f}_2(\boldsymbol{x})\,dx_3 \wedge dx_1 + \tilde{f}_3(\boldsymbol{x})\,dx_1 \wedge dx_2\,,$$

we have

$$(\omega_1 \wedge \widetilde{\omega}_2)(\boldsymbol{x}) = (f_1(\boldsymbol{x})\tilde{f}_1(\boldsymbol{x}) + f_2(\boldsymbol{x})\tilde{f}_2(\boldsymbol{x}) + f_3(\boldsymbol{x})\tilde{f}_3(\boldsymbol{x}))\,dx_1 \wedge dx_2 \wedge dx_3\,.$$

Hence, if $F = (f_1, f_2, f_3)$ and $\widetilde{F} = (\tilde{f}_1, \tilde{f}_2, \tilde{f}_3)$ are the vector fields corresponding to ω_1 and $\widetilde{\omega}_2$, respectively, we see that the vector field determined by $\omega_1 \wedge \widetilde{\omega}_2$ is

$$F \cdot \widetilde{F} = f_1\tilde{f}_1 + f_2\tilde{f}_2 + f_3\tilde{f}_3\,,$$

the "scalar product" of F and $\widetilde{F}$.

If we have a 0-differential form $\omega_0 = f : \mathcal{O} \to \mathbb{R}$, then

$$d\omega_0(\boldsymbol{x}) = \frac{\partial f}{\partial x_1}(\boldsymbol{x})\,dx_1 + \frac{\partial f}{\partial x_2}(\boldsymbol{x})\,dx_2 + \frac{\partial f}{\partial x_3}(\boldsymbol{x})\,dx_3\,,$$

and we know that the corresponding vector field is the "gradient" of f,

$$\nabla f = \left(\frac{\partial f}{\partial x_1}, \frac{\partial f}{\partial x_2}, \frac{\partial f}{\partial x_3} \right).$$

Taking a 1-differential form

$$\omega_1(\boldsymbol{x}) = f_1(\boldsymbol{x})\,dx_1 + f_2(\boldsymbol{x})\,dx_2 + f_3(\boldsymbol{x})\,dx_3\,,$$

we compute

$$d\omega_1(\boldsymbol{x}) = \left(\frac{\partial f_3}{\partial x_2}(\boldsymbol{x}) - \frac{\partial f_2}{\partial x_3}(\boldsymbol{x}) \right) dx_2 \wedge dx_3 +$$

$$+ \left(\frac{\partial f_1}{\partial x_3}(\boldsymbol{x}) - \frac{\partial f_3}{\partial x_1}(\boldsymbol{x}) \right) dx_3 \wedge dx_1 +$$

$$+ \left(\frac{\partial f_2}{\partial x_1}(\boldsymbol{x}) - \frac{\partial f_1}{\partial x_2}(\boldsymbol{x}) \right) dx_1 \wedge dx_2\,.$$

If $F = (f_1, f_2, f_3)$ is the vector field associated with ω_1, we call "curl" of F the vector field corresponding to $d\omega_1$, and we write

$$\nabla \times F = \left(\frac{\partial f_3}{\partial x_2} - \frac{\partial f_2}{\partial x_3}, \frac{\partial f_1}{\partial x_3} - \frac{\partial f_3}{\partial x_1}, \frac{\partial f_2}{\partial x_1} - \frac{\partial f_1}{\partial x_2} \right).$$

Finally, if we consider a 2-differential form

$$\omega_2(\boldsymbol{x}) = f_1(\boldsymbol{x})\, dx_2 \wedge dx_3 + f_2(\boldsymbol{x})\, dx_3 \wedge dx_1 + f_3(\boldsymbol{x})\, dx_1 \wedge dx_2 \,,$$

then

$$d\omega_2(\boldsymbol{x}) = \left(\frac{\partial f_1}{\partial x_1}(\boldsymbol{x}) + \frac{\partial f_2}{\partial x_2}(\boldsymbol{x}) + \frac{\partial f_3}{\partial x_3}(\boldsymbol{x}) \right) dx_1 \wedge dx_2 \wedge dx_3 \,.$$

If $F = (f_1, f_2, f_3)$ is the vector field associated with ω_2, we call "divergence" of F the scalar function corresponding to $d\omega_2$, and we write

$$\nabla \cdot F = \frac{\partial f_1}{\partial x_1} + \frac{\partial f_2}{\partial x_2} + \frac{\partial f_3}{\partial x_3} \,.$$

We will explain later on the physical meaning of curl and divergence.

The properties of the exterior product and those of the exterior differential lead to formulas involving the gradient, the curl, and the divergence. Taking $f : \mathcal{O} \to \mathbb{R}$, $\tilde{f} : \mathcal{O} \to \mathbb{R}$, $F : \mathcal{O} \to \mathbb{R}^3$, and $\tilde{F} : \mathcal{O} \to \mathbb{R}^3$, we have the following formulas:

$$\nabla \times (\nabla f) = 0 \,,$$
$$\nabla \cdot (\nabla \times F) = 0 \,,$$
$$\nabla (f \tilde{f}) = \tilde{f}(\nabla f) + f(\nabla \tilde{f}) \,,$$
$$\nabla \times (f F) = (\nabla f) \times F + f(\nabla \times F) \,,$$
$$\nabla \cdot (f \tilde{F}) = (\nabla f) \cdot \tilde{F} + f(\nabla \cdot \tilde{F}) \,,$$
$$\nabla \cdot (F \times \tilde{F}) = (\nabla \times F) \cdot \tilde{F} - F \cdot (\nabla \times \tilde{F}) \,.$$

The proofs are left to the reader.

12.5   The Integral on an M-Surface

We want to define the notion of integral of an M-differential form $\omega \in \mathcal{F}(\mathcal{O}, \mathbb{R}^N)$ on an M-surface $\sigma : I \to \mathbb{R}^N$, with $1 \le M \le N$. Let

$$\omega(\boldsymbol{x}) = \sum_{1 \le i_1 < \cdots < i_M \le N} f_{i_1,\ldots,i_M}(\boldsymbol{x})\, dx_{i_1} \wedge \cdots \wedge dx_{i_M} \,.$$

Recall, for every $\boldsymbol{x} \in \mathcal{O}$ and every $\boldsymbol{u} \in I$, the $\binom{N}{M}$-dimensional vectors

$$F(\boldsymbol{x}) = \left(f_{i_1,\ldots,i_M}(\boldsymbol{x}) \right)_{1 \leq i_1 < \cdots < i_M \leq N},$$

$$\Sigma(\boldsymbol{u}) = \left(\det \sigma'_{(i_1,\ldots,i_M)}(\boldsymbol{u}) \right)_{1 \leq i_1 < \cdots < i_M \leq N}.$$

If $\sigma(I) \subseteq \mathcal{O}$, then, denoting by "$\cdot$" the Euclidean scalar product in $\mathbb{R}^{\binom{N}{M}}$, we set

$$\int_\sigma \omega = \int_I F(\sigma(\boldsymbol{u})) \cdot \Sigma(\boldsymbol{u}) \, d\boldsymbol{u}.$$

Let us consider the meaning of the given definition in two special cases.

If $M = 1$, then $\sigma : [a, b] \to \mathbb{R}^N$ is a curve and ω is the 1-differential form

$$\omega(\boldsymbol{x}) = f_1(\boldsymbol{x}) \, dx_1 + \cdots + f_N(\boldsymbol{x}) \, dx_N.$$

Then we have $F(\boldsymbol{x}) = (f_1(\boldsymbol{x}), \ldots, f_N(\boldsymbol{x}))$ and

$$\int_\sigma \omega = \int_a^b F(\sigma(t)) \cdot \sigma'(t) \, dt.$$

This quantity will be called the "line integral" of the vector field $F = (f_1, \ldots, f_N)$ along the curve σ, and will be denoted by

$$\int_\sigma F \cdot d\ell.$$

In mechanics, this concept is used, for example, to define the "work" done by a field of forces on a particle moving along a curve.

Example Let us compute the line integral of the vector field $F(x, y, z) = (-y, x, z^2)$ along the curve $\sigma : [0, 2\pi] \to \mathbb{R}^3$, defined by $\sigma(t) = (\cos t, \sin t, t)$:

$$\int_\sigma F \cdot d\ell = \int_0^{2\pi} (-\sin t, \cos t, t^2) \cdot (-\sin t, \cos t, 1) \, dt$$

$$= \int_0^{2\pi} (\sin^2 t + \cos^2 t + t^2) \, dt = 2\pi + \frac{8\pi^3}{3}.$$

If $M = 2$ and $N = 3$, then $\sigma : [a_1, b_1] \times [a_2, b_2] \to \mathbb{R}^3$ is a surface and ω is the 2-differential form

$$\omega(\boldsymbol{x}) = f_1(\boldsymbol{x}) \, dx_2 \wedge dx_3 + f_2(\boldsymbol{x}) \, dx_3 \wedge dx_1 + f_3(\boldsymbol{x}) \, dx_1 \wedge dx_2.$$

Then we have $F(\boldsymbol{x}) = (f_1(\boldsymbol{x}), f_2(\boldsymbol{x}), f_3(\boldsymbol{x}))$, and

$$
\int_\sigma \omega = \int_{a_2}^{b_2} \int_{a_1}^{b_1} \left[f_1(\sigma(u, v)) \det \begin{pmatrix} \frac{\partial \sigma_2}{\partial u}(u, v) & \frac{\partial \sigma_2}{\partial v}(u, v) \\ \frac{\partial \sigma_3}{\partial u}(u, v) & \frac{\partial \sigma_3}{\partial v}(u, v) \end{pmatrix} + \right.
$$

$$
+ f_2(\sigma(u, v)) \det \begin{pmatrix} \frac{\partial \sigma_3}{\partial u}(u, v) & \frac{\partial \sigma_3}{\partial v}(u, v) \\ \frac{\partial \sigma_1}{\partial u}(u, v) & \frac{\partial \sigma_1}{\partial v}(u, v) \end{pmatrix} +
$$

$$
\left. + f_3(\sigma(u, v)) \det \begin{pmatrix} \frac{\partial \sigma_1}{\partial u}(u, v) & \frac{\partial \sigma_1}{\partial v}(u, v) \\ \frac{\partial \sigma_2}{\partial u}(u, v) & \frac{\partial \sigma_2}{\partial v}(u, v) \end{pmatrix} \right] du \, dv
$$

$$
= \int_{a_2}^{b_2} \int_{a_1}^{b_1} F(\sigma(u, v)) \cdot \frac{\partial \sigma}{\partial u}(u, v) \times \frac{\partial \sigma}{\partial v}(u, v) \, du \, dv .
$$

This quantity is called the "surface integral" or "flux" of the vector field $F = (f_1, f_2, f_3)$ through the surface σ and will be denoted by

$$
\int_\sigma F \cdot d\mathcal{S} .
$$

In fluid dynamics, this concept is used, for instance, to define the amount of fluid crossing a given surface in the unit time.

Example Let us compute the flux of the vector field $F(x, y, z) = (-y, x, z^2)$ through the surface $\sigma : [0, 1] \times [0, 1] \to \mathbb{R}^3$, defined by $\sigma(u, v) = (u^2, v, u + v)$:

$$
\int_\sigma F \cdot d\mathcal{S} = \int_0^1 \int_0^1 (-v, u^2, (u + v)^2) \cdot (-1, -2u, 2u) \, du \, dv
$$

$$
= \int_0^1 \int_0^1 (v - 2u^3 + 2u^4 + 6u^3 v + 6u^2 v^2 + 2uv^3) \, du \, dv = \frac{31}{15} .
$$

It is important to analyze how the integral of a differential form ω changes on two equivalent M-surfaces. We recall that $\sigma : I \to \mathbb{R}^N$ and $\tilde{\sigma} : J \to \mathbb{R}^N$ are equivalent if there exists a diffeomorphism $\varphi : A \to B$ such that $\sigma(\boldsymbol{u}) = \tilde{\sigma}(\varphi(\boldsymbol{u}))$ for every $\boldsymbol{u} \in A$, and the sets $I \setminus A$, $J \setminus B$ are negligible.

Definition 12.3 We say that the equivalent M-surfaces σ and $\tilde{\sigma}$ have the "same orientation" if $\det \varphi'(\boldsymbol{u}) > 0$ for every $\boldsymbol{u} \in A$; they have the "opposite orientation" if $\det \varphi'(\boldsymbol{u}) < 0$ for every $\boldsymbol{u} \in A$.

We provide some examples.

Example 1 Given a curve $\sigma : [a, b] \to \mathbb{R}^N$, an equivalent curve with the opposite orientation is, for example, $\tilde{\sigma} : [a, b] \to \mathbb{R}^N$ defined by

$$\tilde{\sigma}(t) = \sigma(a + b - t).$$

Example 2 If σ is regular, an interesting example of an equivalent curve with the same orientation is obtained by considering the function

$$\varphi(t) = \int_a^t \|\sigma'(\tau)\| \, d\tau.$$

Since $\varphi'(t) = \|\sigma'(t)\| > 0$, for every $t \in \,]a, b[$, setting $\iota_1 = \varphi(b)$, we have that $\varphi : [a, b] \to [0, \iota_1]$ is bijective and the curve $\sigma_1 : [0, \iota_1] \to \mathbb{R}^N$, defined as $\sigma_1(s) = \sigma(\varphi^{-1}(s))$, is equivalent to σ. Notice that, for every $s \in \,]0, \iota_1[$, we have that

$$
\begin{aligned}
\|\sigma_1'(s)\| &= \|\sigma'(\varphi^{-1}(s))(\varphi^{-1})'(s)\| \\
&= \left\| \sigma'(\varphi^{-1}(s)) \frac{1}{\varphi'(\varphi^{-1}(s))} \right\| \\
&= \left\| \sigma'(\varphi^{-1}(s)) \frac{1}{\|\sigma'(\varphi^{-1}(s))\|} \right\| = 1.
\end{aligned}
$$

Example 3 Given a surface $\sigma : [a_1, b_1] \times [a_2, b_2] \to \mathbb{R}^3$, an equivalent surface with the opposite orientation is, for example, $\tilde{\sigma} : [a_1, b_1] \times [a_2, b_2] \to \mathbb{R}^3$ defined by

$$\tilde{\sigma}(u, v) = \sigma(u, a_2 + b_2 - v)$$

or by

$$\tilde{\sigma}(u, v) = \sigma(a_1 + b_1 - u, v).$$

Theorem 12.4 *Let $\sigma : I \to \mathbb{R}^N$ and $\tilde{\sigma} : J \to \mathbb{R}^N$ be two equivalent M-surfaces. If they have the same orientation, then*

$$\int_\sigma \omega = \int_{\tilde{\sigma}} \omega \,;$$

if they have opposite orientations, then

$$\int_\sigma \omega = - \int_{\tilde{\sigma}} \omega.$$

Proof We have an M-differential form of the type

$$\omega(\boldsymbol{x}) = \sum_{1 \leq i_1 < \cdots < i_M \leq N} f_{i_1,\ldots,i_M}(\boldsymbol{x})\, dx_{i_1} \wedge \cdots \wedge dx_{i_M}\,.$$

Let $\varphi : A \to B$ be as in the definition of equivalent M-surfaces, such that $\sigma = \tilde{\sigma} \circ \varphi$. By the Change of Variables Theorem 11.43, we have

$$\int_\sigma \omega = \sum_{1 \leq i_1 < \cdots < i_M \leq N} \int_A f_{i_1,\ldots,i_M}(\sigma(\boldsymbol{u}))\, \det \sigma'_{(i_1,\ldots,i_M)}(\boldsymbol{u})\, d\boldsymbol{u}$$

$$= \sum_{1 \leq i_1 < \cdots < i_M \leq N} \int_A f_{i_1,\ldots,i_M}(\tilde{\sigma}(\varphi(\boldsymbol{u})))\, \det(\tilde{\sigma} \circ \varphi)'_{(i_1,\ldots,i_M)}(\boldsymbol{u})\, d\boldsymbol{u}$$

$$= \sum_{1 \leq i_1 < \cdots < i_M \leq N} \int_A f_{i_1,\ldots,i_M}(\tilde{\sigma}(\varphi(\boldsymbol{u})))\, \det \tilde{\sigma}'_{(i_1,\ldots,i_M)}(\varphi(\boldsymbol{u}))\, \det \varphi'(\boldsymbol{u})\, d\boldsymbol{u}$$

$$= \pm \sum_{1 \leq i_1 < \cdots < i_M \leq N} \int_B f_{i_1,\ldots,i_M}(\tilde{\sigma}(\boldsymbol{v}))\, \det \tilde{\sigma}'_{(i_1,\ldots,i_M)}(\boldsymbol{v})\, d\boldsymbol{v} = \pm \int_{\tilde{\sigma}} \omega\,,$$

with positive sign if $\det \varphi' > 0$, negative if $\det \varphi' < 0$. ∎

Remark 12.5 In general, if σ and $\tilde{\sigma}$ are equivalent, we do not necessarily have the equality $|\int_\sigma \omega| = |\int_{\tilde{\sigma}} \omega|$. Indeed, it is not guaranteed that they have the same or opposite orientations. For example, if we consider the two surfaces $\sigma, \tilde{\sigma} : [1, 2] \times [0, 2\pi] \to \mathbb{R}^3$, defined by

$$\sigma(u, v) = \left(\left(\frac{3}{2} + \left(u - \frac{3}{2}\right) \cos \frac{v}{2}\right) \cos v,\right.$$

$$\left(\frac{3}{2} + \left(u - \frac{3}{2}\right) \cos \frac{v}{2}\right) \sin v, \left(u - \frac{3}{2}\right) \sin \frac{v}{2}\right),$$

$$\tilde{\sigma}(u, v) = \sigma\left(u, v + \frac{\pi}{2}\right),$$

it is possible to see that they are both parametrizations of the same set (a Möbius strip), and therefore they are equivalent (the reader is invited to explicitly find a diffeomorphism $\varphi : A \to B$ with the properties required by the definition). On the other hand, if we consider the 2-differential form $\omega(x_1, x_2, x_3) = dx_{12}$, determined by the constant vector field $(0, 0, 1)$, then computation yields

$$\int_\sigma \omega = 0\,, \qquad \int_{\tilde{\sigma}} \omega = -3\sqrt{2}\,.$$

We now consider the important case where $M = N$.

Theorem 12.6 *Let $M = N$; if σ is regular and injective on $\overset{\circ}{I}$ with $\det \sigma' > 0$, and ω is of the type*

$$\omega(\boldsymbol{x}) = f(\boldsymbol{x})\, dx_1 \wedge \cdots \wedge dx_N \,,$$

then

$$\int_\sigma \omega = \int_{\sigma(I)} f \,.$$

Proof By Corollary 10.25, we see that σ induces a diffeomorphism between $\overset{\circ}{I}$ and $\sigma(\overset{\circ}{I})$. Since both the boundary of I and its image through σ are negligible (Lemma 11.38), by the Change of Variables Theorem 11.43, we have

$$\int_\sigma \omega = \int_I f(\sigma(\boldsymbol{u}))\, \det(\sigma'(\boldsymbol{u}))\, d\boldsymbol{u}$$

$$= \int_{\overset{\circ}{I}} f(\sigma(\boldsymbol{u}))\, \det(\sigma'(\boldsymbol{u}))\, d\boldsymbol{u}$$

$$= \int_{\sigma(\overset{\circ}{I})} f = \int_{\sigma(I)} f \,.$$

This completes the proof. ∎

If σ is the identity function, then $\sigma(I) = I$, and instead of $\int_\sigma \omega$, one usually writes $\int_I \omega$. Hence, we have that

$$\int_I f(\boldsymbol{x})\, dx_1 \wedge \cdots \wedge dx_N = \int_I f \,.$$

12.6 Pull-Back Transformation

Consider again an M-differential form $\omega \in \mathcal{F}_M(\mathcal{O}, \mathbb{R}^N)$,

$$\omega(\boldsymbol{x}) = \sum_{1 \le i_1 < \cdots < i_M \le N} f_{i_1,\dots,i_M}(\boldsymbol{x})\, dx_{i_1} \wedge \cdots \wedge dx_{i_M} \,.$$

Let $\phi : \mathcal{V} \to \mathcal{O}$ be a function of class $\mathcal{C}^1$, where $\mathcal{V}$ is an open subset of some $\mathbb{R}^P$. We can then write

$$\phi(\boldsymbol{y}) = (\phi_1(\boldsymbol{y}), \dots, \phi_N(\boldsymbol{y})) \,,$$

with $\boldsymbol{y} = (y_1, \ldots, y_P) \in \mathcal{V}$, and define a new M-differential form $T_\phi \omega \in \mathcal{F}_M(\mathcal{V}, \mathbb{R}^P)$ as

$$T_\phi \omega(\boldsymbol{y}) = \sum_{1 \leq i_1 < \cdots < i_M \leq N} f_{i_1, \ldots, i_M}(\phi(\boldsymbol{y})) \, d\phi_{i_1}(\boldsymbol{y}) \wedge \cdots \wedge d\phi_{i_M}(\boldsymbol{y}) \, ;$$

it is called a "pull-back transformation" through ϕ of ω. Notice that

$$d\phi_{i_1}(\boldsymbol{y}) \wedge \cdots \wedge d\phi_{i_M}(\boldsymbol{y}) = \left(\sum_{j=1}^{P} \frac{\partial \phi_{i_1}}{\partial y_j}(\boldsymbol{y}) dy_j \right) \wedge \cdots \wedge \left(\sum_{j=1}^{P} \frac{\partial \phi_{i_M}}{\partial y_j}(\boldsymbol{y}) dy_j \right)$$

$$= \sum_{j_1, \ldots, j_M = 1}^{P} \frac{\partial \phi_{i_1}}{\partial y_{j_1}}(\boldsymbol{y}) \cdots \frac{\partial \phi_{i_M}}{\partial y_{j_M}}(\boldsymbol{y}) \, dy_{j_1} \wedge \cdots \wedge dy_{j_M} \, .$$

The following three properties are readily verified.

Proposition 12.7 *For any constant $c \in \mathbb{R}$, we have*

$$T_\phi(c\omega) = c \, T_\phi \omega \, .$$

Proposition 12.8 *If $\widetilde{\omega}$ is an $\widetilde{M}$-differential form defined on $\mathcal{O}$, then*

$$T_\phi(\omega \wedge \widetilde{\omega}) = T_\phi \omega \wedge T_\phi \widetilde{\omega} \, .$$

Proposition 12.9 *If, moreover, $M = \widetilde{M}$, then*

$$T_\phi(\omega + \widetilde{\omega}) = T_\phi \omega + T_\phi \widetilde{\omega} \, .$$

Let us now prove some additional properties.

Proposition 12.10 *If $\psi : W \to \mathcal{V}$ and $\phi : \mathcal{V} \to \mathcal{O}$, then*

$$T_\psi(T_\phi \omega) = T_{\phi \circ \psi} \omega \, .$$

Proof By the preceding linearity properties, it will be sufficient to consider the case of a differential form of the type

$$\omega(\boldsymbol{x}) = f_{i_1, \ldots, i_M}(\boldsymbol{x}) \, dx_{i_1} \wedge \cdots \wedge dx_{i_M} \, .$$

Then

$$T_\psi(T_\phi \omega) = \left[(f_{i_1, \ldots, i_M} \circ \phi) \sum_{j_1, \ldots, j_M = 1}^{P} \frac{\partial \phi_{i_1}}{\partial y_{j_1}} \cdots \frac{\partial \phi_{i_M}}{\partial y_{j_M}} \right] \circ \psi \, d\psi_{j_1} \wedge \cdots \wedge d\psi_{j_M} \, .$$

On the other hand,

$$T_{\phi \circ \psi}\omega = (f_{i_1,\ldots,i_M} \circ \phi \circ \psi)\, d(\phi \circ \psi)_{i_1} \wedge \cdots \wedge d(\phi \circ \psi)_{i_M} \, ,$$

and since

$$d(\phi \circ \psi)_{i_k} = d(\phi_{i_k} \circ \psi) = \sum_{j=1}^{P} \left(\frac{\partial \phi_{i_k}}{\partial y_j} \circ \psi \right) d\psi_j \, ,$$

equality then holds.   ∎

Proposition 12.11 *Assume that ϕ is of class C^2. If ω is of class C^1, then $T_\phi \omega$ is too, and*

$$d(T_\phi \omega) = T_\phi(d\omega) \, .$$

Proof Here, too, it is sufficient to consider the case $\omega = f_{i_1,\ldots,i_M}\, dx_{i_1} \wedge \cdots \wedge dx_{i_M}$. We have

$$
\begin{aligned}
d(T_\phi \omega) &= d(f_{i_1,\ldots,i_M} \circ \phi) \wedge d\phi_{i_1} \wedge \cdots \wedge d\phi_{i_M} \\
&\quad + (f_{i_1,\ldots,i_M} \circ \phi)\, d(d\phi_{i_1} \wedge \cdots \wedge d\phi_{i_M}) \\
&= d(f_{i_1,\ldots,i_M} \circ \phi) \wedge d\phi_{i_1} \wedge \cdots \wedge d\phi_{i_M} \\
&= \left[\sum_{m=1}^{N} \left(\frac{\partial f_{i_1,\ldots,i_M}}{\partial x_m} \circ \phi \right) d\phi_m \right] \wedge d\phi_{i_1} \wedge \cdots \wedge d\phi_{i_M} \, .
\end{aligned}
$$

On the other hand,

$$d\omega(\boldsymbol{x}) = \sum_{m=1}^{N} \frac{\partial f_{i_1,\ldots,i_M}}{\partial x_m}(\boldsymbol{x})\, dx_m \wedge dx_{i_1} \wedge \cdots \wedge dx_{i_M} \, ,$$

hence

$$T_\phi(d\omega) = \sum_{m=1}^{N} \left(\frac{\partial f_{i_1,\ldots,i_M}}{\partial x_m} \circ \phi \right) d\phi_m \wedge d\phi_{i_1} \wedge \cdots \wedge d\phi_{i_M} \, ,$$

and the formula is thus proved.   ∎

Proposition 12.12 *If $\sigma : I \to \mathbb{R}^N$ is an M-surface whose image is contained in $\mathcal{O}$, then*

$$\int_\sigma \omega = \int_I T_\sigma \omega \, .$$

Proof As previously, we just consider the case $\omega = f_{i_1,\ldots,i_M} dx_{i_1} \wedge \cdots \wedge dx_{i_M}$. We have

$$\int_I T_\sigma \omega = \int_I f_{i_1,\ldots,i_M} (\sigma(\boldsymbol{u})) \sum_{j_1,\ldots,j_M=1}^M \frac{\partial \sigma_{i_1}}{\partial u_{j_1}} (\boldsymbol{u}) \ldots \frac{\partial \sigma_{i_M}}{\partial u_{j_M}} (\boldsymbol{u}) \, du_{j_1} \wedge \cdots \wedge du_{j_M}$$

$$= \int_I f_{i_1,\ldots,i_M} (\sigma(\boldsymbol{u})) \det \sigma'_{(i_1,\ldots,i_M)} (\boldsymbol{u}) \, d\boldsymbol{u} = \int_\sigma \omega \,.$$

This completes the proof.   ∎

12.7   Oriented Boundary of a Rectangle

Assume that $\sigma_1 : I_1 \to \mathbb{R}^N$, ..., $\sigma_n : I_n \to \mathbb{R}^N$ are some M-surfaces. We call "gluing" of $\sigma_1, \ldots, \sigma_n$ the n-tuple

$$(\sigma_1, \ldots, \sigma_n) \,.$$

Notice that the elements inside the n-tuple need not be necessarily distinct: We could have that $\sigma_i = \sigma_j$ for some indices $i \neq j$.

We define the integral of an M-differential form ω on the aforementioned gluing by setting

$$\int_{(\sigma_1,\ldots,\sigma_n)} \omega = \int_{\sigma_1} \omega + \cdots + \int_{\sigma_n} \omega \,.$$

We will now use this to define the integral on the oriented boundary of a rectangle I of $\mathbb{R}^{M+1}$, with $M \geq 1$, i.e.,

$$I = [a_1, b_1] \times \cdots \times [a_{M+1}, b_{M+1}] \,.$$

We denote by I_k the rectangle of $\mathbb{R}^M$ obtained from I by the suppression of the kth component, i.e.,

$$I_k = [a_1, b_1] \times \cdots \times [a_{k-1}, b_{k-1}] \times [a_{k+1}, b_{k+1}] \times \cdots \times [a_{M+1}, b_{M+1}] \,.$$

Consider, for every k, the M-surfaces $\alpha_k^+, \beta_k^+ : I_k \to \mathbb{R}^{M+1}$ defined by

$$\alpha_k^+ (u_1, \ldots, \widehat{u_k}, .., u_{M+1}) = (u_1, \ldots, u_{k-1}, a_k, u_{k+1}, \ldots, u_{M+1}) \,,$$

$$\beta_k^+ (u_1, \ldots, \widehat{u_k}, .., u_{M+1}) = (u_1, \ldots, u_{k-1}, b_k, u_{k+1}, \ldots, u_{M+1}) \,,$$

where the meaning of the symbol $\frown$ is to "suppress the underlying variable." Consider, moreover, some M-surfaces $\alpha_k^-, \beta_k^- : I_k \rightarrow \mathbb{R}^{M+1}$, equivalent to α_k^+, β_k^+, respectively, with opposite orientations.

Definition 12.13 We call "oriented boundary" of the rectangle I and denote by ∂I a gluing of the following M-surfaces:

(i) α_k^-, and β_k^+ if k is odd.
(ii) α_k^+ and β_k^- if k is even.

Hence, ∂I is the $(2M + 2)$-tuple given by

$$\partial I = (\alpha_1^-, \beta_1^+, \alpha_2^+, \beta_2^-, \dots, \alpha_{M+1}^+, \beta_{M+1}^-) \quad \text{if } M \text{ is odd},$$

$$\partial I = (\alpha_1^-, \beta_1^+, \alpha_2^+, \beta_2^-, \dots, \alpha_{M+1}^-, \beta_{M+1}^+) \quad \text{if } M \text{ is even}.$$

If ω is an M-differential form defined on a subset $\mathcal{O}$ of $\mathbb{R}^{M+1}$ containing the image of ∂I, we will then have

$$\int_{\partial I} \omega = \sum_{k=1}^{M+1} (-1)^k \int_{\alpha_k^+} \omega + \sum_{k=1}^{M+1} (-1)^{k-1} \int_{\beta_k^+} \omega$$

$$= \sum_{k=1}^{M+1} (-1)^{k-1} \left(\int_{\beta_k^+} \omega - \int_{\alpha_k^+} \omega \right).$$

Let $M = 1$, and consider the rectangle $[a_1, b_1] \times [a_2, b_2]$. Then

$$\partial I = (\alpha_1^-, \beta_1^+, \alpha_2^+, \beta_2^-),$$

where, for example,

$$\alpha_1^- : [a_2, b_2] \rightarrow \mathbb{R}^2, \quad v \mapsto (a_1, a_2 + b_2 - v),$$

$$\beta_1^+ : [a_2, b_2] \rightarrow \mathbb{R}^2, \quad v \mapsto (b_1, v),$$

$$\alpha_2^+ : [a_1, b_1] \rightarrow \mathbb{R}^2, \quad u \mapsto (u, a_2),$$

$$\beta_2^- : [a_1, b_1] \rightarrow \mathbb{R}^2, \quad u \mapsto (a_1 + b_1 - u, b_2).$$

We can visualize geometrically ∂I as the gluing of the sides of the rectangle I oriented in such a way that the perimeter be described in counter-clockwise direction.

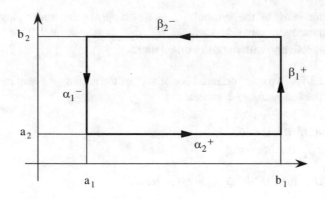

If $M = 2$, we have that

$$\partial I = (\alpha_1^-, \beta_1^+, \alpha_2^+, \beta_2^-, \alpha_3^-, \beta_3^+),$$

where, for example,

$$\alpha_1^- : [a_2, b_2] \times [a_3, b_3] \to \mathbb{R}^3, \quad (v, w) \mapsto (a_1, a_2 + b_2 - v, w),$$

$$\beta_1^+ : [a_2, b_2] \times [a_3, b_3] \to \mathbb{R}^3, \quad (v, w) \mapsto (b_1, v, w),$$

$$\alpha_2^+ : [a_1, b_1] \times [a_3, b_3] \to \mathbb{R}^3, \quad (u, w) \mapsto (u, a_2, w),$$

$$\beta_2^- : [a_1, b_1] \times [a_3, b_3] \to \mathbb{R}^3, \quad (u, w) \mapsto (u, b_2, a_3 + b_3 - w),$$

$$\alpha_3^- : [a_1, b_1] \times [a_2, b_2] \to \mathbb{R}^3, \quad (u, v) \mapsto (a_1 + b_1 - u, v, a_3),$$

$$\beta_3^+ : [a_1, b_1] \times [a_2, b_2] \to \mathbb{R}^3, \quad (u, v) \mapsto (u, v, b_3).$$

In this case, we can visualize ∂I as the gluing of the six faces of the parallelepiped I, each oriented in such a way that the normal unit vector will always be directed toward the exterior.

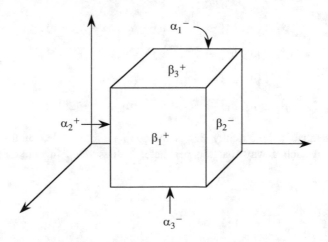

12.8 Gauss Formula

In this section, I will be a rectangle in $\mathbb{R}^N$, with $N \geq 2$. In the following theorem, the elegant **Gauss formula** is obtained.

Theorem 12.14 (Gauss Theorem) *If ω is a $(N-1)$-differential form of class C^1 defined on an open set containing the rectangle I in $\mathbb{R}^N$, then*

$$\int_I d\omega = \int_{\partial I} \omega.$$

Proof We can write ω as

$$\omega(\boldsymbol{x}) = \sum_{j=1}^{N} F_j(\boldsymbol{x})\, dx_1 \wedge \cdots \wedge \widehat{dx_j} \wedge \cdots \wedge dx_N.$$

Then,

$$d\omega(\boldsymbol{x}) = \sum_{j=1}^{N} \sum_{m=1}^{N} \frac{\partial F_j}{\partial x_m}(\boldsymbol{x})\, dx_m \wedge dx_1 \wedge \cdots \wedge \widehat{dx_j} \wedge \cdots \wedge dx_N$$

$$= \sum_{j=1}^{N} (-1)^{j-1} \frac{\partial F_j}{\partial x_j}(\boldsymbol{x})\, dx_1 \wedge \cdots \wedge dx_N.$$

Since the partial derivatives of each F_j are continuous, they are integrable on the rectangle I, and we can use Fubini's Reduction Theorem 11.34 to obtain

$$\int_I d\omega = \sum_{j=1}^{N} (-1)^{j-1} \int_I \frac{\partial F_j}{\partial x_j}(\boldsymbol{x})\, dx_1 \ldots dx_N$$

$$= \sum_{j=1}^{N} (-1)^{j-1} \int_{I_j} \left(\int_{a_j}^{b_j} \frac{\partial F_j}{\partial x_j}(x_1, \ldots, x_N)\, dx_j \right) dx_1 \ldots \widehat{dx_j} \ldots dx_N$$

$$= \sum_{j=1}^{N} (-1)^{j-1} \int_{I_j} \Big[F_j(x_1, \ldots, x_{j-1}, b_j, x_{j+1}, \ldots, x_N) -$$

$$- F_j(x_1, \ldots, x_{j-1}, a_j, x_{j+1}, \ldots, x_N) \Big] dx_1 \ldots \widehat{dx_j} \ldots dx_N,$$

by the Fundamental Theorem. On the other hand, we have

$$\int_{\alpha_k^+} \omega = \sum_{j=1}^{N} \int_{\alpha_k^+} F_j \, dx_1 \wedge \cdots \wedge \widehat{dx_j} \wedge \cdots \wedge dx_N$$

$$= \sum_{j=1}^{N} \int_{I_k} (F_j \circ \alpha_k^+) \det(\alpha_k^+)'_{(1,\ldots,\hat{j},\ldots,N)} \, dx_1 \ldots \widehat{dx_j} \ldots dx_N$$

$$= \int_{I_k} F_k(x_1, \ldots, x_{k-1}, a_k, x_{k+1}, \ldots, x_N) \, dx_1 \ldots \widehat{dx_k} \ldots dx_N \,,$$

since

$$\det(\alpha_k^+)'_{(1,\ldots,\hat{j},\ldots,N)} = \begin{cases} 0 & \text{if } j \neq k \,, \\ 1 & \text{if } j = k \,. \end{cases}$$

Similarly,

$$\int_{\beta_k^+} \omega = \int_{I_k} F_k(x_1, \ldots, x_{k-1}, b_k, x_{k+1}, \ldots, x_N) \, dx_1 \ldots \widehat{dx_k} \ldots dx_N \,,$$

so that

$$\int_{\partial I} \omega = \sum_{k=1}^{N} (-1)^{k-1} \left(\int_{\beta_k^+} \omega - \int_{\alpha_k^+} \omega \right)$$

$$= \sum_{k=1}^{N} (-1)^{k-1} \int_{I_k} [F_k(x_1, \ldots, x_{k-1}, b_k, x_{k+1}, \ldots, x_N) -$$

$$- F_k(x_1, \ldots, x_{k-1}, a_k, x_{k+1}, \ldots, x_N)] \, dx_1 \ldots \widehat{dx_k} \ldots dx_N \,,$$

and the proof is completed.   ∎

Let us now analyze the particular cases $N = 2$ and $N = 3$.
If $N = 2$, then we have a 1-differential form

$$\omega(\boldsymbol{x}) = f_1(\boldsymbol{x}) \, dx_1 + f_2(\boldsymbol{x}) \, dx_2 \,,$$

with the associated vector field $F(\boldsymbol{x}) = (f_1(\boldsymbol{x}), f_2(\boldsymbol{x}))$. The Gauss formula reads
as follows:

$$\int_I \left(\frac{\partial f_2}{\partial x_1} - \frac{\partial f_1}{\partial x_2} \right) dx_1 \wedge dx_2 = \int_{\partial I} f_1 dx_1 + f_2 dx_2 \,,$$

i.e., equivalently,

$$\int_I \left(\frac{\partial f_2}{\partial x_1} - \frac{\partial f_1}{\partial x_2} \right) = \int_{\partial I} F \cdot d\ell \, .$$

If $N = 3$, then we have a 2-differential form

$$\omega(\boldsymbol{x}) = f_1(\boldsymbol{x}) \, dx_2 \wedge dx_3 + f_2(\boldsymbol{x}) \, dx_3 \wedge dx_1 + f_3(\boldsymbol{x}) \, dx_1 \wedge dx_2 \, ,$$

with the associated vector field $F(\boldsymbol{x}) = (f_1(\boldsymbol{x}), \, f_2(\boldsymbol{x}), \, f_3(\boldsymbol{x}))$. The Gauss formula reads as follows:

$$\int_I \left(\frac{\partial f_1}{\partial x_1} + \frac{\partial f_2}{\partial x_2} + \frac{\partial f_3}{\partial x_3} \right) dx_1 \wedge dx_2 \wedge dx_3$$

$$= \int_{\partial I} f_1 dx_2 \wedge dx_3 + f_2 dx_3 \wedge dx_1 + f_3 dx_1 \wedge dx_2 \, ,$$

which is equivalent to

$$\int_I \nabla \cdot F = \int_{\partial I} F \cdot d\mathcal{S} \, .$$

12.9   Oriented Boundary of an M-Surface

In this section, I will be a rectangle in $\mathbb{R}^{M+1}$ and $\sigma : I \to \mathbb{R}^N$ an $(M+1)$-surface.

Definition 12.15 For $1 \le M \le N - 1$, we call "oriented boundary" of σ and denote by $\partial \sigma$ a gluing of the following M-surfaces:

(i) $\sigma \circ \alpha_k^-$ and $\sigma \circ \beta_k^+$ if k is odd.
(ii) $\sigma \circ \alpha_k^+$ and $\sigma \circ \beta_k^-$ if k is even.

Hence, $\partial \sigma$ is the $(2M + 2)$-tuple given by

$$\partial \sigma = (\sigma \circ \alpha_1^-, \sigma \circ \beta_1^+, \sigma \circ \alpha_2^+, \sigma \circ \beta_2^-, \dots, \sigma \circ \alpha_{M+1}^+, \sigma \circ \beta_{M+1}^-) \quad \text{if } M \text{ is odd} \, ,$$

$$\partial \sigma = (\sigma \circ \alpha_1^-, \sigma \circ \beta_1^+, \sigma \circ \alpha_2^+, \sigma \circ \beta_2^-, \dots, \sigma \circ \alpha_{M+1}^-, \sigma \circ \beta_{M+1}^+) \quad \text{if } M \text{ is even} \, .$$

Given an M-differential form ω whose domain contains the image of $\partial\sigma$, we will then have

$$\int_{\partial\sigma}\omega = \sum_{k=1}^{M+1}(-1)^k\int_{\sigma\circ\alpha_k^+}\omega + \sum_{k=1}^{M+1}(-1)^{k-1}\int_{\sigma\circ\beta_k^+}\omega$$

$$= \sum_{k=1}^{M+1}(-1)^{k-1}\left(\int_{\sigma\circ\beta_k^+}\omega - \int_{\sigma\circ\alpha_k^+}\omega\right).$$

Remark 12.16 It is useful to extend the meaning of $\int_{\partial\sigma}\omega$ to the case where $\sigma:[a,b]\to\mathbb{R}^N$ is a curve, with $N\geq 1$, and $\omega = f:\mathcal{O}\to\mathbb{R}$ is a 0-differential form; in this case, we set

$$\int_{\partial\sigma}\omega = f(\sigma(b)) - f(\sigma(a)).$$

Examples As an illustration, consider as usual the case $N=3$. We begin with three examples of oriented boundaries of surfaces, where

$$\partial\sigma = \left(\sigma\circ\alpha_1^-, \sigma\circ\beta_1^+, \sigma\circ\alpha_2^+, \sigma\circ\beta_2^-\right).$$

1. Let $\sigma:[r,R]\times[0,2\pi]\to\mathbb{R}^3$, with $0\leq r < R$, be given by

$$\sigma(u,v) = (u\cos v, u\sin v, 0).$$

Its image is a disk if $r=0$, an annulus if $r>0$. The oriented boundary $\partial\sigma$ is given by a gluing of the following four curves:

$$\sigma\circ\alpha_1^-(v) = (r\cos v, -r\sin v, 0),$$

$$\sigma\circ\beta_1^+(v) = (R\cos v, R\sin v, 0),$$

$$\sigma\circ\alpha_2^+(u) = (u, 0, 0),$$

$$\sigma\circ\beta_2^-(u) = (r+R-u, 0, 0).$$

The first curve has as its image a circle with radius r, which degenerates into the origin in the case where $r=0$. The second has as its image a circle with radius R. Notice, however, that these two circles are described by the two curves in opposite directions. The last two curves are equivalent to each other, with opposite orientations.

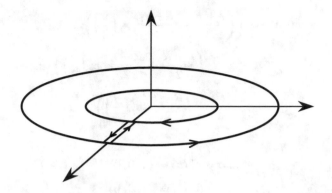

Consider, for example, the vector field $F(x, y, z) = (-y, x, xye^z)$. Then

$$\int_{\partial\sigma} F \cdot d\ell = \int_{\sigma \circ \alpha_1^-} F \cdot d\ell + \int_{\sigma \circ \beta_1^+} F \cdot d\ell$$

$$= \int_0^{2\pi} [-r^2 \sin^2 v - r^2 \cos^2 v]\, dv + \int_0^{2\pi} [R^2 \sin^2 v + R^2 \cos^2 v]\, dv$$

$$= 2\pi (R^2 - r^2).$$

2. Consider the surface $\sigma : [r, R] \times [0, 2\pi] \to \mathbb{R}^3$, with $0 < r < R$, defined by

$$\sigma(u, v) = \left(\left(\frac{r+R}{2} + \left(u - \frac{r+R}{2} \right) \cos \left(\frac{v}{2} \right) \right) \cos v, \right.$$

$$\left(\frac{r+R}{2} + \left(u - \frac{r+R}{2} \right) \cos \left(\frac{v}{2} \right) \right) \sin v,$$

$$\left. \left(u - \frac{r+R}{2} \right) \sin \left(\frac{v}{2} \right) \right),$$

whose image is a Möbius strip. In this case, the oriented boundary is given by a gluing of

$$\sigma \circ \alpha_1^-(v) = \left(\left(\frac{r+R}{2} + \frac{R-r}{2} \cos \left(\frac{v}{2} \right) \right) \cos v, \right.$$

$$- \left(\frac{r+R}{2} + \frac{R-r}{2} \cos \left(\frac{v}{2} \right) \right) \sin v,$$

$$\left. - \frac{R-r}{2} \sin \left(\frac{v}{2} \right) \right),$$

$$\sigma \circ \beta_1^+(v) = \left(\left(\frac{r+R}{2} + \frac{R-r}{2} \cos\left(\frac{v}{2}\right) \right) \cos v, \right.$$

$$\left(\frac{r+R}{2} + \frac{R-r}{2} \cos\left(\frac{v}{2}\right) \right) \sin v,$$

$$\left. \frac{R-r}{2} \sin\left(\frac{v}{2}\right) \right),$$

$$\sigma \circ \alpha_2^+(u) = (u, 0, 0),$$

$$\sigma \circ \beta_2^-(u) = (u, 0, 0).$$

Notice that in this case the last two curves are exactly the same.

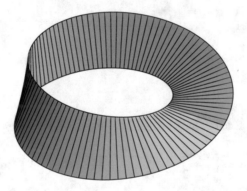

3. Consider the surface $\sigma : [0, \pi] \times [0, 2\pi] \to \mathbb{R}^3$, defined by

$$\sigma(\phi, \theta) = (R \sin\phi \cos\theta, R \sin\phi \sin\theta, R \cos\phi),$$

whose image is a sphere with radius $R > 0$ centered at the origin. In this case, the oriented boundary is given by a gluing of

$$\sigma \circ \alpha_1^-(\theta) = (0, 0, R),$$

$$\sigma \circ \beta_1^+(\theta) = (0, 0, -R),$$

$$\sigma \circ \alpha_2^+(\phi) = (R \sin\phi, 0, R \cos\phi),$$

$$\sigma \circ \beta_2^-(\phi) = (R \sin\phi, 0, -R \cos\phi).$$

Notice that the first two curves are degenerated into one point, while the other two are equivalent to each other, with opposite orientations. Hence, for any choice of a vector field F, we will have $\int_{\partial\sigma} F \cdot d\ell = 0$.

Let us see now an example of an oriented boundary of a volume in $\mathbb{R}^3$. Let $\sigma : [0, R] \times [0, \pi] \times [0, 2\pi] \to \mathbb{R}^3$ be the volume defined by

$$\sigma(\rho, \phi, \theta) = (\rho \sin \phi \cos \theta, \rho \sin \phi \sin \theta, \rho \cos \phi),$$

whose image is the closed ball, centered at the origin, with radius $R > 0$. The oriented boundary $\partial \sigma$ is a gluing of six surfaces,

$$\partial \sigma = (\sigma \circ \alpha_1^-, \sigma \circ \beta_1^+, \sigma \circ \alpha_2^+, \sigma \circ \beta_2^-, \sigma \circ \alpha_3^-, \sigma \circ \beta_3^+),$$

where, e.g.,

$$\sigma \circ \alpha_1^-(\phi, \theta) = (0, 0, 0),$$
$$\sigma \circ \beta_1^+(\phi, \theta) = (R \sin \phi \cos \theta, R \sin \phi \sin \theta, R \cos \phi),$$
$$\sigma \circ \alpha_2^+(\rho, \theta) = (0, 0, \rho),$$
$$\sigma \circ \beta_2^-(\rho, \theta) = (0, 0, -\rho),$$
$$\sigma \circ \alpha_3^-(\rho, \phi) = ((R - \rho) \sin \phi, 0, (R - \rho) \cos \phi),$$
$$\sigma \circ \beta_3^+(\rho, \phi) = (\rho \sin \phi, 0, \rho \cos \phi).$$

Notice that the first surface is degenerated into a point (the origin), the second has as its image the entire sphere, the third and fourth are degenerated in two lines, while the remaining two are equivalent to each other, with opposite orientations. Hence, given a vector field F, we will have

$$\int_{\partial \sigma} F \cdot dS = \int_{\sigma \circ \beta_1^+} F \cdot dS.$$

12.10 Stokes–Cartan Formula

Let us first state the following theorem.

Theorem 12.17 *Let* $f : \mathcal{O} \to \mathbb{R}$ *be a scalar function of class* $\mathcal{C}^1$ *and* $\sigma : [a, b] \to \mathbb{R}^N$ *a curve whose image is contained in* $\mathcal{O}$. *Then*

$$\int_\sigma \nabla f \cdot d\ell = f(\sigma(b)) - f(\sigma(a)).$$

Proof Consider the function $G : [a, b] \to \mathbb{R}$ defined by $G(t) = f(\sigma(t))$. It is of class C^1, and by the Fundamental Theorem we have

$$\int_a^b G'(t) \, dt = G(b) - G(a).$$

Since $G'(t) = \nabla f(\sigma(t)) \cdot \sigma'(t)$, the conclusion follows. ∎

Remark 12.18 The line integral of the gradient of a function f does not depend on the chosen curve itself but only on the values of the function at the two endpoints $\sigma(b)$ and $\sigma(a)$.

Example Given

$$F(x, y, z) = -\left(\frac{x}{[x^2 + y^2 + z^2]^{3/2}}, \frac{y}{[x^2 + y^2 + z^2]^{3/2}}, \frac{z}{[x^2 + y^2 + z^2]^{3/2}} \right)$$

and the curve $\sigma : [0, 4\pi] \to \mathbb{R}^3$ defined by $\sigma(t) = (\cos t, \sin t, t)$, we want to compute the line integral $\int_\sigma F \cdot d\ell$. Observe that $F = \nabla f$, with

$$f(x, y, z) = \frac{1}{\sqrt{x^2 + y^2 + z^2}}.$$

Hence,

$$\int_\sigma F \cdot d\ell = f(\sigma(4\pi)) - f(\sigma(0)) = \frac{1}{\sqrt{1 + 16\pi^2}} - 1.$$

Let us now state the following generalization of the Gauss theorem, where the important **Stokes–Cartan formula** is obtained.

Theorem 12.19 (Stokes–Cartan Theorem—I) *Let* $1 \le M \le N - 1$. *If* ω *is an* M-differential form of class C^1 defined on an open set $\mathcal{O} \subseteq \mathbb{R}^N$, and $\sigma : I \to \mathbb{R}^N$ is an $(M + 1)$-surface whose image is contained in $\mathcal{O}$, then

$$\int_\sigma d\omega = \int_{\partial\sigma} \omega.$$

Proof Since

$$\int_{\sigma \circ \alpha_k^+} \omega = \int_{I_k} T_{\sigma \circ \alpha_k^+} \omega = \int_{I_k} T_{\alpha_k^+}(T_\sigma \omega) = \int_{\alpha_k^+} T_\sigma \omega$$

and

$$\int_{\sigma\circ\beta_k^+}\omega = \int_{I_k} T_{\sigma\circ\beta_k^+}\omega = \int_{I_k} T_{\beta_k^+}(T_\sigma\omega) = \int_{\beta_k^+} T_\sigma\omega\,,$$

we have

$$\int_{\partial\sigma}\omega = \sum_{k=1}^{M+1}(-1)^{k-1}\left(\int_{\sigma\circ\beta_k^+}\omega - \int_{\sigma\circ\alpha_k^+}\omega\right)$$

$$= \sum_{k=1}^{M+1}(-1)^{k-1}\left(\int_{\beta_k^+} T_\sigma\omega - \int_{\alpha_k^+} T_\sigma\omega\right) = \int_{\partial I} T_\sigma\omega\,.$$

If σ is of class $\mathcal{C}^2$, then $T_\sigma\omega$ is of class $\mathcal{C}^1$, and, applying the Gauss formula to $T_\sigma\omega$, we have

$$\int_{\partial I} T_\sigma\omega = \int_I d(T_\sigma\omega)\,.$$

But

$$\int_I d(T_\sigma\omega) = \int_I T_\sigma(d\omega) = \int_\sigma d\omega\,.$$

Hence, we have seen that

$$\int_{\partial\sigma}\omega = \int_{\partial I} T_\sigma\omega = \int_I d(T_\sigma\omega) = \int_\sigma d\omega\,,$$

and the theorem is proved in this case.

If σ is not of class $\mathcal{C}^2$, let $(\sigma_n)_n$ be a sequence of functions as provided by Proposition 11.61. Since they are of class $\mathcal{C}^2$, we know that

$$\int_{\sigma_n} d\omega = \int_{\partial\sigma_n}\omega \qquad \text{for every } n\,.$$

Using Theorem 7.24 we see that

$$\int_\sigma d\omega = \lim_n \int_{\sigma_n} d\omega\,, \qquad \int_{\partial\sigma}\omega = \lim_n \int_{\partial\sigma_n}\omega\,,$$

and the proof is thus completed.   ■

We now concentrate on some corollaries.

The case $M = 1$, $N = 2$. We consider a 1-differential form

$$\omega(\boldsymbol{x}) = F_1(\boldsymbol{x})\,dx_1 + F_2(\boldsymbol{x})\,dx_2\,,$$

and we obtain the **Gauss–Green formula**.

Theorem 12.20 (Gauss–Green Theorem) *Let $F : \mathcal{O} \to \mathbb{R}^2$ be a C^1-vector field and $\sigma : I = [a_1, b_1] \times [a_2, b_2] \to \mathbb{R}^2$ be a surface whose image is contained in $\mathcal{O}$. Then*

$$\int_\sigma \left(\frac{\partial F_2}{\partial x_1} - \frac{\partial F_1}{\partial x_2} \right) dx_1 \wedge dx_2 = \int_{\partial\sigma} F \cdot d\ell\,.$$

Hence, if σ is regular and injective on $\overset{\circ}{I}$, with $\det \sigma' > 0$, then

$$\int_{\sigma(I)} \left(\frac{\partial F_2}{\partial x_1} - \frac{\partial F_1}{\partial x_2} \right) = \int_{\partial\sigma} F \cdot d\ell\,.$$

Example Consider the surface $\sigma : I = [0, 1] \times [0, 2\pi] \to \mathbb{R}^2$ defined by $\sigma(\rho, \theta) = (A\rho \cos\theta, B\rho \sin\theta)$, whose image is an elliptical surface with semiaxes having lengths $A > 0$ and $B > 0$. Take the vector field $F(x, y) = (-y, x)$. Since

$$\frac{\partial F_2}{\partial x}(x, y) - \frac{\partial F_1}{\partial y}(x, y) = 2$$

and (as for the disk)

$$\int_{\partial\sigma} F \cdot d\ell = \int_{\sigma \circ \beta_1^+} F \cdot d\ell\,,$$

the Gauss–Green formula gives us

$$\int_{\sigma(I)} 2\,dx\,dy = \int_0^{2\pi} (-B\sin\theta, A\cos\theta) \cdot (-A\sin\theta, B\cos\theta)\,d\theta = 2\pi AB.$$

We then find the area of the elliptic surface: $\mu(\sigma(I)) = \pi AB$.

The case $M = 1$, $N = 3$. We consider a 1-differential form

$$\omega(\boldsymbol{x}) = F_1(\boldsymbol{x})\,dx_1 + F_2(\boldsymbol{x})\,dx_2 + F_3(\boldsymbol{x})\,dx_3\,,$$

and we obtain the **Kelvin–Stokes formula**.

Theorem 12.21 (Kelvin–Stokes Theorem) *Let $F : \mathcal{O} \to \mathbb{R}^3$ be a C^1-vector field and $\sigma : [a_1, b_1] \times [a_2, b_2] \to \mathbb{R}^3$ a surface whose image is contained in $\mathcal{O}$. Then*

$$\int_\sigma \nabla \times F \cdot dS = \int_{\partial\sigma} F \cdot d\ell \, .$$

Verbally   *The flux of the curl of the vector field F through the surface σ is equal to the line integral of F along the oriented boundary of σ.*

Example Let $F(x, y, z) = (-y, x, 0)$ and $\gamma : [0, 2\pi] \to \mathbb{R}^3$ be the curve defined by $\gamma(t) = (R \cos t, R \sin t, 0)$; we want to compute the line integral $\int_\gamma F \cdot d\ell$. We have already seen how to compute this integral by the direct use of the definition. We now proceed in a different way, as follows. Consider the surface $\sigma : [0, R] \times [0, 2\pi] \to \mathbb{R}^3$ given by $\sigma(\rho, \theta) = (\rho \cos \theta, \rho \sin \theta, 0)$. Observe that $\gamma = \sigma \circ \beta_1^+$, so

$$\int_\gamma F \cdot d\ell = \int_{\sigma \circ \beta_1^+} F \cdot d\ell = \int_{\partial\sigma} F \cdot d\ell = \int_\sigma \nabla \times F \cdot dS \, .$$

Since $\nabla \times F(x, y, z) = (0, 0, 2)$ and

$$\frac{\partial\sigma}{\partial\rho}(\rho, \theta) \times \frac{\partial\sigma}{\partial\theta}(\rho, \theta) = (0, 0, \rho) \, ,$$

we then have

$$\int_\gamma F \cdot d\ell = \int_0^R \int_0^{2\pi} (0, 0, 2) \cdot (0, 0, \rho) \, d\theta \, d\rho = 2\pi R^2 \, .$$

The case $M = 2$, $N = 3$. We consider a 2-differential form

$$\omega(\boldsymbol{x}) = F_1(\boldsymbol{x}) \, dx_2 \wedge dx_3 + F_2(\boldsymbol{x}) \, dx_3 \wedge dx_1 + F_3(\boldsymbol{x}) \, dx_1 \wedge dx_2 \, ,$$

and we obtain the **Gauss–Ostrogradski formula**.

Theorem 12.22 (Gauss–Ostrogradski Theorem) *Let $F : \mathcal{O} \to \mathbb{R}^3$ be a C^1-vector field and $\sigma : I = [a_1, b_1] \times [a_2, b_2] \times [a_3, b_3] \to \mathbb{R}^3$ a volume whose image is contained in $\mathcal{O}$. Then*

$$\int_\sigma \nabla \cdot F \, dx_1 \wedge dx_2 \wedge dx_3 = \int_{\partial\sigma} F \cdot dS \, .$$

Hence, if σ is regular and injective on $\mathring{I}$, with $\det \sigma' > 0$, then

$$\int_{\sigma(I)} \nabla \cdot F = \int_{\partial \sigma} F \cdot dS \,.$$

In Intuitive Terms *The integral of the divergence of the vector field F on the set $V = \sigma(I)$ is equal to the flux of F which exits from V.*

Example We want to compute the flux of the vector field

$$F(x, y, z) = \big([x^2 + y^2 + z^2]x \,, \ [x^2 + y^2 + z^2]y \,, \ [x^2 + y^2 + z^2]z\big)$$

through a spherical surface parametrized by $\eta : [0, \pi] \times [0, 2\pi] \to \mathbb{R}^3$, defined as

$$\eta(\phi, \theta) = (R \sin \phi \cos \theta, \, R \sin \phi \sin \theta, \, R \cos \phi) \,.$$

Recall that $\eta = \sigma \circ \beta_1^+$, where $\sigma : I = [0, R] \times [0, \pi] \times [0, 2\pi] \to \mathbb{R}^3$ is the volume given by

$$\sigma(\rho, \phi, \theta) = (\rho \sin \phi \cos \theta, \, \rho \sin \phi \sin \theta, \, \rho \cos \phi) \,.$$

Hence,

$$\int_\eta F \cdot dS = \int_{\sigma \circ \beta_1^+} F \cdot dS = \int_{\partial \sigma} F \cdot dS = \int_{\sigma(I)} \nabla \cdot F \,.$$

Since $\nabla \cdot F(x, y, z) = 5(x^2 + y^2 + z^2)$, passing to spherical coordinates we have

$$\int_{\sigma(I)} \nabla \cdot F = \int_0^{2\pi} \int_0^\pi \int_0^R (5\rho^2)(\rho^2 \sin \phi) \, d\rho \, d\phi \, d\theta = 4\pi R^5 \,.$$

12.11 Physical Interpretation of Curl and Divergence

The Kelvin–Stokes and the Gauss–Ostrogradski formulas permit us to interpret the physical meaning of the curl and the divergence of a vector field F in $\mathbb{R}^3$. We assume F to be defined on some open set $\mathcal{O} \subseteq \mathbb{R}^3$ and to be continuously differentiable.

Curl Assume at first that $\nabla \times F$ is constant. If we know its direction, we can take a plane orthogonal to it and on this plane a disk with radius $r > 0$ (and hence area πr^2), which we can easily parametrize in polar coordinates, thereby obtaining a 2-surface $\sigma_r : I \to \mathbb{R}^3$. Moreover, we can choose this parametrization in such a way that the normal unit vector ν_{σ_r} has the same direction as $\nabla \times F$. Then

$$\int_{\sigma_r} \nabla \times F \cdot dS = \pi r^2 \|\nabla \times F\|,$$

and, by the Kelvin–Stokes formula,

$$\|\nabla \times F\| = \frac{1}{\pi r^2} \int_{\partial \sigma_r} F \cdot d\ell,$$

providing us the length of the curl of F. We thus see that the length of the curl measures in some way the rotational contribution of the vector field F along the circle $\partial \sigma_r$. This explains the name "curl."

When $\nabla \times F$ is not constant, let us fix a point $\boldsymbol{x}$ in the domain $\mathcal{O}$ and proceed much as we did earlier, taking a disk centered at $\boldsymbol{x}$ with radius r and parametrized by σ_r such that the normal unit vector ν_{σ_r} has the same direction as $\nabla \times F(\boldsymbol{x})$. By continuity, if $r > 0$ is "small," we can think of $\nabla \times F$ as being "almost constant" on this disk. More precisely, we will have that

$$\|\nabla \times F(\boldsymbol{x})\| = \lim_{r \to 0^+} \frac{1}{\pi r^2} \int_{\partial \sigma_r} F \cdot d\ell.$$

We thus interpret physically the intensity of the curl at a point $\boldsymbol{x}$ as a good measure of the rotational motion of the vector field along a circular trajectory centered at $\boldsymbol{x}$ with a very small radius r.

If we do not know the direction of $\nabla \times F(\boldsymbol{x})$ a priori, it can be determined as follows. Assume $\nabla \times F(\boldsymbol{x}) \neq \boldsymbol{0}$. For each plane passing through $\boldsymbol{x}$ we can take a disk centered at $\boldsymbol{x}$ with radius r on this plane and parametrize it by some σ_r, in such a way that the normal unit vector ν_{σ_r} varies continuously with respect to $r > 0$. We compute $\lim_{r \to 0^+} \frac{1}{\pi r^2} \int_{\partial \sigma_r} F \cdot d\ell$, and, from among all these quantities obtained for all those planes, we select the largest one. The plane attaining this maximum quantity will be the one orthogonal to $\nabla \times F(\boldsymbol{x})$, and the unit vector $\lim_{r \to 0^+} \nu_{\sigma_r}$ will indicate the direction of $\nabla \times F(\boldsymbol{x})$.

Divergence   Assume at first that $\nabla \cdot F$ is constant. Let us consider the closed ball $\overline{B}_r$, with radius $r > 0$ (and, hence, volume $\frac{4}{3} \pi r^3$), which we can easily parametrize in spherical coordinates by a 3-surface $\sigma_r : I \to \mathbb{R}^3$ such that $\det J \sigma_r(\boldsymbol{u}) > 0$ for every $\boldsymbol{u} \in \mathring{I}$. Then

$$\int_{\overline{B}_r} \nabla \cdot F = \frac{4}{3} \pi r^3 \nabla \cdot F,$$

whence, by the Gauss–Ostrogradski formula,

$$\nabla \cdot F = \frac{1}{\frac{4}{3} \pi r^3} \int_{\partial \sigma_r} F \cdot dS.$$

We thus see that the divergence provides us a measure of the flux, per unit volume, of the vector field through the spherical surface $\partial \sigma_r$. It will be positive if the flow mostly crosses the surface from the interior toward the exterior, and negative otherwise. This explains the name "divergence."

When $\nabla \cdot F$ is not constant, let us fix a point x in the domain $\mathcal{O}$ and proceed as previously, taking a ball $\overline{B}_r = \overline{B}(x, r)$, centered at x, with radius r, parametrized by some σ_r with $\det J\sigma_r > 0$. By continuity, if $r > 0$ is "small," we can think of $\nabla \cdot F$ as being "almost constant" on this ball. More precisely, we will have that

$$\nabla \cdot F(x) = \lim_{r \to 0^+} \frac{1}{\frac{4}{3}\pi r^3} \int_{\partial \sigma_r} F \cdot d\ell \,.$$

We thus interpret physically the divergence at a point x as a measure of the flux per unit volume of the vector field through a spherical surface centered at x with a very small radius r.

12.12　The Integral on an Oriented Compact Manifold

In this section, we return to the setting of differentiable manifolds. We would like to define the integral of an M-differential form over an M-differentiable manifold. There is a difficulty in doing this, however, due to some problems related to "orientation." Once the definition is given, we will finally be able to obtain a Stokes–Cartan formula also in this setting.

We want to define an orientation for $\mathcal{M}$, which will automatically induce one for $\partial \mathcal{M}$ as well. Given $x \in \mathcal{M}$, let $\sigma : I \to \mathbb{R}^N$ be a local M-parametrization, with $\sigma(0) = x$. Since $\sigma'(u)$ has rank M for every $u \in I$, we have that the vectors

$$\left[\frac{\partial \sigma}{\partial u_1}(u) , \; \ldots \; , \frac{\partial \sigma}{\partial u_M}(u) \right]$$

form a basis for a real vector space of dimension M, which will be called the "tangent space" to $\mathcal{M}$ at the point $\sigma(u)$ and will be denoted by $\mathcal{T}_{\sigma(u)}\mathcal{M}$ (in particular, if $u = 0$, we have the tangent space $\mathcal{T}_x\mathcal{M}$).

Now, once $u \in I$ is considered, the point $\sigma(u)$ will also belong to the images of other M-parametrizations. There can be a $\tilde{\sigma} : J \to \mathbb{R}^N$ such that $\sigma(u) = \tilde{\sigma}(v)$ for some $v \in J$. We have seen how to change the orientation of such a $\tilde{\sigma}$ by a simple change of variables. Hence, we can choose these local M-parametrizations so that the bases of the tangent space $\mathcal{T}_{\sigma(u)}\mathcal{M} = \mathcal{T}_{\tilde{\sigma}(v)}\mathcal{M}$ associated with them are all coherently oriented; this means that the matrix that makes it possible to pass from one basis to the other has a positive determinant. We will refer to such a choice as being "coherent."

A coherent choice of local M-parametrizations is therefore always possible, in a neighborhood of x. But we are interested in the possibility of making a *global* coherent choice, i.e., for *all* possible local M-parametrizations of $\mathcal{M}$. This is not always possible. For example, it can be seen that this cannot be done for a Möbius strip, which is a 2-manifold.

Whenever *all* the local M-parametrizations of $\mathscr{M}$ can be chosen coherently, we say that $\mathscr{M}$ is "orientable." From now on we will always assume that $\mathscr{M}$ is orientable and that a coherent choice of all the local M-parametrizations has been made. We then say that $\mathscr{M}$ has been "oriented."

Once we have oriented $\mathscr{M}$, let us see how it is possible, from that, to define an orientation on $\partial\mathscr{M}$. Given $\boldsymbol{x} \in \partial\mathscr{M}$, let $\sigma : I \to \mathbb{R}^N$ be a local M-parametrization with $\sigma(0) = \boldsymbol{x}$; recall that in this case, I is the rectangle $[-\alpha, \alpha]^{M-1} \times [0, \alpha]$. Since $\partial\mathscr{M}$ is an $(M - 1)$-manifold, the tangent vector space $\mathcal{T}_{\boldsymbol{x}}\partial\mathscr{M}$ has dimension $M - 1$ and is a subspace of $\mathcal{T}_{\boldsymbol{x}}\mathscr{M}$, which has dimension M. Hence, there are two unit vectors in $\mathcal{T}_{\boldsymbol{x}}\mathscr{M}$, which are orthogonal to $\mathcal{T}_{\boldsymbol{x}}\partial\mathscr{M}$. We denote by $v(\boldsymbol{x})$ the one obtained as a directional derivative $\frac{\partial\sigma}{\partial v}(0) = d\sigma(0)\boldsymbol{v}$ for some $\boldsymbol{v} = (v_1, \dots, v_M)$ with $v_M < 0$. At this point, we choose a basis $[v^{(1)}(\boldsymbol{x}), \dots, v^{(M-1)}(\boldsymbol{x})]$ in $\mathcal{T}_{\boldsymbol{x}}\partial\mathscr{M}$ such that $[v(\boldsymbol{x}), v^{(1)}(\boldsymbol{x}), \dots, v^{(M-1)}(\boldsymbol{x})]$ is a basis of $\mathcal{T}_{\boldsymbol{x}}\mathscr{M}$ oriented coherently with the one already chosen in this space. Proceeding in this way for every $\boldsymbol{x}$, it can be seen that $\partial\mathscr{M}$ is thus oriented: We have assigned to it the "induced orientation" from that of $\mathscr{M}$.

Assume now that $\mathscr{M}$, besides being oriented, is *compact*. Given an M-differential form ω defined on an open set $\mathcal{O}$ containing $\mathscr{M}$, we would like to define what we mean by integral of ω on $\mathscr{M}$.

In the case where $\omega|_{\mathscr{M}}$, the restriction of ω to the set $\mathscr{M}$ is zero outside the image of a single local M-parametrization $\sigma : I \to \mathbb{R}^N$, we simply set

$$\int_{\mathscr{M}} \omega = \int_{\sigma} \omega.$$

It can be proved using the same reasoning as in Proposition 11.62 that this is a good definition.

In general, we saw in Theorem 10.39 that for every $\boldsymbol{x} \in \mathscr{M}$ there is a neighborhood A' of $\boldsymbol{x}$ such that $A' \cap \mathscr{M}$ can be M-parametrized by some injective function $\sigma : I \to \mathbb{R}^N$. Moreover, if $\boldsymbol{x} \in \partial\mathscr{M}$, the M-parametrization σ is such that the interior points of a single face of the rectangle I are sent on $\partial\mathscr{M}$.

For every $\boldsymbol{x}$ there is an open ball $B(\boldsymbol{x}, \rho_{\boldsymbol{x}})$ contained in A'. We thus have an open covering of $\mathscr{M}$ made of these open balls. Since $\mathscr{M}$ is compact, by Theorem 4.9 there exists a finite open subcovering, which we denote by $A_1, \dots, A_n$. Let $\phi_1, \dots, \phi_n$ be a partition of unity associated with this covering. Since each $\phi_k \cdot \omega|_{\mathscr{M}}$ is zero outside the image of a single local M-parametrization, we can define the integral of ω on $\mathscr{M}$ as

$$\int_{\mathscr{M}} \omega = \sum_{k=1}^{n} \int_{\mathscr{M}} \phi_k \cdot \omega.$$

It is possible to prove, using the same reasoning as in Proposition 11.64, that such a definition depends neither on the (coherent) choice of the local M-parametrizations nor on the particular partition of unity defined previously.

We can now state the Stokes–Cartan theorem for manifolds.

Theorem 12.23 (Stokes–Cartan Theorem—II) *If ω is an M-differential form of class C^1 defined on an open set $\mathcal{O} \subseteq \mathbb{R}^N$ and $\mathcal{M}$ is an oriented compact $(M + 1)$-manifold contained in $\mathcal{O}$, then*

$$\int_{\mathcal{M}} d\omega = \int_{\partial \mathcal{M}} \omega$$

(provided the orientation on $\partial \mathcal{M}$ is the induced one).

Proof We consider the local parametrizations provided by the preceding argument. Let us first assume that $\omega|_{\mathcal{M}}$ is equal to zero outside the image of a single local M-parametrization $\sigma : I \to \mathbb{R}^N$. Then

$$\int_{\mathcal{M}} d\omega = \int_{\sigma} d\omega = \int_{\partial \sigma} \omega.$$

We now have two possible cases.

Case 1: $\sigma(I) \cap \partial \mathcal{M} = \emptyset$. By the injectivity of σ on an open set containing I, we have that the image of $\partial \sigma$ is contained in the boundary of $\sigma(I)$ in the metric space $\mathcal{M}$. Hence, since it is continuous, we have that ω must be equal to zero on all points of the image of $\partial \sigma$, implying that

$$\int_{\partial \sigma} \omega = 0.$$

On the other hand, since ω is zero on $\partial \mathcal{M}$, we have that

$$\int_{\partial \mathcal{M}} \omega = 0,$$

so the identity is verified in this case.

Case 2: $\sigma(I) \cap \partial \mathcal{M} \neq \emptyset$. We know that the interior points of a single face of I are sent on $\partial \mathcal{M}$. For example, let $\alpha_1^- : I_1 \to \mathbb{R}^M$ be that part of ∂I such that $\sigma(\alpha_1^-(\mathring{I}_1)) \subseteq \partial \mathcal{M}$, indeed the only one. Then $\sigma \circ \alpha_1^- : I_1 \to \mathbb{R}^N$ is a local parametrization of $\partial \mathcal{M}$, and $\omega|_{\partial \mathcal{M}}$ is equal to zero outside its image, whence

$$\int_{\partial \mathcal{M}} \omega = \int_{\sigma \circ \alpha_1^-} \omega.$$

On the other hand, since $\sigma : I \to \mathbb{R}^N$ is injective, we have that the image of $\partial\sigma$ is contained in the boundary of $\sigma(I)$. Hence, since it is continuous, we have that ω must be equal to zero on all points of the images of all the other faces

$$\sigma \circ \beta_1^-, \ \sigma \circ \alpha_2^+, \ \sigma \circ \beta_2^-, \ \ldots$$

Then

$$\int_{\partial\sigma} \omega = \int_{\sigma\circ\alpha_1^-} \omega + \int_{\sigma\circ\beta_1^-} \omega + \int_{\sigma\circ\alpha_2^+} \omega + \int_{\sigma\circ\beta_2^-} \omega + \cdots = \int_{\sigma\circ\alpha_1^-} \omega,$$

showing that even in this case the identity holds.

Consider now the general case. With the previously found partition of unity, $\phi_k \cdot \omega|_{\mathcal{M}}$ is equal to zero outside the image of a single local M-parametrization for every $k = 1, \ldots, n$. Since

$$\sum_{k=1}^n d\phi_k \wedge \omega = d\left(\sum_{k=1}^n \phi_k\right) \wedge \omega = d(1) \wedge \omega = 0,$$

we then have

$$\int_{\mathcal{M}} d\omega = \sum_{k=1}^n \int_{\mathcal{M}} \phi_k \cdot d\omega = \sum_{k=1}^n \int_{\mathcal{M}} d\phi_k \wedge \omega + \sum_{k=1}^n \int_{\mathcal{M}} \phi_k \cdot d\omega$$

$$= \sum_{k=1}^n \int_{\mathcal{M}} d(\phi_k \cdot \omega) = \sum_{k=1}^n \int_{\partial\mathcal{M}} \phi_k \cdot \omega = \int_{\partial\mathcal{M}} \omega,$$

and the proof is completed. ∎

12.13 Closed and Exact Differential Forms

We are now interested in the following problem. Given an M-differential form ω, when is it possible to write it as the exterior differential of another differential form $\widetilde{\omega}$ to be determined?

We assume $1 \le M \le N$ and that $\omega \in \mathcal{F}_M^1(\mathcal{O}, \mathbb{R}^N)$.

Definition 12.24 A M-differential form ω is said to be "closed" if $d\omega = 0$; it is said to be "exact" if there is a $(M-1)$-differential form $\widetilde{\omega}$ such that $d\widetilde{\omega} = \omega$.

Every exact differential form is closed: If $\omega = d\widetilde{\omega}$, then $d\omega = d(d\widetilde{\omega}) = 0$. The converse is not always true.

Example The 1-differential form defined on $\mathbb{R}^2 \setminus \{(0, 0)\}$ by

$$\omega(x, y) = \frac{-y}{x^2 + y^2}\, dx + \frac{x}{x^2 + y^2}\, dy$$

is closed, as is easily verified. Indeed, setting

$$F_1(x, y) = \frac{-y}{x^2 + y^2}\,, \qquad F_2(x, y) = \frac{x}{x^2 + y^2}\,,$$

for every $(x, y) \neq (0, 0)$, we have

$$\frac{\partial F_2}{\partial x}(x, y) = \frac{\partial F_1}{\partial y}(x, y)\,.$$

Let us compute the line integral of its vector field $F = (F_1, F_2)$ on the curve $\sigma :$ $[0, 2\pi] \to \mathbb{R}^2$, defined by $\sigma(t) = (\cos t, \sin t)$:

$$\int_\sigma F \cdot d\ell = \int_0^{2\pi} F(\sigma(t)) \cdot \sigma'(t)\, dt$$

$$= \int_0^{2\pi} (-\sin t, \cos t) \cdot (-\sin t, \cos t)\, dt = 2\pi\,.$$

Assume by contradiction that ω is exact, i.e., that there exists a C^1-function $f : \mathbb{R}^2 \setminus \{(0, 0)\} \to \mathbb{R}$ such that $\frac{\partial f}{\partial x} = F_1$ and $\frac{\partial f}{\partial y} = F_2$. In that case, since $\sigma(0) = \sigma(2\pi)$, we should have

$$\int_\sigma F \cdot d\ell = \int_\sigma \nabla f \cdot d\ell = f(\sigma(2\pi)) - f(\sigma(0)) = 0\,,$$

which contradicts the preceding computation.

The Poincaré Theorem, stated in what follows, says that the situation of the preceding example can never happen if, for example, the open set $\mathcal{O}$ on which the differential form is defined has a particular shape.

Definition 12.25 A set $\mathcal{O}$ is "star-shaped" with respect to a point $\bar{x}$ if, with each of its points x, the set $\mathcal{O}$ contains the whole segment joining x to $\bar{x}$, i.e.,

$$[\bar{x}, x] = \{\bar{x} + t(x - \bar{x}) : t \in [0, 1]\} \subseteq \mathcal{O}\,.$$

For example, every convex set is star-shaped (with respect to any of its points). In particular, a ball, a rectangle, or even the whole space $\mathbb{R}^N$ is star-shaped. Clearly, the set $\mathbb{R}^2 \setminus \{(0, 0)\}$ considered previously is not star-shaped.

Theorem 12.26 (Poincaré Theorem)   *Assume that $\mathcal{O}$, an open subset of $\mathbb{R}^N$, is star-shaped with respect to a point $\bar{x}$. For $1 \leq M \leq N$, an M-differential form ω of class C^1 defined on $\mathcal{O}$ is exact if and only if it is closed.*

To prove the theorem, we need a preliminary result. To simplify the notations, we can assume without loss of generality that

$$\bar{x} = (0, 0, \ldots, 0).$$

Consider now the set $[0, 1] \times \mathcal{O}$, whose elements will be denoted by (t, x), with $x = (x_1, \ldots, x_N)$. Let us define the linear operator K, which transforms a generic M-differential form α defined on $[0, 1] \times \mathcal{O}$ in an $(M - 1)$-differential form $K(\alpha)$ defined on $\mathcal{O}$ in the following way:

(a) If $\alpha(t, x) = f(t, x)\, dt \wedge dx_{i_1} \wedge \cdots \wedge dx_{i_{M-1}}$ (notice here the appearance of the term dt), then

$$K(\alpha)(x) = \left(\int_0^1 f(t, x)\, dt \right) dx_{i_1} \wedge \cdots \wedge dx_{i_{M-1}}.$$

(b) If $\alpha(t, x) = f(t, x)\, dx_{i_1} \wedge \cdots \wedge dx_{i_M}$ (here the term dt does not appear), then $K(\alpha) = 0$.

(c) In all other cases, K is defined by linearity (for each component of a generic M-differential form α, the term dt may or not appear, and the two previous definitions apply).

Moreover, we define the functions $\psi, \xi : \mathcal{O} \to [0, 1] \times \mathcal{O}$ as follows:

$$\psi(x_1, \ldots, x_N) = (0, x_1, \ldots, x_N), \qquad \xi(x_1, \ldots, x_N) = (1, x_1, \ldots, x_N).$$

Lemma 12.27 *If α is an M-differential form of class C^1 defined on $[0, 1] \times \mathcal{O}$, then*

$$d(K(\alpha)) + K(d\alpha) = T_\xi \alpha - T_\psi \alpha.$$

Proof Because of the linearity, it will be sufficient to consider the two cases where the differential form α is one of the two kinds considered in (a) and (b).

(a) If $\alpha(t, x) = f(t, x)\, dt \wedge dx_{i_1} \wedge \cdots \wedge dx_{i_{M-1}}$, then by the Leibniz rule we have

$$d(K(\alpha))(x) = \sum_{m=1}^N \left(\int_0^1 \frac{\partial f}{\partial x_m}(t, x)\, dt \right) dx_m \wedge dx_{i_1} \wedge \cdots \wedge dx_{i_{M-1}};$$

on the other hand,

$$d\alpha(t, \boldsymbol{x}) = \frac{\partial f}{\partial t}(t, \boldsymbol{x})\, dt \wedge dt \wedge dx_{i_1} \wedge \cdots \wedge dx_{i_{M-1}} +$$

$$+ \sum_{m=1}^{N} \frac{\partial f}{\partial x_m}(t, \boldsymbol{x})\, dx_m \wedge dt \wedge dx_{i_1} \wedge \cdots \wedge dx_{i_{M-1}}$$

$$= - \sum_{m=1}^{N} \frac{\partial f}{\partial x_m}(t, \boldsymbol{x})\, dt \wedge dx_m \wedge dx_{i_1} \wedge \cdots \wedge dx_{i_{M-1}},$$

and hence

$$K(d\alpha)(\boldsymbol{x}) = - \sum_{m=1}^{N} \left(\int_0^1 \frac{\partial f}{\partial x_m}(t, \boldsymbol{x})\, dt \right) dx_m \wedge dx_{i_1} \wedge \cdots \wedge dx_{i_{M-1}}$$

$$= -d(K(\alpha))(\boldsymbol{x}).$$

Moreover, since the first component of ψ and of ξ is constant, we have $T_\psi \alpha = T_\xi \alpha = 0$. Hence, the identity is proved in this case.

(b) If $\alpha(t, \boldsymbol{x}) = f(t, \boldsymbol{x})\, dx_{i_1} \wedge \cdots \wedge dx_{i_M}$, we have $K(\alpha) = 0$ and hence $d(K(\alpha)) = 0$; on the other hand,

$$d\alpha(t, \boldsymbol{x}) = \frac{\partial f}{\partial t}(t, \boldsymbol{x})\, dt \wedge dx_{i_1} \wedge \cdots \wedge dx_{i_M} +$$

$$+ \sum_{m=1}^{N} \frac{\partial f}{\partial x_m}(t, \boldsymbol{x})\, dx_m \wedge dx_{i_1} \wedge \cdots \wedge dx_{i_M},$$

and hence

$$K(d\alpha)(\boldsymbol{x}) = \left(\int_0^1 \frac{\partial f}{\partial t}(t, \boldsymbol{x})\, dt \right) dx_{i_1} \wedge \cdots \wedge dx_{i_M}$$

$$= (f(1, \boldsymbol{x}) - f(0, \boldsymbol{x}))\, dx_{i_1} \wedge \cdots \wedge dx_{i_M}.$$

Moreover,

$$T_\xi \alpha(\boldsymbol{x}) = f(1, \boldsymbol{x})\, d\xi_{i_1}(\boldsymbol{x}) \wedge \cdots \wedge d\xi_{i_M}(\boldsymbol{x})$$

$$= f(1, \boldsymbol{x})\, dx_{i_1} \wedge \cdots \wedge dx_{i_M},$$

$$T_\psi \alpha(\boldsymbol{x}) = f(0, \boldsymbol{x})\, d\psi_{i_1}(\boldsymbol{x}) \wedge \cdots \wedge d\psi_{i_M}(\boldsymbol{x})$$

$$= f(0, \boldsymbol{x})\, dx_{i_1} \wedge \cdots \wedge dx_{i_M}.$$

The formula is thus proved in this case as well.   ∎

We are now ready for the proof of the Poincaré Theorem.

Proof By linearity we may assume, for simplicity, that

$$\omega(\boldsymbol{x}) = f_{i_1,\dots,i_M}(\boldsymbol{x})\, dx_{i_1} \wedge \cdots \wedge dx_{i_M}\,.$$

Let $\phi : [0, 1] \times \mathcal{O} \to \mathcal{O}$ be defined by

$$\phi(t, x_1, \dots, x_N) = (tx_1, \dots, tx_N)\,.$$

Consider $T_\phi \omega$, the pullback transformation through ϕ of ω, and set $\widetilde{\omega} = K(T_\phi \omega)$. We want to prove that $d\widetilde{\omega} = \omega$. Since ω is closed, by the linearity of K we have

$$K(d(T_\phi \omega)) = K(T_\phi(d\omega)) = K(T_\phi(0)) = K(0) = 0\,.$$

By the preceding lemma,

$$\begin{aligned}
d\widetilde{\omega} &= d(K(T_\phi \omega)) \\
&= T_\xi(T_\phi \omega) - T_\psi(T_\phi \omega) - K(d(T_\phi \omega)) \\
&= T_\xi(T_\phi \omega) - T_\psi(T_\phi \omega) \\
&= T_{\phi \circ \xi} \omega - T_{\phi \circ \psi} \omega\,.
\end{aligned}$$

Since $\phi \circ \xi$ is the identity function and $\phi \circ \psi$ is the null function, we have that $T_{\phi \circ \xi} \omega = \omega$ and $T_{\phi \circ \psi} \omega = 0$, which concludes the proof.   ∎

Remark 12.28 If $\omega(\boldsymbol{x}) = f_{i_1,\dots,i_M}(\boldsymbol{x})\, dx_{i_1} \wedge \cdots \wedge dx_{i_M}$, then

$$\begin{aligned}
T_\phi \omega(t, \boldsymbol{x}) &= f_{i_1,\dots,i_M}(t\boldsymbol{x})(x_{i_1} dt + t\, dx_{i_1}) \wedge \cdots \wedge (x_{i_M} dt + t\, dx_{i_M}) \\
&= f_{i_1,\dots,i_M}(t\boldsymbol{x})[t^M dx_{i_1} \wedge \cdots \wedge dx_{i_M} + \\
&\quad + t^{M-1} \sum_{s=1}^{M} (-1)^{s-1} x_{i_s} dt \wedge dx_{i_1} \wedge \cdots \wedge \widehat{dx_{i_s}} \wedge \cdots \wedge dx_{i_M}]\,,
\end{aligned}$$

and hence

$$\begin{aligned}
K(T_\phi \omega)(\boldsymbol{x}) &= \\
&= \left(\int_0^1 f_{i_1,\dots,i_M}(t\boldsymbol{x}) t^{M-1} \sum_{s=1}^{M} (-1)^{s+1} x_{i_s}\, dt \right) dx_{i_1} \wedge \cdots \wedge \widehat{dx_{i_s}} \wedge \cdots \wedge dx_{i_M}\,.
\end{aligned}$$

Thus, for a general closed M-differential form

$$\omega(\boldsymbol{x}) = \sum_{1 \leq i_1 < \cdots < i_M \leq N} f_{i_1,\ldots,i_M}(\boldsymbol{x})\, dx_{i_1} \wedge \cdots \wedge dx_{i_M},$$

an $(M-1)$-differential form $\widetilde{\omega}$ such that $d\widetilde{\omega} = \omega$ is given by

$$\widetilde{\omega}(\boldsymbol{x}) = K(T_\phi \omega)(\boldsymbol{x})$$

$$= \sum_{1 \leq i_1 < \cdots < i_M \leq N} \sum_{s=1}^{M} (-1)^{s+1} x_{i_s} \cdot$$

$$\cdot \left(\int_0^1 t^{M-1} f_{i_1,\ldots,i_M}(t\boldsymbol{x})\, dt \right) dx_{i_1} \wedge \cdots \wedge \widehat{dx_{i_s}} \wedge \cdots \wedge dx_{i_M}.$$

We consider now some corollaries that hold true for the case $N = 3$. We will always assume that $\bar{\boldsymbol{x}} = (0, 0, 0)$.

The case $M = 1$. A C^1-vector field $F = (F_1, F_2, F_3)$, defined on an open subset $\mathcal{O}$ of $\mathbb{R}^3$, determines a 1-differential form

$$\omega(\boldsymbol{x}) = F_1(\boldsymbol{x})\, dx_1 + F_2(\boldsymbol{x})\, dx_2 + F_3(\boldsymbol{x})\, dx_3.$$

It is closed if and only if $\nabla \times F = 0$. In this case, the vector field is said to be "irrotational." On the other hand, the vector field is said to be F "conservative" if there is a function $f : \mathcal{O} \to \mathbb{R}$ such that $F = \nabla f$. In that case, f is a "scalar potential" of the vector field F.[1]

Theorem 12.29 *If $\mathcal{O} \subseteq \mathbb{R}^3$ is star-shaped with respect to the origin, then the vector field $F : \mathcal{O} \to \mathbb{R}^3$ is conservative if and only if it is irrotational, and in that case a function $f : \mathcal{O} \to \mathbb{R}$ such that $F = \nabla f$ is given by*

$$f(\boldsymbol{x}) = \int_0^1 F(t\boldsymbol{x}) \cdot \boldsymbol{x}\, dt.$$

Any other function $\tilde{f} : \mathcal{O} \to \mathbb{R}$ which is such that $F = \nabla \tilde{f}$ is obtained from f by adding a constant.

Proof The first part follows directly from the formula in Remark 12.28. Assume now that $F = \nabla f = \nabla \tilde{f}$, and set $g = f - \tilde{f}$. Then $\nabla g = 0$ on $\mathcal{O}$, which is star-shaped with respect to the origin. Using the Fundamental Theorem for the function

---

[1] Beware that in Mechanics it is often the function $-f$ that is called "the potential."

$\mathscr{F}(t) = g(t\boldsymbol{x})$, we have that

$$g(\boldsymbol{x}) = \mathscr{F}(1) = \mathscr{F}(0) + \int_0^1 \mathscr{F}'(t)\,dt = g(\boldsymbol{0}) + \int_0^1 \nabla g(t\boldsymbol{x}) \cdot \boldsymbol{x}\,dt = g(\boldsymbol{0})$$

for every $\boldsymbol{x} \in \mathcal{O}$. Thus, $f - \tilde{f}$ must be constant on $\mathcal{O}$.   ∎

Example Consider the vector field $F(x, y, z) = (2xz+y, x, x^2)$, which, as is easily verified, is irrotational. A scalar potential is then given by

$$f(x, y, z) = \int_0^1 ((2t^2 x^2 z + txy) + txy + t^2 x^2 z)\,dt = xy + x^2 z \,.$$

The case $\underline{M = 2}$. A C^1-vector field $F = (F_1, F_2, F_3)$, defined on an open subset $\mathcal{O}$ of $\mathbb{R}^3$, determines a 2-differential form

$$\omega(\boldsymbol{x}) = F_1(\boldsymbol{x})\,dx_2 \wedge dx_3 + F_2(\boldsymbol{x})\,dx_3 \wedge dx_1 + F_3(\boldsymbol{x})\,dx_1 \wedge dx_2 \,.$$

This is closed if and only if $\nabla \cdot F = 0$. In this case, the vector field is said to be "solenoidal." We say that F has a "vector potential" if there is a vector field $V = (V_1, V_2, V_3)$ such that $F = \nabla \times V$.

Theorem 12.30 *If $\mathcal{O} \subseteq \mathbb{R}^3$ is star-shaped with respect to the origin, then the vector field $F : \mathcal{O} \to \mathbb{R}^3$ has a vector potential if and only if it is solenoidal, and in that case a vector field $V : \mathcal{O} \to \mathbb{R}^3$ for which $F = \nabla \times V$ is given by*

$$V(\boldsymbol{x}) = \int_0^1 t(F(t\boldsymbol{x}) \times \boldsymbol{x})\,dt \,.$$

Any other vector field $\tilde{V} : \mathcal{O} \to \mathbb{R}^3$ such that $f = \nabla \times \tilde{V}$ is obtained from V by adding the gradient of an arbitrary scalar function.

Proof The first part is obtained by applying the formula in Remark 12.28. The second part follows from the fact that if $F = \nabla \times V = \nabla \times \tilde{V}$, then, by the previous theorem, $V - \tilde{V}$ is a conservative vector field, thereby completing the proof.   ∎

Example Consider the solenoidal vector field $F(x, y, z) = (y, z, x)$. A vector potential is then given by

$$V(x, y, z) = \int_0^1 t(ty, tz, tx) \times (x, y, z)\,dt = \frac{1}{3}(z^2 - xy, x^2 - yz, y^2 - xz) \,.$$

The case $M = 3$. A C^1-scalar function f, defined on an open subset $\mathcal{O}$ of $\mathbb{R}^3$, determines a 3-differential form

$$\omega(\boldsymbol{x}) = f(\boldsymbol{x}) \, dx_1 \wedge dx_2 \wedge dx_3 \, .$$

This is necessarily always closed since $d\omega$ is a 4-differential form defined on a subset of $\mathbb{R}^3$.

Theorem 12.31 *If $\mathcal{O} \subseteq \mathbb{R}^3$ is star-shaped with respect to the origin, the function $f : \mathcal{O} \to \mathbb{R}$ is always of the type $f = \nabla \cdot W$, where $W : \mathcal{O} \to \mathbb{R}^3$ is the vector field defined by*

$$W(\boldsymbol{x}) = \left(\int_0^1 t^2 f(t\boldsymbol{x}) \, dt \right) \boldsymbol{x} \, .$$

Any other vector field $\widetilde{W} : \mathcal{O} \to \mathbb{R}^3$ such that $f = \nabla \cdot \widetilde{W}$ is obtained from W by adding the curl of an arbitrary vector field.

Proof The first part follows from Remark 12.28. Concerning the second part, if $\nabla \cdot W = \nabla \cdot \widetilde{W}$, then, by the preceding theorem, $W - \widetilde{W}$ has a vector potential. The proof is thus completed. ∎

Example Consider the scalar function $f(x, y, z) = xyz$. Then a vector field whose divergence is f is given by

$$W(x, y, z) = \left(\int_0^1 t^5 xyz \, dt \right) (x, y, z) = \frac{1}{6}(x^2 yz, xy^2 z, xyz^2) \, .$$

12.14 On the Precise Definition of a Differential Form

We started this chapter with an informal treatment of differential forms. In this section, we will finally provide their precise definition.

Consider, for every positive integer M, the set $\Omega_M(\mathbb{R}^N)$ created by the M-linear antisymmetric functions on $\mathbb{R}^N$ with real values. These are the functions

$$\varphi : \underbrace{\mathbb{R}^N \times \cdots \times \mathbb{R}^N}_{M \text{ times}} \to \mathbb{R} \, ,$$

which assign to each M-tuple $(\boldsymbol{v}^{(1)}, \ldots, \boldsymbol{v}^{(M)})$ of vectors in $\mathbb{R}^N$ a real number

$$\varphi(\boldsymbol{v}^{(1)}, \ldots, \boldsymbol{v}^{(M)}) \, .$$

They need to be linear in each variable, i.e.,

$$\varphi(\ldots, a\boldsymbol{v}^{(j)} + b\boldsymbol{w}^{(j)}, \ldots) = a\,\varphi(\ldots, \boldsymbol{v}^{(j)}, \ldots) + b\,\varphi(\ldots, \boldsymbol{w}^{(j)}, \ldots),$$

and antisymmetric, i.e.,

$$\varphi(\ldots, \boldsymbol{v}^{(i)}, \ldots, \boldsymbol{v}^{(j)}, \ldots) = -\varphi(\ldots, \boldsymbol{v}^{(j)}, \ldots, \boldsymbol{v}^{(i)}, \ldots).$$

Introducing the usual operations $\varphi + \psi$ and $c\varphi$ among functions, with $c \in \mathbb{R}$, the sets $\Omega_M(\mathbb{R}^N)$ may be considered as real vector spaces on $\mathbb{R}$. We also adopt the convention that $\Omega_0(\mathbb{R}^N) = \mathbb{R}$.

If we choose the indices $i_1, \ldots, i_M$ in the set $\{1, \ldots, N\}$, we can define the M-linear antisymmetric function $dx_{i_1, \ldots, i_M}$: It is the function that associates to the vectors

$$v^{(1)} = \begin{pmatrix} v_1^{(1)} \\ \vdots \\ v_N^{(1)} \end{pmatrix}, \ldots, v^{(M)} = \begin{pmatrix} v_1^{(M)} \\ \vdots \\ v_N^{(M)} \end{pmatrix}$$

the real number

$$\det \begin{pmatrix} v_{i_1}^{(1)} & \cdots & v_{i_1}^{(M)} \\ \vdots & \cdots & \vdots \\ v_{i_M}^{(1)} & \cdots & v_{i_M}^{(M)} \end{pmatrix}.$$

Note that whenever two indices coincide, we have the zero function. If two indices are exchanged, the function changes sign. Let us recall the following result from elementary algebra.

Proposition 12.32 *If $1 \leq M \leq N$, the space $\Omega_M(\mathbb{R}^N)$ has dimension $\binom{N}{M}$. A basis is given by $(dx_{i_1, \ldots, i_M})_{1 \leq i_1 < \cdots < i_M \leq N}$. If $M > N$, then $\Omega_M(\mathbb{R}^N) = \{0\}$.*

Proof Let us see that the elements $dx_{i_1, \ldots, i_M}$, with $1 \leq i_1 < \cdots < i_M \leq N$, are linearly independent. Assume that

$$\sum_{1 \leq i_1 < \cdots < i_M \leq N} \alpha_{i_1, \ldots, i_M}\, dx_{i_1, \ldots, i_M} = 0$$

for some real constants $\alpha_{i_1, \ldots, i_M}$. We now fix the indices $1 \leq j_1 < \cdots < j_M \leq N$ and prove that $\alpha_{j_1, \ldots, j_M} = 0$. Applying the preceding sum to the selected vectors of the canonical basis $e_{j_1}, \ldots, e_{j_M}$, we have that

$$
\sum_{1 \leq i_1 < \cdots < i_M \leq N} \alpha_{i_1,\ldots,i_M} \det \begin{pmatrix} \delta_{i_1 j_1} & \cdots & \delta_{i_1 j_M} \\ \vdots & \cdots & \vdots \\ \delta_{i_M j_1} & \cdots & \delta_{i_M j_M} \end{pmatrix} = 0.
$$

(Here δ_{ij} is the Kronecker symbol; it has a value of 1 if $i = j$, otherwise 0.) We see then that, since $1 \leq i_1 < \cdots < i_M \leq N$ and $1 \leq j_1 < \cdots < j_M \leq N$, all the determinants in the preceding sum are equal to 0 except the one with $i_1 = j_1, \ldots, i_M = j_M$, which is equal to 1. Hence, $\alpha_{j_1,\ldots,j_M} = 0$.

It remains to be proved that $\Omega_M(\mathbb{R}^N)$ is generated by the elements $dx_{i_1,\ldots,i_M}$. Let φ be an element of $\Omega_M(\mathbb{R}^N)$. Then,

$$
\varphi(v^{(1)}, \ldots, v^{(M)}) = \varphi\left(\sum_{k_1=1}^{N} v_{k_1}^{(1)} e_{k_1}, \ldots, \sum_{k_M=1}^{N} v_{k_M}^{(M)} e_{k_M} \right)
$$

$$
= \sum_{k_1,\ldots,k_M=1}^{N} \varphi(e_{k_1}, \ldots, e_{k_M}) v_{k_1}^{(1)} \cdots v_{k_M}^{(M)}.
$$

Since φ is antisymmetric, in this sum we can assume that the indices are two by two distinct. Then the sum for $k_1, \ldots, k_M$ going from 1 to N can be determined taking the indices $1 \leq i_1 < \cdots < i_M \leq N$ and considering all their permutations $i_{\sigma(1)}, \ldots, i_{\sigma(M)}$, with $\sigma : \{1, \ldots, M\} \to \{1, \ldots, M\}$. We can thus write

$$
\sum_{k_1,\ldots,k_M=1}^{N} \varphi(e_{k_1}, \ldots, e_{k_M}) v_{k_1}^{(1)} \cdots v_{k_M}^{(M)} =
$$

$$
= \sum_{1 \leq i_1 < \cdots < i_M \leq N} \sum_{\sigma} \varphi(e_{i_{\sigma(1)}}, \ldots, e_{i_{\sigma(M)}}) v_{i_{\sigma(1)}}^{(1)} \cdots v_{i_{\sigma(M)}}^{(M)}.
$$

If we now reorder all terms $e_{i_{\sigma(1)}}, \ldots, e_{i_{\sigma(M)}}$, taking into account that exchanging any two vectors the value of φ changes sign, we obtain

$$
\varphi(v^{(1)}, \ldots, v^{(M)}) = \sum_{1 \leq i_1 < \cdots < i_M \leq N} \varphi(e_{i_1}, \ldots, e_{i_M}) \sum_{\sigma} \varepsilon_\sigma v_{i_{\sigma(1)}}^{(1)} \cdots v_{i_{\sigma(M)}}^{(M)},
$$

where ε_σ denotes the sign of the permutation $\sigma : \{1, \ldots, M\} \to \{1, \ldots, M\}$. By the very definition of the determinant, we thus have that

$$
\varphi(v^{(1)}, \ldots, v^{(M)}) = \sum_{1 \leq i_1 < \cdots < i_M \leq N} \varphi(e_{i_1}, \ldots, e_{i_M}) \det \begin{pmatrix} v_{i_1}^{(1)} & \cdots & v_{i_1}^{(M)} \\ \vdots & \cdots & \vdots \\ v_{i_M}^{(1)} & \cdots & v_{i_M}^{(M)} \end{pmatrix},
$$

i.e.,

$$\varphi = \sum_{1 \le i_1 < \cdots < i_M \le N} \varphi(e_{i_1}, \ldots, e_{i_M}) \, dx_{i_1, \ldots, i_M} \, .$$

The proof is thus completed. ∎

As an example, let us consider the case $N = 3$ and take a closer look at the spaces $\Omega_1(\mathbb{R}^3)$, $\Omega_2(\mathbb{R}^3)$ and $\Omega_3(\mathbb{R}^3)$.

Consider $\Omega_1(\mathbb{R}^3)$, the space of linear functions defined on $\mathbb{R}^3$, with values in $\mathbb{R}$. We denote by dx_1, dx_2, dx_3 the following linear functions:

$$dx_1 : \begin{pmatrix} v_1 \\ v_2 \\ v_3 \end{pmatrix} \mapsto v_1 \, , \quad dx_2 : \begin{pmatrix} v_1 \\ v_2 \\ v_3 \end{pmatrix} \mapsto v_2 \, , \quad dx_3 : \begin{pmatrix} v_1 \\ v_2 \\ v_3 \end{pmatrix} \mapsto v_3 \, .$$

The space $\Omega_1(\mathbb{R}^3)$ has dimension 3, and (dx_1, dx_2, dx_3) is one of its bases.

Consider $\Omega_2(\mathbb{R}^3)$, the space of bilinear antisymmetric functions defined on $\mathbb{R}^3 \times \mathbb{R}^3$, with values in $\mathbb{R}$. It has dimension 3, and a basis is given by $(dx_{1,2}, dx_{1,3}, dx_{2,3})$, where

$$dx_{1,2} : \left(\begin{pmatrix} v_1 \\ v_2 \\ v_3 \end{pmatrix}, \begin{pmatrix} v_1' \\ v_2' \\ v_3' \end{pmatrix} \right) \mapsto \det \begin{pmatrix} v_1 & v_1' \\ v_2 & v_2' \end{pmatrix} = v_1 v_2' - v_2 v_1' \, ,$$

$$dx_{1,3} : \left(\begin{pmatrix} v_1 \\ v_2 \\ v_3 \end{pmatrix}, \begin{pmatrix} v_1' \\ v_2' \\ v_3' \end{pmatrix} \right) \mapsto \det \begin{pmatrix} v_1 & v_1' \\ v_3 & v_3' \end{pmatrix} = v_1 v_3' - v_3 v_1' \, ,$$

$$dx_{2,3} : \left(\begin{pmatrix} v_1 \\ v_2 \\ v_3 \end{pmatrix}, \begin{pmatrix} v_1' \\ v_2' \\ v_3' \end{pmatrix} \right) \mapsto \det \begin{pmatrix} v_2 & v_2' \\ v_3 & v_3' \end{pmatrix} = v_2 v_3' - v_3 v_2' \, .$$

It is useful to recall that

$$dx_{1,1} = dx_{2,2} = dx_{3,3} = 0 \, ,$$

$$dx_{2,1} = -dx_{1,2} \, , \quad dx_{3,1} = -dx_{1,3} \, , \quad dx_{3,2} = -dx_{2,3} \, .$$

Consider $\Omega_3(\mathbb{R}^3)$, the space of trilinear antisymmetric functions defined on $\mathbb{R}^3 \times \mathbb{R}^3 \times \mathbb{R}^3$, with values in $\mathbb{R}$. We denote by $dx_{1,2,3}$ the following trilinear function:

$$
dx_{1,2,3} : \left(\begin{pmatrix} v_1 \\ v_2 \\ v_3 \end{pmatrix}, \begin{pmatrix} v_1' \\ v_2' \\ v_3' \end{pmatrix}, \begin{pmatrix} v_1'' \\ v_2'' \\ v_3'' \end{pmatrix} \right) \mapsto \det \begin{pmatrix} v_1 & v_1' & v_1'' \\ v_2 & v_2' & v_2'' \\ v_3 & v_3' & v_3'' \end{pmatrix}.
$$

Every element of the vector space $\Omega_3(\mathbb{R}^3)$ is a scalar multiple of $dx_{1,2,3}$, hence the space $\Omega_3(\mathbb{R}^3)$ has dimension 1. Recall that

$$
dx_{1,2,3} = dx_{2,3,1} = dx_{3,1,2} = -dx_{3,2,1} = -dx_{2,1,3} = -dx_{1,3,2}
$$

and, when two indices coincide, we have the zero function.

Definition 12.33 Given an open subset $\mathcal{O}$ of $\mathbb{R}^N$, we call "differential form of degree M" (or "M-differential form") a function

$$
\omega : \mathcal{O} \to \Omega_M(\mathbb{R}^N).
$$

If $M \geq 1$, once we consider the basis $(dx_{i_1,\dots,i_M})_{1 \leq i_1 < \cdots < i_M \leq N}$, the components of the M-differential form ω will be denoted by $f_{i_1,\dots,i_M} : \mathcal{O} \to \mathbb{R}$. We will then write

$$
\omega(\boldsymbol{x}) = \sum_{1 \leq i_1 < \cdots < i_M \leq N} f_{i_1,\dots,i_M}(\boldsymbol{x}) \, dx_{i_1,\dots,i_M} .
$$

Hence, the M-linear antisymmetric function $\omega(\boldsymbol{x})$ is determined by the $\binom{N}{M}$-dimensional vector

$$
F(\boldsymbol{x}) = \left(f_{i_1,\dots,i_M}(\boldsymbol{x}) \right)_{1 \leq i_1 < \cdots < i_M \leq N} .
$$

A 0-differential form is simply a function defined on $\mathcal{O}$ with values in $\mathbb{R}$.

We can then define the exterior product of two differential forms, as presented in Sect. 12.2. If we consider the particular case of the two constant differential forms

$$
\omega(\boldsymbol{x}) = dx_1 , \quad \widetilde{\omega}(\boldsymbol{x}) = dx_2 , \quad \text{for every } \boldsymbol{x} \in \mathcal{O} ,
$$

we will have that

$$
(\omega \wedge \widetilde{\omega})(\boldsymbol{x}) = dx_{1,2} , \quad \text{for every } \boldsymbol{x} \in \mathcal{O} .
$$

We can then write

$$dx_1 \wedge dx_2 = dx_{1,2} \,.$$

More generally, in view of the associative property of the exterior product, we can write

$$dx_{i_1} \wedge \cdots \wedge dx_{i_M} = dx_{i_1,\ldots,i_M} \,.$$

The informal definition given in Sect. 12.1 is now completely justified.

Bibliography

References Cited in the Book

1. R.G. Bartle, *A Modern Theory of Integration* (American Mathematical Society, Providence, 2001)
2. Z. Buczolich, The g-integral is not rotation invariant. Real Analy. Exch. **18**, 437–447 (1992/1993)
3. R. Henstock, Definitions of Riemann type of the variational integrals. Proc. London Math. Soc, **11**, 402–418 (1961)
4. T.W. Körner, *Fourier Analysis* (Cambridge University Press, Cambridge, 1989)
5. J. Kurzweil, Generalized ordinary differential equations and continuous dependence on a parameter. Czechoslov. Math. J. **7**, 418–449 (1957)
6. M. Spivak, *Calculus on Manifolds* (Benjamin, Amsterdam, 1965)

Books on the Kurzweil–Henstock Integral

1. R.G. Bartle, *A Modern Theory of Integration* (American Mathematical Society, Providence, 2001)
2. A. Fonda, *The Kurzweil–Henstock Integral for Undergraduates. A Promenade along the Marvelous Theory of Integration* (Birkhäuser, Basel, 2018)
3. R.A. Gordon, *The Integrals of Lebesgue, Denjoy, Perron, and Henstock* (American Mathematical Society, Providence, 1994)
4. R. Henstock, *Theory of Integration* (Butterworths, London, 1963)
5. R. Henstock, *The General Theory of Integration* (Clarendon Press, Oxford, 1991)
6. J. Kurzweil, *Nichtabsolut Konvergente Integrale* (Teubner, Leipzig, 1980)
7. J. Kurzweil, *Henstock–Kurzweil Integration: Its Relation to Topological Vector Spaces* (World Scientific, Singapore, 2000)
8. S. Leader, *The Kurzweil–Henstock Integral and its Differentials* (Dekker, New York, 2001)
9. P.Y. Lee, *Lanzhou Lectures on Henstock Integration* (World Scientific, Singapore, 1989)
10. P.Y. Lee, R. Vyborny, *The Integral. An Easy Approach after Kurzweil and Henstock* (Cambridge University Press, Cambridge, 2000)
11. T.Y. Lee, *Henstock–Kurzweil Integration on Euclidean spaces* (World Scientific, Singapore, 2011)
12. J. Mawhin, *Analyse: Fondements, Techniques, Evolution* (De Boeck, Bruxelles, 1979–1992)
13. R.M. McLeod, *The Generalized Riemann Integral* (Mathematical Association of America, Washington, 1980)
14. E.J. McShane, *Unified Integration* (Academic, New York, 1983)

© The Author(s), under exclusive license to Springer Nature Switzerland AG 2023   425
A. Fonda, *A Modern Introduction to Mathematical Analysis*,
https://doi.org/10.1007/978-3-031-23713-3

15. W.F. Pfeffer, *The Riemann Approach to Integration* (Cambridge University Press, 1993)
16. W.F. Pfeffer, *Derivation and Integration* (Cambridge University Press, Cambridge, 2001)
17. Ch. Swartz, *Introduction to Gauge Integrals* (World Scientific, Singapore, 2001)

Some Textbooks on Exercises

1. M. Gémes, Z. Szentmiklóssy, *Mathematical Analysis - Exercises I*, Eötvös Loránd University (Typotex Publishing House, Budapest, 2014). (Free pdf available online)
2. M. Gémes, Z. Szentmiklóssy, *Mathematical Analysis - Problems and Exercises II*, Eötvös Loránd University (Typotex Publishing House, Budapest, 2014). (Free pdf available online)
3. T. Radożycki, *Solving Problems in Mathematical Analysis, Part I. Sets, Functions, Limits, Derivatives, Integrals, Sequences and Series* (Springer, Berlin, 2020)
4. T. Radożycki, *Solving Problems in Mathematical Analysis, Part II. Definite, Improper and Multidimensional Integrals, Functions of Several Variables and Differential Equations* (Springer, Berlin, 2020)
5. T. Radożycki, *Solving Problems in Mathematical Analysis, Part III. Curves and Surfaces, Conditional Extremes, Curvilinear Integrals, Complex Functions, Singularities and Fourier Series* (Springer, Berlin, 2020)
6. P. Toni, P.D. Lamberti, G. Drago, *100+1 Problems in Advanced Calculus A Creative Journey through the Fjords of Mathematical Analysis for Beginners* (Springer, Berlin, 2022)

Index

A

Absolute value, 26
Additivity of the integral, 184, 304, 319, 354
Adherent point, 36, 93
Almost everywhere, 312
Arccosine, 139
Arcsine, 139
Arctangent, 140
Area
    of a set, 307
    of a surface, 364, 365, 373

B

Banach space, 101
Boundary
    of a manifold, 299
    of a set, 38
Bounded, 14, 95, 104
Bounded from above, 14
Bounded from below, 14

C

Canonical basis, 49
Cauchy product, 214
Cauchy sequence, 100
Chain rule, 276
Chebyshev inequality, 312
Closed
    differential form, 411
    set, 36
Closure, 36
Cluster point, 64, 93
codomain of a function, xix
Compact set, 95
complementary of a set, xviii
Complete additivity of the integral, 318, 354
Complete additivity of the measure, 308
Complete metric space, 100

Complex conjugate, 27
Components of a function, 48
Convergence
    absolute, 207
    in norm, 207
    pointwise, 103
    uniform, 103
Countable, 22
counterimage of a set, xxi
Criterion
    asymptotic comparison, 208
    comparison, 208
    condensation, 210
    Leibniz, 212
Cross product, 30, 382
Curl, 383
Curve, 288
Cylindrical coordinates, 348

D

De l'Hôpital's rule, 147, 149, 150, 153
De Morgan rules, xv, xviii
Derivative, 127, 135, 273
    nth, 136
    second, 136
Diffeomorphism, 286
difference of sets, xviii
Difference quotient, 127
Differentiable, 127, 255, 259
Differentiable manifold, 299
    orientable, 409
    oriented, 409
Differential, 255, 272
Differential form, 377, 422
    closed, 411
    exact, 411
Direction, 256
Directional derivative, 256

Distance
  Euclidean, 32
  in a metric space, 33
  in $\mathbb{R}^N$, 32
Divergence, 384
domain of a function, xix

E
Equivalent M-surfaces, 361
Euler formula, 221
Euler's number, 85
Exact differential form, 411
Exponential, 53
Exterior differential, 380
Exterior product, 378

F
Factorial, 5
Flux, 386
Formulas:
  binomial, 11
  change of variables, 69, 335, 339, 343
  Euler, 221
  Gauss, 395
  Gauss–Green, 404
  Gauss–Ostrogradski, 405
  Kelvin–Stokes, 404
  Stokes–Cartan, 402
  Taylor
    with integral remainder, 181
    with Lagrange remainder, 154, 266
Fourier series, 226
Function, xix
  analytic, 160
  arccosine, 139
  arcsine, 139
  arctangent, 140
  bijective, xx
  bounded, 104
  circular, 57
  of class C^n or C^n-functions, 263
  of class C^n or C^n-functions, 273
  of class C^2 or C^2-functions, 263
  of class C^1 or C^1-functions, 273
  of class C^1 or C^1-functions, 259
  concave, 144
  continuous, 42, 43
  convex, 141
  cosine, 57
  decreasing, 51
  derivative, 135
  differentiable, 127, 255

Dirichlet, 45
  even, xx
  exponential, 53
  hyperbolic, 60
  hyperbolic cosine, 60
  hyperbolic sine, 60
  hyperbolic tangent, 61
  increasing, 51
  injective, xx
  integrable, 165
  invertible, xxi
  L-integrable, 236
  logarithm, 53
  monotone, 51
  odd, xx
  R-integrable, 187
  sine, 57
  strictly concave, 144
  strictly convex, 143
  strictly decreasing, 51
  strictly increasing, 51
  strictly monotone, 51
  surjective, xx
  tangent, 59

G
Gauge, 163, 303
Geometric series, 204
Gluing, 392
Gradient, 383
graph of a function, xix

H
Harmonic series, 205
Hessian matrix, 263
Homothety, 346

I
Image of a M-surface, 288
image of a set, xx
Induced orientation, 409
Induction, 3
Inequality
  Bernoulli, 6
  Schwarz, 31
  triangle, 28, 32, 33
Infimum, 15
Integrable function:
  Kurzweil–Henstock, 165, 303
  Lebesgue, 236, 304
  Riemann, 187, 304

Integral
    of a function, 166, 303
    line integral, 385
    surface integral, 386
Integral function, 173
Integration
    by parts, 177
    by substitution, 179
Interior, 35
Internal point, 35
intersection of sets, xviii
Interval, 19
Irrotational, 416
Isolated point, 35

J
Jacobi Identity, 30
Jacobian matrix, 273

L
Lagrange multipliers, 294
Leibniz rule, 321, 354
Length of a curve, 363, 365, 373
Limit, 63
Line integral, 385
L-integrable, 236, 304
Lipschitz continuous, 334
Local diffeomorphism, 286
Logarithm, 53

M
Mathematicians:
    Banach, 101
    Bernoulli, 6
    Bolzano, 50, 95
    Cantor, 20
    Cartan, 402, 410
    Cauchy, 100, 146, 214
    Cesaro, 226
    Chebyshev, 312
    de l'Hôpital, 147
    De Morgan, xv, xviii
    Descartes, 34
    Dini, 270
    Dirichlet, 45, 228
    Euler, 85
    Fejer, 227
    Fermat, 136, 267
    Fourier, 226
    Fubini, 326, 331, 355
    Gauss, 395, 404, 405
    Green, 404
    Heine, 99
    Henstock, 165, 233, 235
    Hess, 263
    Jacobi, 30, 273
    Kelvin, 404
    Kurzweil, 165
    Lagrange, 138, 266, 294
    Lebesgue, 353
    Leibniz, 179, 212, 321, 354
    Levi, 241, 352
    Lipschitz, 334
    Mertens, 214
    Napier, 85
    Ostrogradski, 405
    Pascal, 10
    Peano, 3
    Poincaré, 413
    Riemann, 302, 304
    Rolle, 137
    Saks, 233, 235
    Schwarz, 31
    Stokes, 402, 404, 410
    Taylor, 154, 160, 181, 266
    Weierstrass, 95, 98
Maximum
    of a function, 98
    of a set, 14
M-dimensional measure, 365
Mean Value Theorem, 279
Measurable, 307, 351
Measure, 307, 351
    M-dimensional, 365
    M-superficial, 365
Metric space, 33
Minimum
    of a function, 98
    of a set, 14
M-manifold, 299
Modulus, 26
M-parametrizable, 291
M-superficial measure, 365
M-surface, 288

N
Napier's constant, 85
Negligible, 310
Neighborhood, 35
Nonoverlapping, 301, 318
Norm, 30
Normal unit vector, 289
Normed vector space, 32

O

Open
   set, 35
Order relation, 12
Orientation
   induced, 409
Oriented boundary
   of a M-surface, 397
   of a rectangle, 393
Orthogonal vectors, 30

P

Parallelogram identity, 32
Parametrizable, 291
Parametrization, 291
Partial derivative, 256
Partition of unity, 371
Point
   adherent, 36, 93
   cluster, 64, 93
   internal, 35
   isolated, 35
Polar coordinates, 347
Potential
   scalar, 416
   vector, 417
Primitivable function, 172
Primitive of a function, 172
Product
   cross, 30
   scalar, 29
product of sets, xviii
Projection, 47
Projection of a set, 331, 355
Proof by induction, 6
Pull-back, 390

R

Ratio test, 210
Rectangle, 288, 301
Recursion, 4
Reflection, 345
Regular M-surface, 288
Riemann sum, 162, 302
R-integrable, 187, 304
Root test, 209
Rotation, 346

S

Scalar product, 29, 383
Second derivative, 136

Section of a set, 331, 355
Separation property, 14
Sequence
   Cauchy, 100
sequence, xix
Series, 203
   Cauchy product, 214
   Fourier, 226
   geometric, 204
   harmonic, 205
   Taylor, 160
Sets:
   bounded, 14, 95
   bounded from above, 14
   bounded from below, 14
   closed, 36
   compact, 95
   countable, 22
   measurable, 307, 351
   M-parametrizable, 291
   negligible, 310
   nonoverlapping, 318
   open, 35
   parametrizable, 291
   star-shaped, 412
Sign permanence, 66
Solenoidal, 417
Space
   Banach, 101
   compact, 95
   metric, 33
   normed, 32
Spherical coordinates, 349
Square root, 18
Square roots in $\mathbb{C}$, 25
Star-shaped set, 412
Subsequence, 94
Summation, 5
Supremum, 15
Surface, 288
Surface integral, 386

T

Tagged partition, 301
   δ-fine, 303
Tangent, 59
   line, 127
   plane, 289
   space, 408
   unit vector, 288
Taylor polynomial, 154
Taylor series, 160
Telescopic sum, 8

Theorems:
    Bolzano, 50
    Bolzano–Weierstrass, 95
    Cantor, 20
    Cauchy, 146
    Change of Variables, 335, 343, 358
    Cousin, 163, 303
    de l'Hôpital, 147, 149, 150, 153
    Dirichlet, 228
    Fejer, 227
    Fermat, 136, 267
    Fubini, 326, 329, 331, 355
    Fundamental Theorem, 170, 172, 200
    Hake, 247
    Heine, 99
    Implicit Function, 270, 280, 285
    Lagrange, 138
    Lebesgue, 246, 353
    Levi, 241, 352
    Local Diffeomorphism, 286
    Mean Value, 279
    Poincaré, 413
    Rolle, 137

    Saks–Henstock, 233
    Schwarz, 261
    Sign Permanence, 66
    Squeeze, 67
    Stokes–Cartan, 410
    Weierstrass, 98
Transformation of a differential form, 390
Translation, 344
Triangle inequality, 28, 32, 33

U
union of sets, xviii
Unit vector, 256

V
Vector field
    irrotational, 416
    solenoidal, 417
Volume, 291
    of a set, 307